Environmental Science, Engineering and Technology

Species Diversity and Extinction

ENVIRONMENTAL SCIENCE, ENGINEERING AND TECHNOLOGY

Additional books in this series can be found on Nova's website at:

https://www.novapublishers.com/catalog/index.php?cPath=23_29&seriesp=Environmental+Science%2C+Engineering+and+Technology

Additional e-books in this series can be found on Nova's website at:

https://www.novapublishers.com/catalog/index.php?cPath=23_29&seriespe=Environmental+Science%2C+Engineering+and+Technology

ENVIRONMENTAL SCIENCE, ENGINEERING AND TECHNOLOGY

SPECIES DIVERSITY AND EXTINCTION

GERALDINE H. TEPPER
EDITOR

Nova Science Publishers, Inc.
New York

Copyright © 2010 by Nova Science Publishers, Inc.

All rights reserved. No part of this book may be reproduced, stored in a retrieval system or transmitted in any form or by any means: electronic, electrostatic, magnetic, tape, mechanical photocopying, recording or otherwise without the written permission of the Publisher.

For permission to use material from this book please contact us:
Telephone 631-231-7269; Fax 631-231-8175
Web Site: http://www.novapublishers.com

NOTICE TO THE READER

The Publisher has taken reasonable care in the preparation of this book, but makes no expressed or implied warranty of any kind and assumes no responsibility for any errors or omissions. No liability is assumed for incidental or consequential damages in connection with or arising out of information contained in this book. The Publisher shall not be liable for any special, consequential, or exemplary damages resulting, in whole or in part, from the readers' use of, or reliance upon, this material. Any parts of this book based on government reports are so indicated and copyright is claimed for those parts to the extent applicable to compilations of such works.

Independent verification should be sought for any data, advice or recommendations contained in this book. In addition, no responsibility is assumed by the publisher for any injury and/or damage to persons or property arising from any methods, products, instructions, ideas or otherwise contained in this publication.

This publication is designed to provide accurate and authoritative information with regard to the subject matter covered herein. It is sold with the clear understanding that the Publisher is not engaged in rendering legal or any other professional services. If legal or any other expert assistance is required, the services of a competent person should be sought. FROM A DECLARATION OF PARTICIPANTS JOINTLY ADOPTED BY A COMMITTEE OF THE AMERICAN BAR ASSOCIATION AND A COMMITTEE OF PUBLISHERS.

LIBRARY OF CONGRESS CATALOGING-IN-PUBLICATION DATA
SPECIES DIVERSITY AND EXTINCTION /EDITOR, GERALDINE H. TEPPER.
   xiv, 448 p. : ill. ; 26 cm.
   Includes bibliographical references and index.
   ISBN 978-1-61668-343-6 (hardcover)
   1. Species diversity. 2. Biodiversity conservation. 3. Extinction (Biology). I. Tepper, Geraldine H.
   (OCoLC)ocn496965603

2010284083

Published by Nova Science Publishers, Inc. † *New York*

CONTENTS

Preface		vii
Chapter 1	Mexican Threatened Cacti: Current Status and Strategies for Their Conservation *M. S. Santos-Díaz, E. Pérez-Molphe, R. Ramírez-Malagón, H. G. Núñez-Palenius and N. Ochoa-Alejo*	1
Chapter 2	Extinction of Cytheroidean Ostracodes (Crustacea) in Shallow-Water around Japan in Relation to Environmental Fluctuations since the Early Pleistocene *Hirokazu Ozawa*	61
Chapter 3	New Frontiers in Genome Resource Banking *Joseph Saragusty and Amir Arav*	111
Chapter 4	The Diversity of Cypriniforms throughout Bangladesh: Present Status and Conservation Challenges *Mostafa A. R. Hossain and Md. Abdul Wahab*	143
Chapter 5	Basic Ecological Information about the Threatened Ant, *Dinoponera lucida* Emery (Hymenoptera: Formicidae: Ponerinae), Aiming its Effective Long-Term Conservation *Amanda Vieira Peixoto, Sofia Campiolo and Jacques Hubert Charles Delabie*	183
Chapter 6	Exploring Microbial Diversity – Methods and Meaning *Lesley A. Ogilvie and Penny R. Hirsch*	215
Chapter 7	Spatial Assemblages of Tropical Intertidal Rocky Shore Communities in Ghana, West Africa *Emmanuel Lamptey, Ayaa Kojo Armah and Lloyd Cyril Allotey*	239
Chapter 8	Pollution and Diversity of Fish Parasites: Impact of Pollution on the Diversity of Fish Parasites in the Tisa River in Slovakia *Vladimíra Hanzelová, Mikuláš Oros and Tomáš Scholz*	265

Chapter 9	Freshwater Endemics in Peril: A Case Study of Species *Echinogammarus Cari* (Amphipoda: Gammaridae) Threatened by Damming *Krešimir Žganec, Sanja Gottstein, Nina Jeran, Petra Đurić and Sandra Hudina*	297
Chapter 10	Threatened Temperate Plant Species: Contributions to Their Biogeography and Conservation in Mexico *Isolda Luna-Vega, Othón Alcántara Ayala and Raúl Contreras-Medina*	317
Chapter 11	Genetic Variability in *Caiman Latirostris* (Broad- Snouted Caiman) (Reptilia, Alligatoridae). Contributions to the Sustainable use of Populations Recovered from the Risk of Extinction *P.S. Amavet, J.C. Vilardi, R. Markariani, E. Rueda, A. Larriera and B.O. Saidman*	341
Chapter 12	The Aquatic Plant Species Diversity in Large River Systems *Dragana Vukov and Ružica Igić*	361
Chapter 13	Changes in Plant Species Diversity around the Copper Plant in Slovakia after Pollution Decline *Viera Banásová, Anna Lackovičová and Anna Guttová*	383
Chapter 14	Crustacean Zooplankton Biodiversity in Chilean Lakes: Two View Points for Study of Their Regulator Factors *Patricio De los Ríos, Luciano Parra and Marcela Vega*	405
Chapter 15	Restoration of Propagation and Genetic Breeding of a Critically Endangered Tree Species, *Abies beshanzuensis* *Shunliu Shao and Zhenfu Jin*	415
Index		433

PREFACE

Species diversity is an index that incorporates the number of species in an area and also their relative abundance. Since species diversity is central to a large amount of ecological theory, its accurate measurement is key to understanding community structure and dynamics. Consequently, during the course of evolution, species have always gone extinct; however, the rate of extinction has increased in recent decades by as much as one hundred fold, some owing to environmental impact, mainly due to human activities. The authors of this book present and review important data on biodiversity and species extinction.

Chapter 1- Cactaceae is an American plant family found from Canada down to Argentina. Cacti have evolved anatomical and physiological adaptations, which allow them to grow and thrive under desert conditions. Therefore, cacti are a main part of the arid and semiarid landscape. The greatest cacti diversity, for genus and species (63 and 669, respectively), is located in Mexico, where approximately 78% of cacti is endemic. Cacti have been used since pre-hispanic times for food, medicines, fodder, and raw material. Furthermore, cacti are considered as one of the most important ornamental plants nowadays, given that they have beautiful flowers and low water requirements. Unfortunately, the meaningless exploitation, poachers and habitat destruction (for agriculture, grazing, housing development, etc.) have posed cacti on an unstable situation, near extinction. Pressure increases as time goes by, since more human developments are found everyday. Currently, more than 250 Mexican cacti are considered as threatened species. Cacti, usually, have a long life cycle and low growth rates, which are prone conditions for vulnerability. One of the main hindrances for cacti conservation is that they have a low multiplication rate, mostly in those species that are not asexually propagated. Cacti sexual multiplication has a low efficiency, and sometimes cacti seeds are very scarce. The precedent cacti multiplication dilemma has caused that enough cacti plants for reforestation are not easily available. In this context, *in vitro* plant tissue culture techniques are a feasible alternative to effortlessly propagate numerous cacti. These techniques use plant fragments, under lab and axenic conditions, to massively propagate cacti plants. The *in vitro*-obtained cacti can be hardened under greenhouse conditions, where they grow as wild type cacti. Plant tissue culture methods allow the fast asexual multiplication of cacti in a short time and in a reduced space, even starting from a scarce supply of plant material. Several successful evidences on cacti propagation using plant tissue culture protocols can be found in the literature; consequently these techniques may be valuable tools to overcome the cacti extinction.

Chapter 2- This chapter overviews extinction events of shallow-water benthic species in multiple families of Cytheroidean ostracodes (Crustacea) in shallow-water areas of the Japanese Islands since the early Pleistocene in the late Cenozoic, and their cause in relation to fluctuations of oceanic environments. The author also refers to the possibility of ostracode extinctions in the near future in shallow-water areas around Japan.

Firstly, this chapter summarizes the faunal changes of species of four selected families in the Japan Sea during the Pleistocene, related to oceanographic environments of shallow-water areas. Tolerance ranges of salinity for these now-extinct species are inferred to have been narrower than those of most extant related-species that live in open water as well as in brackish inner-bays, based on the mode of fossil occurrences in Pleistocene strata at the eastern Japan Sea coast. During glacial periods since 1 Ma, especially in the middle Pleistocene, the salinity decreased in shallow-water areas because low sea levels resulted in the closure of the shallow and narrow straits around this marginal sea. This salinity decrease in shallow-water areas would have caused the extinction of these now-extinct species.

Secondary, the author reviews recent investigations for extinction and disappearance (i.e., regional extinction) events of representative inner-bay species in multiple families along the Japan Sea and Pacific coasts of central Japan during the middle–late Pleistocene, related to coastal environmental fluctuations. Their disappearances along the Pacific coast and extinctions along the Japan Sea coast are hypothesized to have been caused by the drastic changes in salinity and dissolved oxygen due to glacio-eustatic cycles in the middle Pleistocene, with replacement by other bay-dominant taxa.

Finally, this chapter briefly refers to the possibility for the declining, disappearance and extinction of extant relict-cryophilic species in the near future caused by global climatic warming in shallow-water areas around the Japanese island of Hokkaido along the Japan Sea, Okhotsk Sea, and Pacific coasts. Most of these cryophilic species, which favor water temperature conditions around or less than 5°C during the winter, will decline and might disappear from shallow-water areas from around this island within the next 130–250 years, if the global climate warming continues at its present rate.

Chapter 3- During the course of evolution species has always gone extinct; however, the rate of extinction has increased in recent decades by as much as one hundred fold. *In situ* preservation should be supported by *ex situ* efforts like captive breeding, supplemented by assisted reproductive technologies (ART) and the establishment of genome resource banks (GRB).

Semen cryopreservation protocols have been developed for many species but many others proved challenging. Apparently, specie-specific protocols for semen collection and cryopreservation should be developed. The authors and others (Jewgenow et al., 1997; Saragusty et al., 2006) have shown that post mortem semen collection and cryopreservation from endangered species, even hours after death, can save valuable genes. To reduce costs of liquid nitrogen storage and maintenance, the authors have developed large volume cryopreservation technique. Additionally, with this cryopreservation technique they showed that samples can be thawed, used and the balance refrozen (Arav et al., 2002b; Saragusty et al., 2009b). Other mid- and long-term preservation techniques are currently under exploration, including freeze-drying and electrolyte-free preservation.

Oocytes cryopreservation has proved much more challenging than sperm and, even today, success rate is very low. Vitrification is gaining the lead in this field. However, as we have recently demonstrated (Yavin and Arav, 2007), identifying the delicate balance between multiple factors is imperative for success. Alternatively, oocytes collected ante- or post mortem can be fertilized *in vitro* and cryopreserved as embryos.

Ovary freezing is a new technology developed for human fertility preservation in women that undergo cancer treatment. In recent publication we (Arav et al., 2010) documented the longest ovarian function for up to 6 years after whole organ cryopreservation in sheep. The authors have shown endocrine cyclisity and production of normal oocytes and embryos. This strategy could benefit endangered species if allogeneic or even xeno-transplantation after whole ovary or ovarian tissue cryopreservation could be done.

Cloning animals is currently limited to few species and has a low success rate. Nevertheless, despite this limitation, it would seem pragmatic to initiate storage of somatic cells with an eye to future improvements in nuclear transfer efficiency. However, a major obstacle to the establishment of GRB is the cost associated with the long-term maintenance of cell lines in liquid nitrogen. The authors have shown recently (Loi et al., 2008a; Loi et al., 2008b) the capacity of freeze-dried somatic cells, which were held at room temperature for 18 months to maintain their nuclear integrity, and subsequently be used for nuclear transfer and produce viable embryos.

In the following pages, we will review the various aspects of gametes, embryo and tissue preservation for prospective utilization in ART.

Chapter 4- Bangladesh is endowed with a vast expanse of inland openwaters characterised by rivers, canals, natural and man-made lakes, freshwater marshes, estuaries, brackish water impoundments and floodplains. The potential fish resources resulting from these are among the richest in the world; in production, only China and India outrank Bangladesh. The inland openwater finfish fauna is an assemblage of ~267 species, the diversity of which is attributed to the habitats created by the Bengal Delta wetlands and the confluence of the Brahmaputra, Ganges and Jamuna rivers that flow from the Himalayan Mountains into the Bay of Bengal.

The indigenous fish fauna of Bangladesh's inland openwaters, however, are dominated by the cypriniforms - 87 species under 35 genera. Although representatives are rarely encountered in brackish waters, certain species have adapted to some of the country's most extreme freshwater environments.

There are, however, serious concerns surrounding the slow decline in the condition of openwater fish stocks which have been negatively impacted upon through a series of natural and anthropogenic induced changes. These include disturbances resulting from water management programmes including the large scale abstraction of water for irrigation and the construction of water barrages and dams, human activity resulting in the overexploitation of stocks, the unregulated introduction of exotic stocks and pollution from industry. Also, natural phenomena, regular flooding etc cause rivers to continually change course creating complications of soil erosion or oversiltation of waterways. As a consequence, many Bangladeshi species are either critically endangered or extinct. The biodiversity status of many of these have now changed from that listed in the IUCN Red Book almost a decade ago.

Assessment is based primarily on the study of specimens maintained in the Fish Museum and Biodiversity Center (FMBC) of Bangladesh Agricultural University and through surveys conducted over the last ten years. The threat to inland openwater biodiversity is countrywide, but that facing cypriniform species is severe. More than 15% of cypriniforms appear to have disappeared; only one or two individuals of a further 20% of species have been found in the last ten years.

The needs of Bangladesh's poor fisher community to eat what they catch and lack of a legal legislative framework means the situation can only worsen. Hope, however, is offered through several new conservation initiatives including the establishment of fish sanctuaries at strategic points in rivers and floodplains, concerted breeding programmes and the maintenance of captive stocks and cryogenically stored materials.

Chapter 5- The giant ants of the *Dinoponera* genus belong to a convergent group of ants in which there is no morphologically specialized caste of reproducing females and reproduction is done by fertilized workers known as gamergates. The *Dinoponera* genus is endemic of South America. The Brazilian Atlantic rain forest native species, *Dinoponera lucida* Emery, is included on the Brazilian official red list, because of its habitat loss and fragmentation and peculiarities of its biology. Population ecology and biological cycle studies were carried out from August, 2004 to July, 2005, in five forest areas in the states of Bahia and Espirito Santo, Brazil. A range of information was accumulated, with the purpose of providing strong arguments for further implantation of an effective conservation plan for the species: i) ant nest architecture: dynamics of ant nest populations, suggesting colony division per fission, as already seen in other species, was observed; ii) aggregate distribution of ant nests was defined and is explained by the particular reproductive biology of the ant; iii) ant nest colonization by other terrestrial arthropods was studied; and iv) the ant foraging behavior (activity time, prey categories) was studied. Such information is necessary to justify the long-term effort necessary to implant an effective conservation plan for this ant.

Chapter 6- Since species diversity is central to a large amount of ecological theory, its accurate measurement is key to understanding community structure and dynamics. Techniques to assess the diversity of macro-organisms are well established, but when it comes to micro-organisms many challenges still remain. The rapidly developing suite of molecular microbial community analysis methods, from high throughput sequencing to DGGE, has provided unprecedented insights into microbial community structure, revealing unappreciated levels of diversity. However, the analysis of microbial community profiling data is still in its infancy. Diversity indices and analysis methodology used for macro-organisms are often adopted for microbial community profiling data *en masse*. But no comprehensive analysis of their suitability for micro-organisms has been carried out. Here we review the currently available profiling techniques and posit an analysis framework that will facilitate the translation of this data into credible assumptions about microbial diversity and community structure.

Chapter 7- The Ghana's rocky shore is unique (i.e., being flanked by several kilometres of sandy beaches and backed by several river bodies) along the Gulf of Guinea from Cote d'Ivoire to western Cameroon (West Africa). This suggests the existence of specific biotopes of species assemblages on spatial scales. The variability of environmental factors has been highlighted as the main causes of variations in the rocky intertidal communities. This study tested species assemblage patterns between three geomorphic rocky zones (i.e., western, central and eastern shores of Ghana), and quantified the influence of abiotic factors on the assemblage patterns. The study was carried out from December 2003 to January 2004 (best period of daytime good low tides). In describing the species spatial assemblages, four random sites were located within each geomorphic zone. Further, four belt transects were randomly laid from the lower to upper shores of each site and along which a continuous quadrat (1 square meter) was placed to estimate species percentage cover (macroalgae) and abundance of epibenthic fauna. The species data was standardized before submitted to statistical analyses. Also two replicate water samples were taken at each site for the analyses of nutrients (i.e., nitrate and phosphate), while dissolved oxygen concentration, salinity, water and ambient temperatures were measured at two spots of each site. All together, 86 taxa were found comprising 57 macroalgae and 29 epibenthic fauna. Species assemblage patterns indicated significant ($p=0.001$) differences between western and central, as well as western and eastern shores. In general, the species assemblage was dominated by macroalgae, which showed a spatial declivity from west to east shores as opposed to epibenthic fauna. Taxon cumulative dominance decreased from east to west shores indicating high species diversity in the latter. However, spatial differences in the abundance of 24 widespread species influenced the assemblage patterns. Suites of abiotic variables notably nitrate, dissolved oxygen, salinity, and water temperature explained significant variations in Shannon-Wiener species diversity (58.53%), Margalef's species richness (61.49%), and number of species (71.54%). Canonical correspondence analysis showed significant response of individual taxa to nitrate and dissolved oxygen. The results suggest the influence of river flow on the species assemblages. These findings have important consequences for biodiversity and ecosystem functioning as well as conservation.

Chapter 8- An extensive survey of helminth parasites of 1,316 freshwater fish of 31 species from two aquatic ecosystems with different level of environmental pollution in southeastern Slovakia was carried out and a total of 31 gastrointestinal helminths (Trematoda – 11 species, Cestoda – 14, Acanthocephala – 3 and Nematoda – 3) have been found. The Tisa River has been heavily polluted with cyanides and heavy metals after a series of ecological disasters in 2000, whereas its tributary, Latorica River, is less anthropogenically impacted. Even though the fish communities were qualitatively similar (Czekanowski-Sørensen similarity Index - ICS = 81%) and the number of fish examined was approximately the same (676 and 640) in both localities, species richness of helminths and diversity of host-parasite associations were two times lower in the more polluted Tisa River. Helminth communities were also much less abundant in the Tisa River. Based on ICS = 48.8% and the Percentage similarity Index (PI) = 19.5%, the helminth communities were qualitatively and quantitatively different in the two rivers. Four species, the aspidogastrean *Aspidogaster limacoides* Diesing, 1835, the acanthocephalan *Pomphorhynchus tereticollis* (Rudolphi, 1809), and tapeworms *Atractolytocestus huronensis* Anthony, 1958 and *Khawia sinensis* Hsü, 1935 were reported for the first time in Slovakia. Both tapeworms found in wild common carp *Cyprinus carpio carpio* L. have been introduced recently into the Tisa River and their further dissemination to other regions throughout the Danube River basin is probable. Their morphology is briefly described and compared with representatives of other European populations of common carp.

Chapter 9- Freshwaters, especially running waters, have been impacted globally by suite of pressures among which pollution, overexploitation, physical alternation and damming, water abstraction and introduction of non-native species have caused the most severe degradations. Damming of rivers and creation of impoundments have been the most important causes of habitat loss and hydrological alternation in running waters. Moreover, as many freshwater organisms have restricted geographical distributions, any modification of their habitats can have a great impact on species and cause reduction or complete loss of specific local biodiversity.

Echinogammarus cari (S. Karaman 1931) is Croatian endemic species whose presently known distribution is restricted to only 25 km of watercourse length in the upper canyon part of the Gojačka Dobra River and its two tributaries. After the completion of a 52.5 m high dam at the end of the canyon part of the Gojačka Dobra in 2010, about 60% of presently known distribution area of *E. cari* will be lost and the species will become endangered with great probability of extinction.

This study was conducted to determine the extent of *E. cari* distribution in the Dobra River and its tributaries. Also, some aspects of its ecology like its microdistribution and relationship to physicochemical parameters were examined. Hydrological and physicochemical conditions were examined in detail in order to establish the pre-damming state in this river system. Accordingly, impacts of the damming are predicted and possible conservation measures for the species are proposed and discussed. We argue that studies of biodiversity in freshwaters should be more focused on endemic species as they are the most likely to be threatened by damaging activities. This fact can be used as the strong argument for the conservation of extremely endangered freshwater ecosystems.

Chapter 10- Many species of vascular plants inhabiting the Mexican temperate forests are threatened by continuous environmental impact, mainly due to human activities. Some are recorded in some risk category in the Mexican official publication named 'Norma Oficial Mexicana 059' (NOM-059-ECOL-2001), in the IUCN red lists (the World Conservation Union) or in CITES (Convention on International Trade in Endangered Species). Distribution maps of 31 threatened species of vascular plants were generated or updated with information obtained from institutional databases, herbarium specimens, field work, and specialized literature; with this information, we analyzed species richness, endemism and distributional patterns. Mexican territory was divided using a grid system based on a chart index (scale 1:50,000 composed system by grids of 15' x 20'). Factors that represent a threat for these species, to their current habitats and natural populations are discussed, and some strategies are proposed to prevent their extinction. Regarding conservation, we evaluated their geographic distribution based on current Mexican National Protected Areas (NPAs) and Mexican Priority Terrestrial Regions for Conservation (PTRs). We also suggest the urgent incorporation of some of these species in the recent IUCN Red Data List of Threatened Species (2009) and in the NOM-059. Most of the grid-cells with high diversity and endemism are located in eastern Mexico, and are especially associated with mountain landscapes. Among the factors that represent a threat, we concluded that the main problem for all species is habitat destruction, due to increase in agricultural areas, forest exploitation, and animal husbandry, followed by expansion of human settlements, illicit extraction, and traffic of plants. Most of the species studied require special policies for their conservation due to problems that affect their natural populations. Conservation strategies include: (1) demographic and ecological studies of these species; (2) collection of seeds and vegetative propagation for ex situ conservation in botanical gardens; (3) targeting of specific areas where these species inhabit, in order to include them in future conservation plans. With few exceptions, these species are underrepresented in the current Mexican System of Natural Protected Areas, especially those with restricted distributions. Only six species are represented in five or more NAPs, allowing their populations to be protected. In the case of PTRs, most of the species listed are included, but these areas are not official nor are they regulated by federal laws.

Chapter 11- Genetic population analysis using molecular markers is probably the most important issue in conservation genetics and today it is a very useful tool for the study of species subjected to sustainable use. *Caiman latirostris* (broad-snouted caiman) is one of the two crocodilian species cited for Argentina. Their wild populations were drastically reduced in the 1950s and 1960s due to commercial hunting and intense alteration of their habitat, and *C. latirostris* was included in the Appendix I of CITES. Since 1990, management plans that use ranching system (harvest of wild eggs, captive rearing and reintroduction to nature) began in Argentina. Through these management activities, Argentine caiman populations were numerically increased and transferred to the Appendix II of CITES that allows the regulated trade of their products. Genetic population studies are being developed together with these sustainable use plans because genetic monitoring is considered essential in management program execution. This chapter includes genetic population studies about broad-snouted caiman in Santa Fe province, Argentina. Analysis related to variability, differentiation and genetic structure were carried out through isozyme electrophoresis, RAPD markers, and quantitative traits. Furthermore, paternity studies were conducted using microsatellite markers. The obtained results indicate that the broad-snouted caiman populations analyzed have low to intermediate genetic variability values, a significant population differentiation, and a high phenotypic variability for some of the morphometric traits studied. In addition, we found indications that *C. latirostris* mating system could include multiple paternity behavior, since we found more than one paternal progenitor in at least one of the families analyzed. Although the utility and broad applicability of molecular studies are widely accepted, this approach should be complemented by population analyses conducted by means of traditional methods such as morphometry, cytogenetics, ecology, and ethology to get a deeper biological knowledge of the species. To increase the efficiency in the use of natural resources the development of suitable legal guidelines as well as their effective implementation becomes very important to protect wildlife.

Chapter 12- This chapter relates to the survey of aquatic plants – hydrophytes in sections of Danube and Tisa rivers located in Central European part of Serbia, as well as in two small lowland rivers Zasavica and Jegrička. The methodology applied in the survey is in accordance to the Water Framework Directive of the European Council. The main aim of the survey was to list the plant species and to estimate their abundance according to the five-level descriptive scale, in each survey unit. The survey unit in studied rivers was the river kilometer, where the starting and ending point of river km is marked on the river bank with the navigation sign. In small rivers, which are not used for navigation, the survey units were of different length, and their beginning and ending points were the artificial objects on their banks (bridges, farms, etc.) The list of recorded plant species and their estimated abundance in survey unit was the base for calculation of diversity parameters: species richness, Shannon diversity index, and Evenness, in each survey unit. The next step in the study was the spatial analyzes of diversity parameters. Since the survey units are the continual sections along the river, it was possible to observe the longitudinal (upstream-downstream) as well as the latitudinal (backwaters-main channel) trends in diversity. The studied plants belong to the ecological group that contains limited number of species in the given eco-region; the main aim of the study was to test the indicative capacity of their diversity parameters. Analyzes of the aquatic plant species diversity showed that the species richness, diversity index, and evenness, have the great potential to indicate the hydrological conditions of the river.

Chapter 13- Air pollution is a factor that modifies the naturalness of ecosystems, including species diversity. The greatest source of emissions in the region of Krompachy (East Slovakia) was the copper smelter, which has been functioning more than 100 years. Air pollution induced large degradation process, e.g. vegetation decline or die back of a number of plant species. The long-term monitoring of vascular plant, bryophyte and lichen diversity around the smelter started in 1986. The plant diversity decreased rapidly due to the extreme pollution, mostly during the peak in 1987–1992. Over the last 15 years the amount of the pollution decreased. Currently the levels of SO_2 decreased under the limit, high concentrations of Cu, Zn, As and low pH is persisting in the soil. Achieved air quality standards due to the reduction of pollutants in the last years reflect in the first positive changes in the environment. Regeneration trend implies successful secondary succession towards increased species diversity and cover.

Chapter 14- The crustacean zooplankton in Chilean lakes is characterized mainly by their low species number and high predominance of calanoids copepods in comparison to daphnids cladocerans. This is a different pattern in comparison to North American lakes that have abundant cladocerans populations and high number of crustacean species.

The aim of the present study is do an analysis of published information for Chilean lakes and ponds. The information was analyzed using two view points: a) the first step consisted in a Principal Component Analysis considering geographical features, conductivity, trophic status, crustacean species number and calanoid copepod relative abundances. b) the second step consisted in an application of null model co-ocurrence for determine the existence of potential structures or random distribution in species for different groups of water bodies.

The results revealed that the main regulator factors would be the oligotrophy and conductivity, because the high species number are observed in mesotrophic status and low to moderate conductivity, that corresponded to Patagonian oligo-mesotrophic lakes and ponds. In another side, low species number was observed in oligotrophic and/or high conductivity lakes and ponds. The results of null models revealed the presence of regulator factors in one simulation, whereas the other two simulations denoted random in species associations, although this result does not agree with PCA analysis, the cause would be the presence of species repeated in many sites. These results about regulators factors of species diversity are similar with observations for Argentinean Patagonian lakes.

Chapter 15- *Abies beshanzuensis*, Pinaceae, is a geographically distinct tree species in China, which had grown widely at middle to south coastal mountainous area in China during the Riss Ice Age (130–180kBP). However, the population of the species reduced drastically being due to climate change, natural disaster and human activities after the last ice age (Würm: 15–70kBP). There were only 7 wild individuals in 1963, while in 1998 only 3 individuals were discovered at about 1700 m elevation of Mt. Baishanzu in Qingyuan, Zhejiang province, China. *A. beshanzuensis* was approved as one of 12 critically endangered plant species in the world in 1987 by the Species Survival Commission (SSC) of the International Union for Conservation of Nature and Natural Resources (IUCN). Though the efforts to conserve the species by people and many scientists have been performed, it has been recognized that the species has lost natural reproductive ability due to the loss of genetic diversity. It is urgent subject in particular to develop techniques for restoration of propagation and genetic breeding based on the genetic diversity of remaining trees. We developed techniques to produce seeds and seedlings with restoration of high propagation by finding the most suitable tree species for grafting based on taxonomic and genetic information of neighborhood species, and promotion of efflorescence using the most suitable gibberellin.

In: Species Diversity and Extinction
Editor: Geraldine H. Tepper, pp. 1-59

ISBN: 978-1-61668-343-6
© 2010 Nova Science Publishers, Inc.

Chapter 1

MEXICAN THREATENED CACTI: CURRENT STATUS AND STRATEGIES FOR THEIR CONSERVATION

M. S. Santos-Díaz [1], *E. Pérez-Molphe* [2], *R. Ramírez-Malagón* [3], *H. G. Núñez-Palenius* [4] *and N. Ochoa-Alejo* [4]

[1]Facultad de Ciencias Químicas, Universidad Autónoma de San Luis Potosí, Av. Manuel Nava 6; 78210-San Luis Potosí, SLP, México
[2]Departamento de Química-Centro de Ciencias Básicas, Universidad Autónoma de Aguascalientes, Av. Universidad 940; 20100-Aguascalientes, Ags., México
[3]División de Ciencias de la Vida, Campus Irapuato-Salamanca, Universidad de Guanajuato, Exhacienda El Copal s/n, Apartado Postal 311; 36500-Irapuato, Gto., México
[4]Departamento de Ingeniería Genética de Plantas, Cinvestav-Unidad Irapuato, Km 9.6 libramiento norte carretera I rapuato-León; 36821-Irapuato, Gto., México

ABSTRACT

Cactaceae is an American plant family found from Canada down to Argentina. Cacti have evolved anatomical and physiological adaptations, which allow them to grow and thrive under desert conditions. Therefore, cacti are a main part of the arid and semiarid landscape. The greatest cacti diversity, for genus and species (63 and 669, respectively), is located in Mexico, where approximately 78% of cacti is endemic. Cacti have been used since pre-hispanic times for food, medicines, fodder, and raw material. Furthermore, cacti are considered as one of the most important ornamental plants nowadays, given that they have beautiful flowers and low water requirements. Unfortunately, the meaningless exploitation, poachers and habitat destruction (for agriculture, grazing, housing development, etc.) have posed cacti on an unstable situation, near extinction. Pressure increases as time goes by, since more human developments are found everyday. Currently, more than 250 Mexican cacti are considered as threatened species. Cacti, usually, have a long life cycle and low growth rates, which are prone conditions for vulnerability. One of the main hindrances for cacti conservation is that they have a low multiplication rate, mostly in those species that are not asexually propagated. Cacti sexual multiplication has a low efficiency, and sometimes cacti seeds are very scarce. The precedent cacti multiplication dilemma has caused that enough cacti plants for reforestation are not easily available. In this context, *in vitro* plant tissue culture techniques are a feasible alternative to effortlessly propagate numerous cacti. These techniques use plant fragments, under lab and axenic conditions, to massively propagate cacti plants. The *in vitro*-obtained cacti can be hardened under greenhouse conditions, where they grow as wild type cacti. Plant tissue culture methods allow the fast asexual multiplication of cacti in a short time and in a reduced space, even starting from a scarce supply of plant material. Several successful evidences on cacti propagation using plant tissue culture protocols can be found in the literature; consequently these techniques may be valuable tools to overcome the cacti extinction.

INTRODUCTION

It is estimated that 70% of worldwide biodiversity is located in 15 countries around the globe. For these countries, known as mega-diverse due to their great ecosystem and species diversity, their nature heritage represents a paramount responsibility in order to sustain their natural resources for the coming generations (SEMARNAT, 2002).

México is renowned as an immense-biological-diversity country, since almost all worldwide-vegetation types are found within. Ten percent of the entire world plant diversity is found in México. Rzedowski (1991) estimated that this country had 22,800 vascular plant species, and among them 21,000 were flowering plants. Because of this, México is ranking worldwide as third in plant diversity and species number. Moreover, it was calculated that 54.2% of Mexican plants are endemic; therefore, México is ranking sixth around the globe in that aspect. Given that Mexico holds numerous climates, presence of humans for more than 30,000 years, various plant species, and assorted orographic conditions, then the plant endemism, domestication and evolution have flourished (Lépiz and Rodríguez-Guzmán, 2006). Likewise, México has registered 246 plant-flowering families out of 422; similarly, from 12,200 plant genera known around the world, 2,642 thrive in this country (Villaseñor, 2001).

As it was previously pointed out, México has 246 plant-flowering families, and one of the most outstanding is the Cactaceae. Unfortunately, this plant family is highly threatened because their natural populations have been affected by mankind developments, wild zones have been converted to agricultural and/or livestock land, and human collection to sell cacti nationally and globally have been increased (Sánchez-Mejorada, 1982). Hernández-Oria et al., (2007) have mentioned that Cactaceae constitute a dissimilar plant group due to both its great morphology and taxonomy diversity. All cacti species are almost exclusive from America, and its principal diversification center is México. It is estimated that there are 63 cacti genera and more than 669 species (Guzmán et al., 2003), from which more than 70% are endemic to México, distributed predominantly in the arid and semiarid areas, which cover almost half of the country.

Cacti Taxonomy

Cacti classification may be listed as follows (Guzmán et al., 2003):

Kingdom: Plantae
Division: Magnoliophyta
Class: Magnoliopsida
Order: Cariophyllales
Family: Cactaceae

Four subfamilies are established within Cactaceae (Hunt, 2006):

(a) Pereskiodeae. Shrubs and small trees can be found in this subfamily. Stems, branches and leaves are succulent type, but not completely developed. The areoles bear thorns, but not glochids. Flowers are either single or inflorescent, with short stalks. Fruit is a berry with black seeds. In Mexico, the genus *Pereskia* represents this subfamily.
(b) Maihuenioideae. Cacti belonging to this subfamily are shrubs with lawn-like growth, showing only C_3 metabolism, bearing small-curly leaves on short roll-like succulent stems. They usually have three thorns per areola, and flowers are single and terminal. Some fruits are fleshy and have small scales. Their glossy-seeds are rounded with sizes between 3 and 4 mm. This cacti subfamily is found only in South America.
(c) Opuntioideae. This taxonomic group is considered as the most isolated one within the cacti family. Several plant grown-patterns are found in this subfamily; arborescent, shrubby or creeping forms have been reported. Stems may be cylindrical, clava-shape, almost globular or cladode. Stems may be scarcely or highly branched. Leaves are not perennial with small cylindrical-subulate-limbs. The areoles might be from circular to elliptical, bearing thorns, felts and hairs. The most conspicuous characteristic of this subfamily is the presence of glochids, which are present on stem-, flower-, and fruit-areoles. Flower, one per areola, is sessile and opens during the day or sunset. Fruit may be dry or fleshy, with dark-colored and disk-shaped seeds, reproducing freely; the embryo is curved with full-formed cotyledon and perisperm. The *Opuntia, Pereskiopsis* and *Nopalea* genera represent this subfamily in Mexico.
(d) Cactoideae. This subfamily shows the greatest diversity and numerous species within Cactaceae. Cacti belonging to this subfamily may be small, arborescent, terrestrial or epiphyte. Their stems have one single shaft with ramifications, which may be globular, cylindrical, oblong or cladode. Stems have salient parts named aristae, which frequently bear ribs. The space between aristae is called groove, devoid of pods and prickles, but it might have hairs, wool-like material, thorns and bristles. These structures show assorted arrays, colors and forms. Flowers are single, generally one per areola, sessile with radial symmetry or zygomorphic, with variable sizes. In well-developed cacti, older tubers are located at the bottom of the stem, whereas young tubers are found near the shoot tip. Areoles are situated on the cacti-top, and may have scales or not. Seeds are black or darkish, and exceedingly variable regarding size, shape and ornamentation. This subfamily is widely distributed in the

American continent, thriving under dissimilar ecological conditions, reaching a great number of endemic species (Barthott and Hunt, 1993; Guzmán-Cruz, 1997; Anderson, 2001). This subfamily is appropriately represented in Mexico by six tribes: Cacteae, Cereeae, Echinocereeae, Hylocereeae, Pachycereeae and Rhipsalideae (Guzmán et al., 2003).

Cacti Distinctiveness

Plants belonging to Cactaceae developed particular physiological, morphological and anatomical characteristics, which allowed them to endure extreme drought conditions. One of the most important physiological adaptations is the Crassulacean Acid Metabolism, which permits the CO_2 fixation during the night, reducing substantially the water loss due to stomata opening (Gibson and Nobel, 1986; Bravo-Hollis and Sánchez-Mejorada, 1991). The cacti-anatomical modifications consisted mostly of waxy epicuticle, thick and multiple cuticle with depressed stomata, hypodermis with collenchyma, and development of pith and cortex with mucilage cells at cortex (Terrazas and Mauseth, 2002), letting cacti to reduce the plant's total area that is exposed to evapotranspiration, and therefore accumulating water. Due to the last characteristics, cacti are commonly known as succulent plants.

Cacti are perennial and xerophytes, bearing roots, stems, flowers, fruits and seeds. However, two main *sui generis* characteristics are areoles and thorns, which are modified nodes and leaves, respectively (Bravo-Hollis, 1978; Gibson and Nobel, 1986; Bravo-Hollis and Sánchez-Mejorada, 1991; Anderson 2001; Altesor and Ezcurra, 2003). Several cacti show long root systems (up to 15 meters, such as in 'saguaro') and are able to accumulate significant amount of water. Some roots can be fleshy, branched, extended and superficial, or axonomorph, similarly as a carrot (Dubrovsky and North, 2002). For instance, *Ariocarpus* (Britton and Rose 1963; Bravo-Hollis and Sánchez-Mejorada 1991), *Aztekium* (Porembski, 1996), *Leuchtenbergia* (Britton and Rose, 1963), and *Lophophora* (Dubrovsky and North, 2002) genera have succulent roots. Moreover, *Pachycereus pringlei* roots can reach up to 18 cm in diameter. The succulence may be developed within, either the secondary xylem such as in *Maihuenia patogonica, Nyctocereus serpentinus, Opuntia macrorhiza, O. marenae, Pereskia humboldtii, Pterocactus tuberosus* and *Tephrocactus tuberosus russelii*, or the cortex such as in *Neoevansia diguetii* and *Peniocereus greggii* (Gibson, 1978; Dubrovsky and North, 2002). Succulent roots can achieve impressive sizes, e.g. 60 cm wide, 20 cm long and 56 kg weight such as those of *Peniocereus greggii* (Britton and Rose, 1963). Usually, cacti-root capacity to stock up water is lower compared to stem. However, some succulent roots are able to store water up in numbers similar to stem's parenchyma (Dubrovsky and North, 2002). In addition to water, diverse cacti are able to accumulate starch, such as *Wilcoxia poselgeri* and *W. tamaulipensis* (Loza-Cornejo and Terrazas, 1996).

The adventitious roots, which are used by cacti as an asexual reproduction system, may be originated from either the areoles, such as in *Opuntia* and *Nopalea*, or veins in the stem, such as in *Hylocereus* and *Selenicereus*, among others (Bravo Hollis, 1978; Pizzetti, 1987; Bravo-Hollis and Scheinvar, 1995). In some cacti, for example in *Stenocereus gummosus* and *Ferocactus peninsulae*, the root-apical-meristem growth stops soon after the germination, allowing the secondary root development. This is an advantageous characteristic under desert conditions, since more secondary roots mean more water accumulation and an easier plant establishment (Dubrovsky et al., 1998). Aerial roots are rare in cacti; nevertheless, *Stenocereus gummosus*, which thrives in the Sonoran Desert, is able to produce them (Dubrovsky, 1997, 1999).

According to the cacti-root growth pattern, three types may be described: 1) a principal root without or scarce secondary roots, such as in *Lobivia* and *Lophophora* genera; 2) a root system composed of a taproot and horizontal, subsurface lateral roots and/or adventitious roots, as occurs for most columnar cacti and species of *Ferocactus*; and 3) there is no a visible principal root but roots of different lengths, with small cacti species tending to have numerous branched roots directly beneath the shoot, and larger cacti species having long subsurface roots extended some length from the shoot, as seen in several species of *Opuntia* (Gibson and Nobel, 1986).

Cacti stems are succulent and store water, as well as provide the photosynthetic tissues, given that leaves are reduced to thorns (Nobel and Hartsock, 1983; Bravo-Hollis and Sánchez-Mejorada, 1991; Terrazas and Mauseth, 2002; Loza-Cornejo et al., 2003). Cacti phyllotaxy is highly complex, since internodes are enormously compressed, giving place to original phyllotaxies, such as in *Opuntia pilifera*. The greatest level of internode shortening is observed in the globular-cacti species, where a better efficiency, regarding areoles and thorn packed in, is found (Altesor and Ezcurra, 2003).

The cacti-stem growth pattern can be either globular, cylindrical, candelabrum, columnar, oblong or cladode. *Opuntia* spp. stems are arborescent, bushy, or creeping and flattened, and produce mucilage (hydrocoloids), which optimize water retention. *Pachycereus pecten-aborigenum, Carnegiea gigantea, Pachycereus weberi* and *Neobuxbaumia mezcalensis* stems are arborescent, more or less branchy, even columnar, with succulent and thick branches bearing lengthways ribs. In less-arid places, for instance deciduous tropical forest, cacti are able to growth having big arborescent stems, such as in *Pereskia* genus. The species belonging to this genus stand a well-defined shaft, numerous branches and leaves with distinct limb. These species are considered as the most ancient, among cacti. Along the xerophytes coppice, *Mammillaria, Coryphantha, Stenocactus, Ferocactus* and *Echinocactus* genera with globular stems, are particularly abundant. The spiral disposition of tubers is evident in globular cacti, although it may be present additionally in columnar or flattened cacti. This symmetrical-growth type has a mathematical proportion, which is known as Fibonacci series (0, 1, 1, 2, 3, 5, 8, 13, 31...), normally established within the plant kingdom (Altesor and Ezcurra, 2003). For instance, the rib number in *Ferocactus latispinis* follow the Fibonacci series, e.g. seedlings have three ribs, middle plants have five ribs, more developed plants have eight, and finally, older cacti have thirteen ribs. However, *Neobuxbaumia tetezo*, which is a columnar cactus, do not always follow the Fibonacci series. Once a plant develops a longitudinal arrangement of areoles, there is no adaptative reason why it should show ribs in a Fibonancci numbers (Altesor and Ezcurra, 2003).

The cacti ribs play an important role for drought resistance; when the plant is losing water, the stem shrinks following its ribs pattern. When the rain arrives, and the cactus is absorbing and accumulating water inside the stem, the plant structure is conserved due to the ribs (Terrazas and Mauseth, 2002). The rib number in cacti are very diverse, from two up to hundred. Ribs may be straight, spiral or sinuous. The tubers or nipples may have diverse shapes; for instance, flattened, conical, triangle or pyramidal (Gibson and Nobel, 1986; Arreola-Nava, 1997).

The barrel- and columnar-shaped cacti stem faces the predicament of growing erect and losing the mechanical resistance due to succulence. All columnar cacti have bundle sheaths, which form separate rods, under the ribs, giving support to the whole plant. Moreover, the separate rods let the plant change the volume without difficulty, and accumulate water. Using this system, some columnar cacti are able to absorb water up to 10% of their masses in just four days after a rainfall (Gibson and Nobel, 1986; Bravo-Hollis and Sánchez-Mejorada, 1991). This cacti morphology has ecological costs. The space between ribs is exposed to predators, such as woodpecker birds, which bore the stem to build their nests, or herbivores that eat the parenchyma and destroy the whole column. Also, there are mechanical constraints while the cacti is growing; therefore, is very frequent to observe broken columnar cacti in the panorama wilderness (Zavala-Hurtado and Díaz-Solis, 1995). On the other hand, the globular-characteristic stem shape allows the light rotation within the stem, increasing substantially the light harvesting (Bravo-Hollis and Sánchez-Mejorada, 1991; Bravo-Hollis and Scheinvar, 1995; Anderson, 2001; Terrazas and Mauseth, 2002; Loza-Cornejo et al., 2003). In some cacti species, the stem base suffers a lignification process during aging, however, the stem does not reach a true woody structure, and rather it is a spongy tissue similar to cork (Pizzeti, 1987).

As it was previously stated, the *Pereskia* and *Pereskiopsis* ancient genera bear true leaves, with limb and petiole, whereas in other cacti the petiole has been reduced to a tuber, and the limb is observed as a small scale. The areoles are modified buds; most of them are developed at the tuber's axilla, while others are located near the shoot tip. Areoles have two growing-zones; in the superior one the flowers are developed, while in the inferior zone the meristems, called spinuliferous, give origin to thorns. In *Opuntia* and *Nopalea* genera, the areoles are able to produce glochids, new stems, and a cotton-similar material (Barthlott and Hunt, 1993; Arreola-Nava, 1997; Anderson, 2001). The cacti areole phyllotaxy is strongly related to the internal distribution of bundle sheaths (Altesor and Ezcurra, 2003).

Cacti spines protect the plant from direct sunlight and allow the plant body keeping near an air layer, which diminishes sudden temperature changes, conditions often found in the desert. Likewise, cacti thorns are not connected to the internal tissues, they are coming from the epidermis, therefore are superficial structures. Cacti thorns are classified according to their shape, direction, position, and ornamentation. 1) For their shape, thorns are sort them out as acicular, subulate, feathered, papyraceous and hooked. 2) For their direction, thorns such as divergent, porrect, curved, reflected, pectinated, retrorse and ascendant can be found. 3) For their ornamentation, thorns are grouped as smoothed, ringed, and striated and pod (Britton and Rose, 1963; Bravo-Hollis, 1978; Arreola-Nava, 1997). Two types of thorns may be found in the same areole: the radial at the exterior, which are thin and numerous, and the central, which are thick and scarce. Thorns can function as water collectors as well, since they have the ability of conducting water from morning-dew toward the areoles and then to the phloem within the plant. Additionally, thorns protect cacti from being consumed by desert animals, which seek cacti stems as water and food source. In some *Ferocactus, Opuntia* and *Mammillaria* species, there are gland-thorns that exude nectar (Britton and Rose, 1963; Gibson and Nobel, 1986; Bravo-Hollis and Sánchez-Mejorada, 1991).

A unique and distinctive cacti characteristic is their flower type. Cacti flowers are very attractive due to their forms, sizes and colors. Most cacti flowers are single; they form on the superior part of the areoles, but just in a few cases, flowers are originated from the tuber. Cacti flowers are hermaphrodite, except in rare cases where the stigma or anther atrophies, the flower is unisexual. Cacti-flower structures are plenty diverse in size, number, form and color, depending on each species, but environmental conditions, such as length and frequency of raining or pollination style, can alter the flower structure. *Neolloydias, Turbinicarpus* and *Echinocereus* hold astonishing beautiful flowers; similarly, epiphyte cacti bear flowers with amazing sizes, such as 30 cm length (Bravo-Hollis and Sánchez-Mejorada, 1991; Barthlott and Hunt, 1993; Arreola-Nava, 1997; Anderson, 2001).

The cacti-reproductive phenology is exceedingly varied. The globular cacti, such as *Mammillaria* and *Turbinicarpus*, start their reproduction cycle at the beginning of winter, and last for more than five months (Martínez et al., 2001; Sotomayor et al., 2004). In contrast, *Cereus, Myrtillocactus, Pachycereus, Pilosocereus, Selenicereus* and *Stenocereus*, all of them columnar cacti, reproduce during the fall-winter period. Some species of the genera *Carnegiea, Cephalocereus, Escontria, Hylocereus, Lophocereus, Neobuxbaumia, Pachycereus, Polaskia* and *Stenoceresus* reproduce during the spring and summer seasons (Weiss et al., 1994; Steenberg and Lowe, 1977; Casas et al., 1999; Ruiz et al., 2000; Esparza-Olguin et al., 2002).

Most cacti flowers have a short-life period, e.g. some days or even few hours. Having a short flowering period is a great and advantageous characteristic, since the evapotranspiration through the petals is highly reduced. Some cacti flowers are sensitive to light; therefore, several cacti flowering during day- or night-specific times. The nocturnal flowers are always white, with some light-yellow or red tones. The diurnal flowers are white, purple, yellow orange or green, and sometimes different colors are combined giving place to iridescent views. Usually, cacti areole generates one flower; nevertheless, in rare conditions up to ten flowers can be produced by a single areole, such as by *Myrtillocactus geometrizans* (Britton and Rose, 1963; Bravo-Hollis and Schneivar, 1995; Arreola-Nava, 1997).

Cacti flower is typically bell-shaped; nonetheless, cylinder-, pot- and trumpet-shaped flowers are commonly found. Since petals and sepals are not clearly separated, the resultant structures are called tepals. They are separated among themselves forming a dialipetal corolla. Cacti stamens are highly variable in number, from 20 up to 4,000 per flower, depending on cacti species. In numerous species, mostly night flowering, there is a nectary between the ovary and the stamens-base. The nectar is used to attract insects and other pollinators (Bravo-Hollis and Sánchez Mejorada, 1991; Gibson and Nobel, 1986; Anderson, 2001).

Cacti fruit may be either dry or juicy, and vastly miscellaneous in shape, with yellow, green, red or purple color. The seed number per fruit is extremely dissimilar, and it depends on plant-age, fruit size, as well as the flower-number produced. For instance, columnar cacti are able to generate more than 1,000 seeds per fruit, whereas the globular ones generally produce less than 100. Barrel-shaped cacti, cylindropuntias and platyopuntias have fruits, which yield between ten and two hundred seeds (Britton y Rose, 1963; Parker, 1987; León de la Luz and Domínguez-Cadena, 1991). Cacti fruit may be dehiscent, opening through a pore, operculum or split when mature but still juicy, allowing seed dispersion. Likewise, cacti fruit may be indehiscent, opening only once the fruit is dry. Externally, cacti fruit may be either cottony, such as in *Selenicereus*, thorny, such as in *Peniocereus*, scaly, such as in *Hylocereus*, or naked, such as in *Mammillaria*. Most cacti fruits are edible and tasty (Gibson and Nobel, 1986; Bravo-Hollis and Scheinvar, 1995; Arreola-Nava, 1997; Rojas-Aréchiga and Vázquez-Yanes, 2000).

Cacti seeds may be ovoid or reniform, thick or flattened, smooth or corrugated, and bear spots or tubers (Rojas-Aréchiga and Vázquez-Yanes, 2000). Cacti seed are commonly small, with sizes between 0.5 and 5 mm, the seed coat may be yellowish, brown, black or reddish, usually hard, and glossy or dull. Seeds may show dormancy, at different extents, or not (Barthlott and Hunt, 2000; Arias and Terrazas, 2004; Ayala-Cordero et al., 2004; Flores et al., 2005; Orozco-Segovia et al., 2007). Several environmental and physiological conditions, such as temperature, salinity, light, acidity, seed size, seed weight and origin, may influence germination in experimental conditions (Arredondo-Gómez and Camacho-Morfín, 1995; Vega-Villasante et al., 1996; Rojas-Aréchiga et al., 1997; Rojas-Aréchiga and Vázquez-Yanes, 2000; Flores and Briones, 2001; Ayala-Cordero et al., 2004; Ramírez-Padilla and Valverde, 2005; Loza-Cornejo et al., 2008). In the wilderness, the most important factors that determine cacti seed germination are moisture availability, temperature and light, being the first the utmost, since water is the most limited environmental factor in the deserts. However, the embryo developmental stage, seed age, plant hormones and inhibitors within the seed coat may also influence cacti seed germination (Rojas-Aréchiga and Vázquez-Yanes, 2000). Hydration-dehydration cycles help cacti seeds out to germinate easily and accumulate more biomass. This phenomenon, known as "seed-hydration-memory" can be found in the columnar cacti *Carnegiea gigantea*, *Pachycereus pecten-aboriginum*, *Stenocereus gummosus* and *S. thurberi*, as well as in *Ferocactus peninsulae* (Dubrovsky, 1996).

Cacti seeds have the ability to survive in the soil for several months, giving place to seed banks in the wild. *Ferocactus wisliseni* seeds are capable to remain viable for at least 18 months; similarly, *Escontria chiotilla* seeds wait for 12 months for optimal conditions to germinate; and finally, *Mammillaria magnimamma* seeds survive for more than one year in the nature. Likewise, it has been reported that seeds from several *Mammillaria* species remain trapped to the stem, creating natural seed banks (Gibson and Nobel, 1986; Bowers, 2000; Rodríguez-Ortega and Escurra, 2001; Godínez-Álvarez et al., 2003).

The majority of cacti grow at a truly slow pace. For instance, in the wilderness *Carnegia gigantea* and *Stenocereus thurberi* have growth rates as low as eight to ten centimeters per year, whereas some *Pellecyhpora*, *Ariocarpus*, *Coryphanta* and *Mammillaria* species are capable to grow just one millimeter per year! (Nobel, 1988; Zavala-Hurtado and Díaz-Solis, 1995; Tufenkian, 1999). Nonetheless, some cultivated cacti species have growth patterns similar to commercial plants, e.g. *Opuntia ficus-indica* is able to produce 9.95 and 0.527 tons of fresh and dry weight per hectare, respectively (Ruiz-Espinoza et al., 2008). It has been suggested that the sluggish growth showed by some cacti in the wild is an outcome of the adverse environmental conditions, rather than the long cell cycles that cacti have (Nobel, 1988; Dubrovsky et al., 1998).

The time to achieve the reproductive maturity in cacti is assorted. Some cacti species reach their reproductive maturity between two and three years after germination, but others such as *Mammillaria magnimamma* and *Opuntia engelmanni* achieve their first reproductive cycle at four and nine years, respectively (Bowers, 1996). Moreover, several cacti species start their reproductive cycle after an extemely long vegetative period, e.g. *Carnegiea gigantea* starts its reproductive cycle 33 years after germination (Steenbergh and Lowe, 1977), *Cephalocereus columna-trajani* after 70 years (Zavala-Hurtado and Díaz-Solis, 1995) and *Neobuxbaumia macrocephala* after 90 years (Esparza-Olguin et al., 2002). One of the longest cacti life cycle reported is for the saguaros, which are able to live up to 200 years in their natural habitat (Bowers, 1996; Pierson y Turner, 1998).

Traditional Uses of Cacti

In México, indigenous people used cacti in different ways since prehispanic times. The first evidences about their uses were found in Tehuacán and Puebla, and were dated as back as 6,500 to 10,000 years ago (Callen, 1966). Furthermore, some semi-fossilized cacti remains from *Echinocactus grandis* were found in Palo Blanco and Venta Salada in Puebla state (2,000-1,500 B.C.), suggesting that natives used this cactus as foodstuff (Gónzález-Quintero, 1976).

Content	O. ficus-indica (Cultivated)[1]	O. robusta (Wild)[1]	FAO[2]
Protein	16.0a	10.8b	--
Lipid	0.1a	0.1a	--
Fiber	10.8b	10.0b	--
Ash	18.0a	16.8a	--
Carbohydrate	55.1	62.2	--
Energy content	263.6	273.3	--
Arginine	4.7± 0.01	5.6 ±1.8	--
Phenylalanine	2.9 ± 0.2	4.4 ±0.9	--
Tyrosine	2.9 ± 0.2	4.4 ±0.9	1.9 [3]
Histidine	2.4 ± 0.2	3.5 ±0.9	1.6
Isoleucine	4.0 ± 0.1	4.7 ±0.6	1.3
Leucine	5.4 ± 0.0	6.3 ±0.7	1.9
Lysine	3.5 ± 0.0	4.1 ±0.8	1.6
Cysteine	1.0 ± 0.0	1.1 ±0.4	1.7 [4]
Tryptophan	Nd	Nd	0.9
Valine	4.3 ± 0.0	4.7 ±0.2	1.3
Asparagine	6.4 ± 0.0	6.6 ±0.3	--
Glutamine	4.3 ± 0.4	5.5 ±0.2	--
Alanine	12.3 ± 2.3	8.2 ±0.7	--
Glycine	3.5 ± 0.3	4.7 ±1.0	--
Proline	5.3 ± 0.1	6.0 ±1.7	--
Serine	1.4 ± 0.1	3.2 ±0.7	--
Methionine	--	--	--

Figure 1. *O. ficus indica* and *O. robusta* proximal analysis (mg/100 g flour) (modified from Barba de la Rosa, 2007). [1]Different letters between columns are statistically significant at ($p<0.05$). [2]FAO required values. [3]Tyrosine and phenylalanine requirements. [4]Methionine plus cysteine requirements. (- -) Not determined.

The first report about cacti used by humans came from the "Historia de las Indias Occidentales" (History of West Indian Land) written by Gonzalo Hernández de Oviedo and Valdés in 1535, where figures of *Cereus* spp. and *Opuntia* spp. were depicted. In the Mendocino (1541), Badiano (1552) and Florentino (1580) codices illustrations on how cacti plants affected aborigine's way of life in Mexico are also shown. Additionally, the cacti significance is manifested by the hieroglyphic found in those codices and the "Historia de las Indias de la Nueva España (1581)". Of particular interest are the hieroglyphics regarding the Tenochtitlán (main Aztec city; currently Mexico City), which means "Cactus on a stone", foundation (1325). Tells the legend that Aztecs should establish their capital city in a place where an eagle, stood on a cactus (*Opuntia* spp.) devouring a snake, could be came across. This iconography is represented nowadays in the National Coat of Arms of Mexico's Flag.

Cacti are one of most important groceries for Mexican rural and urban (in less grade) people (Russell and Felker, 1987). The "Consejo Nacional para la Cultura y las Artes" (National Council for Arts and Culture; CONACULTA) has more than 35 publications on indigenous cooking and recipes and most of them include cacti, served in many ways (CONACULTA, 2009). The young cladode ("nopalitos") from *Opuntia* spp. and *Nopalea* spp. genera ("nopales verduleros") are the most used, typically grilled, boiled, and stuffed, or in pickles and salads (Almanza, 1999; Buenrostro and Barros, 2000). According to Flores (1995), Mexico's "nopalitos" production for that year was more than 600,000 tons.

The *Opuntia* spp. ("nopal verdulero") proximal analysis has revealed that "nopalitos" contain lipids, proteins and carbohydrates, as well as minerals such as K, Ca, Mg, Mn, Fe, Zn, Cu, and B (Figure 1). *Opuntia* spp. cladodes have a very low amount of sodium, meaning that "nopalitos" consumption is good for human health. The "nopalitos" water and fiber (soluble and insoluble) content is high, similarly to some vegetables, such as spinach and broccoli, and several fruits, such as grape and mango (Orona-Castillo et al., 2004; Anaya-Pérez, 2006; Ramírez-Tobías et al., 2007). Cellulose, lignin and some type of hemicelluloses are the main components; these polymers are related to water, minerals, vitamins and bile acid absorption capacity of "nopalitos" insoluble fiber. Gums, mucilage, pectins and a few remnant hemicelluloses are the foremost substances in the soluble fiber. This soluble fiber seems to reduce the blood cholesterol and glucose in humans. The "nopalitos" total fiber content varies depending on *Opuntia* species. For example, *O. robusta* has between 10 to 18% and *O. ficus-indica* holds between 7-14% dry weights. "Nopalitos" have several vitamins as well, such as niacin, riboflavin, ascorbic acid and thiamine, and antioxidant substances, such as carotenoids (Pérez-Quilantán and Hurtado-González, 2004; Anaya-Pérez, 2006; Sáenz et al., 2006).

The *Opuntia* spp. chemical composition is different between wild and cultivated species (Barba de la Rosa, 2007). The *O. ficus-indica* and *O. robusta* proximal analysis showed that *O. ficus-indica* had a greater protein amount, whereas lipid, fiber and carbohydrate levels were similar in both species. Nevertheless, the essential amino acid content, in general, was higher in the wild species (Figure 1). An interesting and important result was that essential amino acid content for *O. ficus-indica* and *O. robusta* was superior compared to other foods, according to the FAO nutritional requirements (Sáenz et al., 2006).

A quite number of research studies have been carried out in order to use high-value "nopal" (*Opuntia* spp.) sub-products in the food industry, such as yogurt, mayonnaises, flours and bases for cake, flan and ice cream (Saenz et al., 2002; Linaje-Treviño and De la Fuente-Salcido, 2007; Aurea et al., 2007; Moreno-Álvarez et al., 2009; El-Samahy et al., 2009).

Other cacti edible tissues are the fruits, and their nutritional value is mainly because they contain high water, sugars, amino acids, fats, vitamins, micronutrients and antioxidants (Domínguez-López 1995; Lamghari et al., 1996; Kuti, 2004; Piga, 2004). Cacti fruits are known under different names, depending on each species and typical characteristics, such as "tunas", "xoconostles", "garambullos", "pitayos", "pitayas", "jiotillas", "teteches" and "biznagas". *Opuntia* fruits are fleshy and edible, and if they are sweet are known as "tunas", while if they are acidic, they are recognized as "xoconostles". Cacti fruits can be eaten raw, boiled or dried, but they can be processed as well to obtain marmalade, jelly, pickle, honey, wine, vinegar, ice cream, juice, puree, and a jelly named "Tuna cheese" (López-González et al., 1997; Saénz, 2000; Pérez-Quilantán and Hurtado-González, 2004; Cerezal and Duarte, 2004; Moßhammer et al., 2006). The "biznaga" flowers are a delicacy known as "cabuches", and the pulp-stem is used to elaborate a traditional candy known as "acitrón" (Bravo-Hollis and Schneivar, 1995; Arreola-Nava, 1997).

Cacti fruit commercialization is extremely limited since they are liable to decay or spoil easily (Piga et al., 1996; Arnaud-Viñas et al., 1997; Moßhammer et al., 2007). Interestingly, "pitayas" demand has increased in the last five years at great pace as much in the producer countries, such as Colombia, Costa Rica, Nicaragua, Vietnam and Mexico, as in consumer ones, such as in USA and Europe, where "pitayas" are considered exotic fruits (Le Bellec et al., 2006).

Cacti seeds are a rich source of several important healthy compounds; among them, free carbohydrates or arabinogalactanes, proteins such as albumin, lipids (linolenic, palmitic and oleic acids), sterols (β-sitosterol and campesterol), and vitamin E are found. Several studies have been carried out to develop the chemical processes to obtain the cacti-seed oils, but all of them have been only at analytical level; more research must be done in order to take the chemical process to the industrial plane (Ramadan and Mörse, 2003; Moßhammer et al., 2006).

The "nopal" (*Opuntia* spp.) plant is used as livestock fodder. The thorny "nopal" cladode, are first slightly burned to remove the thorns. In Mexico and South-USA, cattle and goats are feed with 30 to 40 kg/day and 6 to 8 kg/ day of burned "nopal", respectively. The inclusion of "nopal" cladode as livestock food reduces the feeding costs between 48 and 65% (Flores-Valdéz and Aranda-Osorio, 1997; Aranda-Osorio et al., 2008); besides, "nopal" provides pectins, water and mineral salts (potassium, calcium and magnesium) to livestock (Anaya-Pérez, 2006; Pérez-Quilantán and Hurtado-González, 2004).

Opuntia groves are also used to culture a "nopales"-plague caused by the insect *Dactylopius coccus*. This pest is commonly named "grana cochinilla","cochinilla grana", "cochinilla del carmín" or "nocheztli". The female insect produce a pigment, carminic acid (E-120, C.I. 75470, Natural Red 004), that gives different color tones (violet, orange, red, gray and black), depending on how it is processed. Since carminic acid is not toxic for humans, it is widely used in the food, cosmetic, photograph and fiber industries, as well as in science as a dye agent. In México, the "grana cochinilla" costs $100 USD per kilogram, representing an interesting income for *Opuntia* growers; due to this reason, "grana cochinilla" culture has received an important support from governments (Bravo-Hollis and Schneivar, 1995; Pérez-Sandi and Becerra, 2001).

Other cacti species, such as *Pachycereus marginatus, Acanthoceresus* spp., *Pereskiopsis* spp., and *Cylindropuntia* spp. have been used as live fences, reducing the land delimitation costs (Bravo-Hollis and Schneivar, 1995; Rodríguez-Arévalo et al., 2006).

Pharmacological and Biological Properties of Cacti

Cacti have been used as medicines to treat several diseases since prehispanic centuries. A bunch of cacti show numerous pharmacological properties such as diuretic, laxative, cardiotonic, analgesic, astringent and antiparasitic. Some cacti species have hallucinogenic properties as well, being the most studied, *Lophophora williamsii* and *L. diffusa,* usually both known as "peyote". More than 56 different alkaloids have been extracted from *L. williamsii*, most of them, isoquinolines and phenethylamines derivatives. The mescaline alkaloid is the main active principle in *L. williamsii*, whereas pellotine is predominant in *L. diffusa* (Agurell, 1969; Helmlin et al., 1992). Mescalines and other phenethylamines have been found in *Opuntia cylyndrica, Trichocereus pachanoi, Pereskia* spp., *Pereskiopsis* spp. and *Islaya* spp. (Doetsch et al., 1980). Moreover, hordenin-type, N-methyltyramine and tyramine alkaloids with hallucinogenic, stimulant and narcotic activities have been isolated from *Turbinicarpus* spp. and *Ariocarpus* spp. (McLaghlin, 1969; Braga and McLaghlin, 1969; Starha et al., 1999a; Starha et al., 1999b). It is well known that mescaline and its analogous inhibit the neuron-muscular cholinergic transmission, since they block the acetylcholine liberation, as well as affect the K^+ conductance (Ghansah et al., 1993).

In *Stenocereus, Epiphyllum, Myrtillocactus, Notocactus, Opuntia, Pereskia, Hylocereus, Lophocereus, Echinopsis* and *Neolloydia* species more than 20 terpenes and sterols have been detected, being the β-sitosterol the most abundant (Djerassi et al., 1957; Knight et al., 1966; Salt et al., 1987). Since phytosteroid structure (β-sitosterol, 24-methylcholesterol, estigmasterol) is very similar to human estrogens, plant steroids might exhibit a similar action as animal steroids, and they could be used to treat some menopause disorders or as birth control agent (Adlercreutz and Mazur, 1997).

Opuntia has been the most studied genus among cacti regarding their pharmacological properties. Aztecs used "nopal" to treat dental infections, tonsillitis and bruises, fractures, burnings and inflammation, and even to induce parturition (Bravo-Hollis and Scheinvar, 1995). These medical properties have been recently evaluated from a scientific approach finding that *Opuntia* spp. has multiple activities, and among others, it is useful to treat gastric disorders (Galati et al., 2001; Vázquez-Ramírez et al., 2006), improves platelet function assisting to control vascular alterations (Wolfram et al., 2003), decreases triglyceride, cholesterol and LDL blood levels minimizing arterioeschlerosis development and other cardiovascular disorders (Li et al., 2005), reduces inflammation and has analgesic characteristics (Ahmed et al., 2005; Saleem et al., 2006), modulates intracellular calcium level and T-cell activation (Aires et al., 2004), and lessens neuronal damage in mouse cortical cell during *in vitro* culture (Kim et al., 2006).

It is well known that "nopal" has hypoglycemic properties, but the exact mechanism by which it is induced is not well understood. Since "nopal" fiber and pectin content is high, it has been proposed that "nopal" reduces the glucose absorption at the intestine. Likewise, it has been shown that "nopal" improves the sensitivity to insulin within the peripheric tissues (Ibañez-Camacho and Meckes-Lozoya, 1983; Frati et al., 1988; Castañeda et al., 1997; Bwititi et al., 2000). Depending on *Opuntia* species and via of administration, the hypoglycemic activity of "nopales" can change. For instance, *O. lindheimeri, O. ficus indica* and *O. robusta* extracts decreased the blood glucose levels in rats when the extracts were injected by intravenous application, but not when they were orally applied. It has been suggested that the hypoglycemic activity is triggered when the extracts are metabolized at the liver (Enigbokan et al., 1996). Equally, polysaccharide-type compounds, with hypoglycemic activity, have been isolated from *O. ficus indica* and *O. streptacantha* (Alarcón-Aguilar et al., 2003).

Cacti extracts have shown cytotoxic activity as well. *O. ficus-indica* fruits extracts have been found to increase apoptosis, and to inhibit the ovary-cancerous, vesicle, cervical and immortal epithelial cell growth. The inhibition has been shown to be doses-time dependent. Cacti-fruit extracts seem to affect cancerous cells due to an increase in G1 phase and a reduction in G2 and S phases, and tumor growth is suppressed, since annexin IV and VEGF genes are possibly modulated (Zou et al., 2005). Similarly, *Pereskia bleo* leaf-extracts have exhibited cytotoxic activity against five different cancerous cells (Abdul Malek et al., 2009).

Cacti are excellent sources of pigments. For instance, *Hylocereus, Mammillaria, Opuntia, Rhodocactus, Cleistocactus, Stenocereus, Pilosocereus* genera are betalain producers (Piatelli and Imperato 1969; Wybraniec and Novak-Wydra, 2007; Wybraniec et al., 2007; Bohm, 2008). Betanidin 5-O-(6′-O-malonyl)-β-soforoside, commonly known as mammillarinine, has been isolated from several *Mammillaria* species (Wybraniec and Nowak-Wydra, 2007). Likewise, betanidin, phyllocactin, betanin, neobetanin, indicaxantin, vulgaxantin I and II and miraxantin II have been obtained from various *Opuntia* species (Piatelli and Imperato, 1969; Piga, 2004). Betalains are water-soluble complex-organic molecules with two or three nitrogen atoms, giving different colors, such as yellow and orange (betaxantines), and red and purple (betacyanins). Betalains are utilized in food industry because they are natural pigments, so they are used to give color to yogurts, ice creams, jellies, sodas, dressings and juices (Stintzing and Carle, 2004). Furthermore, betalains hold antioxidant activity, being able to scavenging free radicals (Reynoso et al., 1997; Stintzing et al., 2005). Similarly to fruit, *in vitro* cell cultures from cacti are excellent betalain sources (Santos-Díaz et al., 2005). These biotechnological systems have several advantages compared to cultivated fruits; among them, pigments are produced and isolated under controlled conditions, betalain synthesis and accumulation can be increased by manipulating the medium characteristics, and the production system can be escalated to industrial level (Charlwood et al., 1990).

Vitamins such as α-tocopherol (vitamin E) and ascorbic acid (vitamin C), enzymes such as catalase (Lee et al., 2002; Piga, 2004), flavonoids such as isorhamnetine 3-O- (6'-O-E-feruloyl)-neohesperidoside (6R)-9,10-dihydroxi-4,7-megastigmadien-3-ona-9-O-β-D- gluco-piranoside and (6S)-9,10-dihydroxyl-4,7-megastigmadien-3-ona-9-O-β-D-glucopiranoside, quercetine-and isorhamnetine-glucoside, isoquercitrine, 3-*O*-metil-quercetine, kaempferol, kaempferide and quercetine have been isolated from *Opuntia* cladodes (Ahmed et al., 2005; Saleem et al., 2006). Due to its antioxidant properties and calcium- and magnesium-high content, "nopal" is considered a nutraceutic food (Piga, 2004). "Nopal" is used to elaborate a great variety of commercial products; for example, gastritis-treatment milk shakes, high-fiber content powders for obesity care and constipation treatment, and shampoos, ointments and creams to avoid hair lose, products for treatment of bruises and reduction of wrinkles, respectively, can be found (Sáenz et al., 2006; Tovar-Puente, 2008). *Nopalea cochenillifera, Ariocarpus kotschoubeyanus* and *A. retusus* extracts have been observed to display activity against several microorganisms, such as *Candida albicans, Salmonella enterica, Escherichia coli, Shigella flexneri, Staphylococcus aureus* and *Bacillus cereus* (Gómez-Flores et al., 2006; Rodríguez et al., 2006). In the same way, *Myrtillocactus geometrizans* extracts have been found to be active against the maize-plague worm *Spodoptera frugiperda*, and *Tenebrio molitor*, an insect that attacks stored-grains (Céspedes et al., 2005). On the other hand, "nopal" plants have been shown to be capable of removing and accumulating Cr, Cd, and Pb from soil and water, so there's the possibility to use "nopal" as a bioremediation system. *Opuntia* spp. mucilage has been employed to remove Pb, As and F as well (Castillo-Bravo et al., 2003; Álvarez-Jiménez et al., 2008; López et al., 2008).

Cacti Origin and Distribution

All cacti species are almost exclusively from the American continent, except a few *Rhipsalis* (Barthlott and Hunt, 1993) and *Opuntia* (Bravo-Hollis and Sánchez-Mejorada, 1991) species. They can be found from Canada (59° North latitude) down to the Argentinean-Chilean Patagonia (52° South latitude), and from the coast dunes up to 5,100 m.a.s.l. such as in Perú (Benson, 1982; Bravo-Hollis and Scheinvar, 1995).

It is considered that cacti are a plant group that evolved in the last 60 to 80 million years from non-succulent plant species with well-developed leaves, C_3 photosynthesis and belonging to the Caryophyllales order. Cactaceae are related to Phytolacaceae, Aizoaceae and Didieriaceae families and it has been suggested that cacti originated in South America subtropical area (Gibson and Nobel, 1986). Until now, it has not been found any plant fossil related to Cactaceae; therefore, their evolution should be studied using geobotanic, compared anatomy and morphology, physiology and molecular biology. Nevertheless, phylogenetic relations of cacti are still controversial. Molecular researches have demonstrated that *Pereskia* and *Mammillaria* genera seem not to be monophyletic; it means that they do not have a common ancestor (Butterworth and Wallace, 2004; Edwards et al., 2005). Moreover, it has been proposed that Opuntioideae subfamily might be the origin of Cactaceae, whereas the Rhipsalidoideae subfamily could have ancient members of subfamily Cactoideae (tribes Browningieae, Cereae, Echinocereae, Hylocereae, Pachycereae, Rhipsalideae, Trichocereae and Notocacteae, except Blossfeldia tribe) leaving out the globular cacti (Cacteae tribe) from North America (Crozier, 2004). However, most authors consider that they are a monophyletic group (Britton and Rose, 1963; Bravo-Hollis, 1978; Gibson and Nobel, 1986; Guzmán-Cruz, 1997), and that asseveration has been recently supported by nuclear, mitochondrial and chloroplast DNA analyzes (Butterworth and Edwards, 2008).

Cacti include, roughly, 100 genera and 1,500 species (Barthlott and Hunt 1993; Hunt, 1999). The great cacti diversity might have resulted from morphological, physiological and anatomical adaptations to several arid and semi-arid areas that appeared during climate changes. Other factors that perhaps could have contributed to the cacti diversity may be the hybridization and increase in ploidy level. For example, *Hylocereus* crosses between diploid members have produced aneuploid, triploid, pentaploid and hexaploid hybrids (Tel-Zur et al., 2004). Similarly, triploid, tetraploid and hexaploid *Opuntia*, tetraploid *Echinocereus*, and tetraploid, hexaploid and octaploid *Mammillaria* members have been described (Pinkava, 1985; Karle et al., 2002; Boyle and Idnurm, 2003; Parks and Boyle, 2003).

The most important cacti diversity center is located in Mexico, with 63-73 genera and 669-737 species (Dávila-Aranda, 2001; Guzmán et al., 2003; Ortega-Baes and Godínez-Álvarez, 2006), followed by Argentina, Bolivia, Brazil and Perú as diversity poles. A significant positive relationship between both, the total and endemic species, and the area of countries has been observed. Despite this fact, the cactus diversity found in Mexico, Argentina, Peru, Bolivia, Chile and Costa Rica is higher than that expected according to their area (Ortega-Baes and Godínez-Álvarez, 2006).

All worldwide cacti species are located in only 24 countries. Even more, 94% of cacti species are sited exclusively in ten countries, and three groups may be formed, depending on the diversity patterns; Mexico, Argentina, Perú, Bolivia, Chile and Costa Rica can be grouped in one, Paraguay and Cuba can be the second, and Brazil and USA may be the last one (Ortega-Baes and Godínez-Álvarez, 2006).

In Mexico, the greatest cacti incidence has been found in the arid and semi-arid regions. The richest Mexican-cacti areas are: the Chihuahuan Desert Region (CDR), the Sonoran Desert, the Tehuacán-Cuicatlán Valley, and the Metzititlán-Zimapán Valley (Hernández and Godínez, 1994; Hernández and Bárcenas, 1995; Hernández et al., 2004; Hernández and Gómez-Hinostrosa, 2005; Godínez-Álvarez and Ortega-Baes, 2007).

In México the CDR includes the Chihuahua, Coahuila, Nuevo León, Tamaulipas, Zacatecas, San Luis Potosí, Guanajuato, Querétaro and Hidalgo states, and in USA, Arizona, Nuevo Mexico and Texas states. The CDR area embraces 507,000 km^2. This zone is protected from the Sonoran Desert influence by the Sierra Madre Mountain range, and includes quite a few non-desert regions. Most of CDR area has elevations between 1,100-1,500 m.a.s.l., and its annual average temperature is 18.6°C. The highest temperatures are attained in the lowest regions, such as the intermountain furrows (Medellín-Leal, 1982; Muldawin, 2002; Hernández et al., 2004).

Considering the CDR endemic cacti distribution, three sub-regions may be defined: 1) the main region, known as the CDR main part that comprises the arid and semi-arid plains; 2) the meridional sub-region that includes the Queretano-Hidalguense arid zone, the Guanajuato dry land, and the southeast San Luis Potosi region; and 3) the eastern sub-region that involves the Jaumave and Aramberri Valleys (Hernández et al., 2004).

The CDR contains 39 genera with approximately 324 species and five hybrids. The most common genera, depending on species numbers, are: *Mammillaria* (with 79 species), *Opuntia* (with 46 species), *Coryphantha* (with 36 species) and *Echinocereus* (with 30 species) (Hernández et al., 2004). It is well accepted that CDR is the Cacteae-tribe diversity center. Most cacti living in that zone are small and globular, with a little number of them being either arborescent or with barrel shape; nevertheless, a lot of *Opuntia* spp. is observed as well (Hernández and Gómez-Hinostrosa, 2005). For instance, *Opuntia rastrera* can reach populations as numerous as 4,000 plants per hectare or more (Mandujano et al., 2001). The *Platyopuntias* proliferation might be related to various factors, such as an efficient seed-dispersion system, lack of predators and diseases, drought, or selective pasturing to other cacti, among others (Burger and Louda, 1995; Jordan and Nobel, 1982; Bowers, 2005). Other cacti species inhabit specific ecosystems, such as *Ariocarpus* spp., *Astrophytum* spp., *Aztekium* spp., *Epithelantha* spp., *Geohintonia* spp., *Leuchtenbergia* spp., *Lophophora* spp., *Obregonia* spp., *Pelecyphora* spp., *Stenocactus* spp., *Strombocactus* spp. and *Turbinicarpus* spp. (Hernández and Bárcenas, 1996; Gómez-Hinostrosa and Hernández, 2000; Hernández et al., 2004; Sotomayor et al., 2004).

The Sonoran Desert covers some regions of Arizona and California in USA, and most of Baja California peninsula and Sonoran state areas in México. Sonoran Desert area is around 300,000 km^2. Raining season is unpredictable and extremely scarce; temperature differences between seasons are tremendously significant with very hot summers and very cold winters, including snowing and freezing. There are numerous small and arborescent cacti, although columnar cacti are the most representatives, being *Carnegiea, Pachycereus* and *Stenocereus* genera the majority (Yeaton and Cody, 1979; McAuliffe, 1984; Turner et al., 1995).

The Tehuacán-Cuicatlán Valley, between Puebla and Oaxaca states, and the Metzititlán-Zimapán Valley, between Hidalgo and Querétaro states, represents the southern area from North American deserts (Medellín-Leal, 1982). The Tehuacán-Cuicatlán Valley embraces a 10,000 km^2 area and holds 81 cacti species (Dávila et al., 2002). The humid zones of Puebla, Oaxaca, Morelos and Guerrero states generally influence climate in this region; therefore, the

Tehuacán-Cuicatlán Valley has mild temperatures all the way through the year. This zone is considered a speciation center, mostly for columnar and endemic cacti such as *Polaskia chende, Mammillaria zephyranthoides* and *Ferocactus flavovirens*. Up to 1,800 individuals per hectare of columnar cacti have been counted in Zapotilán region. *Cephalocereus columna-trajani, Mitrocereus fulviceps, Myrtillocactus* spp., *Neobuxbaumia* spp., *Pachycereus* spp., *Stenocereus* spp., *Polaski* spp. and *Escontria chiotilla* are the most common species found in that Valley (Arias et al., 1997; Valiente-Banuet and Godinez-Álvarez, 2002; Hernández and Gómez-Hinostrosa, 2005).

Taking into consideration the cacti species distribution and diversity in descending order by Mexican states, are found San Luis Potosí, Coahuila, Nuevo León, Oaxaca, Zacatecas, Tamaulipas and Sonora (Figure 2) (Godínez-Álvarez and Ortega-Baes, 2007). The greatest cacti densities have been detected in the Huizache region (San Luis Potosí), with 75 native species, and in its boundaries *Mammillaria crinita* subsp. *leucantha* and *Pellecyphora aseelliformis* have been found. A single portion of the Huizache area (114 km^2) contains 41 cacti species, comparable to Tehuacán-Cuicatlán Valley, where 76 cacti species have been described; however, the Tehuacán-Cuicatlán Valley is three times bigger than the Huizache region. The majority of Huizache-region cacti species is similar to CDR; nevertheless, cacti species from Queretano-Hidalguense area have also been found (Hernández et al., 2001; Arredondo-Gómez et al., 2001).

It has been proposed that the great Huizache-cacti diversity might be explained by the combined effect of three ecological and biogeographic factors: 1) the mild climate resulting from not having severe seasons; 2) the heterogeneity in microclimates and cacti species; and 3) Huizache region is enclosed by Queretana-Hidalguense, Tula-Jaumave and CDR zones (Hernández et al., 2001). Other zones with elevated cacti densities are located within Coahuila state, particularly in Cuatro Ciénegas region, with 48 species, and La Paila region, with 44 species (Hernández and Bárcenas, 1996; Hernández et al., 2004; Gómez-Hinostrosa and Hernández, 2000).

Recent studies, carried out with 660 cacti species in 32 Mexican states, have confirmed that 512 cacti species (78%) are endemic (Figure 2). The greatest endemic cacti numbers have been detected in San Luis Potosí, Oaxaca and Zacatecas. On the contrary, Quintana Roo and Campeche have the lowest number of cacti species and endemic ones (Godínez-Álvarez and Ortega-Baes, 2007).

After evaluating the effect of several factors such as altitude, raining, temperature, aridness and area (km^2) on Mexican cacti diversity and endemism, it has been concluded that only dryness positively correlated with cacti diversity and endemism rate, whereas the area size completely correlated with richness number species (Godínez-Álvarez and Ortega-Baes, 2007). Additionally, it has been theorized that cacti biodiversity and endemism, particularly within CDR, might be due to the fact that the area was a cacti refugee center during the Pleistocene climate changes, allowing the cacti isolation and diversification (Rzedowski, 1991; Hernández and Bárcenas, 1995, 1996; Gúzman et al., 2003; Hernández and Gómez-Hinostrosa, 2005; Ortega-Baes and Godínez-Álvarez, 2006).

On the other hand, soil type plays an important role regarding the observed cacti distribution and endemism rate (Johnston, 1977; Powell and Turner, 1977). For instance, within the CDR were found obligated gypsophile cacti species, such as *Aztekium riterri, A. hintonii, Geohintonia mexicana* and *Turbinicarpus zaragozae* (Hernández and Gómez-Hinostrosa, 2005). The cacti species selectivity, depending on soil type, was described in the calcifuge *Ferocactus histrix* and the calcicole *Echinocactus platyacanthus* distribution. Both species grow in similar weather and topographic conditions, and their distribution areas are boundaries, but there are no any overlapped areas reported so far (Del Castillo, 1996; Del Castillo and Trujillo, 1997).

Mexican state	Genus	Species	Endemic species	Endemism %	Threatened species	Threatened species (%)
San Luis Potosí	33	151	115	76	69	46
Coahuila	24	126	71	56	53	42
Nuevo León	30	119	71	60	61	51
Oaxaca	32	118	97	82	29	25
Zacatecas	26	112	86	77	37	33
Tamaulipas	31	105	57	54	47	45
Sonora	21	100	42	42	24	24

Figure 2. Summary of Mexican endemic and threatened cacti species. Adapted from Godínez-Álvarez and Ortega-Baes (2007).

Current Status of Mexican Cacti

Cactaceae is one of the most threatened families within the plant kingdom. According to SEMARNAT (Acronym for Secretaría del Medio Ambiente y Recursos Naturales – Natural Resources and Environmental Ministry) official list, published in 2002, 285 cacti species are considered in risk. Almost 40 cacti species are comprised within the CITES appendix I (Convention on International Trade of Endangered Species), which seeks to supervise wildlife (plants and animals), its commercialization, avoiding survival risks. Due to its commercial significance all Cactaceae family is included in appendix II (Hunt, 1999; Luthy, 2001). CITES-appendix I consists of all wild extinction-endangered species that are not allowed to be extracted from their natural habit for commercial purposes, and their global interchange is only allowed under precise conditions, such as scientific trade off or CITES-certificated nursery specimens. CITES-appendix II embraces species that are not currently considered as extinction endangered, but they could easily fall under this term, if their collect in the wild and commercialization is not supervised. Within appendix III are included all wildlife species that might be vulnerable at least in one CITES-country members, who has requested CITES assistance to control their commercialization. There are no any cacti species considered in this category (CITES, 1990). Correspondingly, 48 Mexican cacti (Figure 3) are included in the "Red Data Book of the International Union for the Conservation of Nature" or IUCN (IUCN, 2009) at different threatened degrees. *Mammillaria guillauminiana* is considered as wild extinct cactus, 15 cacti species are in critical extinction endangered, seven are extinction endangered, 20 are catalogued as vulnerable, three are near of being threatened and two are threatened-free, at least currently. *Turbinicarpus* genus is highly threatened with five species in the category of extinction endangered, five species in vulnerable status and two categorized as near to be threatened. *Turbinicarpus* genus is endemic to México, and is habitually found in Coahuila, Guanajuato, Hidalgo, Querétaro, Nuevo León, San Luis Potosí, Tamaulipas and Zacatecas states. Twenty taxa are situated in San Luis Potosí state, where 75% of

Turbinicarpus species are endemic, of which 37% are microendemic (found in no more than one or two locations) (Anderson et al., 1994; Sotomayor et al., 2004).

Bearing in mind the cacti distribution with some degree of threatening, the Mexican states that have the greatest numbers are: San Luis Potosí with 69 species, Nuevo León with 61 species, and Coahuila with 53 species (Figure 2).

Species	Status	Population trend
Acharagma aguirreanum	CE	Decreasing
Ariocarpus agavoides	V	Decreasing
Ariocarpus bravoanus	V	Stable
Ariocarpus kotschoubeyanus	NT	Decreasing
Ariocarpus scaphirostris	V	Decreasing
Astrophytum asterias	V	Decreasing
Coryphantha hintoniorum	V	Decreasing
Coryphantha maíz-tablasensis	E	Decreasing
Coryphantha pycnacantha	E	Decreasing
Cumarinia odorata	V	Decreasing
Echinocactus grusonii	CE	Decreasing
Epiphyllum phyllanthus	LC	In alert, since its natural habit is under destruction
Lophophora diffusa	V	Decreasing
Mammillaria albicoma	E	Decreasing
Mammillaria albiflora	CE	Decreasing
Mammillaria anniana	CE	Decreasing
Mammillaria aureilanata	V	Decreasing
Mammillaria berkiana	CE	Decreasing
Mammillaria duwei	E	Decreasing
Mammillaria guelzowiana	CE	Increasing
Mammillaria guillauminiana	EC	--
Mammillaria herrerae	CE	Decreasing
Mammillaria luethyi	E	Unknown
Mammillaria mathildae	V	Decreasing
Mammillaria microhelia	V	Decreasing
Mammillaria pennispinosa	E	Decreasing
Mammillaria rettigiana	V	Decreasing
Mammillaria sanchez-mejorada	CE	Decreasing
Mammillaria schwarsii	CE	Decreasing
Mammillaria weingartiana	V	Decreasing
Obregoni denegrii	V	Decreasing
Opuntia chaffei	CE	Decreasing
Opuntia megarrhiza	E	Decreasing
Pereskia aculeata	LC	--
Thelocactus hastifer	V	Decreasing
Turbinicarpus alonsoi	CE	Decreasing
Turbinicarpus gielsdorfianus	CE	Decreasing

Turbinicarpus hoferi	CE	Decreasing
Turbinicarpus horripilus	V	Decreasing
Turbinicarpus laui	V	Decreasing
Turbinicarpus lophophoroides	V	Decreasing
Species	Status	Population trend
Turbinicarpus mandragora	CE	Decreasing
Turbinicarpus pseudomacrochele	V	Decreasing
Turbinicarpus pseudopectinatus	V	Decreasing
Turbinicarpus schmiedickeanus	NT	Decreasing
Turbinicarpus swobodae	CE	Decreasing
Turbinicarpus valdezianus	V	Decreasing
Turbinicarpus viereckii	NT	Decreasing

Figure 3. Mexican cacti included in the Red Data Book of the International Union for the Conservation of Nature (modified from IUCN, 2009). LC= Least concern; NT= Near threatened; V= Vulnerable; E= Endangered; CE= Critically endangered.

It has been clearly pointed out that the local level risk of several cacti species can change suddenly to higher menace categories, since more natural habit is destroyed and more poacher activity is going on, giving place to alarming numbers of threatened cacti species (Arredondo-Gómez et al., 2001; Martínez-Ávalos and Jurado, 2005; Ortega-Baes and Godínez-Álvarez, 2006).

The cacti taxa exhibiting the highest extinction-risk degree have showed extremely low population densities. For instance, in the Queretan desert, *Echinocactus grusonii*, *Echinocereus schmollii*, *Mammillaria crinita*, *M. herrerae*, *M. longimamma* and *Turbinicarpus hastifer* disclosed less than 50 individuals; *Ariocarpus kotschoubeyanus*, *Astrophytum ornatum*, *Echinocactus platyacanthus*, *Ferocactus histrix* and *Lophophora diphusa* exposed low densities (between 50 and 500 specimens); or resembling *T. pseudomacrochele* and *Mammillaria microhelia*, which has shown low-seedling numbers in the wild, reducing severely the total recruitment to increase the population. Likewise, the *Acharagma* spp., *Aztekium* spp., *Pelecyphora* spp. and *Turbinicarpus* spp. seedling recruitment is erratic and atypical within the CDR (Anderson 2001; Godínez-Álvarez et al., 2003; Sotomayor et al., 2004; Hernández and Gómez-Hinostrosa, 2005; Hernández-Oria et al., 2007).

The reasons to explain the low density of cacti populations observed in the wild, comprise inherent characteristics to cactus physiology, as well as ecosystem disturbances. Within the first group, reproductive, phenological and ecological limitations may be:

1. The highest habit-specificity and edaphic-specialization shown by cacti communities strongly influence their distribution in the wild. Such as is the case of *Cephalocereus columna-trajani, Escontria chiotilla, Neobuxbamia tetetzo, Pachycereus fulviceps, P. Pringlei, Stenocereus gummosus* and *S. thurberi* (Parker, 1991; Hernández and Godínez, 1994; Valiente-Banuet et al., 1995; Contreras and Valverde, 2002; Valiente-Banuet and Godínez-Álvarez, 2002; Zavala-Hurtado and Valverde, 2003).

2. Nursing requirements, which provide protection against predators, increase nutrients availability, and more important, supply shade for seedlings, are a constraint. For

example, *Echinocactus platyacanthus* (5 cm) seedlings need their association (nursing) to *Agave macroacantha, Castela tortuosa* and *Acacia coulteri* in order to survive. Similarly, *Neobuxbaumia tetetzo* and *Cephalocereus hoppenstedtii* (columnar cacti), and *Coryphantha pallida, Mammillaria colina* and *Mammillaria casoi* (globular cacti) establishment was facilitated by a nursing process in the Zapotitlán Valley. The most beneficial effect of nursing is the solar reduction, e.g. temperature differences at noon between a sunny and a shaded (by nursing) area at the soil level can reach up to 16°C (Valiente-Banuet et al., 1991; Mandujano et al., 2002; Godínez-Álvarez et al., 2003; Castro-Cepero et al., 2006; Hernández-Oria et al., 2006; Jiménez-Sierra et al., 2007).

3. Slow-growth rates and long-reproductive periods of cacti species affect their distribution patterns, since community and population density depends on the relationship between cacti sprouting *vs* mortality (Gibson and Nobel, 1986; Zavala-Hurtado and Díaz-Solis, 1995; Godínez-Álvarez et al., 2003).
4. Low seed production, seed dormancy and low-vigor seed reduce harshly cacti density (Barthlott and Hunt, 2000; Rojas-Aréchiga and Vázquez-Yanes, 2000; Flores et al., 2005; Mandujano et al., 2005; Flores et al., 2008).
5. The presence of auto-sterility phenomenon avoids seed formation (Boyle and Idnurm, 2001).

Several authors agree that the most important reason for cacti population reduction in the wild is the natural-habit devastation, followed by the excessive collect, mostly by poachers (Anderson et al., 1994; Hernández and Bárcenas, 1996; Bárcenas-Luna, 2003; Sotomayor et al., 2004; Hernández and Gómez-Hinostrosa, 2005; Martorell and Peters, 2005; Hernández-Oria et al., 2007). The environmental disturbance by human activities may be categorized into two groups: 1) urban growth, deforestation, overgrazing and agriculture; and 2) opening mining areas, and road and highway constructions. These human activities are increasing everyday, since human population is increasing as well; therefore, more cacti species are expected to fall in the threatened category. Furthermore, it has been extensive documented that overgrazing by horses, cattle and goats, increase desertification and cacti population declination (Hernández et al., 1999; Martorell and Peters, 2005). Predatory actions by animals have increased recently, causing cactus communities to be in peril. For example, Steenberg and Lowe (1969) described that no more than 0.001% of *Carnegiea gigantea* seeds survived to predators such as ants, birds and mammals. Correspondingly, the *T. pseudomacrochele* and *A. kotschoubeyanus* natural habit in the Queretan desert, is reducing severely due to opening mining (Hernández-Oria et al., 2007), and in the Huizache region a similar case is happening with *T. schmiedickeanus* subpopulation (Sotomayor et al., 2004; Hernández-Oria et al., 2007).

It is well known that poachers collect cacti in the wild in order to satisfy the national- and worldwide-cacti market demand. The illegal-cacti extraction is carried out by well-organized bands that work periodically and systematically, mostly at cacti-flowering point since they are much more easily find in that period. Juvenil- and seed-cacti collectors alter their population structure, holding back the species permanence (Anderson et al., 1994; Hernández and Godínez, 1994; Gómez-Hinostrosa and Hernández, 2000; Arredondo-Gómez et al., 2001; Robbins, 2002; Bárcenas-Luna, 2003; Martínez-Ávalos and Jurado, 2005). *Mammillaria guillauminiana* is extinct-categorized in the wild since 1997 due to the excessive collection (www.endangeredspeciesinternational.org). The worldwide-cacti market pose a strong demand for wild individuals, e.g. within San Luis Potosí, the 70% of *Turbinicarpus* species are constantly threatened by the pressure of the international market demand, whereas 16% of that threatened population has been caused by the cactus use for traditional medicine (Arredondo-Gómez et al., 2001; Sotomayor et al., 2004).

The Mexican and Dutch governments seized from 1996 to 2000 more than 5,100 cacti specimens belonging to 75 species, supposedly coming from CDR area. From those cacti, 1,000 were *Mammillaria*, 639 were *Ariocarpus*, and 558 *Ferocactus* species (Bárcenas-Luna, 2003).

Genus	Species for national market	Species for international market	MRP
Ariocarpus	5	6	100
Astrophytum	5	4	100
Aztekium	2	2	100
Cephalocereus	1	1	20
Coryphantha	5	37	68.5
Echinocactus	2	5	83.3
Echinocereus	4	35	59.3
Epithelantha	1	2	100
Escobaria	-	23	92
Ferocactus	8	5	17.2
Geohintonia	1	1	100
Isolatocereus	1	1	100
Hamatocactus	-	1	100
Leuchtenbergia	1	1	100
Lophophora	-	2	100
Mammillaria	34	93	53.8
Myrtillocactus	1	1	25
Neolloydia	1	1	100
Obregonia		1	100
Opuntia	2	47	13.8
Pelecyphora	2	2	100
Peniocereus	-	1	5.6
Sclerocactus	2	19	95
Stenocactus	2	7	70
Stenocereus	1	3	12.5
Strombocactus	1	1	100
Thelocactus	3	12	100
Turbincarpus	8	20	87

Figure 4. National and international market availability of cacti. MRP: Market Representation Percentage. Modified from Bárcenas-Luna (2003).

Cacti are vastly appreciated worldwide due to their beautiful flowers and/or stem distinctiveness. The highest cacti demands are found in Germany, Austria, Sweden, Switzerland, Belgium, USA, France, Holland, Italy and Japan. Cacti can reach extraordinary prices in those countries; for instance, *Pellecyphora strobiliformis* and *Strombocactus disciformis* can cost up to $150 US dollars each (Fuller, 1985), whereas *Geohintonia mexicana* and *Aztekium hintonii* can be sold by $2,000 USD (Franco-Martínez, 1997).In México, numerous Federal regulations and laws have been dictated in order to protect wild life. For example, the Norma Oficial Mexicana (Mexican Official Norm) NOM-059-ECOL-1999 and also the NOM-059-ECOL-2001 were established with the aim of controlling the collection and commercialization of protected species. México pledged to CITES in 1996, and the Instituto Nacional de Ecología (INE) (National Institute of Ecology) was named as the administrative and scientific authority for CITES. Thanks to that pledge to CITES it is possible to demand the devolution of Mexican seized cacti either in Mexico or around the globe. Usually, detained cacti remain, either temporally or in definitive form, in research or high education institutions to preserve them. Likewise, several federal ministries are involved to preserve cacti in the wild, among them: a) the Natural Resources and Environmental Ministry (Secretaría del Medio Ambiente y Recursos Naturales, SEMARNAT, www.semarnat.org.mx) whose main goal is to foster the protection, restoration and conservation of natural resources and ecosystems, and natural services and goods, to propitiate the sustainable development and use of them; b) the Environmental Management and Ecology Ministry (Secretaría de Ecología y Gestión Ambiental, SEGAM, www.segam.org.mx), which is the administrative authority, depending from the State Executive Power, encharged to formulate, conduct and evaluate the Mexican states environmental policy; and c) the Federal Procurement to Environmental Protection (Procuraduría Federal de Protección al Ambiente, PROFEPA, www.profepa.org.mx), which has the responsibility to secure the necessary actions before administrative, judicial and legislative authorities to accomplish an efficient regulation of environmental justice. Likewise, the General Law for the Ecologic Equilibrium and the Environmental Protection is focused on preservation and restoration of the ecologic equilibrium, as well as on the protection of the environment within the Mexican territory (www.diputados.gob.mx/LeyesBiblio/pdf/148.pdf). Similarly, the Wild Life General Law was dictated with the responsibility of conserving and obtaining a sustainable profit of wildlife and their habits inside the Mexican territory (www.semarnat.gob.mx/leyesynormas/ Leyes del sector/vidasilvestre). On the other hand, the National Commission for the Biodiversity (CONABIO www.conabio.org.mx) was established to promote and coordinate actions for the Mexican biodiversity knowledge and its sustainable use.

Nevertheless, the Law and Official Norm are not enough to save the wildlife, if they are not applied accordingly to as they were formulated. Environmental impact studies are very important, advised by expert panels, in order to dictate the conservation actions and strategies to apply mitigation procedures. It is necessary to increase and reinforce the propagation and conservation steps on threatened cacti species and also to elevate the vigilance to avoid the extraction and exportation of illegal plants. Some actions to preserve and increase the cacti populations include the *in situ* and *ex situ* protection and conservation strategies, such as the Protected Natural Areas, where the environmental resources are used under a sustainable approach under ecological and technical criteria.

The National Comission of Protected Natural Areas (Comisión Nacional de Áreas Naturales Protegidas, CNANP) currently manages 166 Federal Natural Areas, which means more than 23 million of hectares. The Mexican Protected Natural Areas, including arid and semi-arid ecosystems are listed in Figure 5 (Melo and Alfaro, 2007;

www.conanp.gob.mx/q_anp.html). Paradoxically, being San Luis Potosí the Mexican state with the greatest cacti diversity and richness, it has only one state reserve named Reserva Real de Guadalcázar. For this reason, it does not appear within the oficial list of CNANP, and it is extremely urgent to modify the Reserva Real de Guadalcázar status to Biosphere Reserve category, giving the immense cacti number species living in that region (Hernández et al., 2001). Using a complementary analysis it is possible to define in which Mexican states the conservation efforts must focus (Godínez-Álvarez and Ortega-Baes, 2007). This will allow preserving specific cacti groups by means of creating Protected Natural Areas. So, the Cacteae tribe conservation (Cactoidea subfamily) could look out San Luis Potosí, Coahuila and Nuevo León states, whereas the Pachycereae tribe preservation should focus on Jalisco, Hidalgo and Oaxaca states (Godínez-Álvarez and Ortega-Baes, 2007; Ortega-Baes and Godínez-Álvarez, 2006). Other conservation strategies might take place in Baja California Sur and Sonora. Pereskioidea and Opuntioidea subfamilies preservation may occur in the majority of Mexican states, with a previous selection by complementary analysis.

The *ex situ* conservation strategies for cacti consist of: 1) maintaining cacti specimens within botanical gardens, such as Jardín Botánico del Instituto de Biología de la UNAM (www.ibiologia.unam.mx/jardin/), Jardín Botánico Regional de Cadereyta "Ing. Manuel González de Cosío" (www.concyteq.edu.mx/Jardin%20Botanico/paginaprincipal.htm), and Jardín Botánico "El Charco del Ingenio" *(*www.elcharco.org.mx/*)*, 2) cacti propagation in nurseries, and 3) micropropagation.

State	Name	Area (ha)
Baja California	El Vizcaíno	2,493,091
Baja California Sur	Parque Nacional de Bahía de Loreto	206,581
Coahuila	Áreas de Protección de Flora y Fauna Silvestres de Cuatro Ciénegas	84,347
Coahuila	Area de protección de flora y Fauna de Ocampo	344,238
Coahuila	Maderas del Carmen	208,381
Chihuahua	Cañón de Santa Elena	277,210
Chihuahua	Área de Protección de Flora y Fauna de Ocampo	344,238
Durango	Reserva de la Biosfera de Mapimí	342,388
Hidalgo	Barranca de Metztitlán	96,043
Hidalgo	ANP de Tula de Allende	100
Nuevo León	Parque Nacional Cumbres	177,396
Oaxaca-Puebla	Tehuacan-Cuicatlán	490,187
Querétaro	ANP El Cimatario	2,448
San Luis Potosí[1]	Reserva Estatal Real de Guadalcázar	225,000
Tamaulipas	El Cielo	144,530
Sonora	El Pinacate y Gran Desierto de Altar	714,557
Sonora	Reserva Especial de la Biosfera Cajón del Diablo	147,000

Figure 5. Mexican Protected Natural Areas with arid and semi-arid ecosystems. (www.conanp.gob.mx).
[1] www.slp.gob.mx.

Propagation and Conservation by Conventional Methods

Since wild cacti are subjected to an enormous national and international demand, they are exceedingly endangered. Sixty-six years ago, Mexican cacti protection, as natural resource, officially started. During that period, Federal government, states, counties, NGOs, scientists and cacti-lovers have collaborated to improve a theoretical and practical paradigm in order to preserve most cacti (Bárcenas, 2006). Likewise, in the richness and threatened cacti ecosystems, such as Sonora and Baja California deserts, several governmental actions have taken place; among them, Natural Protected Areas and Mexican Biosphere Reserves have been declared, such as the Vizcaíno (Northern Baja California Sur, California Gulf and Colorado River delta areas), the El Pinacate and Altar Grand Desert, and Organ Pipe Cactus and Cabeza Prieta (Northwest Sonora and South Arizona). These protected areas embrace 18,000 km^2 in USA and 68,000 km^2 in México, meaning 17% of the total wild area. In 1988, the Biosphere Reserve "El Vizcaíno" was created, covering 2.5 million hectares, being the biggest Mexican Biosphere Reserve. Between 1993 and 1998, four more Natural Protected Areas were founded, all of which include 2.9 million hectares (Riley and Riley, 2005).

In order to conserve most cacti plant resources it is imperative to account with a legal framework to save the Country's rights over those plant assets. Likewise, that legal framework must contribute to norm and control the *in situ* and *ex situ* cacti conservation throught Botanic Gardens, Natural Protected Areas, commercial and private collections. Using these legal figures, significant data should be collected and organized, such as national and regional cacti inventories, propagation and conservation strategies, education and scientific activities, among others. At the end, all these legal frameworks should allow the cacti propagation, conservation and dissemination, as well as the reduction on cacti-threatened species numbers.

Legal framework

With the aim of contributing to find a solution of two main national priorities, accelerated biological richness lose and human quality life deterioration, the Mexican Government, throught SEMARNAT, established the "Programa de Conservación de la Vida Silvestre y Diversificación Productiva en el Sector Rural 1997-2000" (Program of Wildlife Conservation and Productive Diversification for the Rural Sector 1997-2000). This program combines a series of interconnected strategies with economic, social, legal and environmental issues, looking for a permanent, extensive and committed participation of all society segments. This program has also two principal strategies: 1) Recuperation and Conservation of Priority Species, and 2) Unit Systems for the Conservation, Management and Sustainable Improvement of Wildlife (USCMSIW).

In order to develop the Recuperation and Conservation Priority strategy, wild species are selected depending on several facts, such as: 1) if they are categorized in some threatened level; 2) if they are suitable to be recovered and managed; 3) if their protection helps to conserve other species; 4) if they are important species; and 5) if they pose a cultural or economic interest. A fundamental part of this strategy is the creation of Technical Consultive Committees Specialized by species, supported by the competent authority, increasing in this way the social-shared responsibility to manage, conserve and take the most from wildlife. The second strategy consisting of create USCMSIW is a cooperation scheme that aims to promote the development of production alternatives compatible to wildlife care. This system is formed once a "Unidad de Conservación, Manejo y Aprovechamiento Sustentable de la Vida Silvestre (UMA)" (Unit of Conservation, Management and Sustainable Improvement of Wildlife) is established. UMA's concept is based on the creation of sustainable, legal and viable opportunities to profit from wildlife, but avoiding its destruction, endangering and annihilation (Instituto Nacional de Ecología, 2007).

On the other hand, the CITES is an international agreement and started its actions on June 1st. 1975. Up to now, 155 countries, including Mexico, have signed CITES agreement (Benítez and Dávila, 2002). CITES covers wildlife species under three appendixes, which were explained with anteriority. Within appendix I *Ariocarpus, Astrophytum, Aztekium, Coryphantha, Disocactus, Echinocereus, Escobaria, Mammillaria, Melocactus, Obregonia, Pachycereus, Pediocactus, Pelecyphora, Sclerocactus, Strombocactus, Turbinocarpus* and *Uebelmannia* Mexican cacti genera are found. Within the appendix II are the rest of wildlife cacti species, including their seeds (Benitez and Dávila, 2002). Each CITES country member design, under CITES regulation, one administrative authority, which in México case is the "Dirección General de la Vida Silvestre" (Wildlife General Direction) belonging to SEMARNAT, and one scientific authority, which in México is the CONABIO (Benítez and Dávila, 2002).

Generally speaking, plant species conservation may be divided into two approaches: *in situ* and *ex situ*. Natural Reserve establishments can assist both systems as well.

In situ conservation

The *in situ* conservation is achieved when plant material is protected right in the natural spot. This conservation has several advantages over the *ex situ* one, among them: 1) it is possible to maintain the genetic material and keep the natural processes by which diversity is originated; 2) it is feasible to preserve the genetic variation; and 3) it is possible to carry on the conservation of numerous plant species in just one single place.

Gómez-Hinostroza and Hernández (2000) have carried out a study in the Chihuahuan Desert to identify cacti species diversity. They have made a botanic collection in 80 different places within that Mexican state, finding 54 cacti species. Similarly, they have found that these cacti species were not equally distributed along the Chihuahua state, and that most cacti species concentration was at the Southwest state area, e.g. Mier and Noriega region. From all collected species, 82% were endemic, 6% had a much-localized distribution, and 19 species out of 54 were extinction-endangered. For that reason, these authors proposed that these cacti species as well as their habitats should be *in situ* conserved, since García de la Cruz and Cruz-Alvarado (1999) commented that according to the "Diario Oficial de la Federación (1994)" (Official Newspaper of the Federation) an important number of cacti species are extinction-threatened. The principal factors that diminish cacti-species survival, are: 1) natural habit destruction; 2) over-collection for national and international markets; and 3) lack of an efficient law enforcement and legislation to protect them. In order to give an answer to this problem and to maintain cacti diversity, it is extremely important to design permanent strategies to propagate and preserve cacti plant material, *in situ* and *ex situ*, even more, considering that cacti population recovery is very slow.

Pierson and Turner (1998) have indicated that, even in natural protected environments, cacti multiplication is sluggish. For instance, a study on "saguaro" cactus (*Carnegia gigantea*) in Arizona, has demonstrated that despite its wild population in 700 hectares has been protected by fences, the cactus number has duplicated after 85 years. One of the Mexican places with elevated priority to be *in situ* conserved is the Queretan semi-arid region, because this area lodges the greatest worldwide-diversity cacti communities, most of them categorized as threatened species. Hernández-Oria et al. (2007) have evaluated diverse potential factors that might contribute to perturb cacti populations. Their results have suggested that high impact reasons are the lost and modification of natural habits. They have also established that thicket protection is an important feature to avoid cacti disturbance.

A pilot nursery assay has been carried out within the botanic garden "Jardín Botánico Regional de Cadereyta" (Queretaro state) aiming to support the threatened-cacti conservation. Young students from rural communities have participated reproducing extinction-endangered cacti species. The propagated cacti species are: *Astrophytum ornatum* ("biznaga burra"), *Echinocactus grusonii* ("biznaga dorada"), *Echinocereus schmollii* ("organito"), *Mammillaria herrerae* ("bolita de hilo") and *Thelocactus hastifer* ("biznaguita"). In just few months, their propagation work generated more than 560 cacti individuals from those species with appropriated sizes and elevated survival likelihood. These positive results enhance the possibility to improve the extinction-endangered cacti propagation. In the same manner, the National Institute of Agricultural and Livestock Research (Federal government) (Instituto Nacional de Investigaciones Agrícolas y Pecuarias-INIFAP) has developed several projects in San Luis Potosí state involving five private and six social organizations from the rural sector, such as cooperatives ("organizaciones ejidales") to stimulate the productive chain for cacti propagation and commercialization (Arredondo-Gómez, 2007).

A similar propagation and re-establishment in the wild study for commercial purposes has been established for *Ferocactus pilosus* (cacti species considered under Special Protection Official Norm -NOM-059-SEMARNAT-2001) in the "Catorce" county. Two different started-plant materials were used: 1) seedling-suckers (20-40 cm height) from adult plants, and 2) three-years-old greenhouse-seedlings (8 cm height). Both seedling-types were sown at a 2,500 plant-density per hectare. After three years in the wild, cacti populations were evaluated to assess the survival rate, finding 97% survival for both seedling-types. This study has demonstrated that it is possible the establishment of *F. pilosus* in the wild using nursery-propagated cacti (Arredondo-Gómez et al., 2007).

All those previous results on *in situ* cacti propagation enlighten an important issue that must be considered in order to propagate threatened cacti species, e.g. the human factor. Mankind has posed cacti-wild populations under pressure, since human activities have devastated wild habits. It is essential and urgent to merge with responsibility, human activities within *in situ*-cacti conservation programs.

Ex situ conservation

The main goal of *ex situ* plant conservation is to maintain and propagate, out of their natural habitat for long term, threatened-cacti populations under environmental-controlled conditions. With this approach it is possible to support the *in situ* conservation as well. Germplasm banks and wildlife management centers are considered as *ex situ* plant conservation systems. These systems emerged as a complementary alternative for *in situ* conservation programs, in order to preserve the genetic plant material and to avoid the plant genetic erosion. With the purpose of achieve the *ex situ* conservation, seed banks, *in vitro* germplasm, gene banks, field collections and botanic gardens have been established (Benítez, 2001).

In Mexico, the *ex situ* cacti conservation has been mostly carried out in 40 botanic gardens (registered before the SEMARNAT creation), research federal centers, public and private universities, commercial nurseries (authorized by SEMARNAT), and in a number of private nurseries from cacti-lovers. Among the Mexican botanic gardens, the most significant are:

Jardín Botánico del Instituto de Biología de la UNAM (Botanic Garden of Biology Institute from UNAM, Mexico city) (www.ibiologia.unam.mx/jardin/). This is the most prestigious Mexican botanic garden; it has more than 600 cacti species and close to 4,000 specimens. All cacti species are meticulously categorized, propagated and subjected to research. This botanic garden is not only a cacti conservation center, but it is a teaching and diffusion cactus-knowledge institution. Each cactus depicts a sign that describes its mainly essential characteristics and geographic origin. This Mexican garden is the most frequently visited.

Jardín Botánico "El Charco del Ingenio", A. C. San Miguel Allende, Guanajuato state (Botanic Garden "El Charco del Ingenio") (www.elcharco.org.mx/). This botanic garden has more than 500 cacti species, with diverse sizes and forms, such as columnar, globular, arborescent and creeping, among others. The biggest cacti found in that garden are *Cephalocereus senilis, Pachycereus marginatus, Stenocereus eruca, Ferocactus histrix, Echinocactus grusonii, Echinocactus platyacanthus, Peniocereus serpentines*, and *Opuntia, Echinocereus, Selenicereus,* and *Acanthocereus* genera.

El Jardín Botánico Regional de Cadereyta, Querétaro state (Regional Botanic Garden of Cadereyta) (www.concyteq.org.mx/jardin.html). The goal of this garden is to protect cacti species, mostly those belonging to Tolimán quadrant (Chihuahuan Desert, specifically "Queretano-Hidalguense region"). This garden keeps 17 extinction-endangered cacti species, such as *Ariocarpus kotschoubeyanus* (Lem.), *Astrophytum ornatum* (Britton & Rose), *Echinocactus grusonii* (Hildm.), *Echinocactus platyacanthus* (Link & Otto), *Echinocereus schmollii* (Weing.), *Ferocactus histrix* (Lind.), *Lophophora diffusa* (Croizat), *Mammillaria crinita* subsp. *painteri* (Rose ex Quehl), *Mammillaria herrerae* (Werderm.), *Mammillaria longimamma, Mammillaria microhelia* (Werderm.), *Mammillaria parkinsonii* (Ehrenb.), *Mammillaria schiedeana* subsp. *dumetorum* (J.A. Purpus), *Strombocactus disciformis* (Britton & Rose), *Thelocactus hastifer* (Werderm. & Boed.), *Turbinicarpus pseudomacrochele* subsp. *pseudomacrochele* (Backeb.) *Buxb. & Backeb* and *Stenocactus sulphureus* (A. Dietr.).

Jardín Botánico del INIFAP, Campo Experimental Todos Santos, Baja California Sur state (Botanic Garden from INIFAP, Experimental Field Todos Santos) (http://intranet.inifap.gob.mx/cgi-bin/pagina_web/campos_nueva_pn.cgi?uaa=108). This garden is located 80 km from La Paz city, and it is registered within the Mexican Association of Botanic Gardens. This botanic garden has more than 50% of threatened-wild cacti that are in the Cactaceae Checklist (1992).

Vivero y Jardín Botánico Comisión Federal de Electricidad (CFE), "Cactus del Noreste". Address: Calle Orión No. 300, San Nicolás de los Garza, Nuevo León state *(Nursery and Botanic Garden of National Electricity Commission "Northeast Cactus")* (http://www.cfe.gob.mx/QuienesSomos/sustentabilidad/Documents/Vivero%20C actus%20del%20Noreste.pdf). This nursery was established as a consequence of CFE's building and construction work that led to the *Digitostigma caput-medusae* (medusa head) discovery by CFE workers in June 2001. Medusa-head cactus distribution is extremely restricted, which causes that this plant is highly vulnerable. The CFE has also conducted the rescue of other threatened cacti, and their reproduction by seeds and vegetative parts. Afterward, propagated cacti are re-introduced to their natural habitat.

Figure 6. Several Mexican cacti species propagated and maintained at the University of Guanajuato cactus collection. a) *Ferocactus echidne*; b) *Ferocactus emoryi*; c) *Echinocactus grusonii*; d) *Opuntia erinacea*; e) *Mamillaria longimamma*; f) *Lophophora williamsii*; g) *Myrtillocactus geometrizans*; h) *Mammillaria hahniana* subsp. *woodsii*; i) *Ferocactus histrix*; j) *Mammillaria bocasana*; k) *Stenocactus crispatus*; l) *Echinocactus grusonii*.

Jardín Botánico Regional "Cassiano Conzatti" del CIIDIR Oaxaca state. (Regional Botanic Garden "Cassiano Conzatti" from CIIDIR) (http://www.cidiroax.ipn.mx/index.php?option=com_search&searchword=conzatti). Located in Santa Cruz Xoxocotlan city, Oaxaca state, this botanic garden has 218 collects, 34 families, 89 genera and 175 cacti species, from which four are extinction-threatened, seven are threatened and ten are under special protection.

In more than some few public and private Universities there are cacti *ex situ* conservation programs as well (UNAM, 1993). By the cacti specimen number stand out, it should be mentioned UNAM with 3,772; Ecology Institute of Jalapa in Veracruz state, with 304; the Autonomous University Antonio Narro de Saltillo in Coahuila state, with 296; the Center of Scientific Research of Yucatán in Mérida, Yucatán state with 118; and the University of Guanajuato in Irapuato city, Guanajuato state, with 170 cacti species and 900 specimens. Some cacti species propagated and conserved at the University of Guanajuato program are depicted in Figure 6.

A number of Mexican nurseries authorized by SEMARNAT, conserve and commercialized a variety of cacti species. Among them may be found the following:

Cultivadores de Cactus de S.A. de C.V., (Cacti Growers of México Inc.) (http://www.cactusmex.pue-mx.com). This nursery is located in Tlaxcalancingo, Puebla state, at 2,300 m.a.s.l. This nursery manage and commercialize more than 600 cacti varieties from more than 400 cacti species, and also sells cacti seeds from 307 species. Because of that, it is considered as the biggest company for propagation and commercialization of cacti species in Mexico.

Viveros GDV, (Nurseries GDV) (www.viverosgdv.com/). This nursery is located in the Charapendo County, Michoacán state. This nursery propagates and commercializes more than 50 cacti species; some of them are from México, but some others are from South America and the Caribbean region.

Cacti propagation

Cacti propagation may be achieved by asexual (vegetative) or sexual methods. Most of columnar and globular cacti, and *Opuntia* genus are asexually propagated by stem or cladode segments and by suckers that are separated from mother plants. However, a great number of cacti species are propagated merely by seeds.

Vegetative propagation. Columnar cacti species include around 170 species, from which 80 are found in México and 12 of them are regularly utilized as live fences or cultivated in groves and backyards (Bravo-Hollis, 1978). Although most of these columnar cacti are competent of being propagated by seeds, vegetative propagation is easier, faster and more reliable. Spring is the best season for vegetative propagation of columnar cacti when the cacti water content is low. Columnar cacti have "arms" that are stem fragments joined by fibrous material, and they are used as propagation cuttings (30 cm in length or more). *Stenocereus thurberi* ("pitaya"), *Stenocereus stellatus* ("tunillo") and *Escontria chiotilla* ("jiotilla") propagation is achieved by using one meter cuttings, and these species are especially important columnar cacti for rural communities, such as in Oaxaca state. López-Gómez et al.

(2000) have studied "pitaya", "tunillo" and "jiotilla" vegetative propagation by using cuttings from cacti "arms". Their results showed that "pitaya" and "tunillo" can be easily propagated utilizing 0.5 m long "arms", but the apical part must be removed to break the apical dominance, and then the cutting should be planted vertically. It seems that apical dominance is stronger in "pitaya" than in "tunillo". The results for "jiotilla" propagation have been ambiguous, and the authors have concluded that more research should be done for this columnar cactus in order to determine the best vegetative propagation method.

The best cacti phenotypes in the wild and nurseries are selected for vegetative propagation. For *Stenocereus stellatus, S. pruinosus* and *Pachycereus hollianus* vegetative propagation, stems or branches are cut transversally giving place to 1 or 1.5 m length cuttings (Casas, 2002). On the other hand, for *Polaskia chende, P. chichipe, Escontria chiotilla* and *Myrtillocactus schenkii*, stems or branches are cut into around 40 cm length sections (Casas, 2002). Cuttings are left under sunlight for two weeks to facilitate cicatrisation of the cut-area in order to avoid microorganism infections after planting. Later on, cactus cuttings are sown on trenches to which goat's manure has been previously added. The plantation takes place at the end of April or beginning of May, few weeks before the raining season starts (Casas, 2002). Vegetative propagation of cylindroopuntias ("chollas") is carried out with stem segments that are derived from donor plants and then they are rooted. This propagation method has been successfully applied to *Opuntia acanthocarpa, O. bigelovii, O. echinocarpa* and *O. ramosissima* in the Northwest Sonoran Desert and South Mojave areas (Bobich and Nobel, 2001). From these four species, only *O. bigelovii* has been successfully reproduced in the wild; likewise, it has been the only cacti species that has shown a great rooting capacity and weak joints between segments, facilitating the propagation process. What's more, *O. bigelovii* has stems with the greatest spine numbers per segment, which is an advantageous characteristic, since this property increases the probability of vegetative dispersion by animals (Bobich and Nobel, 2001). It has been suggested that those cacti species that are vegetative-propagated have weak joints between segments and possess easy-rooting capacity. In general, the link strength between segments is related to its diameter; therefore, a higher diameter means a stronger link. Link strength also depends on the presence or absence of fibers and parenchyma cell numbers (Bobich and Nobel, 2001).

Andrade et al. (2006) have studied the effect of sunlight on stem growth in *Hylocereus undatus* at Yucatán state (Mexico). These authors have tested the influence of 25, 36, 48 and 90% of sunlight treatment for 55 weeks. Their results showed that stem segments that received 36 and 48% of sunlight increased 67% their length compared to those that received 25 and 90% of sunlight. It was also observed that a sunlight irradiance of 25% was not enough to accomplish an efficient photosynthesis, and at 90% there might be a photosynthesis-saturation process.

De Andrade and Martins (2007) tested a number of factors, such as cutting source and type, and cicatrisation time on the vegetative propagation of *Hylocereus undatus*. They used propagation stems from three sources (cuttings from adult plants, young stems from rooted cuttings and from young plants coming from seeds), and three scar-timing periods (0, 7 and 15 days after being cut). After 30 days of incubation, rooting, root volume, principal root length (cm), dry and fresh root weight (g), and number and size of new sprouts (cm) were evaluated. The best vegetative propagation response was obtained from stems with 0 days of cicatrisation time, independently of the stem source. Consequently, it is recommended that *Hylocereus undatus* plantation must be carried out just after the stems are cut.

Segment-stem juvenility is other essential factor that contributes to cacti vegetative propagation rate and success (Cavalcante and Martins, 2008). Juvenile stems are the preferred organ source for propagation when the goal is to obtain cacti for ornamental purposes, whereas older stems should be used when the aim is the propagation of cacti material for fruit production. Cavalcante and Martins (2008) studied the effects of *Hylocereus undatus* juvenility stage on rooting. They found that cutting's position on mother plant has a quantitative effect on rooting; e.g. juvenil (top's mother plant) cuttings showed a 35% more rooting compared to adult (bottom's mother plant) cuttings. Root density, area, dry weight and length were higher for juvenile cuttings than older ones as well. These results demonstrated, at least for *Hylocereus undatus,* that juvenility plays an important role for vegetative propagation. Other cacti such as *Cylindropuntia fulgida* var. *mamillata* are easily propagated by almost any type of cutting during their entire life cycle independently of their position in the mother plant (Bobich, 2005). Most *Opuntia ficus-indica* varieties are propagated through cladode, commonly known as "penca". Several cladodes are used as cuttings when cactus plant material can be easily available. On the contrary, if cactus availability is scarce a cladode fragment bearing an areole may create individual cuttings. In general, cladodes from the plant bottom are used as cuttings, avoiding those that have already developed corky epidermis. Once each cladode is cut in several pieces, they are left under shade for 15 days to let cicatrisation to occur (Manzanero-Medina and Flores-Martínez, 2008). In addition to columnar and cladode cacti, the globular cacti, such as *Mammillaria oteroi* and *Mammillaria kraehenbuehlii,* that are usually propagated by seeds, can be also vegetative-propagated (Manzanero-Medina and Flores-Martínez, 2008).

Sexual propagation. In order to preserve cacti biodiversity, it is of paramount importance to design reliable, affordable and efficient strategies with the aim of propagating and preserving cacti plant material by indefinite periods, mostly considering that cacti wild populations are in truth time-consuming to recover by themselves (Pierson and Turner, 1998). Seeds have frequently been chosen to achieve sexual cacti propagation and preservation; however, cacti seeds may show dormancy, restricting their easy germination. Therefore, cacti seed germination studies are required in order to attain their efficient propagation (Rojas-Aréchiga and Vázquez-Yanes 2000). Fortunately, official and private Botanic Gardens are excellent sources for cacti seeds, and the application of dormancy-breaker treatments, such as keeping seeds at 3-4 °C in dry atmospheres, is frequently utilized at commercial cacti propagation level.

Rojas-Aréchiga and Arias (2007) commented that most of seed germination cacti studies have been focused on the effect of several factors (light, temperature, water, mechanical and chemical scarification, plant hormone addition, maturation period, use of mother plant, seed size and age, soil volume, water potential, "rareness" and germination rate correlation, and dormancy) on different cacti species. Elevated seed dormancy (low germination rate) and cactus attractiveness can make a cactus species considered as "rareness". This term means that those cacti that are highly demanded by the market might have a higher threatening-species level (Ramírez-Padilla and Valverde, 2005). Correlation between species "rareness" and seed dormancy level has been studied by Ramírez-Padilla and Valverde (2005) in *Neobuxbaumia macrocephala*, *N. tetetzo* and *N. mezcalaensis*. These authors applied some treatments (different temperatures, immersion into HCl, darkness and diverse water potentials) to cacti seeds from those species. Interestingly, the most "rare" species, e.g. *N.*

macrocephala, showed the lowest seed germination rate. *N. mezcalaensis* (the most common) had photoblastic germination, and the water potential negatively affected the seeds.

Regarding seed size, it is well known that seeds from most cacti species weight less than one milligram. This is an advantage since each gram of cacti seeds holds more than 1,000. Nevertheless, this may be a drawback as well, giving that predators are able to consume high amount of cacti seeds with no trouble. Loza-Cornejo et al. (2008) studied the *Escontria chiotilla, Myrtillocactus geometrizans, Neobuxbaumia mezcalaensis, N. multiareolata, Pachycereus grandis* and *Stenocereus queretaroensis* seed germination in order to test if cacti seed size might have an effect on germination rate. These authors found that *E. chiotilla* and *M. geometrizans* seeds, which are smaller and lighter than other cacti species, germinated 15 days later than *Neobuxbaumia mezcalaensis, N. multiareolata, Pachycereus grandis* and *Stenocereus queretaroensis* seeds, showing phylogenetic differences. Similarly, Sánchez-Salas et al. (2006) reported that *Astrophytum myriostigma* seed germination is affected by seed size. In controversy with Loza-Cornejo et al. (2008) results, small seeds showed higher germination percentages than the control for all treatments except scarification, for which germination was indistinguishably low for both seed size categories. Germination speed varied in response to seed size and treatment x size interaction. Small seeds germinated faster (3.8 seeds/day) than large seeds (1.6 seeds/day). In conclusion, small seeds showed higher and faster germination than large seeds. In general, small seeds might have higher germination percentages as a result of faster water uptake.

Seed age is another important factor to be considered for cacti propagation (Sánchez-Salas et al., 2006). Cacti seed viability depends on cacti species and plant growing conditions. For instance, Sánchez-Salas et al. (2006) studied *Astrophytum myriostigma* seed germination using four-years-old seeds, and they found that small seeds had a 91% germination rate, whereas big ones depicted a 39%. Likewise, Manzanero-Medina and Flores-Martínez (2008), using *Mammillaria oteroi* and *Mammillaria kraehenbuehlii*, determined that seed age also affects the germination rates. Their results showed germination rates for *Mammillaria oteroi* of 77%, 97% and 65% at four, six and fifteen month-old seeds, respectively. In *Mammillaria kraehenbuehlii*, seed germination was 71-78%, 92%, 45% and 88% for control, acid, darkness and darkness + acid treatments, respectively. These authors concluded that acid treatment reduced from 30 to 10 days the germination process.

Flores and Jurado (2008) reported the effect of seed age and putrescine (dormancy breaker) addition on seed germination of *Turbinicarpus lophophoroides* and *Turbinicarpus pseudopectinatus*. Both cacti species are currently under special protection and produce dormant seeds. One to four-years-old seeds treated with three putrescine levels (0.0, 0.1 and 100 µM) were used in this study. Regarding seed age, *Turbinicarpus lophophoroides* had the greatest germination rate (66%) using four-years-old seeds, whereas for *T. pseudopectinatus*, one-year-old seeds showed the greatest germination percentage (48%). Putrescine reduced germination rates compared to control treatments in both species. Flores-Martínez and Medina (2008) reported that *Mammillaria huitzilopochtli* did not exhibit seed dormancy, since a 90% germination rate (seven days after being wetted) was observed when one-year-old seeds were utilized; nonetheless, germination percentage decreased to the extent they aged. *M. huitzilopochtli* seeds were not able to germinate efficiently, when they were kept at room temperature for more than two years. Their germination rate was below 14%.

Sulphuric acid has been used to overcome dormancy in cacti seeds (Álvarez-Aguirre and Montaña, 1997; Mandujano et al., 2005; Sánchez-Salas et al., 2006). Sulphuric acid dissolves the hard and impermeable seed coat, assisting root protrusion. When acids are used to disrupt mechanically seed dormancy, the treatment is known as chemical scarification. For instance, Álvarez-Aguirre and Montaña (1997) evaluated the germination of *Cephalocereus chrysacanthus, Cephalocereus hoppenstedtii, Ferocactus latispinus, Stenocereus stellatus* and *Wilcoxia viperina* seeds subjected to chemical scarification, among others. All those Mexican cacti are endemic species and they are used as ornamental plants at their juvenile stages. Mechanical and chemical scarification treatments, as well as a control were tested. *C. chrysacanthus* had the lowest germination rate (36%), while the other four species showed a 79% germination average. Chemical scarification and control were not significantly different, but mechanical scarification was deleterious for germination, since the germination rate was 50% lower than the control. Survival rate, after one year, was 65% for *W. viperina*, and 18.3% on average for the rest of cacti species.

On the other hand, Mandujano et al. (2005) investigated the germination of mature seeds of *Opuntia rastrera* using sulphuric acid, water imbibition and mechanical scarification as pre-treatments. *O. rastrera* seeds require at least one year in order to mature, so that the 50% will be able to germinate. Similarly, Altare and Trione (2006) established that *Opuntia ficus-indica* seeds, as many *Opuntia* species, showed a low germination rate, mostly due to their lignificated integuments. Their main goal was to speed the germination process by means of using several physical and chemical agents. Sulphuric acid treatment, followed by incubation of seeds in a H_2O_2 diluted solution under photoperiod induced the greatest germination rates in the shortest time.

Light is other essential factor that determines cacti seed germination (Zúñiga et al., 2005). It is far and wide known that some cacti species display photoblastic (a positive germination seed process under light) or photodormant (a negative germination seed effect under light) seed behavior (Zúñiga et al., 2005). For example, Zúñiga et al. (2005) studied the spatial distribution of *Lophophora diffusa*, a Mexican endemic endangered-threatened cactus species in Queretaro state (México), using SADIE (**S**patial **A**nalysis by **D**istance **I**ndic**E**s). The analysis of *L. diffusa* spatial distribution and its association to shrub dominant species, as well as the microclimate conditions (light, temperature and humidity) where *L. diffusa* thrives protected by nursery plants, revealed that *L. diffusa* and some bushes showed an aggregated distribution with patches and gaps. *Lophophora diffusa* thriving was positively associated to arborescent vegetation, especially to *Larrea tridentata* and *Acacia sororia*, but it was negatively associated to *Celtis pallida* and *Myrtillocactus geometrizans*. The microclimate evaluation indicated that *C. pallida* canopy significantly reduced the light and temperature received by *L. diffusa*, compared to the other arborescent plant species. The authors concluded that *C. pallida* canopy reduced light availability to *L. diffusa* seeds and seedlings, diminishing cactus establishment.

Flores et al. (2006) determined the seed photodormancy behavior of 28 cacti species from the Chihuahuan Desert region. Eleven cacti seed species, out of 28, produced non-dormant seeds under light presence, with germination rates superior to 70%; therefore being photoblastic.

Water potential is another evaluated factor regarding cacti seed germination. Guillen and Benítez (2009) compared seed germination rates in the columnar cacti *Stenocereus pruinosus, Polaskia chichipe, Myrtillocactus schenckii* and *Polaskia chende*. *S. pruinosus* and *P. chichipe* seed germination was significantly affected under low water potential, whereas *M. schenckii* seeds had the best germination rate at -0.2 MPa. Surprisingly, *P. chende* seeds germinated well, irrespectively of the used treatment. Seed germination, from wild and cultivated plants, did not show any significant differences. Therefore, it seems that each cactus species needs specific water potential in order to achieve its greatest germination rate.

If seed cacti germination is going to be used as cacti propagation and preservation method, it is important to consider not only their germination rate, but also to obtain the best seedling development pattern. For instance, De Andrade and Martins (2008) investigated the volume-size substrate influence on *Hylocereus undatus* seedling development and they observed that cactus survival and seedling development was positively correlated to volume-size cell, since the greater volume-size used, the higher increase in survival and bigger seedlings were obtained.

Biotechnological Approaches for Cacti Propagation and Conservation

One of the main barriers for cacti conservation is that some species are highly difficult to be massively propagated either sexually or asexually. This fact possesses a greater challenge in the case of cacti species where asexual propagation is totally impaired or significantly reduced. Seed germination is the only propagation option for those species. However, propagation by seeds may have some of the following inconveniences: 1) several cacti species show sexual incompatibility process, due mainly to a severely diminished cacti population in the wild and consequently a decreased cross pollination; 2) seed germination rates are extremely low because of seed dormancy, e.g. a number of *Turbinicarpus* threatened-species had a 8% germination rate (Flores et al., 2005; Flores et al., 2008); and 3) the initial cacti-seedling development from seeds is very slow, and the survival percentage is exceptionally low due to wild predators and low-drought tolerance (Nobel, 1998). Because of these reasons some cacti species are extraordinarily complicated to be propagated by conventional methods, and the marketed specimens consist mostly from illegal wild gathering. Within this context, biotechnology is an invaluable tool for a fast, efficient, reliable and easy approach for the massive cacti propagation. Plant Tissue Culture Methods (PTCM) are biotechnological systems to maintain, develop and manipulate plant cells, tissues, organs or complete individuals under controlled and artificial conditions (Nuñez-Palenius et al., 2006). PTCM are carried out under axenically (microorganism-free environment) conditions. This methodology is based on inoculated plant tissues, previously superficial-disinfected, into artificial media cultures that include all plant-required nutrients. These media contain macro and micronutrients, a carbon source (mostly sucrose), and organic compounds such as vitamins, amino acids and myo-inositol. Agar, Gelrite® or other gelling agents may be added if the culture medium is going to be semi-solid. Furthermore, plant growth regulators or phytohormones (auxins, cytokinins, gibberellins, among others) are usually included in the culture media to stimulate plant cell differentiation, de-differentiation or multiplication processes, and finally to produce callus growth, organ formation (shoots, roots or somatic embryos) or development of pre-existing buds.

One of the utmost attractive PTCM's opportunities is to generate *de novo* plants from somatic tissues by culturing *in vitro* plant segments. These *in vitro* plants are artificially produced, but afterward, they may be adapted and transferred to soil where they generally develop and grow as normal plants. This multiplication process is known as micropropagation or massively *in vitro* propagation. This technique allows the multiplication of unlimited plant numbers, in short time and reduced spaces, without needing huge amounts of seed or vegetative plant material to start the process. In addition, this is a clonally propagation system, since new plants are genetically identical to progenitor plant material, being suitable this technique to be successfully applied to elite plants, as well as to obtain pathogen-free plants.

The cacti micropropagation system involves four main stages (Murashige, 1974):

Stage I. Axenic culture establishment

This stage consists of explant selection (plant segment that is going to be cultured) and its surface-sterilization to start the axenic culture. In theory, any plant segment containing an elevated alive-cell number might be used as explant. In regard to cacti, cultures may be initiated from germinated seeds, previously surface-sterilized, on artificial media (Figure 8a). In this case, the resulting axenic cacti seedlings are utilized directly as explants for the next micropropagation stages. Other possibility is to use adult plant tissues as explant sources, such as those containing areoles from which new shoots might be developed (Figure 8b); these new shoots may be further utilized as explants as well. Nevertheless, the surface-disinfection process may be not so easy, since cacti bear spines, thick cuticle and irregular surfaces that usually complicate the disinfectant penetration, and consequently difficulting destruction of contaminant microorganisms. Besides, juvenile cacti tissues from seedlings respond better than those from older plants. Explant surface-disinfection is carried out by a series of rinses with detergent (fungicides might be also added), followed by ethanol and sodium or calcium hypochlorite treatments. Sanitizer type and exposition time must be empirically determined for each cacti species and explant types. If explant contamination persists despite the antiseptic treatment, antibiotics or commercial plant sterilizers should be tested. It must be considered that disinfectants distress the plant tissue as well; therefore, a dilemma should be deciphered each time in order to find an optimal treatment for each explant type. How to remove all microorganism-contamination, but not killing and maintaining viable the plant tissue? That issue has to be addressed for each cactus species and explant type.

Stage II. Propagule multiplication

This stage is where massive plant propagation takes place, obtaining a great number of new plants or shoots from tiny explants. There are three plant multiplication or *in vitro* regeneration systems:

(a) Propagation from meristems

This system generates new shoots from pre-existing plant meristems; consequently, it does not entail cell de-differentiation and re-differentiation processes, such as in organogenesis and somatic embryogenesis routes. Plant propagation from meristems is based in the fact that because apical and axillary meristems (within buds) are undifferentiated growing zones, they have the ability to form new shoots. When plant buds are *in vitro* cultured, new shoots emerge easily, then they are rooted and new plants are produced. Regarding cacti condition, axillary buds are contained within a special structure named areole; subsequently, this system is known as "areole activation". This plant propagation scheme requires cytokinin addition to culture media at different doses depending on cactus species; the most commonly used cytokinins are: 6-benzylaminopurine (BAP) also known as 6-benzyladenine (BA), 6-(γ,γ-dimethylallylamino) purine (2iP), and kinetin (6-furfury laminopurine). Recently, meta-topolin [6-(3-hydroxybenzylamino) purine] was used for *Browningia candelaris in vitro* mass propagation with superior results regarding the number of regenerated shoots, compared to other cytokinins (Sánchez-Morán and Pérez-Molphe-Balch, 2007). For efficient areole activation, it is important to find the most appropriated cytokinin type and concentration for each cactus species, in view of the fact that the response might be variable (Figure 8c). The lowest cytokinin level that gives place to a good areole activation must be used, since higher cytokinin levels might induce negative responses, such as callus growth and hyperhydricity, commonly found in *in vitro* cacti cultures (Elias-Rocha et al., 1998). By areole activation, somaclonal variation and genetic alterations in regenerated plants are usually absent or are found at low levels compared to other regeneration systems. This is because there is no callus growth involved, only pre-existing meristem activation (Vyscot and Jára, 1984).

(b) Organogenesis

This plant regeneration method involves *de novo* organ formation from cultured explants. Hence, a cell re-differentiation process takes place in the plant tissue. Adventitious roots and shoots may be formed by this regeneration system. Adventitious shoots are bud-like structures, capable of being rooted and form a complete plant. Unlike propagation by meristems, shoots from organogenesis are not derived from pre-existing meristems; rather they are formed from somatic cells by de-differentiation and re-differentiation processes. Cacti organogenesis can be induced by cytokinins alone or combined with auxins. Direct (no callus-mediated plant formation) and indirect (callus-mediated plant formation) organogenesis may be present in cacti *in vitro* cultures. Indirect organogenesis was observed in *Turbinicarpus laui* (Mata-Rosas et al., 2001), *Selenicereus megalanthus* (Pelah et al., 2002) and *Notocactus magnificus* (de Medeiros et al., 2006) *in vitro* cultures. Indirect organogenesis may be highly productive due to the great number of shoots produced from each multiplication cycle. However, this propagation system involves a callus formation event; because of that, regenerated cacti might have somaclonal variation or genetic changes (Oliveira et al., 1995). Therefore, indirect organogenesis pathway is not recommended when genetic integrity is a goal.

The areole activation system is greatly desirable when wild cacti species must be propagated and conserved devoid of genetic alterations. Because of that, areole activation has broadly been used as the most widespread system for cacti regeneration (Figure 7). An additional advantage of this system is that secondary stem proliferation could be stimulated giving place that more shoots per initial explants can be obtained. This event happens when the initial induced shoots are kept in the same medium, and the primary-shoot areole will produce more shoots (Figure 8d). For instance, in *Pelecyphora aselliformis* and *P. strobiliformis* cultures, up to 128 and 136 shoots per explant, respectively, can be obtained due to secondary stem proliferation (Pérez-Molphe-Balch and Dávila-Figueroa, 2002).

Cacti species	Reported achievement	Reference
Erytrorhipsalis pilocarpa, Epiphyllum grandiflorum, E. phyllanthus, Hylocereus calcaratu, Mammillaria elongata, Opuntia polyacantha, Rhipsalis teres, Weberocereus biolleyi	Tested several auxin:cytokinin combinations, accomplishing shoot production in some species.	Johnson and Emino (1979)
Astrophytum myriostigma, Mammillaria carmenae, M. prolifera, Trichocereus spachianus	*In vitro* propagation using areole as explants and auxin:cytokinin combinations.	Vyscot and Jára (1984)
Leuchtenbergia principis	*In vitro* propagation using auxin:cytokinin combinations.	Starling (1985)
Opuntia amyclaea	*In vitro* propagation by areole activation using cytokinins.	Escobar et al. (1986)
Ferocactus acanthodes	*In vitro* propagation using auxin:cytokinin combinations.	Ault and Blackmon (1987)
Mammillaria haageana, M. san-angelensis	*In vitro* propagation of almost extinct species by areole activation using cytokinins.	Martínez-Vázquez and Rubluo (1989)
Aztekium ritteri	*In vitro* culture with low propagation rate.	Rodríguez-Garay and Rubluo (1990)
Escobaria missourensis, E. robbinsorum, Mammillaria wrightii, Pediocactus bradyi, P. despaini, P. knowltonii, P. paradinei, P. winkleri, Sclerocactus mesae-verdae, S. spinosior, Toumeya papyracantha	*In vitro* propagation using cytokinins alone or combined with low auxin levels.	Clayton et al. (1990)
Sulcorebutia alba	*In vitro* propagation by areole activation using cytokinins.	Dabekaussen et al. (1991)
Mediocactus coccineus	*In vitro* propagation and somatic embryogenesis.	Infante (1992)
Cereus peruvianus	Shoot generation - callus tissue.	Oliveira et al. (1995)
Turbinicarpus pseudomacrochele	Somatic embryogenesis from adult tissues.	Torres-Muñóz and Rodríguez-Garay (1996)
Mammillaria candida	*In vitro* propagation by areole activation using cytokinins.	Elias-Rocha et al. (1998)
Astrophytum myriostigma, Cephalocereus senilis, Coryphantha clavata, C. durangensis, C. radians, Echinocactus platyacanthus, Echinocereus dubius, E. pectinatus, Echinofossulocactus sp., Ferocactus hamatacanthus, F. histrix F. latispinus, F. pilosus, Mammillaria candida, M. craigii, M. Formosa, M. obscura, M. sphacelata, M. uncinata, Nyctocereus serpentinus, Stenocactus coptonogonus	*In vitro* propagation by areole activation using cytokinins alone or combined with low auxin levels.	Pérez-Molphe-Balch et al. (1998)
Coryphantha elephantidens	Shoot generation -callus tissue.	Bhau (1999).
Turbinicarpus laui	Shoot generation -callus tissue.	Mata-Rosas et al. (2001)
Mammillaria elongata	Shoot generation -callus tissue.	Papafotiou et al. (2001)

Figure 7. (Continued).

Acharagma aguirreana, Astrophytum ornatum, Coryphantha elephantidens, Ferocactus flavovirens, Mammillaria bocasana, M. oteori, Pachycereus schottii, Pilosocereus chrysacanthus, Stenocereus stellatus, Thelocactus hexaedophorus	*In vitro* propagation by areole activation using cytokinins.	Castro-Gallo et al. (2002)
Pelecyphora aselliformis, P. strobiliformis	*In vitro* propagation by areole activation using cytokinins.	Pérez-Molphe-Balch and Dávila-Figueroa (2002)
Carnegiea gigantea, Pachycereus pringlei, Stenocereus thurberi	*In vitro* propagation by areole activation using cytokinins.	Pérez-Molphe-Balch et al. (2002)
Hylocereus undatus	*In vitro* propagation by areole activation using cytokinins.	Mohamed-Yasseen (2002)
Escobaria minima, Mammillaria pectinifera, Pelecyphora aselliformis	*In vitro* propagation by areole activation using cytokinins.	Giusti et al. (2002)
Opuntia ellisiana	*In vitro* propagation using cytokinins and auxins.	Juárez and Passera (2002)
Selenicereus megalanthus	Propagation by organogenesis.	Pelah et al. (2002)
Ariocarpus kotschoubeyanus	Propagation by organogenesis and somatic embryogenesis.	Moebius-Goldammer et al. (2003)
Pelecyphora aselliformis	Propagation by organogenesis.	Santos-Díaz et al. (2003)
Mammillaria gracillis	Shoot generation of callus tissue.	Poljuha et al. (2003)
Astrophytum asterias, Echinocactus grusonnii, Coryphantha werdermannii, Echinocereus adustus, E. delaetii, E. ferreirianus, Epithelantha micromeris, Ferocactus cylindraceus, Morangaya pensilis	*In vitro* propagation by areole activation using cytokinins.	Lizalde et al. (2003)
Rhipsalidopsis (R. gaertneri X R. rosea), Schlumbergera (S. russeliana X S. truncata)	*In vitro* propagation using cytokinins and auxins by organogenesis.	Sriskandarajah and Serek (2004)
Turbinicarpus laui, T. lophophoroides, T. pseudopectinatus, T. schmiedickeanus subsp. Flaviflorus, T. schmiedickeanus, subsp. Klinkerianus, T. schmiedickeanus subsp. Schmiedickeanus, T. subterraneus, T. valdezianus	*In vitro* propagation by areole activation using cytokinins.	Dávila-Figueroa et al. (2005)
Notocactus magnificus	Shoot generation callus tissue.	De Medeiros et al. (2006)
Opuntia ficus-indica	Somatic embryogenesis.	Ferreira-Gomes et al. (2006)
Mammillaria bocasana, M. densispina, M. hahniana, M. hutchinsoniana, M. orcutii, M. pectinifera, M. perbella, M. picta, M. rhodantha, M. zephyranthoides	*In vitro* propagation using cytokinins and auxins.	Ramírez-Malagón et al. (2007)
Browningia candelaris	*In vitro* propagation by areole activation using cytokinins.	Sánchez-Morán Pérez-Molphe-Balch (2007)
Echinocereus knippelianus, Echinocereus schmollii, Mammillaria carmenae, M. carmenae f. rubrisprina, M. herrerae, M. theresae, Melocactus curvispinus, Escontria chiotilla, Polaskia chichipe	*In vitro* propagation by areole activation using cytokinins.	Retes-Pruneda et al. (2007)

Figure 7. In vitro propagation of different cacti species including some threatened species

Figure 8. An assortment of *in vitro* propagated Mexican cacti species. a) *In vitro* germination of *Coryphantha poselgeriana;* b) areole sprout from *Backebergia militaris* adult tissue; c) shoot formation from *Mammillaria herrerae* areole treated with 6 mg L^{-1} 2iP (top), 4 mg L^{-1} 2iP (middle) and 1 mg L^{-1} BA (bottom) (notice dissimilar responses to each treatment); d) *Turbinicarpus subterraneus* secondary proliferation; e) *Ariocarpus kotschoubeyanus* embryogenic callus showing different embryo stages; f) *Turbinicarpus pseudomacrochele* subsp. *pseudomacrochele* rooted shoots; g) *in vitro* propagated threatened cacti species developing under greenhouse; h) *in vitro* propagated *Ferocactus histrix* five-months later of being transplanted to the wild.

(c) Somatic embryogenesis

Somatic cells under specific *in vitro* cultures are able to undergo a series of genetic, biochemical and physiological events that resembles the zygotic embryogenesis process. By this "somatic pathway", named somatic embryogenesis, it is possible to generate plant embryos under *in vitro* culture (Figure 8e). This plant regeneration route is the most convenient for massive plant production. This is because somatic embryos are capable to germinate and form a complete plant similarly as zygotic embryos do, with the exemption that somatic embryogenesis is an asexual propagation via, and the formed embryos are genetically identical to the somatic progenitor cell. Similarly as organogenesis, somatic embryogenesis may be direct or indirect (previous callus formation). Indirect somatic embryogenesis is more commonly used, and in order to obtain callus formation, synthetic auxins are added to the culture media. Later on, auxins are removed from culture media to facilitate the somatic embryo differentiation. During somatic embryo development, a series of intermediate stages, correspondingly as zygotic embryogenesis, are observed. For dicotyledonous plants, the developmental intermediate stages are: globular, heart and torpedo shape. In regard to cacti, a somatic embryogenesis system has been reported for *Ariocarpus kotschoubeyanus* (Moebius-Goldammer et al., 2003). Similarly as indirect organogenesis, the main drawback for indirect somatic embryogenesis is the formation of callus growth that might induce somaclonal variation or genetic variations during *in vitro* culture.

Stage III. Shoot growth and rooting

In general, direct and indirect organogenesis produces tiny plants devoid of roots. Likewise, direct and indirect somatic embryogenesis generates small embryos. Both, somatic embryos and shoots are unable to survive under *ex vitro* conditions. In this third stage, the main goals are rooting the shoots, and then stimulate the rooted shoots and embryos to grow (Figure 8f). Separating the new shoots from the original explants and transferring them to a diluted medium culture, with low amounts of auxins and/or activated charcoal, may induce the rooting process. Alternatively, transferring the new shoots directly to the final substrate, adding previously a commercial rooting substance to the shoot, and maintaining them under a high relative humidity may also induce rooting. Although, plant mortality can be substantially increased by this system, a micropropagation step is saved, thus reducing considerably time and costs. In the cacti particular case, it has been observed that residual exogenous cytokinin in shoots from the multiplication step can restrain the rooting process (Pérez-Molphe-Balch and Dávila-Figueroa, 2002). This effect has been more frequently observed when BA or other synthetic cytokinins are used, since there are no natural mechanisms in the plant to metabolize them. In these cases, activated charcoal may reduce the negative-cytokinin effect on rooting (Hemphill et al., 1998).

Stage IV. Plant hardening

Plants generated through PTCM require a hardening process in order to survive *ex vitro* conditions (Figure 8g). This is because during *in vitro* culture they are exposed to high humidity conditions, and their mechanisms to avoid water lose are not normally developed. For example, an *in vitro* plant possesses a weak-formed cuticle, and its close-stomata system is atrophied. Likewise, during *in vitro* culture carbon is normally supplied through sucrose, and then plants are not autotrophic, and their leaves are thinner and with lower chlorophyll content than normal plants. For all these reasons, hardening process must be as gradually as achievable, regarding humidity and light intensity, to avoid *in vitro* plants to suffer a water-shock. At the same time, it is imperative to let *in vitro* plants to develop a normal photosynthetic apparatus. Moreover, *in vitro* plants grow under axenic conditions, so their defense mechanisms against pathogens are also not well developed. During hardening is exceedingly recommended to work under hygienic circumstances as much as possible. For instance, all medium culture residues, rich in nutrients, attached to *in vitro* plants must be carefully removed. Lastly, a paramount aspect to consider is the substrate used for the *in vitro* seedlings transplantation; the substrate should be low in organic matter content and should have an excellent drainage, mostly for cacti. A mixture of commercial substrate for potting and sand (1:1 v/v) will work for cactus transplanting. Additionally, if *in vitro* rooted cacti are allowed to dry at room temperature for three days before transferring them to the final substrate, the survival rate will be substantially increased (Ramírez-Malagón et al., 2007).

The massive *in vitro* plant propagation through PTCM has been successfully applied to diverse Mexican and American cacti species and varieties (Figure 7 and 8). In fact, it would be possible to assess that Cactaceae is a plant family with excellent propagation results through biotechnology. For instance, Retes-Pruneda et al. (2007) have reported that *in vitro* cultures of *Echinocereus knippelianus, E. schmollii, Mammillaria carmenae* and *M. theresae* are able to produce an average of 8.9, 13.5, 7.7 and 8.2 shoots per explant, respectively. These values are obtained during a 50 days incubation cycle. Nonetheless, taking into account that this plant multiplication system is exponential rather than lineal, then in five culture cycles (250 days) it would be expected to achieve 55,840, 448,403, 27,067 and 37,074 shoots per one single explant from *Echinocereus knippelianus, E. schmollii, Mammillaria carmenae* and *M. theresae*, respectively. These excellent propagation efficiency rates would be impossible to attain using any conventional plant propagation method.

Moreover, PTCM offers other advantage for cacti species, e.g. by using this biotechnological system, the juvenile period can be drastically reduced, which is the utmost episode for susceptibility to predation or droughtness-death. This has been demonstrated by Malda et al. (1999), who studied the *in vitro*-cultured cactus growth and CAM activity (CO_2 capture is carried out at night). These authors discovered that *Coryphanta minima* plants obtained and maintained under *in vitro* conditions grow seven times more than their counterparts under similar *ex vitro* conditions. Plant hormones and sugar added to the medium culture have a strong influence on this increment, as well as the high relative humidity kept in the plant culture container. However, the most remarkable fact is that gas-exchange and malic acid daily-fluctuation analyzes revealed a photosynthetic net augmentation, measured as carbon assimilation, in the *in vitro* cultured plants. Unlike *ex vitro* CAM plants, *Coryphanta minima in vitro* plants carry out CO_2 capture in the presence of light and also in darkness. Thus, it would be possible to assume that *in vitro* cultured cacti simultaneously achieve C3 and CAM photosynthesis, increasing substantially their biomass. Once those *in vitro* cultured cacti are transferred to *ex vitro* conditions, they start their CAM system, yet with a superior size compared to the *ex vitro* counterparts.

Despite all advantages that PTCM offer for cactus propagation and conservation, the amount of cacti species that has been subjected to these biotechnological techniques is still low compared to the total worldwide cacti species number. The reasons for this is that each cactus species, or even at subspecies level, respond differently to the same PTC protocol at each propagation step. These differences belong to the plant hormone type and level, the gelling type and concentration, and the organic compounds added to the culture medium, such as activated charcoal. On the other hand, diverse drawback phenomena are present during the *in vitro* cactus culture, such as hyperhydricity, oxidation and callus overgrowth, among others. Hence, particular micropropagation systems have to be developed for each cacti taxon (Hubstenberger et al., 1992; Giusti et al., 2002).

An important aspect of *in vitro* cultured cactus that must be extensively studied is the development and successful adaptation to the wilderness. This stage is of paramount importance, since it is not worthy to obtain thousands of *in vitro* propagated cacti plants, and once they are transplanted to the wild, none or low numbers of them survive; therefore, it is urgent and necessary to test quite a lot of transplanting conditions in order to obtain the best adaptation environment for each cacti species. Among the factors to be tested are fertilization and watering regimens, light intensity and quality, temperatures, type and amount of substrate, and environmental relative humidity. Other attracting factor to be examined is the use of mychorrizae during cacti transplanting to soil. Although there are no reports on mychorrizae utilization during *in vitro*-cacti transplant to soil, notorious positive effects of this approach have been found in other plant groups (Rai, 2001).

A successful protocol for cacti reintroduction to the wilderness has been established for *in vitro* propagated *Ferocactus histrix* (Figure 8h). It was observed that 45% of transferred cacti were able to survive in the wild after one year without any maintenance care (Flores-Balderas and Pérez-Molphe-Balch, unpublished results). This was only achieved after a series of ecological studies were carried out to determine the most favorable areas for cacti adaptation, and the evaluation of some factors such as humidity, terrain slop, presence of nursing plants and cardinal orientation (Flores-Balderas and Pérez-Molphe-Balch, unpublished results).

Secondary Metabolite Production by *in Vitro* Cacti Cultures

Cacti, in their natural habitat, are capable of producing a horde of secondary metabolites, and it has been observed that *in vitro* cultured cacti are also competent to produce those secondary metabolites. For this reason, PTCM can be applied effectively to study and produce diverse secondary metabolites in cacti. For example, *Cereus peruvianus* callus cultures were found to generate different alkaloids (Braz de Oliveira and Da Silva Machado, 2003). *Mammillaria candida* callus cultures were able to produce betaxanthins and betacyanins, pigments commonly known as betalains, and their production is affected by the presence of auxins and cytokinins in the culture medium. Moreover, betalain production can be substantially increased when osmotic stress (abiotic) or fungi extracts (biotic) are applied to *in vitro* cacti cultures (Santos-Díaz et al., 2005).

A fascinating alternative for secondary-metabolite production by *in vitro* cacti cultures is the generation and culture of genetically transformed roots. *Agrobacterium rhizogenes*, a pathogenic gram-negative bacterium, genetically transforms susceptible plants by transferring genes that alter the plant phenotype and induce the hairy root syndrome. Genetically transformed roots have important biotechnological applications for secondary metabolite production, since these roots are able to generate a great amount of cellular mass in short time, and these organs are easily cultured. Additionally, unlike callus cultures, transformed roots are differentiated tissues; therefore, they keep their ability to synthesize and accumulate a bunch of secondary metabolites. In this sense, González-Díaz et al. (2006) evaluated 65 Mexican-cacti species, belonging to 22 genera, regarding their susceptibility to be transformed by *Agrobacterium rhizogenes* strain A4. *In vitro* cacti tissues were used to achieve the transformation event. Thirty-four cacti species out of 65 were capable of generating transformed roots, and they were tested for their transformed nature by histochemical and molecular assays. It was found, at least for some cacti species, that the secondary-metabolite production profiles from normal and transformed cacti were almost identical. This study demonstrated that quite a few cacti species are susceptible of being transformed by *Agrobacterium rhizogenes*, and they are able of producing transgenic roots. On the other hand, transgenic roots conserve their ability to synthesize and accumulate secondary metabolites in the same way as the non-transgenic tissues do. These results open the opportunity to utilize this biotechnological approach to study and manufacture secondary metabolites at industrial level in short time, and without neither collecting Mexican cacti species in the wild nor risking their preservation.

CONCLUSION

Cacti are very attractive due to the ornamental value, low maintenance cost, and resistance to different pests, diseases and abiotic factors such as water limitation. However, they usually have slow growth rates and long life cycles. For all these reasons Cactaceae is one of the most endangered-extinction threatened plant groups worldwide. Overcollection and wild disturbance dramatically contribute to this problem. Nonetheless, there is a hope to avoid their extinction, through the establishment of protected areas and methods for their preservation and conservation. Cacti species are usually vegetatively propagated or by seeds, but they are also relatively easy to manage by Plant Tissue Culture Methods. However, more economic support and research is needed in order to develop and achieve reliable micropropagation protocols for each cacti species. The application of biotechnology to cacti propagation is not only desirable if we want to reduce the extinction-risk of diverse cacti species, but it must be also used to expand the rational profit of this plant family. It is imperative to take to commercial schemes those reliable cacti propagation protocols in order to reduce the human collection of cacti species in the wild.

REFERENCES

Abdul-Malek, S. N., Shin. S. K., Abdul-Wahab. N. & Yaacob, H. (2009). Cytotoxic components of *Pereskia bleo* (Kunt) DC (Cactaceae) leaves. *Molecules*, *14*, 1713-1724.

Ault, J. R. & Blackmon, W. J. (1987). *In vitro* propagation of *Ferocactus acanthodes* (Cactaceae). *HortScience*, *22*, 126-127.

Adlercreutz, H. & Mazur, W. (1997). Phyto-oestrogens and Western diseases. *Ann. Med.*, *29*, 95-120.

Agurell, S. (1969). Cactaceae alkaloids. *Lloydia*, *32*, 206-216.

Ahmed, M. S., El-Tanbouly, N. D., Islam, W. T., Sleem, A. A. & El-Senousy, A. S. (2005). Antiinflammatory flavonoids from *Opuntia dilleni* (Ker-Gawl) Haw. Flowers growing in Egypt. *Phytother. Res.*, *19*, 807-809.

Aires, V., Adote, S. & Hichami, A. (2004). Modulation of intracellular calcium concentrations and T cell activation by prickly pear polyphenols. *Mol. Cell Biochem*, *260*, 103-110.

Alarcón-Aguilar, F. J., Valdés-Arzate, A., Xolalpa-Molina, S., Banderas-Dorantes, T., Jiménez-Estrada, M., Hernández-Galicia, E. & Román-Ramos, R. (2003). Hypoglycaemic activity of two polysaccarides isolated of *Opuntia ficus-indica* and *O. streptacantha*. *Proc. West. Pharmacol. Soc.*, *46*, 139-142.

Almanza, R. A. (1999). *Cocina indígena y popular. Recetario guanajuatense del xoconostle"*. México, D.F.: Comisión Nacional para la Cultura y las Artes (CONACULTA).

Altesor, A. & Ezcurra, E. (2003). Functional morphology and evolution of stem succulence in cacti. *J. Arid Environ.*, *53*, 557-567.

Altare, M. & Trione, S. (2006). Stimulation and promotion of germination in *Opuntia ficus-indica* seeds. *J. Professional Assoc. Cactus Develop.*, *8*, 91-100.

Álvarez-Aguirre, C. & Montaña, C. (1997). Germinación y supervivencia de cinco especies de cactáceas del valle de Tehuacan: implicaciones para su conservación. *Acta Bot. Mex*, *40*, 43-58.

Álvarez-Jiménez, E., Morales-Jiménez, Y., Goretti, B., Murillo-Gómez, S., Cruz-Jiménez, G., Correño-Alonso, J. & Hernández-Castillo, J. (2008). Remoción de plomo del agua de fuentes de abastecimiento para comunidades de San Luis de la Paz, Guanajuato, utilizando mucílago de nopal y hueso molido en reactores de mezcla completa. VII Congreso Internacional, XIII Congreso Nacional, III Congreso Regional de Ciencias Ambientales. Junio 2008, *Ciudad Obregón*, Sonora, México.

Anaya-Pérez, M. A. (2006). History of the use of *Opuntia* as forage in México. In: Cactus (*Opuntia* spp.) as forage. FAO Plant Production and Protection Paper 169. (http://www.fao.org/Waicent/Faoinfo/Agricult/Agp/Agpc/doc/pasture/CACTUS.HTM).

Anderson, E. F. (2001). *The cactus family*. Portland, OR: Timber Press.

Anderson, F. E., Arias, M. & Taylor, N. P. (1994). *Threatened cacti of México*. Kew: Royal Botanic Gardens.

Andrade, J. L., Rengifo, E., Ricalde, M. F., Simá, J. L., Cervera, J. C. & Vargas-Soto G. (2006). Light microenvironments, growth and photosynthesis for pitahaya (*Hylocereus undatus*) in an agrosystem of Yucatán, Mexico. *Agrociencia*, *40*, 687-697.

Aranda-Osorio, G., Florez-Valdez, C. A. & Cruz-Miranda, E. F. (2008). Inclusion for cactus pear cladodes in diets for finishing lambs in Mexico. *J. Professional Assoc. Cactus Develop.*, *10*, 49-55.

Arias, S., Gama, S. & Guzmán, U. (1997). *Cactaceae A.L. Juss. Flora del Valle de Tehuacán-Cuicatlán*. 14. México , D.F.: Instituto de Biología, UNAM.

Arias, S. & Terrazas, T. (2004). Seed morphology and variation in the genus *Pachycereus* (Cactaceae). *J. Plant Res.*, *117*, 277-289.

Arnaud-Viñas, R., Santiago-García, P. & Bautista, P. B. (1997). Agroindustria de algunos frutos. In *Suculentas mexicanas: Cactáceas*, (79-95). México, D.F.: Conabio, Semarnap, Unam, Cvs Publicaciones.

Arredondo-Gómez, A. (2007). El sistema producto cactáceas en San Luis Potosi. Campo Experimental San Luis. Cirne-Inifap. *San Luis Potosí*, México. Folleto para productores No. 46.

Arredondo-Gómez, A. & Camacho-Morfín, F. (1995). Germinación de *Astrophytum myriostigma* (Lemaire) en relación con la procedencia de las semillas y la temperatura de incubación. *Cact. Suc. Mexicanas*, 40, 34-38.

Arredondo-Gómez, A., Sánchez-Barra, R. & Martínez-Méndez, M. (2007). Ensayo de plantación de *Ferocactus pilosus* (biznaga roja o cabuchera) en San Luis Potosí. Campo Experimental San Luis. CIRNE-INIFAP. *San Luis Potosí*, México. Folleto para productores No. 48.

Arredondo-Gómez, A., Sotomayor, M., Sánchez-Barra, F. R. & Martínez-Méndez, M. (2001). *Cactáceas amenazadas de extinción en el estado de San Luis Potosí* (Publicación especial No. 4, disco compacto). San Luis Potosí, S.L.P.: SAGARPA, INIFAP.

Arreola-Nava, H. J. (1997). Formas de vida y características morfológicas. In *Suculentas mexicanas: Cactáceas*, (27-35). México, D.F.: CONABIO, SEMARNAP, UNAM, CVS Publicaciones.

Ayala-Cordero, G., Terrazas, T., López-Mata, L. & Trejo, C. (2004). Variación en el tamaño y peso de la semilla y su relación con la germinación en una población de *Stenocereus beneckei*. *Interciencia*, 29, 692-697.

Aurea, B. N., Gónzalez-Cruz, L., Ramírez-ortiz, M. E., Güemes-Vera, N. & Godínez-Roldan, S. Utilization of the emulsifying capacity of the Nopal (*Opuntia robusta* Wendl) for the obtaining of a dressing type mayonnaise. *Institute of Food Technologist Annual Meeting*, 2007, Chicago, Illinois, 28 July-1 August.

Barba de la Rosa, P. (2007). Estudio integral de las propiedades del nopal. Informe Técnico. FOMIX-GTO.

Bárcenas, R. T. (2006). Comercio de cactáceas mexicanas y perspectivas para su conservación. CONABIO. *Biodiversitas*, 68, 11-15.

Bárcenas-Luna, R. T. Chihuahuan Desert cacti in Mexico: An assessment of trade, management, and conservation. Priorities. Part II. 2003. Available from URL: http://www.worldwilfe.org/species/attachments/cactus_report_part2.pdf.

Barthlott, T. & Hunt, D. (1993). In K., Kubitzki, J. Rohwer, & V. Bittrich, (Eds.), *Cactaceae: The family and genera of vascular plants. II Dicotyledons*, (161-197). New York, NY: Springer Verlag.

Barthlott, W. & Hunt, D. (2000). *Seed-diversity in the Cactaceae subfamily Cactoideae*. Richmond, VA: Remous Ld.

Benítez, I. S. (2001). Conservación de recursos fitogenéticos *ex situ*. In *Estrategia en recursos fitogenéticos para los países del Cono Sur*. Programa Cooperativo para el Desarrollo Tecnológico Agroalimentario y Agroindustrial del Cono Sur. (PROCISUR). Montevideo, Uruguay.

Benítez, H. & Dávila, P. (2002). Las cactáceas mexicanas en el contexto de la CITES. CONABIO. *Biodiversitas*, 40, 8-11.

Benson, L. (1982). *The cacti of United States and Canada*. Stanford, CA: Stanford University Press.

Bhau, B. S. (1999). Regeneration of *Coryphantha elephantidens* (Lem.) Lem. (Cactaceae) from root explants. *Sci. Hortic.-Amsterdam*, 81, 337-344.

Bobich, E. G. (2005). Vegetative reproduction, population structure, and morphology of *Cylindropuntia fulgida* var. *mamillata* in a desert grassland. *Int. J. Plant Sci.*, *166*, 97-104.

Bobich, E. G. & Nobel, P. S. (2001). Vegetative reproduction as related to biomechanics, morphology and anatomy of four cholla cactus species in the Sonoran Desert." *Ann. Bot.*, *87*, 485-493.

Böhm, H. (2008). *Opuntia dillenii* an interesting and promising cactaceae taxon. *J. Professional Assoc. Cact. Develop.*, *10*, 148-170.

Bowers, J. E. (1996). Growth rate and life span of a pricklypear cactus, *Opuntia engelmannii*, in the northern Sonoran Desert. *Southwest. Nat*, *41*, 315-318.

Bowers, J. E. (2000). Does *Ferocactus wislizeni* (Cactaceae) have a between-year seed bank?. *J. Arid Environ.*, *45*, 197-205.

Bowers, J. E. (2005). Influence of climatic variability on local population dynamics of a Sonoran Desert *Platyopuntia*. *J. Arid Environ.*, *61*, 193-210.

Boyle, T. H. & Idnurm, A. (2001). Physiology and genetics of self-incompatibility in *Echinopsis chamaecereus* (Cactaceae). *Sexual Plant Reprod*, *13*, 323-327.

Boyle, T. H. & Idnurm, A. (2003). Intergeneric hybridization between *Schlumbergera* Lem. and *Hatiora* Britt. & Rose. (Cactaceae). *J. Am. Soc. Hort. Sci.*, *128*, 724-730.

Braga, D. L. & McLaughlin, J. L. (1969). Cactus alkaloids. V. Isolation of hordenine and N-methyltyramine from *Ariocarpus retusus*. *Planta Medica*, *17*, 87-94.

Bravo-Hollis, H. (1978). *Las cactáceas de México (Vol. I)*. México, D.F.: UNAM.

Bravo-Hollis, H. & Sánchez-Mejorada H. (1991). *Las cactáceas de México (Vol. II)*. México, D.F.: UNAM.

Bravo-Hollis, H. & Scheinvar, L. (1995). *El interesante mundo de las cactáceas*. México, D.F.: *Consejo Nacional de Ciencia y Tecnología*, Fondo de Cultura Económica.

Braz de Oliveira, A. J. & da Silva Machado, M. F. P. (2003). Alkaloid production by callus tissue cultures of *Cereus peruvianus* (Cactaceae). *Appl. Biochem. Biotechnol*, *104*, 149-155.

Britton, N. L. & Rose, J. N. (1963). *The Cactaceae*. Toronto, ON: General Publishing Company Ltd.

Buenrostro, M. & Barros, C. (2000). Recetario del nopal de Milpa Alta, D.F. y Colima. *Colección de cocina indígena y popular*, No. 48. México, D.F.: CONACULTA.

Burger, J. C. & Louda, S. M. (1995). Interaction of diffuse competition and insect herbivory in limiting brittle pricklypear cactus *Opuntia fragilis* (Cactaceae). *Am. J. Bot*, *82*, 1558-1566.

Butterworth, C. & Edwards, E. (2008). Investigating *Pereskia* and the earliest divergences in Cactaceae. *Haseltonia*, *14*, 46-53.

Butterworth, C. A. & Wallace, R. S. (2004). Phyllogenetic studies of *Mammillaria* (*Cactaceae*)- insights from chloroplast sequence variation and hypothesis testing using the parametric bootstrap. *Am. J. Bot*, *91*, 1086-1098.

Bwititi, P., Musabayane, C. T. & Nhachi, C. F. B. (2000). Effects of *Opuntia megacantha* on blood glucose and kidney function in streptozotocin diabetic rats. *J. Ethnopharmacol*, *69*, 247-252.

Callen, E. O. (1966). Analysis of the Tehuacán coprolites. In D. S. Byers, (Ed.), *The prehistory in Tehuacán Valley. I. Environment and Subsistence*, (261-289). Austin, TX: University of Texas.

Casas, A. (2002). Uso y manejo de cactáceas columnares mesoamericanas. Conabio. *Diversitas*, *40*, 18-23.

Casas, A, Valiente-Banuet, A, Rojas-Martínez, A. & Dávila, P. (1999). Reproductive biology and the process of domestication of the columnar cactus *Stenocereus stellatus* in central Mexico. *Amer. J. Bot*, *86*, 534-542.

Castañeda-Andrade, I., González-Sánchez, J. & Frati-Munari, A. C. (1997). Hypoglycemic effect of an *Opuntia streptacantha* Lemaire dialysate. *J. Professional Assoc. Cactus Develop.*, *2*, 73-75.

Castillo-Bravo, A., Netzahaual-Nava, M., Pérez-Pérez, F. & Guevara-García, J. A. (2003). Estudio de la absorción de metales pesados por la planta de nopal *Opuntia ficus indica*. III *Congreso Iberoamericano de Física y Química Ambiental*, Tlaxcala, México. 6 al 10 de Octubre.

Castro-Cepero, V., Eyzaguirre-Pérez, R. & Ceroni-Stuva, A. (2006). Survival of *Melocactus peruvianus* Vaupel and *Haageocereus pseudomelanostele* subsp. *aureispinus* (Rauh & Backeberg) Ostolaza, plants at Umarcata Hill. Chillon River Valley, Lima. *Ecol. Aplicada*, *5*, 61-66.

Castro-Gallo, I. A., Meza-Rangel, E., Pérez-Reyes, M. E. & Pérez-Molphe-Balch, E. (2002). Propagación *in vitro* de 10 especies mexicanas de cactáceas. *Sci. Naturae*, *4*, 5-24.

Cerezal, P. & Duarte, G. (2004). Sensory influence of chemical additives in peeled cactus pears *(Opuntia ficus-indica* (L.) in syrup conserved by combined methods. *J. Professional Assoc. Cactus Develop.*, *6*, 102-119.

Cavalcante, I. H. L. & Martins, A. B. G. (2008). Effect of juvenility on cutting propagation of red pitaya. *Fruits*, *63*, 277-283.

Céspedes, C. L., Salazar, J. R., Martínez, M. & Aranda, E. (2005). Insect growth regulatory effects of some extracts and sterols from *Myrtillocactus geometrizans* (Cactaceae) against *Spodoptera frugiperda* and *Tenebrio molitor*. *Phytochemistry*, *66*, 2481-2493.

Charlwood, V., Charlwood, K. & Molina-Torres J. (1990). Accumulation of secondary compounds by organized plants cultures. In V. B. Charlwood, & M. J. Rodhes, (Eds.), *Secondary products from plant tissue culture*, (167-200). Oxford: Clarendon Press.

Chávez-Martínez, J., Hernández-Oria, J. G. & Sánchez-Martínez, E. (2007). Documentación de factores de amenaza para la flora cactológica del semidesierto Queretano. *Bol. Nakari*, *18*, 89-95.

CITES (Convention on International Trade in Endangered Species). (1990). *Appendix I and II*. Washington, DC.: U.S. Fish and Wildlife Service.

Clayton, P. W., Hubstenberger, J. F., Phillips, G. C. & Butler-Nance, S. A. (1990). Micropropagation of members of the Cactaceae subtribe Cactinae. *J. Am. Soc. Hortic. Sci.*, *115*, 337-343.

Comisión Nacional de Áreas Naturales Protegidas (CNAP). Agust (2009). Available from URL: http:// www.conanp.gob.mx/q_anp.html.

Comisión Nacional para el Conocimiento y uso de la Biodiversidad (CONABIO). May 2008. Available from URL: http://www.conabio.gob.mx

CONABIO (2009) (www.conabio.gob.mx).

CONACULTA (2009) (http://www.culturaspopularesindigenas.gob.mx).

Contreras, C. & Valverde, T. (2002). Evaluation of the conservation status of a rare cactus (*Mammillaria crucigera*) through the analysis of its population dynamics. *J. Arid Environ.*, *51*, 89-102.

Crozier, B. S. (2004). Subfamilies of Cactaceae Juss., including *Blossfeldioideae* subfam. nov. *Phytologia, 86*, 52-64.

Dabekaussen, M. A. A., Pierik, R. L. M., van der Laken, J. D. & Hoek Spaans, J. (1991). Factors affecting areole activation *in vitro* in the cactus *Sulcorebutia alba* Rausch. *Sci. Hortic.-Amsterdam, 46*, 283-294.

Dávila-Aranda, P. (2001). Lista taxonómca de Cactaceae en México. Informe final del proyecto Q045; CONABIO (www.conabio.gob.mx).

Dávila, P., Arismendi, M. S., Valiente-Banuet, A., Villaseñor, J. L., Casas, A. & Lira, R. (2002). Biological diversity in the Tehuacán-Cuicatlán Valley. *Biodivers. Conservat, 11*, 421-442.

Dávila-Figueroa, C. A., De la Rosa-Carrillo, M. L. & Pérez-Molphe-Balch, E. (2005). *In vitro* propagation of eight species or subspecies of *Turbinicarpus* (Cactaceae). *In Vitro Cell. Dev. Biol.-Plant, 41*, 540-545.

De Andrade, R. A. & Martins, A. B. G. (2007). Influence of the material source and the cicatrize time in vegetative propagation of red dragon fruit (*Hylocereus undatus* Haw). *Rev. Brasileira Fruticult, 29*, 183-186.

De Andrade, R. A. & Martins, A. B. G. (2008). Development of seedlings of red pitaya (*Hylocereus undatus* Haw) in different substrate volumes. *Acta Sci.-Agron, 30*, 697-700.

Del Castillo, R. F. (1996). Sobre la naturaleza calcífuga y calcícola en cactáceas: un ensayo. *Cact. Suc. Mexicanas, 41*, 3-11.

Del Castillo, R. F. & Trujillo, S. (1997). Sobre la naturaleza calcífuga y calcícola en cactáceas. II. Comparaciones de germinación y establecimiento en *Echinocactus platyacanthus* y *Ferocactus histrix. Cact. Suc. Mex, 42*, 51-59.

Djerassi, C. (1957). Cactus triterpens. *J. Am. Chem. Soc., 26*, 330-352.

de Medeiros, L. A., Roberval Cassia, S. R., Gallo, L. A., de Oliveira, E. T. & Payao Dematte, E. S. (2006). *In vitro* propagation of *Notocactus magnificus*. *Plant Cell Tiss. Organ Cult, 84*, 165-169.

Doetsch, P. W., Cassady, J. M. & McLaughlin, J. L. (1980). Cactus alkaloids: XL. Identification of mescaline and other β-phenethylamines in *Pereskia, Pereskiopsis* and *Islaya* by use of fluorescamine conjugates. *J. Chromatogr. A, 189*, 79-85.

Domínguez-López, A. (1995). Review: use of the fruits and stems of the prickly pear cactus (*Opuntia* spp.) into human food. *Food Sci. Tech. Int., 1*, 65-74.

Dubrovsky, J. G. (1996). Seed hydration memory in Sonoran Desert cacti and its ecological implication. *Am. J. Bot, 83*, 624-632.

Dubrovsky, J. G. (1997). Determinate primary root growth in *Stenocereus gummosus* (Cactaceae), its organization and role in lateral root development. In A. Altman, A. & Y. Waisel, Y. (Eds.). *Biology of root formation and development*, (13-20). New York, NY: Plenum Press.

Dubrovsky, J. G. (1999). Desarrollo de sistema radicular durante la ontogénesis de plantas del género *Stenocereus* (Cactaceae). In E. Pimienta-Barrios (Ed.). *El pitayo en Jalisco y especies afines en México*, (133-146). Guadalajara, Jal.: Universidad de Guadalajara, Fundación Produce Jalisco.

Dubrovsky, J. G., Contreras-Burciaga, L. & Ivanov, V. B. (1998). Cell cycle duration in the root meristem of Sonoran Desert Cactaceae as estimated by cell-flow and rate-of-cell-production methods. *Ann. Bot, 81*, 619-624.

Dubrovsky, J. G. & North, G. B. (2002). Root structure and function. In P. S. Nobel, (Ed.). *The cacti: Biology and uses*, (41-56). Berkeley, CA: University of California Press.

Edwards, E. J., Nyffeler, R. & Donoghue, M. J. (2005). Basal cactus phylogeny: implications of *Pereskia* (Cactaceae) praphyly for the transition to the cactus life form. *Am. J. Bot, 92*, 1177-1188.

Elias-Rocha, M. A., Santos-Díaz, M. S. & Arredondo-Gómez, A. (1998). Propagation of *Mammillaria candida* (Cactaceae) by tissue culture technique. *Haseltonia, 6*, 96-101.

El-Samahy, S. K., Youssef, K. M. & Moussa-Ayoub, T. E. (2009). Producing ice cream with concentrated cactus pear pulp: a preliminary study. *J. Professional Assoc. Cactus Develop., 11*, 1-12.

Enigbokan, M. A., Felder, T. B. & Thompson, J. O. (1996). Hypoglycaemic effects of *Opuntia ficus-indica* Mill., *Opuntia lindheimeri* Engelm and *Opuntia robusta* Wendl. in streptozotocin-induced diabetic rats. *Phytotherapy Res., 10*, 379-382.

Escobar, A, H. A., Villalobos, A, V. M. & Villegas M, A. (1986). *Opuntia* micropropagation by axillary proliferation. *Plant Cell Tiss. Organ Cult, 7*, 269-277.

Esparza-Olguin, L., Valverde, T. & Vilchis-Anaya, E. (2002). Demographic analysis of a rare columnar cactus (*Neobuxbaumia macrocephala*) in the Tehuacán Valley, México. *Biol. Con., 103*, 349-359.

Ferreira-Gomes, F. L. A., Fernandes-Heredia, F., Barbeta e Silva, P., Facó, O. & de Paiva Campos, F. A. (2006). Somatic embryogenesis and plant regeneration in *Opuntia ficus-indica* (L.) Mill. (Cactaceae). *Sci. Hortic.-Amsterdam, 108*, 15-21.

Flores, C. A. (1995). Cactus pear and nopalito production in Mexico. *Proc. Professional Assoc. Cactus Develop., 1*, 27-44.

Flores J., Arredondo A. & Jurado E. (2005). Comparative seed germination in species of *Turbinicarpus*: an endangered cacti genus. *Nat. Areas, J. 25*, 183-187.

Flores, J. & Briones, O. (2001). Plant life-form and germination in a Mexican inter-tropical desert: effects of soil water potential and temperature. *J. Arid Environ., 47*, 485-497.

Flores, J. & Jurado, E. (2006). Effect of light on germination of seeds of Cactaceae from the Chihuahuan Desert, Mexico. *Seed Sci. Res., 16*, 149-155.

Flores, J. & Jurado, E. (2008). Breaking seed dormancy in specially protected *Turbinicarpus lophophoroides* and *Turbinicarpus pseudopectinatus* (Cactaceae). *Plant Species Biol., 23*, 43-46.

Flores, J., Jurado, E. & Jiménez-Bremont, J. F. (2008). Breaking seed dormancy in specially protected *Turbinicarpus lophophoroides* and *Turbinicarpus pseudopectinatus* (Cactaceae). *Plant Species Biol., 23*, 43-46.

Flores-Martínez, A. & Medina, G. I. M. (2008). Seed age germination responses and seedling survival of an endangered cactus that inhabits cliffs. *Nat. Areas J, 28*, 51-57.

Flores-Valdez, C. A. & Aranda-Osorio, C. (1997). *Opuntia*-based ruminant feeding systems in Mexico. *J. Professional Assoc. Cactus Develop., 2*, 3-8.

Franco-Martínez, I. S. (1997). Legislación y conservación. In *Suculentas mexicanas: Cactáceas*, (101-111). México, D.F.: Conabio, Semarnap, Unam, Cvs Publicaciones.

Frati, A. C., Gordillo, B. E., Altamirano, P. & Ariza, R. (1988). Hypoglycemic effects of *Opuntia steptacantha* Lemaire in NIDDM. *Diabetes Care, 11*, 63-66.

Fuller, D. (1985.) U.S. Cactus and succulent bussines moves toward propagation. *World-Wildlife Fund –US, 6(2)*, 1-13.

Galati, E. M., Monforte, M. T., Tripodo, M. M., D'Aquino, A. & Mondello, M. R. (2001). Antiulcer activity of *Opuntia ficus-indica* (L.) Mill. (Cactaceae): ultrastructural study. *J Ethnopharmacol, 76*, 1-9.

García de la Cruz, R. F. & Cruz-Alvarado, L. F. (1999). Método alternativo para la conservación de brotes de *Mammillaria cándida*, cactácea en peligro de extinción. Memoria del VIII Congreso Nacional y VI Internacional sobre conocimiento y aprovechamiento del nopal. Universidad Autónoma de San Luis Potosí. 6-10 de Sept. 1999. *San Luis Potosí*, México, 126-127.

Ghansah, E., Kopsombut, P., Malleque, M. A. & Brossi, A. (1993). Effects of mescaline and some of its analogs on cholinergic neuromuscular transmission. *Neuropharmacology, 32*, 169-74.

Gibson, A. C. (1978). Structure of *Pterocactus tuberosus,* a cactus geophyte. *Cactus Succ. J, 50*, 41-43.

Gibson, A. C. & Nobel, P. S. (1986). *The cactus primer.* Cambridge, MA: Harvard University Press.

Giusti, P., Vitti, D., Fiocchetti, F., Colla, G., Saccardo, F. & Tucci, M. (2002). *In vitro* propagation of three endangered cactus species. *Sci. Hortic.-Amsterdam, 95*, 319-332.

Gobierno de San Luis Potosí (http://www.slp.gob.mx).

Godínez-Álvarez, H. & Ortega-Baes, P. (2007). Mexican cactus diversity: environmental correlates and conservation priorities. *Bol. Soc. Bot. México, 81*, 81-87.

Godínez-Álvarez, H., Valverde, T. & Ortega-Baes, P. (2003). Demographic trends in the Cactaceae. *Bot. Rev., 69*, 173-203.

Gómez-Flores, R., Tamez-Guerra, P., Tamez-Guerra, R., Rodríguez-Padilla, C., Monreal-Cuevas, E., Hauad-Marroquin, L. A., Córdova-Puente, C. & Rangel-Llanas, A. (2006). *In vitro* antibacterial and antifungal activities of *Nopalea cochenillifera* Pad extracts. *Am. J. Infect. Disease, 2*, 1-8.

Gómez-Hinostrosa, C. & Hernández, H. M. (2000). Diversity, geographical distribution and conservation of Cactaceae in the Mier y Noriega región, México. *Biodivers. Conserv, 9*, 403-418.

González-Díaz, M. C., Pérez-Reyes, M. E. & Pérez-Molphe-Balch, E. (2006). *In vitro* analysis of susceptibility to *Agrobacterium rhizogenes* in 65 species of Mexican cacti. *Biol. Plant, 50*, 331-337.

González-Quintero, L. (1976). Las cactáceas subfósiles de Tehuacán. *Cact. Suc. Mex., 17*, 3-15.

Guillen, S. & Benítez, J. (2009). Seed germination of wild, *in situ*-managed, and cultivated populations of columnar cacti in the Tehuacan-Cuicatlan Valley, Mexico. *J. Arid Environ., 73*, 407-413.

Guzmán-Cruz, L. U. (1997). Grupos taxónomicos. In *Suculentas mexicanas: Cactáceas,* (37-41). Mexico, D.F.: Conabio, Semarnap, Unam, CVS Publicaciones.

Guzmán, U., Arias, S. & Dávila, P. (2003). *Catálogo de cactáceas mexicanas.* México, D.F.: UNAM, CONABIO. Proyecto AP003.

Helmlin, H. J., Bourquin, D. & Brenneisen, R. (1992). Determination of phenylethylamines in hallucinogenic cactus species by high performance liquid chromatography with photodiode-array detection. *J. Chromatogr, 623*, 381-385.

Hemphill, J. K., Maier, C. G. A. & Chapman, K. D. (1998). Rapid *in-vitro* plant regeneration of cotton (*Gossypium hirsutum* L.). *Plant Cell Rep., 17*, 273-278.

Hernández, H. M. & Bárcenas, R. T. (1995). Endangered cacti in the Chihuahuan Desert: I. Distribution patterns. *Conserv. Biol.*, *9*, 1176-1188.

Hernández, H. M. & Bárcenas, R. T. (1996). Endangered cacti in the Chihuahuan Desert: II. Biogeography and conservation. *Conserv. Biol.*, *10*, 1200-1209.

Hernández, H. M. & Godínez H. (1994). Contribución al conocimiento de las cactáceas mexicanas amenazadas. *Acta Bot. Mex.*, *26*, 33-52.

Hernández, H. M. & Gómez-Hinostrosa. (2005). Cactus diversity and endemism in the Chihuahuan Desert region. In J. E. L., Cartron, G. Ceballos, & R. S. Felger, (Eds.), *Biodiversity, ecosystems and conservation in Northern Mexico*, (264-275). New York, NY: Oxford University Press.

Hernández, H. M., Gómez-Hinostrosa, C. & Bárcenas, R. T. (2001). Diversity, spatial arrangement, and endemism of Cactaceae in the Huizache area, a hot spot in the Chihuahuan Desert. *Biodivers. Conserv*, *10*, 1097-1112.

Hernández, H. M., Gómez-Hinostrosa, C. & Goettsch, B. (2004). Checklist of Chihuahuan Desert Cactaceae. *Harvard Papers Bot*, *9*, 11-26.

Hernández, L., Barral, H., Halffter, G. & Sánchez, S. (1999). A note on the behavior of feral cattle in the Chihuahuan Desert of Mexico. *Appl. Anim. Behav. Sci.*, *63*, 259-267.

Hernández-Oria, J. G, Chávez-Martínez, R. & Sánchez-Martínez E. (2006). Estado de conservación de *Echinocereus schmollii* (Weing.) N. P. Taylor en Cadereyta de Montes, Querétaro, México. *Cact. Suc. Mex*, *51*, 68-95.

Hernández-Oria, J. G., Chávez-Martínez, R., Sánchez-Martínez, E. (2007). Factores de riesgo en las cactáceas amenazadas de una región semiárida en el sur del desierto Chihuahuense, México. *Interciencia*, *32*, 728-734.

Hubstenberger, J. F., Clayton, P. W. & Phillips, G. C. (1992). Micropropagation of cacti (Cactaceae). In Y. P. S. Bajaj, (Ed.), *Biotechnology in agriculture and forestry: High-Tech and Micropropagation IV*, (Vol. 20, 49-68). Berlin-Heilderberg: Springer-Verlag.

Hunt, D. (1999). CITES Cactaceae Checklist; Royal Botanic Gardens/International Organization for Succulent Plant Study. Kew, UK.

Hunt, D. (2006). *New cactus lexicon*. Milborne Port: DH Books.

Ibañez-Camacho, R. & Meckes-Lozoya, M. (1983). Effect of a semipurified product obtained from *Opuntia streptacantha*, L. (a cactus) on glycemia and triglyceridemia of rabbit. *Arch. Invest. Med*, *14*, 437-443.

Infante, R. (1992). *In vitro* axillary shoot proliferation and somatic embryogenesis of yellow pitaya *Mediocactus coccineus* (Slam-Dyck). *Plant Cell Tiss. Organ Cult*, *31*, 155-159.

Instituto Nacional de Ecología. (2007). Conservación, manejo y aprovechamiento sustentable de la vida silvestre. SEMARNAT.

Instituto Nacional de Ecología. (2009). Los desiertos de Sonora y Baja California Sur. www.ine.gob.mx/ueajei/publicaciones/artnopubli/368/asilvestres.html.

IUCN, version 2009.1 (www.iucnredlist.org/search).

Jiménez-Sierra, C., Mandujano, M. C. & Eguiarte, L. E. (2007). Are populations of the candy barrel cactus (*Echinocactus platyacanthus*) in the desert of Tehuacán, México at risk? Population projection matrix and life table response analysis. *Biol. Conserv*, *125*, 278-292.

Johnson, J. L. & Emino, E. R. (1979). Tissue culture propagation in the cactaceae. *Cactus & Succulent J*, *51*, 275-277.

Johnston, M. C. (1977). Brief resume of botanical, including vegetational features of the Chihuahuan Desert Region with special emphasis o the uniqueness. In R. H. Wauer, R. H. & D. H. Riskind, (Eds.), *Transactions of the Symposium on the Biological Resources of the Chihuahuan Desert Region United States and Mexico*, (335-359). Washington, DC: National Park Service.

Jordan, P. W. & Nobel, P. S. (1982). Height distributions of two species of cacti in relation to rainfall, seedling establishment, and growth. *Bot. Gaz, 143*, 511-517.

Juárez, M. C. & Passera, C. B. (2002). *In vitro* propagation of *Opuntia ellisiana* Griff. and acclimatization to field conditions. *Biocell, 26*, 319-324.

Karle, R., Parks, C. A., O'Leary, M. C. & Boyle, T. H. (2002). Polyploidy-induced changes in the breeding behavior of *Hatiora x graeseri* (Cactaceae). *J. Am. Soc. Hortic. Sci., 127*, 397-403.

Kim, J. H., Park, S. M. & Ha, H. J. (2006). *Opuntia ficus-indica* attenuates neuronal injury in *in vitro* and *in vivo* models of cerebral ischemia. *J. Ethnopharmacol, 104*, 257-262.

Knight, J. C, Wilkinson, D. I. & Djerassi C. (1966). The structure of the cactus sterol macdougallin (14α-Methyl-Δ^8-cholestene-3β,6α-Diol). A novel link in sterol biogenesis. *J. Chem. Soc., 88*, 790-798.

Kuti, J. O. (2004). Antioxidant compounds from four *Opuntia* cactus pear fruit varieties. *Food Chem., 85*, 527-533.

Lamghari, E. L., Kossori, R., Villaume, C., El Bootani, E., Suavaire, Y. & Méjean, L. (1996). Composition of pulp, skin and seeds of prickly pear fruit (*Opuntia ficus indica sp.*). *Plant Foods Hum. Nutr., 52*, 263-270.

Le Bellec, F., Vaillant, F. & Imbert, E. (2006). Pitahaya (*Hylocereus* sp). *Fruits, 61*, 237-250.

Lee, S. P., Lee, S. K. & Ha, Y. D. (2002). Alcohol fermentation of *Opuntia ficus* fruit juice. *J. Food Sci. Nutr., 5*, 32-36.

León de la Luz, J. L. & Domínguez-Cadena, R. (1991). Evaluación de la reproducción por semilla de la pitaya agria (*Stenocereus gummosus*) en Baja California Sur, México. *Acta Bot. Mex, 14*, 75-87.

Lépiz, R. I. & Rodríguez-Guzmán E. (2006). los Recursos Fitogenéticos de México. Segundo Informe Nacional sobre los Recursos Filogenéticos para la Alimentación y la Agricultura de México. *Sociedad Mexicana de Fitogenética*. 192.

Ley General del Equilibrio Ecológico y la Protección al Ambiente, May (2008). Available from URL: http://www.diputados.gob.mx/LeyesBiblio/pdf/148.pdf.

Li, C. Y., Cheng, X. S. & Cui, M. Z. (2005). Regulative effect of *Opuntia* powder on blood lipids in rats and its mechanism] Zhongguo Zhong. *Yao Za Zhi, 30*, 694-696.

Linaje-Treviño, M. S., De la Fuente-Salcido, N. M. (2007). Evaluación sensorial de yogurt adicionado con nopal (*Opuntia ficus-indica*) como alimento funcional. IX Congreso de Ciencia de los Alimentos y V Foro de Ciencia y Tecnología de Alimentos. 31 de mayo, *Monterrey, Nuevo León*, México.

Lizalde, V. H. J., Pérez-Molphe-Balch, E., Dávila, F. C. A. & Pérez, R. M. E. (2003). Propagación de nueve especies de cactáceas mexicanas por técnicas de cultivo de tejidos. *Sci. Naturae, 5*, 21-31.

López-Gómez, R., Díaz-Pérez, J. C. & Flores-Martínez, G. (2000). Propagación vegetativa de tres especies de cactáceas: Pitaya (*Stenocereus griseus)*, tunillo (*Stenocereus stellatus*) y jiotilla (*Escontria chiotilla*). *Agrociencia, 34*, 363-367.

López-González, J. J., Fuentes-Rodríguez, J. M. & Rodríguez-Gámez, A. (1997). Prickly Pear Fruit Industrialization (*Opuntia streptacantha*). *J. Professional Assoc. Cactus Develop.*, *2*, 169-175.

López, G. B., L., Murillo-Gómez, S., Álvarez-Jiménez, E., Morales-Rodríguez, Y., Cruz Jiménez G, Coreño Alonso, J., Hernández-Castillo, D. (2008). Remoción de arsénico en agua de abastecimiento humano utilizando mucílago de nopal y hueso molido en reactores de mezcla completa. VII Congreso Internacional y XIII Congreso Nacional, III Congreso Regional de Ciencias Ambientales. Junio 2008. Ciudad Obregón, Sonora, México.

Loza-Cornejo, S. & Terrazas, T. (1996). Anatomía del tallo y de la raíz de dos especies de *Wilcoxia* Britton & Rose(Cactaceae) del Noroeste de México. *Bol. Soc. Bot. México*, *59*, 13-23.

Loza-Cornejo, S., Terrazas, T., López-Mata, L. & Trejo, C. (2003). Características morfo-anatómicas y metabolismo fotosintético en plántulas de *Stenocereus queretaroensis* (Cactaceae): su significado adaptativo. *Interciencia*, *28*, 83-89.

Loza-Cornejo, S., López-Mata, L. & Terrazas, T. (2008). Morphological seed trait and germination of six species of *Pachycereeae* (Cactaceae). *J. Professional Assoc. Cactus Develop.*, *10*, 71-84.

Lutty, J. M. (2001). The Cacti of CITES Appendix I. *CITES Management Authority of Switzerland*, Bern.

Malda, G., Backhaus, R. A. & Martin, C. (1999). Alterations in growth and crassulacean acid metabolism (CAM) activity of *in vitro* cultured cactus. *Plant Cell Tiss. Organ Cult*, *58*, 1-9.

Mandujano, M. C., Flores-Martínez, A., Golubov, J. & Escurra, E. (2002). Spatial distribution of three globose cacti in relation to different nurse-plant canopies and bare areas. *Southwest. Nat*, *47*, 162-168.

Mandujano, M. C., Montaña, C., Franco, M., Golubov, J. & Flores-Martínez, A. (2001). Integration of demographic annual variability in a clonal desert cactus. *Ecology*, *82*, 344-359.

Mandujano, M. C., Montaña, C. & Rojas-Aréchiga, M. (2005). Breaking seed dormancy in *Opuntia rastrera* from Chihuahuan desert. *J. Arid Environ.*, *62*, 15-21.

Manzanero-Medina, G. I. & Flores-Martínez A. (2008). Conservación y uso sostenible de cactáceas globosas endémicas de Oaxaca. XXI Reunión Nacional de Jardines Botánicos Con el tema: Especies prioritarias para los jardines botánicos mexicanos. Jardín Botánico Regional, Cassiano Conzatti. CIIDIR Oaxaca. 25 de febrero 2008. Oaxaca, Oax., 47.

Martínez, D., Flores-Martínez, A., López, F. & Manzanero, G. (2001). Aspectos ecológicos de *Mammillaria oteroi* Glass & R. Foster en la región mixteca de Oaxaca. *Cact. Suc. Mex*, *46*, 32-39.

Martínez-Ávalos, J. G. & Jurado, E. (2005). Geographic distribution and conservation of Cactaceae from Tamaulipas, México. *Biodiv. Cons.*, *14*, 2483-2506.

Martínez-Ávalos, J. G., Suzán, H. & Salazar, C. A. (1993). Aspectos ecológicos y demográficos de *Ariocarpus trigonous* (Weber) Schumann. *Cact. Suc. Mex.*, *38*, 30-38.

Martorell, C. & Peters, E. (2005). The measurement of chronic disturbance and its effects on the threatened cactus *Mammillaria pectinifera*. *Biol. Cons.*, *124*, 197-207.

Mata-Rosas, M., Monroy-De la Rosa, M. A., Moebius-Goldamer, K. & Chávez-Avila, V. M. (2001). Micropropagation of *Turbinucarpus laui* Glass *et* Foster, an endemic and endangered species. *In Vitro Cell. Dev. Biol.-Plant*, *37*, 400-404.

Martínez-Vázquez, O. & Rubluo A. (1989). *In vitro* mass propagation of the near-extinct *Mammillaria san-angelensis* Sánchez-Mejorada. *J. Hortic. Sci.*, *64*, 99-105.

McAuliffe, J. R. (1984). Prey refugia and distributions of two Sonoran desert cacti. *Oecologia*, *65*, 82-85.

McLaghlin, J. L. (1969). Identification of hordenine and n-methyltyramine in *Ariocarpus fissuratus* varieties *fissuratus* and *lloydii. Lloydia*, *32*, 392-94.

Medellín-Leal, F. (1982). The Chihuahuan desert. In G. L. Bender, (Ed.), *Reference handbook on the desserts of North America*. Westport, CN: Greenwood Press.

Melo, C. & Alfaro, G. (2007). Mapa áreas naturales protegidas federales de México, Escala 1:4 000 000, Sección Medio Ambiente. *Nuevo Atlas Nacional de México*. México, D. F.: Instituto de Geografía, UNAM.

Moßhammer, M. R., Stintzing, F. C. & Carle, R. (2006). Cactus pear fruits (*Opuntia* spp): a review of processing technologies and current uses. *J. Professional Assoc. Cactus Develop.*, *8*, 1-25.

Moßhammer, M. R., Rohe, M., Stintzing, F. & Reinhold, C. (2007). Stability of yellow-orange cactus pear (*Opuntia ficus-indica* L. Mill cv. Gialla) betalains as affected by the juice matrix and selected food additives. *Eur. Food Res. Tech.*, *225*, 21-32.

Moebius-Goldammer, K. G., Mata-Rosas, M. & Chávez-Avila, V. (2003). Organogenesis and somatic embryogenesis in *Ariocarpus kotschoubeyanus* (Lem.) K. Schum. (Cactaceae), an endemic and endangered Mexican species. *In Vitro Cell. Dev. Biol.-Plant*, *39*, 388-393.

Mohamed-Yasseen, Y. (2002). Micropropagation of Pitaya (*Hylocereus undatus* Britton et Rose). *In Vitro Cell. Dev. Biol.-Plant*, *38*, 427-429.

Moreno-Álvarez, M. J., Hernández, R., Belén-Camacho, D. R, Medina-Martínez, C. A., Ojeda-Escalona, C. E. & García-Pantaleon, D. M. (2009). Making of bakery products using composite flours: wheat and cactus pear *(Opuntia boldinghii* Britton *et* Rose) stem (cladodes). *J. Professional Assoc. Cactus Develop*, *1*, 78-87.

Muldawin, E. H. (2002). Some floristic characteristics of the northern Chihuahuan Desert: A search for its northern boundary. *Taxon*, *51*, 453-462.

Murashige, T. (1974). Plant propagation through tissue culture. *Annu. Rev. Plant Physiol.*, *25*, 135-166.

Nobel, P. S. (1988). *Environmental biology of agaves and cacti*. New York, NY: Cambridge University Press.

Nobel, P. (1998). *Los incomparables agaves y cactos*. México, D.F.: Editorial Trillas.

Nobel, P. S. & Hartsock, T. L. (1983). Relationships between photosynthetically active radiation, nocturnal acid accumulation, and CO_2 uptake for a crassulacean acid metabolism plant, *Opuntia ficus-indica. Plant Physiol.*, *71*, 71-75.

Nuñez-Palenius, H. G., Cantliffe, D. J., Klee, H. H., Ochoa-Alejo, N., Ramírez-Malagón, R. & Pérez-Molphe, E. (2006). Methods in plant tissue culture. In K., Shetty, G., Paliyath, A. Pometto, & R. Levin, (Eds.), *Food Biotechnology* (2nd Edition; pp. 553-601). Boca Raton, FL: CRC Taylor & Francis.

Oliveira, S. A., Pires da Silva Machado, M. F., Prioli, A. J. & Aparecida M. C. (1995). *In vitro* propagation of *Cereus peruvianus* Mill. (Cactaceae). *In Vitro Cell. Dev. Biol.-Plant* 31, 47-50.

Orona-Castillo, I., Cueto-Wong, J. A., Murillo-Amador, B., Santamaría-César, J., Flores-Hernández, A., Valdez-Cepeda, R. D., García-Hernández, J. L. & Troyo-Diéguez, E. (2004). Mineral extraction of green prickly pear cactus under drip irrigation. *J. Professional Assoc. Cactus Develop*, 6, 90-101.

Orozco-Segovia, A., Márquez-Gúzman, J., Sánchez-Coronado, M. E., Gamboa de Buen, A., Baskin, J. M. & Baskin, C. C. (2007). Seed anatomy and water uptake in relation to seed dormancy in *Opuntia tomentosa* (Cactaceae, Opuntioideae). *Ann. Bot*, 99, 581-592.

Ortega-Baes, P. & Godínez-Álvarez, H. (2006). Global diversity and conservation priorities in the Cactaceae. *Biodiv. Conserv*, 15, 817-827.

Papafotiou, M., Balotis, G. N., Louka, P. T. & Chronopoulos, J. (2001). *In vitro* plant regeneration of *Mammillaria elongata* normal and cristate forms. *Plant Cell Tiss. Organ Cult*, 65, 163-167.

Parker, K. C. (1991). Topography, substrate and vegetation patterns in the northern Sonoran Desert. *J. Biogeogr*, 18, 151-163.

Parker, K. C. (1987). Seed crop characteristics and minimum reproductive size of organ pipe cactus (*Stenocereus thurberi*) in southern Arizona. *Madroño*, 34, 294-303.

Parks, C. A. & Boyle, T. H. (2003). Variation in ploidy level, fertility, and breeding behavior in cultivated *Schlumbergera* (Cactaceae). *Acta Hortic*, 623, 341-349.

Pelah, D., Kaushik, R. A., Mizrahi, Y. & Sitrit, Y. (2002). Organogenesis in the vine cactus *Selenicereus megalanthus* using thidiazuron. *Plant Cell Tiss. Organ Cult*, 71, 81-84.

Pérez-Molphe-Balch, E., Pérez-Reyes, M. E., Villalobos-Amador, E., Meza-Rangel, E., Morones-Ruiz, L. R. & Lizalde-Viramontes, H. J. (1998). Micropropagation of 21 species of Mexican cacti by axillary proliferation. *In Vitro Cell. Dev. Biol.-Plant*, 34, 131-135.

Pérez-Molphe-Balch, E. & Dávila-Figueroa, C. A. (2002). *In vitro* propagation of *Pelecyphora aselliformis* Ehrenberg and *P. strobiliformis* Werdermann (Cactaceae). *In Vitro Cell. Dev. Biol.-Plant*, 38, 73-78.

Pérez-Molphe-Balch, E., Pérez-Reyes, M. E., Dávila-Figueroa, C. A. & Villalobos-Amador, E. (2002). *In vitro* propagation of three species of columnar cacti from the Sonoran desert. *HortScience*, 37, 693-696.

Pérez-Quilantán, L. M. & Hurtado-González, M. (2004). *El nopal Opuntia sp.: Aprovechamiento integral de sus alimentos derivados y componentes funcionales.* Tamaulipas, Tamps.: Universidad Autónoma de Tamaulipas.

Pérez-Sandi, M. & Becerra, R. (2001). Nocheztli: el insecto del rojo carmín. Boletín Bimestral de la Comisión Nacional para el Conocimiento y Uso de la Biodiversidad. Biodiversitas No. 36: 2-7. (available from URL: http://www.conabio.gob.mx/otros/biodiversitas/doctos/pdf/biodiv36.pdf.

Piatelli, M. & Imperato, F. (1969). Betacyanins of the family Cactaceae. *Phytochemistry*, 8, 1503-1507.

Pierson, E. A. & Turner, R. M. (1998). An 85-year study of saguaro (*Carnegiea gigantean*) demography. *Ecology*, 79, 2676-2693.

Piga, A. (2004). Cactus pear: a fruit of nutraceutical and functional importance. J. Professional Assoc. *Cactus Develop.*, 6, 9-22.

Piga, A., D'Aquino, S., Agabbio, M. & Schirra, M. (1996). Storage life and quality attributes of cactus pears cv "Galia" as affected by packaging. *Agric. Med*, *126*, 423-427.

Pinkava, D. J. (1985). Chromosome numbers in some cacti of Western North America. *Systematic Bot*, *10*, 471-483.

Pizzetti, M. (1987). *Guía de cactus*. Barcelona: Grijalbo.

Poljuha, D., Balen, B., Bauer, A., Ljubešić, N. & Krsnik-Rasol, M. (2003). Morphology and ultrastructure of *Mammillaria gracillis* (Cactaceae) in in vitro culture. *Plant Cell Tiss. Organ Cult*, *75*, 117-123.

Porembski, S. (1996). Functional morphology of *Aztekium ritteri* (Cactaceae). *Bot. Acta*, *109*, 167-171.

Powell, A. M. & Turner, B. L. (1977). Aspects of the plant biology of the gypsum outcrops of the Chihuahuan Desert. In R. H. Wauer, & D. H. Riskind, (Eds.), *Transactions of the Symposium on the Biological Resources of the Chihuahuan Desert Region, United States and Mexico*, (315-325). Washington, DC: National Park Service.

Rai, M. K. (2001). Current advances in mycorrhization in micropropagation. *In Vitro Cell. Dev. Biol.-Plant*, *37*, 158-167.

Ramadan, M. F. & Mörse, J. T. (2003). Oil cactus pear (*Opuntia ficus-indica* L.). *Food Chem.*, *82*, 339-345.

Ramírez-Malagón, R., Aguilar-Ramírez, I., Borodanenko, A., Pérez-Moreno, L., Barrera-Guerra, J. L., Nuñez-Palenius, H. G. & Ochoa-Alejo, N. (2007). In vitro propagation of ten threatened species of *Mammillaria* (Cactaceae). *In Vitro Cell. Dev. Biol.-Plant*, *43*, 660-665.

Ramírez-Padilla, C. A. & Valverde, T. (2005). Germination responses of three congeneric cactus species (*Neobuxbaumia*) with differing degrees of rarity. *J. Arid Environ.*, *61*, 333-343.

Ramírez-Tobías, H. M, Reyes-Agüero, J. A., Pinos-Rodríguez, J. M. & Aguirre-Rivera, J. R. (2007). Efecto de la especie y madurez sobre el contenido de nutrientes de cladodios de nopal. *Agrociencia*, *41*, 619-626.

Retes-Pruneda, J. L., Valadez-Aguilar, M. L., Pérez-Reyes, M. E. & Pérez-Molphe-Balch, E. (2007). Propagación in vitro de especies de *Echinocereus, Escontria, Mammillaria, Melocactus* y *Polaskia* (Cactaceae). *Bol. Soc. Bot. México*, *81*, 7-16.

Reynoso, R., García, F. A. & González-de Mejía, E. (1997). Stability of betalain pigments from a cactacea fruit. *J. Agric. Food Chem.*, *45*, 2884-2889.

Riley, L. & Riley, W. (2005). *Nature's strongholds: The world's great wildlife reserves*. Princeton, NJ: Princeton University Press.

Robbins, C. (2002). Cactus conundrum: Traffic examines the trade in Chihuahuan Desert cacti. *Traffic North America*, *1*, 2-3.

Rodríguez, R. G., Treviño, J. F., Morales, M. E., Oranday, A., Verde M. J & Rivas, J. C. (2006). Preliminares fisicoquímicos y actividad bactericida de *Ariocarpus kotschoubeyanus* (Lem.) K. Schum. y *Ariocarpus retusus* Scheidw. (Cactaceae). *Bol. Soc. Latin. Caribe Cact. Suc.*, *3*, 8-10.

Rodríguez-Arévalo, I., Casas, A., Lira, R. & Campos, J. (2006). Uso, manejo y procesos de domesticación de *Pachycereus hollianus* (Weber) Buxb. (Cactaceae), en el Valle de Tehuacán-Cuicatlán, México. *Interciencia*, *31*, 677-685.

Rodríguez-Garay, B. & Rubluo, A. (1990). In vitro morphogenetic responses of the endangered cactus *Aztekium ritteri* (Boedeker). *Cactus & Succulent J*, *64*, 116-119.

Rodríguez-Ortega, C. E. & Escurra, E. (2001). Distribución espacial en el hábitat de *Mammillaria pectinifera* y *Mammillaria carnea* en el Valle de Zapotitlán Salinas, Puebla, México. *Cactac. Suc. Mex*, *45*, 4-14.

Rojas-Aréchiga, M. & Arias, S. (2007). Avances y perspectivas en la investigación biológica de la familia Cactaceae en México. *Bol. Soc. Latin. Caribe Cact. Suc.*, *4*, 1-3.

Rojas-Aréchiga, M., Orozco-Segovia, A. & Vázquez-Yanes, C. (1997). Effect of light on germination of seven species of cacti from the Zapotitlán Valley in Puebla, Mexico. *J. Arid Environ.*, *36*, 571-578.

Rojas-Aréchiga, M. & Vázquez-Yanes, C. (2000). Cactus seed germination: a review. *J. Arid Environ.*, *44*, 85-104.

Ruiz, A., Santos, M., Cavalier, J. & Soriano, P. J. (2000). Estudio fenológico de cactáceas en el enclave seco de la Tatacoa, Colombia. *Biotropica*, *32*, 397-407.

Ruiz-Espinoza, F. H., Alvarado-Mendoza, J. F., Murillo-Amador, B., García-Hernández, J. L., Pargas-Lara, R., Duarte-Osuna, J. D., Beltrán-Morales, F. A. & Fenech-Larios L. (2008). Rendimiento y crecimiento de nopalitos de cultivares de nopal (*Opuntia ficus-indica*) bajo diferentes densidades de plantación. *J. Professional Assoc. Cactus Develop.*, *10*, 22-35.

Russell, C. E. & Felker, P. (1987). The pricklypears (*Opuntia* sp., Cactaceae): A source of human and animal food in semiarid regions. *Econ. Bot*, *41*, 433-445.

Rzedowski, J. (1991). El endemismo en la flora fanerógamica mexicana: una apreciación analítica preliminar. *Acta Bot. Mex.*, *15*, 47-64.

Sáenz, C. (2000). Proccesing technologies: an alternative for cactus pear *(Opuntia* spp.) fruits and claodes. *J. Arid Environ.*, *46*, 209-225.

Sáenz, C., Berger, H., Corrales-García, J., Galletti, L., García-de Cortazar, V., Higuera, I. C., Mondragón, A., Rodríguez-Félix, E., Sepúlveda, M. & Varnero. T. (2006). Utilización agroindustrial del nopal. Boletín de Servicios Agrícolas de la FAO # 162.

Sáenz, C., Sepúlveda, E., Pak, N. & Vallejos, X. (2002). Uso de fibra dietética de nopal en la formulación de un polvo para flan. *Alan*, *52*, 387-392.

Saleem, M., Kim, H. J., Han, C. K., Jin, C. & Lee, Y. S. (2006). Secondary metabolites from *Opuntia ficus-indica* var. *saboten*. *Phytochemistry*, *67*, 1390-1394.

Salt, T. A., Tocker, J. E. & Adler, J. H. (1987). Dominance of Δ^5- sterols in eight species of the Cactaceae. *Phytochemistry*, *26*, 731-733.

Sánchez-Mejorada, H. (1982). México's problems and programmes monitoring trade in common and endangered cacti. *Cact. Succ. J. Gr. Brit*, *44*, 36-38.

Sánchez-Morán, M. R. & Perez-Molphe-Balch, E. (2007). Propagación *in vitro* de *Browningia candelaris* (Cactaceae) usando meta-topolina. *Bol. Soc. Latin. Caribe Cact. Suc.*, *4*, 17-19.

Sánchez-Salas, J., Flores, J. & Martínez-García, E. (2006). Efecto del tamaño de semilla en la germinación de A*strophytum myriostigma* Lemaire. (cactaceae), especie amenazada de extinción. *Interciencia*, *31*, 370-375.

Santos-Díaz, M. S., Méndez-Ontiveros, R., Arredondo-Gómez, A. & Sántos-Díaz, M. L. (2003). *In vitro* organogenesis of *Pelecyphora aselliformis* Erhenberg. *In Vitro Cell. Dev. Biol.-Plant*, *39*, 480-484.

Santos-Díaz, M. S., Velásquez-García, Y. & González-Chávez, M. M. (2005). Producción de pigmentos por callos de *Mammillaria candida* Scheidweiler (Cactaceae). *Agrociencia*, *39*, 619-626.

Secretaría de Ecología y Gestión Ambiental (SEGAM). (2009). Agust, Available from URL: http://segam.org.mx

Secretaría de Medio Ambiente y Recursos Naturales (SEMARNAT). (2005). March (http://www.semarnat.gob.mx/saladeprensa/sistesisdeprensanacional/Pages/Síntesis008.aspx).

Semarnat (Secretaría del Medio Ambiente y Recursos Naturales). (2002). Acceso a recursos genéticos y distribución justa y equitativa de beneficios. Reunión de Ministros de los Países Megadiversos. *Cancún*, México - Febrero 16-18, 2002. 15.

Emarnat. (2002). Norma Oficial Mexicana NOM-059-ECOL-2001, Protección ambiental-Especies nativas de México de flora y fauna silvestres-Categorías de riesgo y especificaciones para su inclusión, exclusión o cambio-Lista de especies en riesgo. Secretaria de Medio Ambiente y Recursos Naturales. Diario Oficial de la Federación. México, D.F.

Sotomayor, M., Arredondo-Gómez, A., Sánchez-Barra, F. & Martínez-Méndez, M. (2004). *The genus Turbinicarpus in San Luis Potosí*. Venegono: Cactus & Co.

Sriskandarajah, S. & Serek, M. (2004). Regeneration from phylloclade explants and callus cultures of *Schlumbergera* and *Rhipsalidopsis*. *Plant Cell Tiss. Organ Cult*, 78, 75-81.

Startha, R., Chybidziurová, A. & Lacný, Z. (1999a). Alkaloids of the genus *Turbinicarpus* (Cactaceae). *Biochem. System. Ecol.*, 27, 839-841.

Starling, R. (1985). *In vitro* propagation of *Leuchtenbergia principis*. *Cactus & Succulent J.* 57, 114, 115.

Startha, R., Chybidziurová, A. & Lacný, Z. (1999b). Constituents of *Turbinicarpus alonsoi* Glass & Arias (Cactaceae). *Acta Univers. Palack. Olomucensis*, 38, 71-73.

Steenbergh, W. & Lowe, C. (1969). Critical factors during the first years of life of the saguaro *(Cereus giganteus)* at the Saguaro National Monument, Arizona. *Ecology*, 50, 825-834.

Steenbergh, W. & Lowe, C. (1977). Ecology of the saguaro: II Reproduction, germination, establishment, growth and survival of the young plant. Washington, DC: *Department of the Interior*, National Park Service.

Stintzing, F. C. & Carle, R. (2004). Functional properties of anthocyanins and betalains in plants, food, and in human nutrition. *Trends Food Sci. Tech.*, 15, 19-38.

Stintzing, F. C., Herbach, K., M., Mosshammer, M. R., Carle, R., Yi, W., Sellapan, S., Akoh, C. C., Bunch, R. & Felker, P. (2005). Color, betalain pattern, and antioxidant properties of cactus pear (*Opuntia* spp.) clones. *J. Agric. Food Chem.*, 53, 442-451.

Tel-Zur, N., Abbo, S., Bar-Zvi, D. & Mizrahi, Y. (2004). Genetic relationships among *Hylocereus* and *Selenicereus* vine cacti (Cactaceae): evidence from hybridization and cytological studies. *Ann. Bot.*, 94, 527-534.

Terrazas, T. & Mauseth, J. D. (2002). Shoot anatomy and morphology. In P. S. Nobel, (Ed.), *The cacti: biology and uses*, (23-40). Berkeley, CA: California University Press.

Torres-Muñóz, L. & Rodríguez-Garay, B. (1996). Somatic embryogenesis in the threatened cactus *Turbinicarpus pseudomacrochele* (Buxbaum & Backeberg). *J. Professional Assoc. Cactus Dev.*, 1, 36-38.

Tovar-Puente, A. (2008). Nuevos productos y subproductos a base de nopal (alimenticios y medicinales) una alternativa de producción sustentable para las zonas rurales de México. VII Simposium-Taller "Producción y Aprovechamiento del Nopal en el Noreste de México", *Monterrey*, Nuevo León.

Tufenkian, D. (1999). *Ariocarpus* easy to grow. *Cactus Succ. J*, 71, 210-205.

Turner, R. M., Bowers J. E. & Burgess, T. L. (1995). *Sonoran Desert plants: An ecological atlas*. Tucson, AZ: University of Arizona Press.

UNAM (Universidad Nacional Autónoma de México). (1993). Base de datos de colecciones de cactáceas de Norte y Centroamérica. *Ser. Bot*, *64*, 87-94.

Valiente-Banuet, A., Bolongaro-Crevenna, A., Briones, O., Ezcurra, E., Rosas, M., Núñez, H., Barnard, G. & Vázquez, E. (1991). Spatial relationship between cacti and nurse shrub in a semi-arid environment in central México. *J. Veget. Sci.*, *2*, 15-20.

Valiente-Banuet, A., Dávila, P., Ortega, R. J., Arizmendi, M. C., Leon, J. L., Breceda, A. & Cancino, J. (1995). Influencia de la evolución de una pendiente de piedemonte en una vegetación de cardonal de *Pachycereus pringlei* en Baja California Sur, México. *Inv. Geogr. Bol.*, *3*, 101-113.

Valiente-Banuet, A. & Godínez-Álvarez, H. (2002). Population and community ecology. In P.S. Nobel (Ed.), *Cacti: Biology an uses*, (91-108). Berkeley, CA: University of California Press.

Vázquez-Ramírez, R., Olguín-Martínez, M., Kubli-Garfias, C. & Hernández-Muñoz, R. (2006). Reversing gastric mucosal alterations during ethanol-induced chronic gastritis in rats by oral administration of *Opuntia ficus-indica* mucilage. *World J. Gastroenterol*, *12*, 4318-4324.

Vega-Villasante, F., Nolasco, H., Monataño, C., Romero-Schmidt, H. & Vega-Villasante, E. (1996). Efecto de la temperatura, acidez, iluminación, salinidad, irradiación solar y humedad sobre la germinación de semillas de *Pachycereus pecten-aboriginum* "cardón barbón" (Cactaceae). *Cact. Suc. Mex.*, *41*, 51-61.

Villaseñor, J. L. (2001). La flora de México en el umbral del siglo XXI: ¿qué sabemos y hacia dónde vamos? Manuscrito inédito. Conferencia Magistral del XV Congreso Mexicano de Botánica en Querétaro.

Vyscot, J. & Jára, Z. (1984). Clonal propagation of cacti through axillary buds *in vitro*. *J. Hortic. Sci.*, *59*, 449-452.

Weiss, J., Nerd, A. & Mizrahi, Y. (1994). Flowering behavior and pollination requirements in climbing cacti with fruit crop potential. *HortScience*, *29*, 1487-1492.

Wolfram, R., Budinsky, A., Efthimiou, Y., Stomatopoulos, J., Oguogho, A. & Sinzinger, H. (2003). Daily prickly pear consumption improves platelet function. *Prostag. Leukotr. Essent. Fatty Acids*, *69*, 61-66.

Wybraniec, S. & Novak-Wydra, B. (2007). Mammillarinin: a new malonylated betacyanin from fruits of *Mammillaria*. *J. Agr. Food Chem.*, *55*, 8138-8143.

Wybraniec, S., Nowak-Wydra, B., Mitka, K., Kowalsk, P. & Mizrahi, Y. (2007). Minor betalains in fruits of *Hylocereus* species. *Phytochemistry*, *68*, 251-259.

Yeaton, R. I. & Cody, M. L. (1979). The distribution of cacti along environmental gradients in the Sonoran and Mohave deserts. *J. Ecol.*, *67*, 529-541.

Zavala-Hurtado, J. A. & Díaz-Solis, A. (1995). Repair, growth, age and reproduction in the giant columnar cactus *Cephalocereus columna-trajani* (Karwinski ex Pfeiffer) Schumann (Cactaceae). *J. Arid Environ.*, *31*, 21-31.

Zavala-Hurtado, J. A. & Valverde, P. L. (2003). Habitat restriction in *Mammillaria pectinifera*, a threatened endemic mexican cactus. *J. Veg. Sci.*, *14*, 891-898.

Zou, D., Brewer, M., García, F., Feugang, J. M., Wang, J., Zang, R., Liu, H. & Zou, Ch. (2005). Cactus pear: a natural product in cancer chemoprevention. *Nutr.*, *J. 4*, 25-37.

Zuñiga, B., Malda, G. & Suan, H. (2005). Planta-nodriza interactions in *Lophophora diffusa* (Cactaceae) in a subtropial desert in Mexico. *Biotropica, 37*, 351-356.

Chapter 2

EXTINCTION OF CYTHEROIDEAN OSTRACODES (CRUSTACEA) IN SHALLOW-WATER AROUND JAPAN IN RELATION TO ENVIRONMENTAL FLUCTUATIONS SINCE THE EARLY PLEISTOCENE

Hirokazu Ozawa

Department of Geology, National Museum of Nature and Science, Tokyo, Japan

ABSTRACT

This chapter overviews extinction events of shallow-water benthic species in multiple families of Cytheroidean ostracodes (Crustacea) in shallow-water areas of the Japanese Islands since the early Pleistocene in the late Cenozoic, and their cause in relation to fluctuations of oceanic environments. The author also refers to the possibility of ostracode extinctions in the near future in shallow-water areas around Japan.

Firstly, this chapter summarizes the faunal changes of species of four selected families in the Japan Sea during the Pleistocene, related to oceanographic environments of shallow-water areas. Tolerance ranges of salinity for these now-extinct species are inferred to have been narrower than those of most extant related-species that live in open water as well as in brackish inner-bays, based on the mode of fossil occurrences in Pleistocene strata at the eastern Japan Sea coast. During glacial periods since 1 Ma, especially in the middle Pleistocene, the salinity decreased in shallow-water areas because low sea levels resulted in the closure of the shallow and narrow straits around this marginal sea. This salinity decrease in shallow-water areas would have caused the extinction of these now-extinct species.

Secondary, the author reviews recent investigations for extinction and disappearance (i.e., regional extinction) events of representative inner-bay species in multiple families along the Japan Sea and Pacific coasts of central Japan during the middle–late Pleistocene, related to coastal environmental fluctuations. Their disappearances along the Pacific coast and extinctions along the Japan Sea coast are hypothesized to have been caused by the drastic changes in salinity and dissolved oxygen due to glacio-eustatic cycles in the middle Pleistocene, with replacement by other bay-dominant taxa.

Finally, this chapter briefly refers to the possibility for the declining, disappearance and extinction of extant relict-cryophilic species in the near future caused by global climatic warming in shallow-water areas around the Japanese island of Hokkaido along the Japan Sea, Okhotsk Sea, and Pacific coasts. Most of these cryophilic species, which favor water temperature conditions around or less than 5°C during the winter, will decline and might disappear from shallow-water areas from around this island within the next 130–250 years, if the global climate warming continues at its present rate.

1. INTRODUCTION

Research on the effects of climate changes in the geological past on global biodiversity and ecosystems has yielded key insights for modeling the potential impact of future human-induced climate change on the world's biota (e.g., Cronin, 1999; Cronin and Dwyer, 2003). Studying the biotic responses of past faunal histories related to climatic fluctuations, especially in the late Cenozoic when the climate and oceanographic environments drastically fluctuated, is useful for predicting extinction and survival in response to possible future climatic changes, and has become increasingly important (e.g., Cronin and Raymo, 1997; Cannariato et al., 1999; Cronin et al., 2002).

At the shallow-water areas of the Japanese archipelago in the mid-latitude of the Northern Hemisphere, there is a high diversity of both marine habitats and faunas. Furthermore, this area is one of the most tectonically active on earth, because of its geological setting, i.e., an oceanic-plate subduction zone. This area has an advantage in that we can more easily obtain successive and abundant late Cenozoic fossils from outcrops on the land compared with continental Asia, which is relatively tectonically stable. Therefore, the Japanese Islands are one of the most suitable regions on earth for studying the history of the highly diversified marine fauna, including the extinction, migration, and speciation from late Cenozoic to today, in relation to environmental fluctuations.

The Japan Sea, located in the northwestern Pacific region, is particularly well suited to this type of study. The reason is that it is a semi-closed, mid-latitude (35–45°N), marginal sea (Figure 1), and its oceanographic environment drastically fluctuated during late Cenozoic due to strong influences by glacio-eustatic sea-level changes linked to global climatic oscillations (e.g., Tanimura, 1981; Kanazawa, 1990; Koizumi, 1992; Cronin et al., 1994; Tada, 1994; Kitamura et al., 2001). Additionally, the Japan Sea has a modern, unique oceanographic character, being connected to the East China Sea, Okhotsk Sea and Pacific by four shallow and narrow straits (Figure 1) with a mean depth of ca. 100 m. So its paleoceanographic history is quite distinct from the Pacific side of the Japanese Islands (e.g., Oba et al., 1991; Tada, 1994; Hanagata, 2003; Itaki, 2007; Kamikuri and Motoyama, 2007). The Tsushima Warm Current, a branch of the northward-flowing Kuroshio Current, transports a large amount of heat and countless marine organisms from the south towards the north, and dominates the present-day surface waters of the Japan Sea (Nishimura, 1966; Tanimura, 1981; Ikeya and Cronin, 1993; Ozawa, 2003a; Domitsu and Oda, 2005). During glacial periods, especially in the Pleistocene, the sea level in this sea was low, and the water temperature and salinity in shallow areas were lower than today's, due to a reduced influx of warm and high-salinity water from the south (e.g., Tada, 1994; Ozawa and Kamiya, 2001).

However, there are only a few detailed comparative studies on faunal changes from the Cenozoic to modern times in relation to environmental factors of this sea.

Benthic ostracodes, a type of small, bivalved crustacean, are ideal for such studies, because they are extremely sensitive to environmental changes, and show high regional-endemism due to the lack of a planktic growth-stages during their lifecycles (e.g., Athersuch *et al.*, 1989; Ikeya and Yamaguchi, 1993; Cronin *et al.*, 2002; Frenzel and Boomer, 2005). Their shells have been abundantly found from both late Cenozoic strata on the land at the Japan Sea coast and modern surface sediments in its shallow-water areas (e.g., Hanai, 1957; Cronin and Ikeya, 1987; Okada, 1979; Ikeya and Suzuki, 1992; Irizuki, 1993; Ishizaki *et al.*, 1993; Kamiya *et al.*, 2001). Although the fossil ostracode assemblages from each stratum have individually been reported (e.g., Ishizaki and Matoba, 1985; Tabuki, 1986; Hayashi, 1988; Irizuki, 1989; Ozawa, 1996), there have been no studies summarizing the geographical and geological records of each ostracode species in this area and no discussions on their detailed faunal history except for restricted species of a few genera (Irizuki, 1996; Tsukagoshi, 1996). Additionally, the summary for the modern distribution and habitat characters for ostracode species in the Japan Sea has not been well represented.

Knowledge of the modern distribution of ostracode species in the Japan Sea has greatly increased especially since the latter half of 1990's (e.g., Tsukawaki *et al.*, 1997, 1999; Kamiya *et al.*, 2001; Ozawa and Kamiya, 2001, 2005a; Ozawa *et al.*, 2004a; Ozawa and Tsukawaki, 2008). The modern distribution of ostracode faunas in the Japan Sea is mainly controlled by water-mass properties in summer, rather than by sediment type or water depth (Figure 1; e.g., Ikeya and Suzuki, 1992; Ikeya and Cronin, 1993; Ozawa, 2003a; Tanaka, 2008). Furthermore, the data of fossil occurrences of Plio–Pleistocene ostracodes along the Japan Sea coast have also increased recently (e.g., Ozawa, 2003b, 2007; Ozawa and Kamiya, 2005b, 2008, 2009; Yamada *et al.*, 2005; Irizuki and Ishida, 2007; Irizuki *et al.*, 2007; Ozawa and Ishii, 2008; Ozawa *et al.*, 2008). These recent increases in the data of the recent and fossil ostracode faunas of the Japan Sea and its adjacent areas in middle latitudes provide the opportunity to study the causal relationship between species distribution and fluctuations of environmental factors since the late Cenozoic.

The overall aim of this chapter is to review extinction events of shallow-water benthic species in multiple families of Cytheroidean ostracodes from Japanese coasts in and around the Japan Sea during the late Cenozoic, and their causal relationship to changes of oceanic environments, briefly referring to the possibility of their extinctions in the near future. This overview is based on results of the recent research, mainly by the author and his colleagues during the last decade.

This chapter firstly overviews ostracode faunal changes by showing species occurrences of four selected families along the Japan Sea coast since the early Pleistocene, related to changes of oceanographic environments. Secondly, the author reviews the recent research of extinction and disappearance (=regional extinction) events of representative inner-bay species along the Pacific and Japan Sea coasts of central Japan during the middle–late Pleistocene, related to coastal environmental fluctuations. Finally, this chapter briefly refers to the possibility for the disappearance (=regional extinction) and extinction of extant relict-species of cryophilic ostracodes in the near future around the northern Japan, e.g. shallow-water areas of the Okhotsk Sea, in relation to global climatic warming.

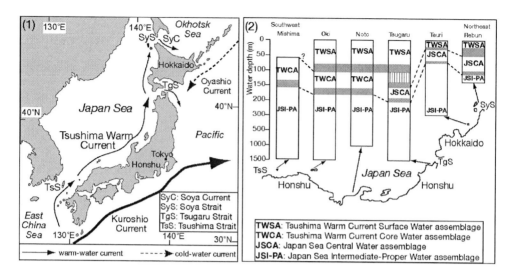

Figure 1. (1) Map of the Japan Sea, showing the locations of the three straits and schematic flow patterns of main ocean currents around the Japanese Islands; (2) Water-depth ranges for ostracode assemblages at six localities of the Japan Sea, modified after Ozawa (2003a). Vertical lines at depths of ca. 100–150 m off the Tsugaru Peninsula indicate an interval with a low occurrence of ostracodes.

2. GENERAL FEATURES OF OSTRACODES: TAXONOMY, ECOLOGY AND MORPHOLOGY

Ostracodes are a class of small crustaceans (Figure 2), with the adult form typically ranging from 0.3–3.0 mm in length. They are generally not so well known to many people due to their small size and not being commercially important in comparison with other crustaceans, e.g., shrimps and crabs. This chapter treats ostracodes of the Order Podocopida in the Subclass Podocopa, which consists of more than 20,000 named-species including both living and fossil ones from all over the world, and is the most diversified taxonomic group in the Class Ostracoda (Horne et al., 2002). The famous bioluminescent animal 'the sea firefly' (called 'umi-hotaru' in Japanese) is an ostracode, but this species belongs to another Subclass, the Myodocopa.

From the Japanese coast and adjacent areas, about 1,000 podocopid species of living and fossil taxa have been described (Ikeya et al., 2003).

Ostracodes occur in most aquatic environments on earth (Figure 3; Benson, 2003) such as the deep-sea at several thousand meters depth, through to shallow-seas on the continental shelf. They also live in rock-pools in the inter-tidal zone, brackish water areas at the river mouths, lagoons and estuaries, freshwater lakes, ponds, irrigated rice fields, and temporary puddles. Most species are benthic throughout their life, and crawl on or through the surface sediment and among aquatic plants, and a number of interstitial species, living between sediment particles, are also known (e.g., Athersuch et al., 1989; Tsukagoshi, 2004).

The most distinctive feature of ostracodes is the calcareous bivalved carapace (or shell), consisting of two valves that totally envelop their soft body, all the appendages (Figure 2). The various types of appendages are protruded between the opened valves for locomotion, feeding, and reproduction. The two valves (termed right and left) are connected with a hinge running along the dorsal margin (Figure 2). The word ostracode (or 'ostracod') is derived from the Greek word 'ostrakon' which means 'a shell'. This carapace or shell has various morphological characters (Figure 2), which allow taxonomic and phylogenetic studies to be made on both living and fossil specimens in detail (e.g., Athersuch *et al.*, 1989; Tsukagoshi, 1996).

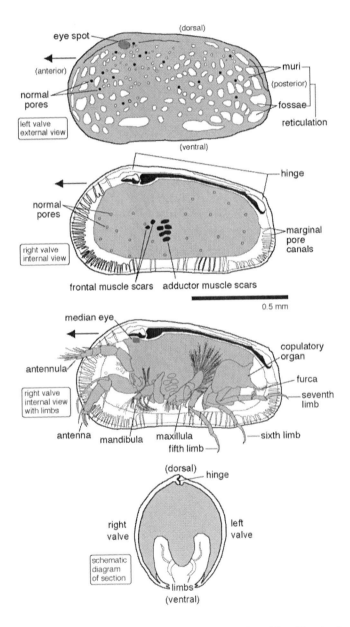

Figure 2. Morphology of podocopian ostracode (Podocopida, Cytheroidea, Hemicytheridae), male, based on *Hemicythere villosa* Sars, 1866, modified from Horne *et al.* (2002).

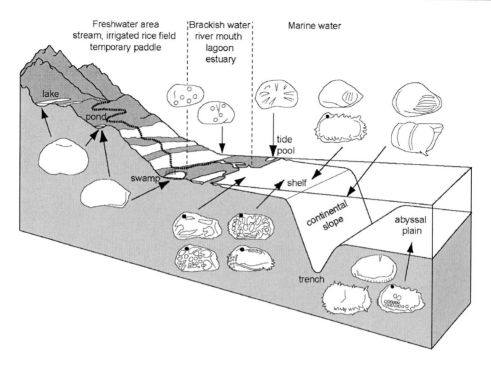

Figure 3. Schematic hypothetical profile of terrestrial aquatic to abyssal habitats of living ostracodes, modified from Benson (2003).

Like other crustaceans such as the decapods, ostracodes grow by molting (ecdysis). The adult stage is termed as A, while juvenile stages are called A-1 (i.e. one stage before the adult), A-2 (i.e. two stages before the adult), etc. In the Podocopida, there are usually eight molt stages between the egg and the adult, and the last molting is the first sexually mature stage. Sexual dimorphism (i.e. morphological differences between males and females) of the carapace and appendages can be recognized especially in the last adult stage, and to a lesser degree in the late juvenile stages.

Species with strongly calcified carapaces, such as most marine species, are relatively easily fossilized, so ostracodes have been abundantly reported in sediments from the Paleozoic Cambrian period (ca. 550 Ma = 550 million years ago) onwards from all over the world. The high-protein (= 'chitinous') soft body and appendages, due to a lack of mineralized parts, are rarely fossilized, except for a few rare cases. Typically an ostracode fossil therefore only consists of the hard carapace, but this is enough for specimens, both recent and fossil, to be identified to the species level based on various morphological characters. In particular, the surface ornamentation (reticulation, fossae, muri, eye tubercle, ridges etc), hinge type, muscle scar morphology, and pore shape and numbers (Figure 2) are very useful in ostracode taxonomy. With living specimens, the morphology of the copulatory organs and appendages are also used for species identification, similar to identification techniques used for other crustaceans, such as decapods. Fossil ostracodes are utilized by paleontologists as important paleo-environmental and stratigraphic (geological age) indices (Cronin *et al.*, 2002; Boomer *et al.*, 2003), and are used in oil and gas exploration (Athersuch *et al.*, 1989).

3: EXTINCTION AND SURVIVAL OF SHALLOW-WATER SPECIES IN JAPAN SEA

At first, this part briefly overviews the modern species-distribution and the condition of their habitat (water temperature and salinity) for the 14 living cryophilic species of three families Hemicytheridae (Hemicytherinae), Cytheruridae and Eucytheridae in and around the Japan Sea coast (Figure 4), mainly based on results of Ozawa *et al.* (2004a) and Ozawa (2006). The reason why examining the water temperature and salinity for habitats herein is that the modern distribution of most species is controlled by the physical-chemical conditions, i.e., water temperature and salinity (Athersuch *et al.*, 1989; Ikeya and Cronin, 1993; Benson, 2003; Frenzel and Boomer, 2005). Then this chapter reviews the fossil occurrence mode of the now-extinct species (= NEX species) from a lower Pleistocene stratum yielding the maximum species number of NEX species, summary for their fossil occurrences from the Japan Sea coast. It also refers to causal factors of their extinction, and the reason for survival of the 14 modern species, mainly based on contents of Ozawa and Kamiya (2005a) and Ozawa (2007). These NEX species are phylogenetically related to the 14 modern species of the above three families, but probably became extinct during the Pleistocene.

This chapter uses the word 'cryophilic species' of ostracodes reported in and around the Japan Sea. This word itself is the same as the word 'cryophilic species' defined by Cronin and Ikeya (1987) to 21 species from the same area, but this word in this chapter does not mean the same species-group as those of Cronin and Ikeya (1987). Herein the word 'cryophilic species' means modern species just in the three families Hemicytheridae (Hemicytherinae), Cytheruridae and Eucytheridae, which mainly inhabit the Japan Sea and have the common ecological feature favoring the condition of the low-water temperature as described below, with their phylogenetically related and extinct species in the same families.

Secondary, this article reviews two example cases for extinction of each one species of other taxonomic groups, Loxoconchidae and Hemicytheridae (Aurilinae), i.e., *Loxoconcha kamiyai* and *Aurila tsukawakii*, in comparison with occurrence mode of their most related-species, based on results of Ozawa and Ishii (2008) and Ozawa and Kamiya (2009). These two species were once endemic to the Japan Sea until the middle Pleistocene.

3.A. Modern Distribution of Four Assemblages

There had been still few data for the modern water-depth and areal distribution of ostracode species from the Japan Sea coast until the former-half of 1990's. So our research group investigated six regions on the sea bottom of continental shelf and slope areas from off southern to northern Japan by multiple research cruises (e.g., Tsukawaki *et al.*, 1997). As a result, we recognized a total of ca. 250 ostracode species from ca. 200 sediment samples. Ozawa (2003a) summarized the ostracode fauna on the continental shelf and slope in the six areas in the northwestern and northeastern Japan Sea, and defined four assemblage types (Figure 1; Table 1).

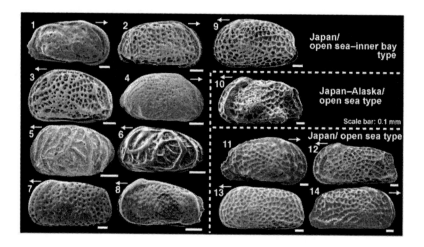

Figure 4. SEM images of fourteen species of the modern cryophilic ostracodes. 1: *Cornucoquimba alata* (Tabuki, 1986), RV. 2: *Finmarchinella nealei* Okada, 1979, RV. 3: *Hemicythere orientalis* Schornikov, 1974, LV. 4: *Howeina camptocytheroidea* Hanai, 1957, RV. 5: *Howeina higashimeyaensis* Ishizaki, 1971, LV. 6: *Howeina leptocytheroidea* (Hanai, 1957), LV. 7: *Johnnealella nopporensis* Hanai & Ikeya, 1991, LV. 8: *Munseyella hatatatensis* Ishizaki, 1966, LV. 9: *Yezocythere hayashii* Hanai & Ikeya, 1991, LV. 10: *Laperousecythere robusta* (Tabuki, 1986), LV. 11: *Baffinicythere ishizakii* Irizuki, 1996, RV. 12: *Baffinicythere robusticostata* Irizuki, 1996, LV. 13: *Daishakacythere abei* (Tabuki, 1986), LV. 14: *Daishakacythere posterocostata* (Tabuki, 1986), RV. Specimens of nos. 1–3, 7, 9, 10–14: Family Hemicytheridae; 4–6: Cytheruridae; 8: Eucytheridae. All specimens are adult valves. Specimens of nos. 1, 4, 6, 7, 11, 12 from Okhotsk Sea; 5 from Pacific; 2, 3, 8–10, 13, 14 from Japan Sea. Arrows indicate anterior. LV: left valve, RV: right valve.

Table 1. Water temperatures, salinities and characteristic species in modern ostracode assemblages of the Japan Sea, modified from Ozawa (2003a). In the assemblage names, an A for 'assemblage' is appended to abbreviations of the water masses

assemblage	temperature (°C)	salinity (‰)	characteristic species
TWSA	15–25	33–34.5	*Aurila spinifera*, *Loxoconcha viva*
TWCA	7–20	34–34.5	*Bradleya* spp., *Acanthocythereis munechikai*
JSI-PA	0–10	around 34	*Acanthocythereis dunelmensis*, *Robertsonites* spp.
JSCA	5–15	around 34	*Laperousecythere robusta*, *Munseyella hatatatensis*

The waters as well stratified especially in summer in this sea, because the winds are weak, and a strong thermocline forms near the sea surface water (e.g., Sudo, 1986). Waters in the Japan Sea are thus largely divided into two types: the Tsushima Warm Current Water occurs at depths shallower than 150 m, and a cold-water mass is present at depths greater than 150 m, e.g., in case of the southern area (e.g., Ozawa, 2003a). These water masses differ in temperature and salinity, although there is some variability in the depth of the boundary between the two water masses among multiple areas. The warm surface current water becomes cooler and heavier due to winter cooling in the northern area, where it sinks and returns southwards as part of the cold water mass (e.g., Sudo, 1986).

Four ostracode assemblages (Table 1) can be correlated to four summer water masses respectively: the Tsushima Warm Current Surface Water (TWS), the Tsushima Warm Current Core Water (TWC), the Japan Sea Intermediate-Proper Water (JSI-P) and the Japan Sea Central Water (JSC), and were named after the water masses they inhabit (Ozawa, 2003a). The assemblages are characterized by ostracode taxa occurring in a specific temperature-salinity range (Table 1).

These assemblages differ in depth distribution, and the depth boundaries between assemblages vary between the northern and southern areas (Figure 1). The JSI-PA is present in the deepest areas, whereas the TWSA occurs at the shallowest depths. The TWCA occurs at around 100 m depth in the three southern locations, and is replaced by the JSCA in the three northern ones. Assemblages differ significantly between the southern and northern areas. At the northern three locations, the TWCA does not occur, and the JSCA is found in a water mass intercalated between the shallowest and warmest water (TWS) and the deepest, coldest water (JSI-P) (Figure 1).

A large percentage of the Tsushima Warm Current flows out to the Pacific through the Tsugaru Strait, while the rest of the warm current flows to the north, off Hokkaido Island (Figure 1). Therefore water temperatures and salinities are lower in shallow areas off Hokkaido than in those off Honshu, especially at 50–100 m depths (Ozawa, 2003a). For example, at the northern two locations, the lower depth limit for the TWSA is 50 m shallower than at the southern two areas, while the upper depth limits for the JSCA and JSI-PA in the north are approximately 50–100 m shallower than for the area south of Tsugaru Strait. As a result, north of Tsugaru Strait, three assemblages live within a narrow depth range (Figure 1).

3.B. Case of Cryophilic Species of Three Families

3.B.1. Habitat conditions of living species

This part briefly mentions the geographical distribution and habitat condition (water temperature–salinity in summer and winter) of the 14 modern cryophilic species, based on results of Ozawa et al. (2004a) and Ozawa (2006).

Among four ostracode assemblages in the modern Japan Sea (Figure 1), this part focuses on one ostracode assemblage JSCA, and specifically on their ecology and relationships with water-temperature and salinity for particular cryophilic species. That is because just one assemblage JSCA includes many cryophilic species in the three families Hemicytheridae (Hemicytherinae), Cytheruridae and Eucytheridae (Figure 4; Ozawa, 2003a). These cryophilic ostracodes are the characteristic fauna in the Plio–Pleistocene Japan Sea, and are characterized by very high-species diversity (Cronin and Ikeya, 1987; Ozawa, 1996). These are so called cold-water or cold-current species in previous studies (e.g., Okada, 1979), because they belong to genera that inhabit higher latitudes than the Japan Sea today (e.g., Tabuki, 1986). These ostracodes had commonly occurred just in the Plio–Pleistocene shallow-marine strata along the Japan Sea coast (Cronin and Ikeya, 1987). Therefore, these species had been vaguely considered to flourish in the cold-water environment in the shallow-sea, especially during glacial periods (Ozawa, 2006). However, after that, due to the repeated warm current inflow during the Pleistocene interglacial periods to this sea (e.g., Tada, 1994;

Kitamura *et al.*, 2001; Ozawa and Kamiya, 2001), it had been inferred that many of these cryophilic species become now extinct at least around the Japanese Islands (Ozawa, 2003a).

However, living specimens with soft parts of these cryophilic species in three representative families, Hemicytheridae (Hemicytherinae), Cytheruridae and Eucytheridae, were recovered from the northern Japan Sea in the latter-half of the 1990's, by our investigation in offshore and inner-bay area (e.g., Ozawa *et al.*, 1999; Tsukawaki *et al.*, 1999). As a result, these ostracodes are now regarded as relict species (e.g., Ozawa, 2003a, 2003b; Ozawa *et al.*, 2004a). On the other hand, extinct, fossil endemic species to the Japan Sea are clarifying to attain more than few score species (Ozawa and Kamiya, 2005a; Ozawa, 2007).

Furthermore, since the 1990's, it has been established that some of these species in the three families, especially the Hemicytheridae (Hemicytherinae), were mis-identified, and the juvenile-adult relationship of some species was unrecognized by some workers (Ozawa *et al.*, 2004a). This led to new genera and species being established since the 1990's (e.g., Hanai and Ikeya, 1991; Irizuki, 1993, 1996). In recent studies, many species have assigned to different genera and have different species names from studies before 1990's.

In these circumstances, to clarify the environmental factor dividing into survival and extinction of cryophilic species in the three families from the shallow-water area of the Japan Sea, firstly the author began to study their ecology such as water temperature and salinity, because habitat characters of living environment for these species had been still not understood well. The author began to re-examine and summarize the modern geographical and water depth distributions of these cryophilic species in three families in and around the Japan Sea, also including results of published literatures before 1990's with re-identifying species. Then Ozawa *et al.* (2004a) identified the specific ranges of water temperature and salinity in their distributional areas both in summer and winter in order to understand the survival factors for the cryophilic species in this marginal sea through the late Cenozoic environmental changes, in terms of the water conditions.

The author focuses on 14 representative extant species of cryophilic ostracodes, especially in the three families Hemicytheridae (Hemicytherinae), Cytheruridae and Eucytheridae (Figure 4). These 14 species were chosen, because they abundantly or commonly occur from many strata of the Pliocene and Pleistocene strata along the Japan Sea coast. Furthermore, they are relatively easy to identify, because of having relatively distinct morphological characters on carapace.

In order to summarize the modern geographical and water depth distributions of the 14 species, the published data from the Japan Sea and its adjacent regions of other sea were checked by the author, adding data of species occurrences obtained by our research cruise (see literatures referred in Ozawa *et al.*, 2004a; Ozawa, 2006). Their modern distribution is summarized in Ozawa *et al.* (2004a) and Ozawa (2006).

To establish the environmental factors of their distributional areas in summer and winter, the mean values of water temperature and salinity in August and February were used. Here, the temperature and salinity for the open sea regions, the average data between 1906–1994 within one degree of latitude–longitude of each depth between 0–500 m (surface, 10, 20, 30, 50, 75, 100, 125, 150, 200, 250, 400, 500 m) from the Japan Oceanographic Data Center (JODC) database (http://www.jodc.go.jp/service_j.htm) were utilized. The mean values at the nearest water depth for each site, summarized in this study, were used, and the data for Alaska region, where lacks JODC data, were covered by Brouwers (1988). For inner bay regions of Japanese coasts, survey reports for the water temperature and salinity in each bay were used (see literatures referred in Ozawa et al., 2004a). Summarizing these data, the plotted graph of water temperature and salinity in summer and winter for each species was shown in Ozawa et al. (2004a).

These data provide new information on the ecology of cryophilic species in the ostracode fauna and their survival through Pleistocene environmental changes. These fourteen species are found from the Japan Sea, Okhotsk Sea, Pacific, and Bo-Hai Sea, and many species are reported from off the northern Japan around the Hokkaido Island (Figure 5; Ozawa et al., 2004a; Ozawa, 2006). Especially, the maximum 14 species are distributed in the open-sea area (continental shelf region) at the northeastern part of the Japan Sea. As a result, their temperature–salinity habitat requirements, based on the summer temperature which shows the largest difference through a year, were divided into three species groups, named after the distribution area's name (Figure 6; Table 2; the author might need to change names of these three groups, if their distributional data are added from many other areas.): (1) Japan/ open sea-inner bay (0–20°C, 30–34‰; 9 species); (2) Japan–Alaska/ open sea (around 5°C, 31–34‰; 1 species); (3) Japan/ open sea (0–20°C, around 34‰; 4 species). According to this division of summer condition, these 14 species are divided into two types, i.e., wide tolerance range for water temperature 0–25°C and small tolerance range for water temperature around 5°C.

On the other hand, the winter temperature–salinity of three species groups falls in a single range of 0–5°C and 30–34‰. Their summer temperature–salinity conditions are characterized by a wide range either for temperature or salinity or both. The summer temperature range of most species reflects the Tsushima Warm Current water in summer. The winter temperature range of all the species corresponds to as the coldest Japan Sea Intermediate–Proper Water through the year. The large temperature range difference between summer and winter is a remarkable character of most species. It is clear that many of these species examined here also live in temperatures as high as 20°C in summer, but are generally cryophilic species as a whole. Therefore the winter low water-temperature (less than 5°C) is considered to be critical for the survival of all of these species.

For the salinity condition, its range for species living both in inner-bay and open-sea is wide especially to low salinity range, comparing with species living only in the open-sea area, and that range is 30–34‰ at the maximum.

Table 2. Summary of species numbers and marine environmental factors in the three groups for the fourteen species, modified from Ozawa et al. (2004a)

ostracode type name	species number	temperature (°C) summer	temperature (°C) winter	salinity (‰) summer	salinity (‰) winter
Japan/ open sea–inner bay	9	5–20	0–5	30–34	30–34
Japan–Alaska/ open sea	1	around 5	0–5	31–34	32–34
Japan/ open sea	4	5–20	0–5	around 34	around 34

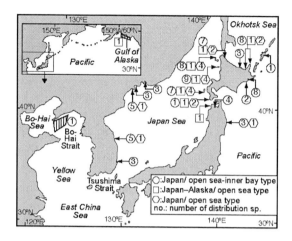

Figure 5. Summary of the geographical distribution and species number in the three types of the modern cryophilic species, modified from Ozawa (2007).

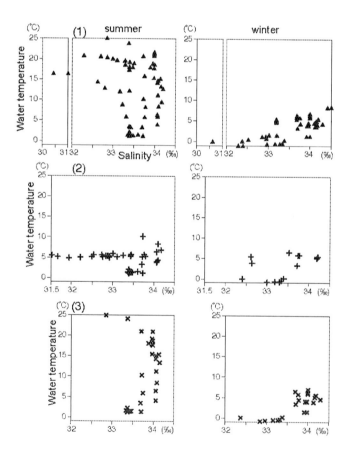

Figure 6. Summary of the water temperature and salinity in summer (August, left graphs) and winter (February, right graphs) for the distributional site of all the species in the three types. One mark means the temperature-salinity data at one site of one water depth in one area, modified from Ozawa et al. (2004a). (1) Japan/ open sea–inner bay type (for 9 species); (2) Japan-Alaska/ open sea type (for 1 species); (3) Japan/ open sea type (for 4 species). The names of three types are the same as in Figure 5.

3.B.2. Paleoecology of now-extinct ('Nex') species

This part briefly explains the probable now-extinct species, which are found just as fossil specimens from Plio–Pleistocene strata at the Japan Sea coast, in the same three families as the modern 14 cryophilic species, based on results of Ozawa and Kamiya (2005a) and Ozawa (2007). Research for the geographical–geological distribution (occurrence age) of these extinct species in detail is significant for clarifying the faunal history of the Japan Sea, because most of them were once endemic to this semi-closed marginal sea. This chapter conveniently calls these fossil species the NEX (= now extinct) species, which mean becoming extinct until today.

We firstly estimated the paleo-habitat of NEX species during glacial periods, i.e., which did they favor the environment of inner-bay or open-sea area. This analysis focuses on 25 now-extinct species in the three families cited above (Table 3). These species were chosen, because they are abundant or common in Plio-Pleistocene strata at its coast. Of these 25 NEX species, 17 are in the Hemicytheridae (Hemicytherinae), five in the Cytheruridae, and three in the Eucytheridae (Figure 7). Their occurrences have no reported from the modern Japan Sea or elsewhere in more than 40 studies prior to 2009.

We first examined fossil occurrences of these species in horizons of the Plio-Pleistocene Omma Formation, which crops out along the Japan Sea coast of central Japan (Figure 8) and is well known for its abundant fossils, including also ostracodes (e.g., Cronin and Ikeya, 1987; Ozawa, 1996). This stratum contains the 17 NEX species especially in horizons correlated to glacial periods dating to about 1.5 Ma under both the inner-bay (two areas) and open-sea (one area) environments, and have the greatest abundance and highest number of NEX species among all Plio–Pleistocene deposits at the Japan Sea coast (Ozawa and Kamiya, 2001, 2005a). These factors led us to analyze and compare ostracode occurrences in several outcrops of the Omma Formation, in order to infer the environments, e.g. inner bay or open sea, where these NE species lived during glacial periods.

Occurrence data for these NEX species of Ozawa (1996) in the shallow-cold water assemblage as defined by Ozawa and Kamiya (2001), were re-examined from horizons dated at about 1.5 Ma in three areas: Okuwa, Kofutamata, and Sakuramachi. This assemblage is one of their three types of assemblages in this formation, and is found from periods correlated to glacial periods. Other assemblages, such as shallow-warm and deep-warm ones in horizons during interglacial periods, were excluded, because the shallow-cold assemblage has the greatest number of NEX species, and other assemblages contain just few individuals of NEX species. The shallow-cold ostracode assemblage contains at Okuwa was found in 24 samples; also present in 18 samples at Kofutamata, and 21 samples at Sakuramachi, and each sample contains around 200 individuals (Ozawa and Kamiya, 2001).

As a result, NEX species in these samples occurred with extant species that are characteristic of modern shallow-open sea (*Finmarchinella hanaii*, *Cytheropteron sawanense*, *Schizocythere ikeyai*) and brackish-inner bay environments (*Bicornucythere bisanensis*) of Japanese coasts (e.g., Ikeya *et al.*, 1992; Ikeya and Shiozaki, 1993; Zhou, 1995; Tsukagoshi and Briggs, 1998). A total of 19 NEX species in three families occur in the Omma Formation within the three study areas (Figure 8; e.g., Ozawa and Kamiya, 2005a). Seventeen species are present only at the Sakuramachi locality in many samples, whereas only three species occur at Okuwa and Kofutamata. *Kotoracythere* sp. is restricted to Okuwa and Kofutamata only, whereas *Laperousecythere* sp. B is more common at Sakuramachi than at Okuwa and

Kofutamata. *Kotoracythere tsukagoshii* is more abundant at Okuwa and Kofutamata than at Sakuramachi.

The maximum percentages of individuals of the extant open-sea species *F. hanaii, C. sawanense* and *S. ikeyai*, the extant inner-bay species *Bicornucythere bisanensis* and the 19 NEX species in each study area in all samples are shown in Figure 8. The open-sea ostracodes *F. hanaii, C. sawanense* and *S. ikeyai* are abundant among the predominantly extant species at Sakuramachi, and *S. ikeyai* occurs only at Sakuramachi (Figure 8). The dominant species at Okuwa and Kofutamata is the characteristically inner-bay species *B. bisanensis* (see Ozawa and Kamiya, 2001, for details). The number of NEX species is higher at the open sea locality (Sakuramachi; 17 species) than at inner-bay environments (Okuwa and Kofutamata; 3 species) (Figure 8).

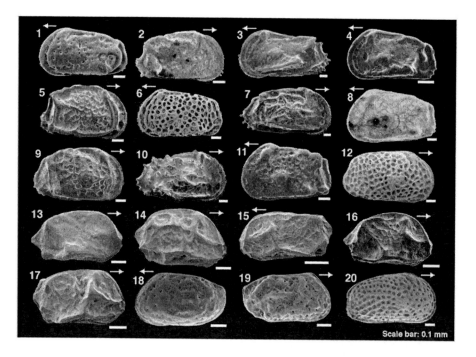

Figure 7. SEM images of now-extinct (NEX) species from the Japan Sea coast in the early Pleistocene Omma Formation, central Japan. 1: *Cornucoquimba* sp. A, LV. 2: *Cornucoquimba* sp. B, RV. 3: *Cornucoquimba* sp. C, LV. 4: *Cornucoquimba* sp. D, LV. 5: *Hemicythere kitanipponica* (Tabuki, 1986), RV. 6: *Johnnealella*? sp., LV. 7: *Laperousecythere* cf. *ishizakii* Irizuki & Matsubara, 1996, RV. 8: *Laperousecythere* sp. A, LV. 9: *Laperousecythere* sp. B, RV. 10: *Laperousecythere* sp. C, RV. 11: *Laperousecythere* sp. D, LV. 12: *Urocythereis*? *gorokuensis* Ishizaki, 1966, RV. 13: *Semicytherura leptosubundata* Ozawa & Kamiya, 2008, RV. 14: *Semicytherura robustundata* Ozawa & Kamiya, 2008, RV. 15: *Semicytherura subslipperi* Ozawa & Kamiya, 2008, LV. 16: *Semicytherura subundata* (Hanai, 1957), RV. 17: *Semicytherura tanimurai,* Ozawa & Kamiya, 2008, RV. 18. *Kotoracythere tsukagoshii* Tanaka, 2002, LV. 19. *Kotoracythere* sp., RV. 20. *Yezocythere* sp., RV. All specimens are adult, except for one juvenile in no. 10. Specimens of nos. 1–20 are from the Omma Formation, central Japan, and only one specimen of no. 12 is from the Kaidate Formation, central Japan. Arrows indicate anterior. LV: left valve, RV: right valve.

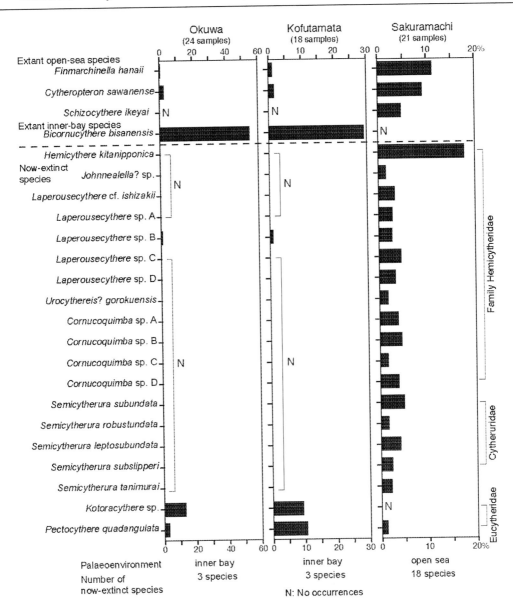

Figure 8. Species occurrences and maximum percentages of individual numbers per sample for now-extinct (NEX) species, and two types (open-sea and inner-bay) of extant species, in the three study areas, modified from Ozawa and Kamiya (2005a).

Shallow, open-sea ostracode assemblages around the Japanese Islands are characterized by a higher species-diversity than those in inner bays (e.g. Ikeya and Suzuki, 1992; Ikeya and Shiozaki, 1993). Shallow, open-sea ostracode assemblages at Sakuramachi, which have the highest number of NEX species, show a Shannon-Weaver species-diversity index averaging 3.54 (Figure 8; Ozawa and Kamiya, 2001). Each sample from this assemblage at Sakuramachi contains ca. 40–70 species, with *F. hanaii, C. sawanense* and *S. ikeyai* dominant (Ozawa and Kamiya, 2001). In contrast, the dominant species at Okuwa and Kofutamata is *B. bisanensis*, which is often commonly found with *Finmarchinella uranipponica* and *Pistocythereis bradyformis* (Ozawa, 1996; Ozawa and Kamiya, 2001). These species inhabit brackish, inner-bay areas of the Japan Sea coast and around the Japanese Islands in low-species-diversity assemblages (e.g. Ikeya and Shiozaki, 1993; Kamiya *et al.*, 2001). This assemblage at Okuwa and Kofutamata contains ca. 20–50 species in each sample, and has a relatively low species-diversity index (average = 2.75) (Figure 8; Ozawa and Kamiya, 2001). These facts with co-occurring fossils of other extant ostracode species and planktic foraminifers from characteristic lithofacies (calcareous sandstone) at Sakuramachi (Takata, 2000), including many NEX species, suggest that the 17 NEX species in the Omma Formation favorably inhabited a shallow, open sea during glacial periods, even though the dominant occurrences of two NEX Eucytheridae species at Okuwa and Kofutamata imply an inner-bay environment.

Table 3. Now-extinct species in three families treated in this chapter, modified from Ozawa and Kamiya (2005a)

Family Hemicytheridae
Cornucoquimba sp. A (= *Cornucoquimba* sp. 1 and *Cornucoquimba* sp. 2 of Ozawa, 1996)
Cornucoquimba sp. B (= *Cornucoquimba* sp. 5 of Ozawa, 1996)
Cornucoquimba sp. C (= *Cornucoquimba* sp. 6 of Ozawa, 1996)
Cornucoquimba sp. D (= *Cornucoquimba* sp. 7 of Ozawa, 1996)
Daishakacythere sp. (= *Daishakacythere* sp. of Irizuki, 1993)
Finmarchinella daishakaensis Tabuki, 1986
Finmarchinella rectangulata Tabuki, 1986
Hemicythere kitanipponica (Tabuki, 1986)
Johnnealella? sp. (= *Johnnealella?* sp. of Ozawa, 1996)
Laperousecythere cf. *ishizakii* (= *Laperousecythere* cf. *ishizakii* of Ozawa, 1996)
Laperousecythere sp. A (= *Patagonacythere* sp. 2 of Ozawa, 1996)
Laperousecythere sp. B (= *Laperousecythere* sp. 2 of Ozawa, 1996)
Laperousecythere sp. C (= *Laperousecythere* sp. 1 of Ozawa, 1996)
Laperousecythere sp. D (= *Laperousecythere?* sp. of Ozawa, 1996)
Normanicythere japonica Tabuki, 1986
Patagonacythere sasaokaensis Irizuki, 1993
Urocythereis? gorokuensis Ishizaki, 1966
Yezocythere sp. (= *Yezocythere* ? sp. of Ozawa, 1996)

Family Cytheruridae
Semicytherura leptosubundata Ozawa and Kamiya, 2008
Semicytherura robustundata Ozawa and Kamiya, 2008
Semicytherura subslipperi Ozawa and Kamiya, 2008
Semicytherura subundata (Hanai, 1957)
Semicytherura tanimurai Ozawa and Kamiya, 2008

Family Eucytheridae
Kotoracythere tsukagoshii Tanaka, 2002
Kotoracythere sp. (= *Kotoracythere* sp. 1 of Ozawa, 1996)
Pectocythere daishakaensis Tabuki, 1986

Table 4. Disappearance of now-extinct (NEX) species per period since 1.5 Ma, modified from Ozawa and Kamiya (2005a). Epochs (A)– (F) are the same as those in Figure 10

variable / period of epoch	A–B	B–C	C–D	D–E	E–F	F–G
a. disappearance of now-extinct species	4	2	5	11	0	1
b. time interval (million of years)	0.3	0.1	0.1	0.1	0.15	0.1
	(1.5)–1.2	1.0–0.9	0.9–0.8	0.5–0.4	0.25–0.1	0.1–0
c. disappearances of now-extinct species per hundred of thousands of years (= 10 x a / b)	1.3	2	5	11	0	1

3.B.3. Faunal history of 'Nex' species during pleistocene

In the next step, the occurrence patterns of NEX and extant cryophilic species in three families were summarized for the 14 strata from upper Pliocene to upper Pleistocene along the Japan Sea coast (Figure 9), in order to understand processes and causal factors for the decreased species number in modern times as compared to that the 1.5-Ma bed, in relation to the Pleistocene environmental changes of this sea.

We examined ca. 60 samples from thirteen formations to check species occurrences, and the chronostratigraphic ranges of these formations are shown in Figure 9 (see reference list for the stratigraphy in Table 5 of Ozawa and Kamiya, 2005a). We have also used data for ostracode occurrences in nine formations from published studies that were illustrated with SEM images to compensate for data of species occurrence (see Table 6 of Ozawa and Kamiya, 2005a). By summarizing occurrences of these species, we also calculated the disappearance rate of NEX species.

Figure 10 represents the summary for occurrences of a total of 23 NEX species, which is 17 species occurring from the Omma Formation, adding with the 6 species in two families Hemicytheridae (Hemicytherinae) and Eucytheridae just from other areas estimated for shallow open-sea species judging from co-occurring extant species as fossils, in each stratum at its coast. In Figure 10, the fourteen strata were divided into the six periods (= Epochs) by depositional ages, and they are arranged from old to young periods from left to right. As a result, species number is high until the Epoch A, but its number decreases as time goes on to the Epoch F, then it becomes just one species *Semicytherura subundata* in the youngest Epoch F of late Pleistocene (ca. 0.1 Ma). To compare with the fossil occurrence mode of NEX species, also occurrence data of the extant 14 cryophilic species are also shown in lower column of Figure 10. According to these data, most extant species are found from many strata at this coast.

The change of species numbers of NEX species during these six periods was shown in Figure 11. According to it, the species number of NEX species represents the maximum until around 1.5 Ma, then their number of species begins to decrease after 1.5 Ma (Figure 11), and many of them disappeared in 0.9–0.4 Ma. The disappearance rate of species per 0.1-million-years is the highest within this period, especially between Epochs D–E (0.5–0.4 Ma; Figure 11; Table 4). Especially in 0.9–0.4 Ma about 70% of NEX species are not found, and probably they became extinct. The rapid decrease in NEX species at that time suggests a major reduction or disappearance of suitable area for their habitats in the Japan Sea, due to several factors.

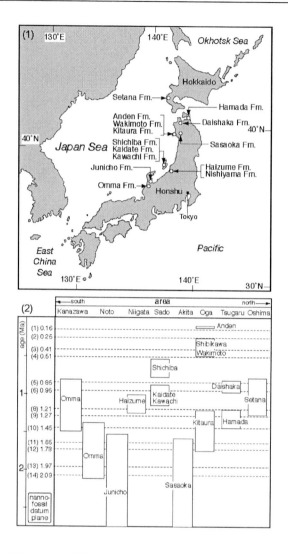

Figure 9. (1) Locations of Pliocene and Pleistocene strata along the Japan Sea coast; (2) Periods of deposition and correlation of Pliocene and Pleistocene formations along the Japan Sea coast, modified from Ozawa and Kamiya (2005a). The ages of calcareous nannofossil datum levels are from Sato et al. (1999). See the caption of Figure 5 of Ozawa and Kamiya (2005a) for compiled each literature of chronological data of formations respectively.

These NEX cryophilic species had obscurely been considered that they favored the cold-water environment in the shallow-sea especially during glacial periods, then after that, due to the beginning of the repeated warm-current inflow during the Pleistocene interglacial periods, it had been inferred that many of them become now-extinct (e.g., Ozawa, 2006). However, based on the detailed investigation of their fossil occurrences by our research, the extinction event at 0.9–0.4 Ma actually occurred after the beginning of interglacial intervals with low-volume influxes of warm water (around 1.5 Ma) and after the beginning of high-volume influxes of warm water (1.2–1.0 Ma) (Figure 11). Their disappearance rate was highest at 0.5–0.4 Ma (Epoch D–E) and second highest at 0.9–0.8 Ma (Epoch C–D; Table 4). There was a time lag of up to 1.1 million years (1.5–0.4 Ma) at the maximum between the extinction of numerous species (−0.4 Ma) and the beginning of increased periods of warm-water influx (1.5 Ma) accompanying interglacial periods (Figure 11).

On the other hand, it is considered that the shallow-sea area under the low-salinity condition was widely found in glacial periods since ca. 1 Ma (Figure 11; e.g., Tada, 1994; Amano, 2004). Judging from the period of the paleoenvironmental fluctuations of this sea, it is appropriate that these NEX species became extinct owing to low-salinity conditions during glacial periods in the Japan Sea, not to warm-current water influx during interglacial periods. It is unlikely that the increases in the water temperature related to the influx of warm currents during interglacial periods were the direct causal factor of extinction. If these temperature-increases had been the direct cause of their extinctions, many more of NEX species would have disappeared during 1.5–1.0 Ma. So this implies that the low-salinity environments in glacial periods at 0.9–0.4 Ma were more likely to be the main cause of extinction than the increases in water temperature at 1.2–1.0 Ma or 1.5 Ma.

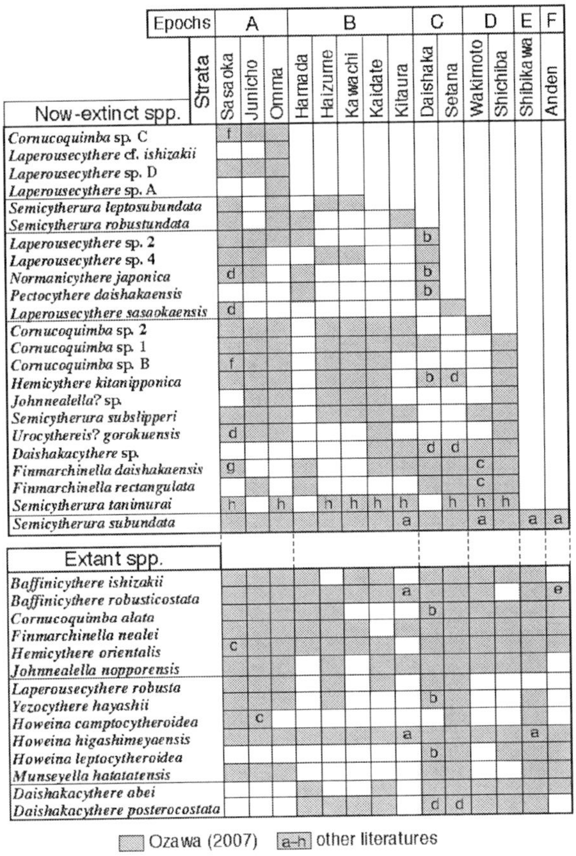

Figure 10. Summary for occurrences of now-extinct (NEX) and extant cryophilic species in Pliocene and Pleistocene strata along the Japan Sea coast, modified from Ozawa and Kamiya (2005a). Age of epochs; (A): −1.5 Ma, (B): 1.2–1.0 Ma, (C): 0.9 Ma, (D): 0.8–0.5 Ma, (E): 0.4–0.25 Ma, (F): 0.1 Ma. Cited literatures for ostracode occurrences; a: Ishizaki and Matoba (1985), b: Tabuki (1986), c: Cronin and Ikeya (1987), d: Irizuki (1993), e: Irizuki (1996a), f: Ishida (unpubl. data), g: Irizuki and Ishida (2007), h: Ozawa and Kamiya (2008). Strata names are the same as in Figure 9. Epochs (A)–(F) are the same as in Table 4.

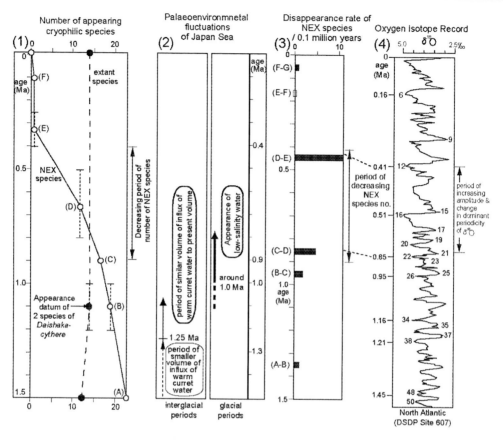

Figure 11. (1) Decrease in the number of now-extinct (NEX) and extant cryophilic species at the Japan Sea since 1.5 Ma; (2) Paleoenvironment of the Japan Sea; (3) Disappearance of now-extinct (NEX) species per 0.1 million years at the Japan Sea; (4): Timing of events compared to oxygen isotope records at North Atlantic DSDP Site 607 (Ruddiman et al., 1989) with ages of the oxygen isotope stages after Sato et al. (1999). (1)–(4) are modified from Ozawa and Kamiya (2005a) and Ozawa (2007). Periods of warm-water influx events in the Japan Sea are from Ozawa and Kamiya (2001). The three bars in graph of (1), B, D and F, show age ranges. Epochs (A)– (F) are the same as in Figure 10 and Table 4.

The salinity condition in the shallow-water area during glacial periods between 1.5 Ma and 0.9–0.4 Ma would be different (Amano, 2004; Ozawa and Kamiya, 2005a). So most NEX species became extinct, because they favored the open-sea environment and could not tolerate the low-salinity condition in glacial periods during 0.9–0.4 Ma, adding with the reduction in area or disappearance of suitable habitat environments in shallow-water, with the closure of shallow-strait connections to other seas, and the resulting lack of refuge areas due to the lowered sea-level.

During 0.9–0.4 Ma, it is considered that the range of sea-level fluctuations became larger than that before 0.9 Ma (e.g., Ozawa and Kamiya, 2005a). Few paleoceanographic data, such as successive records of oxygen isotope, exist for the Japan Sea since 1.5 Ma. The calcium carbonate compensation depth (CCD) in the Japan Sea is shallower than other oceans and seas, so that its complete Pleistocene oxygen isotope records from calcareous foraminifers in deep-sea cores could not be recovered (Tada, 1994). To work around this problem, we have used oxygen isotope records from other oceans in order to infer the paleoceanography of the

Japan Sea. We could correlate records from the Japan Sea to global oceanic records by relying on sea-level fluctuations resulting from the Northern Hemisphere glaciations (e.g., Kanazawa, 1990; Tada, 1994; Kitamura et al., 2001; Ozawa and Kamiya, 2001). When sea-level dropped during glacial low-stands, the influx of warm current water into the Japan Sea, and the salinity in shallow areas, decreased due to a reduction in both the flow's cross-sectional area at the southernmost strait (Paleo-Tsushima Strait) and the volume of the warm current from the south (e.g., Tada, 1994). The influx of high-salinity, warm water through the southern strait into the Japan Sea increased when sea-level rose during interglacial periods.

An overall positive shift in $\delta^{18}O$ during glacial periods at 0.9–0.4 Ma has been recognized in oxygen isotope stages 22–12 in many marine records (Figure 11; stage ages from Sato et al., 1999) including those from DSDP Site 607 in the North Atlantic Ocean (Ruddiman et al., 1989) and ODP Site 806 in the Western Equatorial Pacific (Berger et al., 1994). If these oxygen isotope records are suitable proxies for glacio-eustatic conditions in the Japan Sea, the 0.9–0.4 Ma interval saw a large fall in sea-level during each glacial lowstand, compared to the period around 1.5 Ma (Figure 11). The large sea-level falls during glacial periods at 0.9–0.4 Ma would have caused a stronger decrease in the cross-section of flow through the southern strait into the Japan Sea, compared to glacial periods around 1.5 Ma. During each low-stand period, a low-salinity environment expanded onto the shelf areas of this marginal sea. As a result, the NEX species would have lost much of their suitable habitat during 0.9–0.4 Ma glacial periods. The shallow-strait connection with northern seas would have been very shallow and/ or narrow, and virtually closed during glacial low-stands (or even interglacial periods; e.g., Kitamura et al., 2001). Thus, it would have been difficult for NEX species to seek refuge in adjacent seas during glacial periods, so that they were confined within the Japan Sea. Many species could not survive the very low salinity in glacial periods between 0.9–0.4 Ma, so their species numbers had decreased.

In addition, the regular cycle length for $\delta^{18}O$ fluctuations increased from 40 ky to 100 ky during the Mid-Pleistocene Revolution at 0.9–0.4 Ma, (e.g., Ruddiman et al., 1989; Berger et al., 1994). This change would have caused the longer-term development of low-salinity conditions during each glacial period between 0.9–0.4 Ma in the Japan Sea, compared to the period around 1.5 Ma. This long-term persistence of low-salinity conditions likely was unsuitable for NEX species, and was also one of the factors in their decline.

Therefore the main causal factors of extinction of NEX species are summarized as follows: (A) the reduction in area or disappearance of suitable environments in shallow water, due to the onset of low-salinity conditions, (B) closure of the shallow-strait connections with other seas, and the resulting lack of refuge areas because of the lowered sea-level, and (C) the long-term presence of unsuitable low-salinity environments. Probably most species became extinct during Pleistocene glacial periods at 0.9–0.4 Ma, owing to these three paleoceanographic factors that were related to glacio-eustatic sea-level oscillations.

3.B.4. Reasons for survival of modern species

Since the extinction of NEX species in the shallow open-sea areas during 0.9–0.4 Ma, the modern cryophilic species, originally favoring open-sea areas (i.e., the four species of 'Japan/ open sea' type), colonized and flourished in the shallow-water areas of the Japan Sea (e.g., Figure 11).

The 10 modern cryophilic species of the two types, i.e., 'Japan/ open sea-inner bay' and 'Japan–Alaska/ open sea', can tolerate winter water temperatures of around or less than 5°C, even in areas under low-salinity conditions (Ozawa et al., 2004a). Therefore, during the Pleistocene they could survive in relatively low-salinity conditions, probably in areas of low-water temperatures during the winter in the Japan Sea. After all, cryophilic species, which could adapt to the new appearance of low-salinity conditions during glacial periods, can survive in inner-bay or open-sea areas in or outside of the Japan Sea. Probably species that could not adapt to this new type of environment became extinct.

As described above, most NEX species that favored open-sea areas became extinct in the Japan Sea and other areas. However, the two species of *Daishakacythere* of 'Japan/ open sea' type favor the open sea areas like the NEX species, and inhabit just the shallow-water area of the Japan Sea. There are no reports of occurrences of these two *Daishakacythere* species from areas outside of the Japan Sea, not only as fossils but also as modern records, indicating that these two species are endemic to the Japan Sea since their appearance 1.2 Ma (Ozawa et al., 2004a; Ozawa, 2007). The four species of the 'Japan/ open sea' type, including the two *Daishakacythere* species, inhabit the shallow open-sea off the present Hokkaido Island, under low water-temperature conditions around or less than 5°C in winter and relatively high-salinity condition of around 34‰. How could these four species in the open-sea areas survive within the Japan Sea during glacial periods between 0.9–0.4 Ma when the shallow-sea areas under the low-salinity conditions spread widely within this sea? For this subject, one hypothesis has been represented (e.g., Ozawa, 2006, 2007).

This hypothesis is based on the probable existence of a high salinity environment, similar to the present conditions of around 34‰, in a part of the shallow open-sea areas (from lower shelf to upper continental slope in ca. 100–400 m water depths) under a surface layer of low-salinity water during glacial periods since 0.9 Ma. Paleobiogeographical and paleoecological research on the molluscan fauna (e.g., Amano and Watanabe, 2001; Amano, 2004, 2007) and occurrence reports for radiolarian species in the Japan Sea during glacial periods (Itaki et al., 1996; Itaki, 2007) suggest the existence for this type of water mass in this sea during glacial periods since the early Pleistocene. Species of 'Japan/ open sea' type are dominantly distributed in waters shallower than 100 m depth of the present Japan Sea, but they are also occasionally reported from 100–200 m depths (e.g., Ozawa et al., 2004a, 2004b). These facts suppose that they could survive glacial periods within this sea by inhabiting slightly deeper water areas compared with their modern counterparts.

Species of 'Japan/ open sea' type are not found from the modern inner-bay regions under low-salinity conditions, and there are no reports of occurrences from inner-bay fossil assemblages (e.g., Ozawa, 1996; Ozawa et al., 2004a). One reason is that species of 'Japan/ open sea' type favor sea areas under relatively high-dissolved oxygen condition, like the present open-sea areas in the northeastern Japan Sea (6–9 ml/ l; e.g., Kuwahara, 1990; Ikeya and Cronin, 1993). For example, in Mutsu Bay and Lake Saroma, which are representative inner-bays in modern coasts of northern Japan, the dissolved oxygen in the bottom waters decreases to less than 3 ml/ l during the summer, producing a hypoxia water mass (e.g., Nagamine et al., 1982; Tada and Nishihama, 1988). So probably the open-sea species, favoring high-dissolved oxygen conditions, would not be able to inhabit inner-bay areas during the Pleistocene.

It is considered that the distributional range of water-depths for NEX species and species of the 'Japan/ open sea' type would be slightly different, and the NEX species were restricted in slightly shallower water depth than species of the 'Japan/ open sea' type. That is because fossils of the NEX species have a tendency to co-occur with fossils of extant species inhabiting intertidal-zones and upper-shelf areas in coarser sediments in the stratum, and not with those of the 'Japan/ open sea' type (e.g., Ozawa, 1996; Takata, 2000; Ozawa and Kamiya, 2005a). The difference in the range of the distributional water depth would decide their fate during the Pleistocene. Species of the 'Japan/ open sea' type would be able to survive glacial periods in slightly deeper water areas in the shallow-open seas within the Japan Sea since the early Pleistocene. The NEX species, restricted to shallower environments, might be more influenced by the surface low-salinity water than species of 'Japan/ open sea' type, so these NEX species would become extinct during the Pleistocene in this semi-closed marginal sea (Ozawa and Kamiya, 2005a; Ozawa, 2007).

Furthermore, Ozawa (2007) highlighted that the NEX species and the modern four species of the 'Japan/ open sea' type, first appear in the geological record at different times. Comparing the ages of their first occurrences, most NEX species had appeared during the Pliocene at the latest, and some species such as *Hemicythere kitanipponica* and *Semicytherura subundata* had already appeared in the Miocene (e.g., Irizuki, 1994). In contrast, the two living species of *Daishakacythere* appeared around 1.2 Ma (Ozawa, 2007), so they are the youngest and last-appearing taxonomic group among a total of 37 species in three families covered in this chapter. The period in which the two *Daishakacythere* began to expand their distributions was after 1.0 Ma, almost the same time as the two *Baffinicythere* species started to expand their distributions (Irizuki, 1996a). So these four species had invaded shallow open-sea areas of the Japan Sea replacing NEX species, which were declining and disappearing (finally becoming extinct) from this sea during the Pleistocene (Ozawa, 2006). Since then, these four species are the dominant or common species of the modern ostracode fauna in the northern Japan Sea (Ozawa *et al.*, 1999, 2004a).

Based on the first appearance age and process of expanding distributional area, it looks that the four species of two genera, *Daishakacythere* and *Baffinicythere,* would occupy their ecological niche just becoming vacant for the ostracode fauna during Pleistocene at shallow open-sea areas in this sea, due to extinctions of many NEX species. So they might be able to geographically expand their distributional area within this marginal sea. The factors that cause the expansions of distributions in the shallow open-sea regions for each species of the 'Japan/ open sea' type need to be clarified.

Furthermore, the paleoenvironment during glacial periods of the Japan Sea, especially in the Pleistocene, has often been estimated to consist of anoxic bottom-waters so that most benthic organisms would not be able to survive there (e.g., Oba *et al.*, 1991; Tada, 1994). This common idea for this sea's glacial periods, i.e., 'the sea of death', is based on the research of bottom sediments of deep-sea core samples obtained from 2,000–3,000 m water depths in offshore areas in deep-sea basins. However, in the shallow-sea areas during glacial periods, the number of species of these cryophilic ostracodes in the three families was much higher than that of the modern fauna. So the ecological and paleoceanographic settings of the shallow-sea areas during glacial periods, such as food supply, nutrients, and dissolved oxygen for supporting this high-diversified fauna, might have been different from those of shallow-areas during interglacial periods like today or those of the deep-sea areas during glacial periods. Therefore we need to research the ecological conditions in this sea, with the causal relationship of the species diversity based on examples of ostracode research.

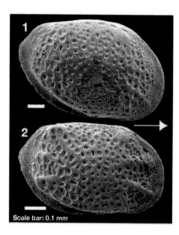

Figure 12. SEM images of two *Loxoconcha* species (left valve). 1: *Loxoconcha kamiyai* Ozawa & Ishii, 2008. 2: *Loxoconcha mutsuense* (Ishizaki, 1971). Specimens of no. 1 from the Omma Formation, central Japan, and no. 2 from the Kaidate Formation, central Japan. Arrows indicate anterior.

3.C. Case of an Extinct Species of *Loxoconcha*

This article overviews extinction of an ostracode species *Loxoconcha kamiyai* (Figure 12), which was once endemic to the Japan Sea, in comparison with the occurrence mode of its most related, extant species *Loxoconcha mutsuense* (Figure 12; Ozawa and Ishii, 2008).

The ostracode genus *Loxoconcha* (Loxoconchidae) is widely distributed in shallow-marine environments from tropical to subarctic regions around the world. This is one of the most diversified genera of ostracodes, and about 600 species belonging to this genus have been identified (e.g., Tanaka and Ikeya, 2002; Horne, 2003). This genus also commonly occurs in and around the Japanese Islands (e.g., Ishizaki, 1968; Hanai *et al.*, 1977; Zhou, 1995), and approximately 40 living and fossil species of *Loxoconcha* have been described (Ikeya *et al.*, 2003). Thus, *Loxoconcha* is one of the most important genera of Japanese ostracodes.

Species of *Loxoconcha* are common in late Cenozoic strata along the Japanese coast of the Japan Sea (e.g., Ishizaki and Matoba, 1985; Tabuki, 1986; Tanaka *et al.*, 2004). These strata contain several undescribed *Loxoconcha* species (Ozawa, 1996), but previous studies had not described these unnamed species as new ones.

Ozawa and Ishii (2008) described a new fossil species of *L. kamiyai* in the Pleistocene stratum from the eastern coast of the Japan Sea, with the first detailed description of its carapace. They discussed the paleo-biogeographical significance for *L. kamiyai*, and assessed its phylogenetic relationship to other related *Loxoconcha* species on the basis of the distribution pattern of pores. The number, distribution, and differentiation of pores (= a kind of ostracode sensory organs; Figure 13) on the ostracode carapace during ontogeny were studied to determine the phylogenetic relationships among species (e.g., Kamiya, 1997). The reconstruction of ostracode phylogeny based on pore analyses was first proposed by Tsukagoshi (1990) for the 14 species in the genus *Cythere*. This work was followed by Irizuki (1993), who studied 21 species in eight genera of hemicytherinae ostracodes, and Ishii *et al.* (2005), who investigated 17 species in the genus *Loxoconcha,* and Sato and Kamiya (2007), who researched 13 species in the genus *Xestoleberis*. Kamiya (1997) named the phylogenetic reconstruction method proposed by Tsukagoshi (1990) "differentiation of distributional pattern of pore (DDP) analysis". The distribution of pores in *L. kamiyai* obtained from the Japan Sea coast was examined with this method, and the results compared to the pore data for other 17 *Loxoconcha* species in Ozawa and Ishii (2008).

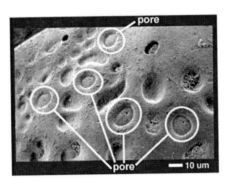

Figure 13. SEM images of pores of *Loxoconcha kamiyai* from the Omma Formation, central Japan.

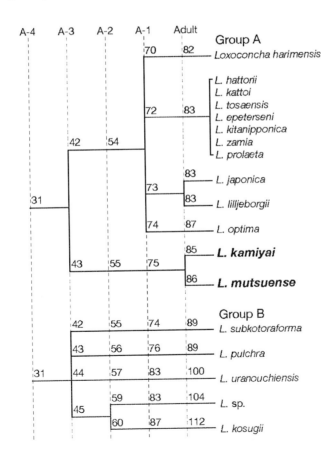

Figure 14. Results of DDP analysis for 18 *Loxoconcha* species, modified from Ishii et al. (2005) and Ozawa and Ishii (2008). Numbers indicate total numbers of pores for each lineage and stage. Trees drawn by hand.

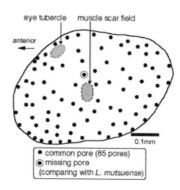

Figure 15. Distributional pattern of pores in adult left valve of *Loxoconcha kamiyai*. Position of one missing pore of this species is determined by comparison with the distributional pattern of pores of *Loxoconcha mutsuense* shown in Ishii *et al.* (2005).

On the basis of the DDP results for its adult and A-1 juvenile stages, *L. kamiyai* is judged to be the species most closely related to *Loxoconcha mutsuense* in the Loxoconchidae (Figure 14; Ozawa and Ishii, 2008). Both species have the same total number of pores on the carapace at the A-1 juvenile stage (75 pores per valve; Figure 14). The difference in total number of pores in the adult stage is just one pore between these two species, which is missing on the central area in *L. kamiyai* (Figure 15).

Based on the oldest fossil record, *L. kamiyai* and *L. mutsuense* first appeared in the Japanese Islands during the Pliocene around 3 Ma. The oldest record of *L. kamiyai* is from the Pliocene stratum at the Japan Sea side of central Japan (Figure 16; Ozawa *et al.*, 2008; ca. 3.5 Ma; age data from Nagamori *et al.*, 2003). The oldest record of *L. mutsuense* is from the Pliocene stratum on the Pacific side of southwestern Japan (Figure 17; Ishizaki, 1983; ca. 3 Ma; age data from Iwai *et al.*, 2006). Therefore it is still difficult to specify which is the ancestral species on the sole basis of their fossil records. However, the genus *Loxoconcha* would primarily have southern origins, because it shows high species diversity in areas affected by the modern warm Kuroshio Current at the western Pacific region in East and Southeast Asia (e.g., Zhou, 1995). So *L. mutsuense*, firstly appearing at the Pacific during Pliocene, might be the probable ancestral species of *L. kamiyai*, distributed only in the Japan Sea (Ozawa and Ishii, 2008).

According to its geographical and geological occurrences (Figure 16), *L. kamiyai* is judged to be a formerly endemic species restricted to the Pliocene–Pleistocene of the Japan Sea. Therefore its origin appears to be the Japan Sea, and this species would become extinct since 0.5 Ma, based on the youngest fossil record from the Pleistocene stratum (Figure 16; age data from Kato *et al.*, 1995). The extinction period of this species is the same as those of more than 20 now extinct, formerly endemic species in the Japan Sea in three families Hemicytheridae (Hemicytherinae), Cytheruridae, and Eucytheridae (e.g., Ozawa and Kamiya, 2005a, 2008) as described above in this chapter. These species would have become extinct, because of low-salinity water at the Japan Sea surface during glacial and corresponding low sea-level periods since the early Pleistocene, related to glacio-eustatic sea level changes. These species probably only inhabited open marine environments in shallow areas, and could not have lived in low-salinity areas such as the brackish inner-bay (Ozawa and Kamiya, 2005b).

Extinction of Cytheroidean Ostracodes (Crustacea) in Shallow-Water around Japan ... 87

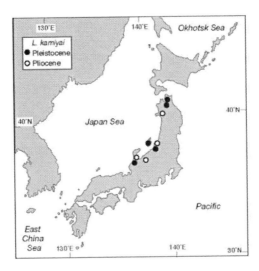

Figure 16. Summary for geographical and geological occurrences of *Loxoconcha kamiyai*, modified from Ozawa and Ishii (2008).

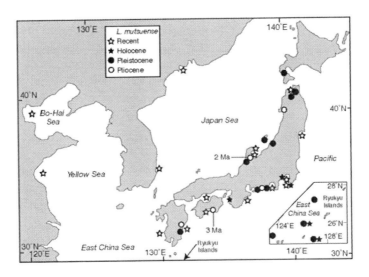

Figure 17. Summary for geographical and geological distribution of *Loxoconcha mutsuense*, based on data of previous studies, modified from Ozawa and Ishii (2008), adding with distributional data of Irizuki *et al*. (2006, 2008a) and Iwatani and Irizuki (2008). See Figures 6 and 7 of Ozawa and Ishii (2008) for compiled literatures of its occurrence data.

Like these extinct species, the paleo-habitat of *L. kamiyai* is also inferred to have been restricted to shallow, open, marine environments on the basis of its fossil occurrence with shallow, open marine ostracodes from Pleistocene strata (Ozawa, 1996). It would also have become extinct during the Pleistocene since 0.5 Ma because of its narrow ecological niche and salinity tolerance (Ozawa and Ishii, 2008).

L. mutsuense first appeared at the Pacific around 3 Ma, and migrated into the Japan Sea around 2 Ma in view of its oldest fossil record on the Japan Sea coast from the Pliocene stratum in Japan (Figure 17; age data from Yamada *et al*., 2002). Its first appearance in the

Japan Sea is more than one million years later than that of *L. kamiyai*. Therefore the origin of *L. mutsuense* appears to be the western Pacific around Japan (Ozawa and Ishii, 2008).

Thereafter, *L. mutsuense* expanded its geographical range, probably by floating on the leaves of marine plants, and became widely distributed in shallow-water areas at the Pacific Ocean, Japan Sea, Yellow Sea, and Bo-Hai Sea until today (Figure 17), because this species is phytal living (Kamiya, 1988). This species is found in both open marine and inner bay environments (e.g., Ishizaki, 1971; Zhou, 1995), where it could survive because of its wide salinity tolerance. This ecological tolerance would have allowed *L. mutsuense* to have the wide distributional area in East Asia than *L. kamiyai* (Ozawa and Ishii, 2008).

3.D. Case of an Extinct Species of *Aurila*

This part reviews the extinction of a species, *Aurila tsukawakii* (Figure 18), which was once endemic to the Japan Sea, in comparison with occurrence mode of its most related, extant species *Aurila* sp. (Figure 18) (Ozawa and Kamiya, 2009).

The genus *Aurila* (Hemicytheridae, Aurilinae) contains more than 150 named-species ranging from Miocene to present, reported from shallow-marine environments worldwide (Schornikov and Tsareva 1995). About 20 living and fossil species of *Aurila* have been described from Japan and its adjacent areas since Miocene (Ikeya *et al.* 2003). This fact makes this genus an important taxonomic group among the ostracode fauna from Japan.

Fossils of *Aurila*, including undescribed species, are commonly found from late Cenozoic shallow-marine strata along the eastern Japan Sea coast (e.g., Kamiya *et al.* 2001). Some species of fossil *Aurila* from this coast had not been examined on the taxonomy in detail, nor described, due to the difficulty of accurately identifying species in this highly-diversified genus that has many species with close morphological similarities to their congeneric species. Therefore Ozawa and Kamiya (2009) described a new species *Aurila tsukawakii* from one Pleistocene stratum at the Japanese coast of the Japan Sea, providing a detailed description of its carapace morphology and preliminary data on the distribution pattern of pores in a particular part.

Of the living and fossil *Aurila*, the carapace morphology of *A. tsukawakii* is most similar to the living undescribed species *Aurila* sp. (Figure 18), which is commonly found in the northeastern Japan Sea off Hokkaido (Figure 19; e.g., Ozawa and Tsukawaki 2008). The general pattern of ornamentation on the carapace surface of both species is more similar than other Japanese species (Figure 18). However, *A. tsukawakii* differs from *Aurila* sp. in plural characters such as carapace outline, shape of ventral sub-marginal ridge (Figure 18), reticulation's shape and pattern in the antero- and postero-median areas, and distribution pattern of two types of pores in the dorso-median area among reticulations A-B-C and C-D-E (Figures 18 and 20, 21; Ozawa and Kamiya, 2009). Other multiple species of *Aurila* from Japan, differing in the carapace outline and surface ornamentation such as *Aurila munechikai* and *Aurila kiritsubo* (Figure 18), show different patterns of the pore distribution in the median area (Ozawa, unpubl. data).

Fossils of *A. tsukawakii* often co-occur with those of ostracode species of the genera *Neonesidea*, *Finmarchinella* and *Schizocythere* in calcareous fine sandstones of the Plio–Pleistocene strata along the eastern Japan Sea coast (e.g., Ozawa 1996). Furthermore, extant species of these genera are commonly found together with *Aurila* species in calcareous sandy sediments of the shallow-open sea area in the northeastern Japan Sea (e.g., Ozawa and Tsukawaki 2008). These data suggest that *A. tsukawakii* inhabited sandy sediments of the shallow-open sea environment in the Plio–Pleistocene Japan Sea.

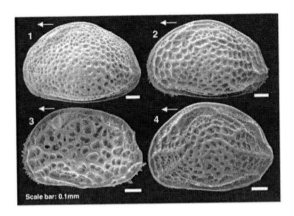

Figure 18. SEM images of four *Aurila* species (left valve). 1: *Aurila tsukawaki* Ozawa & Kamiya, 2009, from the Omma Formation, central Japan. 2: *Aurila* sp. of Ozawa & Kamiya, 2009, from the modern northeastern Japan Sea off Hokkaido. 3: *Aurila kiritsubo* Yajima, 1982, from the Kioroshi Formation, central Japan. 4: *Aurila munechikai* Ishizaki, 1968, from the modern northwestern Pacific coast of the Miura Peninsula, central Japan. Arrows indicate anterior.

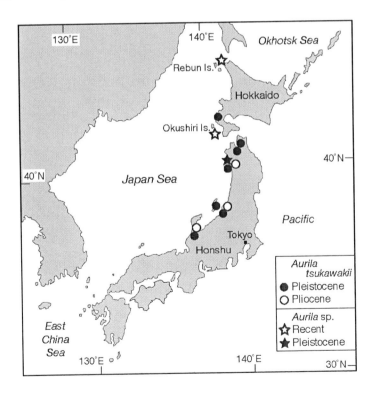

Figure 19. Summary for geographical and geological occurrences of *Aurila tsukawaki* and *Aurila* sp. of Ozawa and Kamiya (2009), modified from Ozawa and Kamiya (2009).

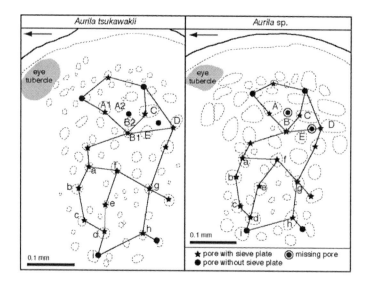

Figure 20. Distributional pattern of pores and reticulations in median area of adult left valve in two species of two *Aurila* species, modified from Ozawa and Kamiya (2009). Reticulations with letters (A–E and a–i) are the same in each of the two species. Positions of the two missing pores of *Aurila* sp. were determined by comparison with *A. tsukawakii*. Areas within reticulations 'a-b-c-d-e-f-a' and 'd-e-f-g-h-i-d' correlate to fields of frontal muscle scars and adductor muscle scars on the internal valve surface. Arrows indicate anterior.

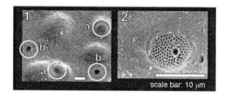

Figure 21. SEM images of pores of *Aurila tsukawakii* from the Omma Formation, central Japan. 1: pores with sieve plate (a) and without sieve plate (b). 2: Close-up view of a pore with sieve plate.

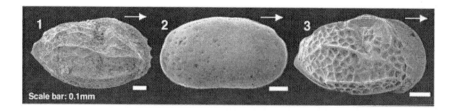

Figure 22. SEM images of selected three bay-taxa from the middle Pleistocene strata at the Pacific coast (right valve). 1: *Neomonoceratina delicata* Ishizaki & Kato, 1976. 2: *Sinocytheridea* sp. (Brady, 1869). 3: *Spinileberis quadriaculeata* (Brady, 1880). Specimens of nos. 1, 2 from the Yabu Formation, central Japan, and no. 3 from the Naganuma Formation, central Japan. Arrows indicate anterior.

Geological and geographical records of *A. tsukawakii* (Figure 19) suggest that this species originated in the eastern Japan Sea during late Pliocene around 3 Ma (age data from Cronin *et al.*, 1994; Irizuki *et al.*, 2007). Until ca. 3 Ma, this sea was still partially closed and colder than the present Japan Sea, because it was not well connected with the East China Sea (e.g., Tada 1994; Irizuki *et al.* 2007). The youngest fossil record suggests that this species was once endemic to this sea and flourished shallow sea areas in the eastern Japan Sea (Figure 19) until the Middle Pleistocene, ca. 0.6 Ma (Ozawa and Kamiya, 2009; age data from Kato *et al.*, 1995). It was likely extinct by the middle Pleistocene (ca. 0.4 Ma), sharing the fate of other more than 20 species of shallow-marine ostracodes of families Hemicytheridae (Hemicytherinae), Cytheruridae, Eucytheridae and Loxoconchidae within this sea (e.g., Ozawa and Kamiya, 2005a, 2008) as stated above. The simultaneous extinction of these species suggests the event of drastic fluctuations in environmental conditions, such as salinity, during glacial periods (Ozawa, 2006, 2007) related to Pleistocene glacio-eustatic changes in sea-level (e.g., Tanimura 1981; Tada 1994; Ozawa and Kamiya 2001; Amano 2004) as described above.

The one congeneric species *Aurila* sp. appeared around the Japanese Islands in late Pleistocene at the latest, based on the oldest fossil record of ca. 0.1 Ma from the northeastern Japan (Figure 19; Ozawa and Kamiya, 2009; age data from Shirai *et al.* 1997). During this period, *Aurila* sp. colonized shallow-water habitats of the eastern Japan Sea that had been vacated previously by the extinction of *A. tsukawakii*. This led to the present situation in which *Aurila* sp. mainly inhabits shallow-open areas of the northeastern Japan Sea (Figure 19; e.g., Ozawa and Tsukawaki 2008).

A similar replacement of one ostracode species with another in shallow-water at the eastern Japan Sea in the Pleistocene has been reported for the genus *Cythere*, whereby *Cythere hanaii* was replaced by *Cythere omotenipponica* in late Pleistocene, ca. 0.1 Ma (Tsukagoshi, 1996). It is likely that the future research will identify more examples of species extinction/ appearance or replacement in various ostracode genera, which will help in the construction of a detailed paleobiogeographic history of the Japan Sea. The accumulation of the data and detailed examination of the ostracode species-composition in both the Plio–Pleistocene and the modern Japan Sea will provide a representative example of faunal changes in marine benthic organisms at mid-latitudes, where repeated drastic changes in the oceanic environment had occurred during late Cenozoic.

4. DISAPPEARANCE AND EXTINCTION OF INNER-BAY SPECIES

This part overviews the extinction and disappearance (= regional extinction) of inner-bay ostracodes at the Japanese coasts of the Pacific and Japan Sea, based on results of mainly Irizuki *et al.* (2005, 2009) and Ozawa (2009 in press).

4.A. Pacific Coast

Ozawa (2009 in press) newly reported the ostracode fauna from the middle Pleistocene Naganuma Formation (ca. 0.5 Ma) near the Tokyo Bay of central Japan. This fauna includes species of representative shallow-bay ostracodes of the modern Pacific coast of central Japan, i.e., *Bicornucythere* species and *Spinileberis quadriaculeata*. However, it does not include two shallow-bay taxa, i.e., *Sinocytheridea* species and *Neomonoceratina delicata* (Figure 22). The latter two taxa are currently distributed from southern Japan to Southeast Asia, although they are abundantly found in fossil assemblages from central Japan just during restricted periods of the middle or late Pleistocene also in the Tokyo Bay area, commonly with other bay ostracodes, e.g., *Spinileberis quadriaculeata* (Figure 22), *Bicornucythere* species and *Pistocythereis bradyformis* (e.g., Yajima, 1982; Ozawa et al., 1995; Irizuki et al., 2008b).

Irizuki et al. (2009) discussed the paleo-biogeography for *N. delicata* in the Southeast–East Asia in detail. According to it, this species migrated from south of the Tokara Strait (Watase's line; one of zoogeographical lines of demarcation) by the MIS (= marine oxygen isotope stage) 11 (ca. 0.4 Ma) through the southern Japan. It also attained to the Tokyo Bay area by the MIS 9 at the latest (e.g., Yajima, 1982; Ozawa et al., 1995; Figure 23), and this occurrence is its farthest record along the Pacific coast from the southern Japan. After that, it is considered that this species became regionally extinct since the MIS 5 north of the Tokara Strait (Figure 23) due to the decrease of water temperature in these coasts and the deepened water depth of this strait, after that it presently lives just south of this strait (Irizuki et al., 2009).

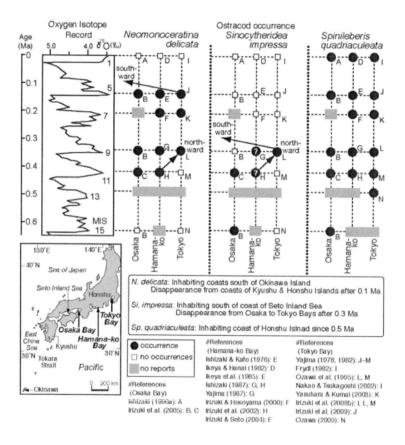

Figure 23. Summary for temporal changes of occurrences of selected three bay taxa at the three bay areas in Honshu Island at the Pacific coast. Oxygen isotope stratigraphy is modified from Yoshikawa and Mitamura (1999). Correlations between Marine oxygen Isotope Stage (MIS) and age of each stratum are referred from Sugiyama et al. (1987), Okazaki et al. (1997), Irizuki et al. (2005) and Nakashima et al. (2008). *Sinocytheridea* species occurs from Hamana-ko Bay area (Ishizaki, 1987), but its occurrence age has not clearly been described from which age (0.4 or 0.3 Ma) in Ishizaki (1987). This figure represents these occurrences of *Sinocytheridea* attached with '?'.

Its occurrence in the Tokyo Bay area for pre-0.4 Ma had not been investigated, because there is just one example of the fossil assemblage in this period of Japanese Islands from one stratum of ca. 0.6 Ma in the Osaka Bay coast (Figure 23; Ishizaki, 1990a, 1990b). Furthermore I did not find *N. delicata* from the Naganuma Formation (ca. 0.5 Ma). These facts suggest that this species firstly appeared at the Tokyo Bay in 0.3 Ma at the latest (Figure 23), and this result is consistent with that of Irizuki *et al.* (2009). One of the reasons for its later appearance in the Tokyo Bay (0.3 Ma), than that of Osaka and Hamana-ko Bays (0.6–0.5 Ma at the latest; Figure 23), is possibly due to the distance from southern Japan and the periods when the migration occurred.

For *Sinocytheridea* species, Irizuki *et al.* (2005) discussed their paleo-biogeography in the Southeast–East Asia in detail. According to it, this species-group migrated from the South China Sea through the southern Japan to the Honshu Island such as Osaka and Hamana-ko Bays. It became regionally extinct in these two bays around 0.4 Ma (Figure 23), and presently inhabits just south of the Seto Inland Sea. In the Tokyo Bay area, its fossil has not been found from many strata since the middle Pleistocene (e.g., Yajima, 1982; Irizuki *et al.*, 2005; Ozawa, 2009 in press), although Ozawa *et al.* (1995) reported just two individuals of *Sinocytheridea* sp. from the middle Pleistocene strata (0.3 Ma) near the Tokyo Bay (Figure 23). So its migration to Tokyo Bay from the south (0.3 Ma) was slightly later, three hundred thousand years at least, than the migration to Osaka Bay (0.6 Ma). This taxon lived in Tokyo Bay during a very limited period around 0.3 Ma, one hundred thousand years later than its presence in Osaka Bay. One of the reasons for its later appearance in Tokyo Bay, than that of Osaka Bay, is possibly due to the distance from southern Japan and its migration periods.

Causal factor of its regional extinction of north of the Seto Inland Sea has been inferred to the splitting of its population, due to the deepened shelf areas between China–Japan that resulted from the transgression around 0.3 Ma (Ishizaki, 1990a, 1990b) and level changes of salinity, dissolved oxygen and food-supply in bay areas by glacio-eustatic cycles, with replacement by *N. delicata* after 0.4 Ma (Irizuki *et al.*, 2005, 2009). These factors might cause its regional extinction also in the case of the Tokyo Bay.

4.B. Japan Sea

Kotoracythere tsukagoshii, *Kotoracythere* sp. and *Yezocythere* sp. (Figure 7) favored a brackish inner-bay environment (e.g., Ozawa and Kamiya, 2005a). That is because in fossil assemblages these species often co-occurred with the brackish inner-bay species of genera *Bicornucythere* and *Spinileberis*, and do not tend to co-occur with open-sea species of genera such as *Cythere* and *Neonesidea*, e.g., from the lower Pleistocene stratum in central Japan (e.g., Ozawa, 1996; Ozawa and Kamiya, 2001). For example, *Yezocythere hayashii* (Figure 4), which has inhabited both inner-bay and open-sea areas since the early Pleistocene at the latest, has a relatively wide range of tolerance for water conditions (Ozawa and Kamiya, 2001; Ozawa *et al.*, 2004a). So it could survive the drastic environmental changes related to glacio-eustatic cycles during the Pleistocene, with more than 10 ostracode species having the same ecological character such as *Munseyella hatatatensis* (Figure 4; Ozawa *et al.*, 2004a; Ozawa and Kamiya, 2005a) as described above. In addition, the broader geographical distribution of e.g., *M. hatatatensis* and *Y. hayashii*, compared to *K. tsukagoshii*, *K.* sp. and *Y.*

sp., would have facilitated its survival (e.g., Figure 24; Ozawa *et al.*, 2004a; Ozawa and Kamiya, 2005a). The latter three species, which favored inner-bay areas, likely became extinct, because it had a narrower tolerance than *M. hatatatensis* and *Y. hayashii* for fluctuations in water conditions. For example, with its narrow geographical distribution, circumstances such as salinity in restricted inner-bay areas related to glacio-eustatic cycles since the middle Pleistocene, would be led to its replacement by other bay-dominant taxa such as *Spinileberis* (e.g., Kamiya *et al.*, 2001).

Irizuki *et al.* (2005) reported the regional extinction of one brackish inner-bay taxon, *Sinocytheridea* species, in the middle Pleistocene (0.3 Ma) along the Pacific coast of central Japan. They hypothesized that its regional extinction was caused by the drastic change in salinity and dissolved oxygen due to glacio-eustatic cycles in the middle Pleistocene, with replacement by other bay-dominant taxa such as *Neomonoceratina delicata*, as described above. It is still uncertain whether the detailed period and causal factor for both extinction events of *Sinocytheridea* species at the Pacific side and of three species at the Japan Sea side in inner-bay areas had been exactly the same or not. Its reason is that the Japan Sea especially in Pleistocene glacial periods was a semi-isolated area, and not well connected with Pacific, due to the low sea-level stand with surrounding for shallow and narrow straits (Figure 1) in a mean depth of ca. 100 m (e.g., Tada, 1994; Ozawa and Kamiya, 2001; Amano, 2004). Also it is that *Sinocytheridea* species still inhabit many inner-bay areas at the western margin of the Pacific between the southern Japan and Southeast Asia (Irizuki *et al.*, 2005), differing from cases of complete extinction of species of the Japan Sea.

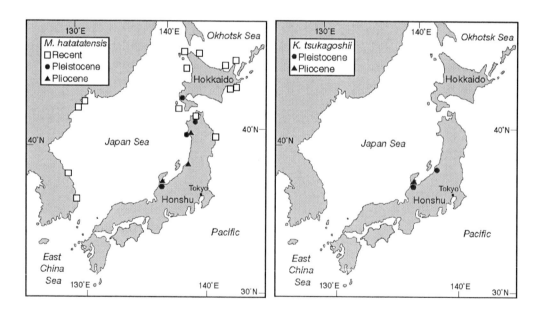

Figure 24. Summary for geographical and geological occurrences of *Munseyella hatatatensis* and *Kotoracythere tsukagoshii* since late Pliocene. Data are cited from Ozawa (1996, 2006, 2007, unpubl. data), Ozawa *et al.* (2004a) and Irizuki *et al.* (2007).

Extinction of Cytheroidean Ostracodes (Crustacea) in Shallow-Water around Japan ... 95

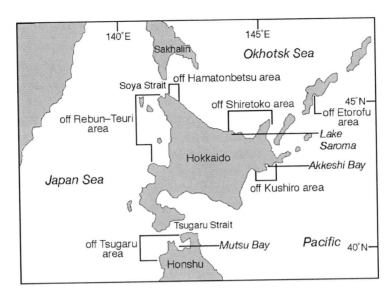

Figure 25. Location map of showing regions for which species distribution data are available for the Okhotsk Sea, Japan Sea and Pacific around the Hokkaido Island.

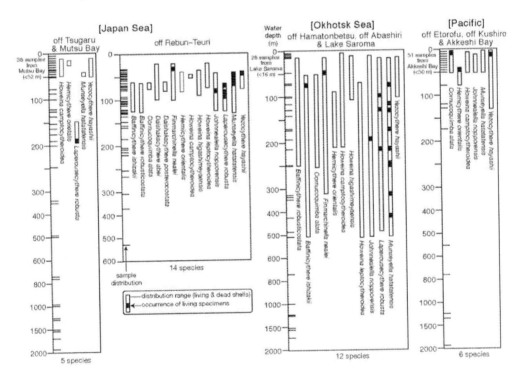

Figure 26. Water depth range of 14 species in the three sea areas of the Okhotsk Sea, Japan Sea and Pacific around Hokkaido, modified from Ozawa (2004), adding the data represented by Ozawa et al. (1999, 2004a, 2004b). Area names are the same as in Figure 25.

5. POSSIBILITY OF SPECIES DISAPPEARANCE AND EXTINCTION AROUND JAPAN NEAR FUTURE

As described in the article 3-A, there is one common ecological character that the three groups of extant 14 cryophilic ostracodes (Figure 4) favor the condition around or less than 5°C in winter (Figure 6) especially shallower than 100 m water depths (Ozawa et al., 2004a). Therefore, due to the probable increase for the water-temperature in shallow-water areas accompanied with the global warming, it is highly possible that the distribution of these cryophilic species would largely change. For example, the 14 species, which is the maximum species number of cryophilic taxa in the East Asia, are found in the shallow-sea area around the Hokkaido Island of northern Japan (Figures 5, 25 and 26).

However, the sea around Hokkaido, such as southwestern Okhotsk Sea, is considered to be the most increasing area for the winter surface water-temperature on earth during the next 100 years in the recent projection, because the winter sea-ice in the Okhotsk Sea is estimated to disappear after 70 years (e.g., Japan Meteorological Agency, 1996). If the global warming continues, the surface water-temperature in winter and summer at the northwestern Pacific region including Okhotsk Sea is projected to show the increase of ca. 2–4°C until A. D. 2100 (Japan Meteorological Agency, 1996, 2003, 2008). For example 3–4°C increase is supposed, if the CO_2 concentration in the air keeps increasing at the rate of 1% increase per a year for the next 100 years (Japan Meteorological Agency, 1996).

A round the Hokkaido Island, the water-temperature ranges for 14 cryophilic ostracodes, such as *Munseyella hatatatensis,* in the shallow sea of summer and winter, were represented by Ozawa et al. (2004a). Especially their study concluded that the winter low water-temperature, less than 5°C, is critical for the survival of these cryophilic species. So these cryophilic taxa could be difficult to inhabit the shallow sea around Hokkaido, and would disappear from there after 130–250 years, if the bottom water-temperature in the northwestern Pacific especially in winter simply keeps increasing at the same rate of the increase for surface water-temperature (e.g., Ozawa, 2004). In this case, most cryophilic species in the Okhotsk Sea would migrate to deeper areas in this sea or other areas neighboring this sea for their suitable temperature condition. However, in the southwestern Okhotsk Sea, there are two water depth-ranges under the condition of low-dissolved oxygen around 1°C through the year in 250–500 m water depths (2–3 ml/ l) and in 500–1,500 m depths (0–1 ml/ l; Figure 27) where is lower than those of the northern Japan Sea (6–9 ml/ l at present) where they live (e.g., Kuwahara, 1990; Ikeya and Cronin, 1993; Abe and Hasegawa, 2002).

So these species would begin to migrate to the Japan Sea or Pacific from the Okhotsk Sea, and to disappear from the Okhotsk Sea. They might become extinct finally, if these species completely lack for refuge areas under the condition less than 5°C in winter off the northern Japan, and low oxygen regions in the Okhotsk Sea keep to exit in deeper than 250 m water depths herein until A. D. 2,100 as the same depths as the present situation. Therefore, we must continue to monitor changes in the areal and water-depth distribution of these cryophilic ostracodes in and around the southwestern Okhotsk Sea with marine environmental fluctuations caused by near-future climate changes.

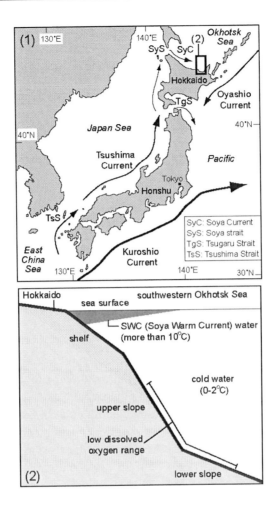

Figure 27. (1) Schematic flow patterns of main ocean currents around the Japanese Islands, with location of examined area in the southwestern Okhotsk Sea by Ozawa (2004); (2) schematic water mass in summer of the southwestern Okhotsk Sea examined by Ozawa (2004), modified from Ozawa (2004).

6. SUMMARY AND FUTURE WORK

The author continues to be interested in the causal relationship between the distribution of the marine biota and factors of the oceanographic environment, and has treated ostracode faunas as the suitable material for this subject. This is because ostracodes have the advantages that we can accurately identify species, and can obtain abundant geological and phylogenetic records based on fossils. Especially, mainly for the Japan Sea with its adjacent sea areas, I have examined the modern geographical and water-depth distribution for species of multiple families with ecological features of their habitats, faunal histories of extinction and survival during glacial–interglacial cycles in the late Cenozoic, and their causal relationship with environmental fluctuations. Based on results from in and around the Japan Sea, important points for analyzing the relationship between modern species distribution and oceanic environments are as follows.

On the basis of the modern distribution of each ostracode species found from the Japan Sea, some species are found not only from around Japanese Islands, but also in areas far from Japan, e.g., the other side of the Japan Sea around Russia, the Bo-Hai Sea, and the Gulf of Alaska. Furthermore, for a specific species, water-depths and topographic characters of habitats, such as the inner-bay or open-sea, are different in each area. Furthermore, the Japan Sea, a semi-closed marginal sea located in mid-latitudes, has largely different characters for the water temperature and water-mass structure in shallow-water areas between summer and winter, than neighboring large oceans such as the Pacific. The conditions of salinity and dissolved oxygen are also different between inner-bay and open-sea areas. Therefore, in order to investigate the causal relationship between the distribution of extant species and modern habitat conditions, we should carefully extract the data of summer and winter water-temperatures for each species in each distributional area or water-depth separately, in case there are species inhabiting the open-waters of marginal seas and inner-bay areas. In these cases, it is not appropriate to examine the mean annual water temperatures and salinities as have been done in previous studies. We must compare these data plotted on one graph, and must find common or different characters for environmental factors.

The method of these analyses itself is very easy and simple. However, we cannot obtain clues to analyze their response when facing environmental fluctuations during the geological past with causal factors for their extinction, survival, and migration, until we compare with these obtained characters for their habitats (i.e., the range for the temperature and salinity, modern distribution of geography and water depths, and range for the dissolved oxygen), and take into consideration their fossil records and phylogenetic data.

On the other hand, the history of the ostracode research in East–Southeast Asia is much shorter than those of Europe–North America, so accumulation data for ostracode faunas and its geological change in East–Southeast Asia are still much smaller than those of Europe–North America (e.g., Tsukagoshi, 1996, 2004). Consequently, there are still many unknown aspects for the faunal history of ostracodes in East–Southeast Asia, e.g., species-diversity and geographical distribution of each species.

This situation holds true for research of the ostracode fauna in the Japan Sea and its adjacent areas. For example, multiple genera of the three families including the 14 cryophilic species described above, such as *Hemicythere, Howeina* and *Semicytherura*, still have many unnamed species reported as fossils from strata of the upper Pliocene and lower Pleistocene in coastal outcrops of the Japan Sea (Ozawa, 2007), although the author does not refer to these in this chapter. It is considered that they had been restrictedly distributed within the Japan Sea until the early Pleistocene at the latest (e.g., Cronin and Ikeya, 1987; Ozawa, 1996). If we describe these probable now-extinct species as new species, and summarize all the fossil records, it will surely clarify that the number and diversity of cryophilic species during the early Pleistocene had been much higher than those of the modern cryophilic species in the Japan Sea, where today a maximum of just 14 species are found off Hokkaido. This poses an interesting question for the shallow-water environments during glacial periods in the Japan Sea, namely, what were the oceanic environmental conditions that had supported the surprisingly much higher diversity ostracode fauna compared with the modern fauna?

Furthermore it is also still unknown, when, where, and how these species speciated, what were the environmental triggers for their speciation, and what led them to becoming endemic species to the Japan Sea. In order to clarify these subjects using ostracodes, we need to also investigate the phylogenetic relationship among these species. We must examine their detailed faunal history, such as their speciation, evolution, flourishing and extinction, by adding more data of fossil occurrences of each species since the Miocene when the ostracodes probably became highly diversified in and around the Japan Sea. Based on results from these analyses at mid-latitudes, the most sensitive region to global-scale environmental fluctuations, we will be able to clarify the extinction, evolution and phylogeny of species of organisms, especially during the late Cenozoic, when the climate and oceanographic environment drastically fluctuated. Furthermore, we will be able to obtain clues to the timing, processes and triggers of species-diversity changes including extinction and speciation for endemic species, the effects of late Cenozoic climate changes on global biodiversity and ecosystems, and the biotic response of past faunas related to drastic climatic fluctuations in geological periods.

We cannot identify the migration routes and periods of appearance, extinction and expanding distributional areas of species over long time-scales, including geological periods, until we analyze the modern geographical distributions and fossil occurrences. We need to collect not only the data for modern species distribution and conditions of their habitats, but also to accumulate fossil data, if we want to solve questions of the ecology of modern species. If we do so, we can obtain abundant knowledge of organisms' responses (e.g., extinction, migration and speciation) and the faunal history at a much longer scale than that treated in ecological studies examining just modern living-species.

Ostracodes are the most abundantly preserved fossilized arthropod and one of the most taxonomically diversified groups, bearing various taxonomically useful characters of the carapace and appendages. In geological periods since the early Paleozoic, more than 10,000 ostracode species at least have appeared and become extinct (e.g., Horne et al., 2002). Facing various types of environmental fluctuations in every aquatic area on earth, some taxa could adapt to them by giving full play to their ecological tolerances, or some of them could not succeed to adapt to them, and became extinct. These facts suggest many aspects for the ecological adaptability and morphological plasticity in each species of organisms during their survival or extinction process, when in facing environmental fluctuations.

The carapace length of most ostracodes is around 1 mm. Its mall size is often an advantage treating in paleontological studies, because large-sized fossils such as molluscan bivalves are often broken up, especially in coring sediments. In some cases, only a small sample is needed to extract many ostracodes, although to get a reasonable sample size of larger fossils takes a lot longer. This is reason why ostracode fossils are used in the oil industry (e.g., Athersuch et al., 1989), but bivalve fossils, which are simply too big-sized, are not there.

However, there are still many unknown facts for the distribution, ecology, and species-diversity of ostracodes worldwide, so possibly the potential of their untapped data has not yet been fully realized. The author expects to be able to deepen our understanding of the history of organisms in the geological past since the early Paleozoic, such as extinction and evolution, with the causal relationship to environment fluctuation on earth, through the research of ostracodes. We must clarify the detailed natural history over a long-time scale of each ostracode species based on continuation of the research and accumulation of data from in and around Japan, globally one of the best areas for this kind of research.

7. Conclusion

1. This chapter overviews extinction events of shallow-water benthic species in multiple families of Cytheroidean ostracodes from Japanese coasts during the Pleistocene in the late Cenozoic, and the causes of oceanic environment changes, while briefly referring to the possibility of their extinctions in the near future, based on results of the recent research.
2. Tolerance ranges of salinity for more than 20 now-extinct species of four selected families from the eastern Japan Sea coast, are inferred to have been narrower than those of most extant related-species, as indicated by fossil and modern occurrences in and around this sea. This is because the extant species have relatively wide tolerance ranges for salinity living in both open-water and brackish inner-bays. Salinity-decreases in shallow-water areas, as a result of lowered sea-levels closing the shallow, narrow straits around this marginal sea during glacial periods, would have caused the extinction of now-extinct species since ca. 1 Ma, especially during the middle Pleistocene.
3. The disappearance (= regional extinction) of two inner-bay taxa from the Pacific coast of central Japan is hypothesized to have been caused by the drastic changes in salinity and dissolved oxygen due to glacio-eustatic cycles in the middle Pleistocene, which resulted in replacement by other bay-dominant species. The three inner-bay species of the Japan Sea likely became extinct during the Pleistocene, because they had narrower tolerance for fluctuations of water conditions in relation to glacio-eustatic cycles compared to their related-species of the same subfamilies. Due to their restricted geographical distributions and changing environmental conditions, such as salinity, in restricted inner-bay areas during glacio-eustatic cycles since the middle Pleistocene, these inner-bay taxa would have been replaced by other species.
4. Recent projections suggest that the shallow-water areas around the Hokkaido Island of northern Japan, e.g., southwestern Okhotsk Sea, are likely to see some of the highest increases in sea surface-water temperatures on earth during the next 100 years. One of the reasons is that the winter sea-ice of the Okhotsk Sea is estimated to disappear within the next 100 years due to global warming. In these areas, especially in the Okhotsk Sea, extant relict-species of cryophilic ostracodes, which favor water-temperature conditions of less than 5°C in winter, might begin to decline and disappear from shallow-water areas around Hokkaido region of northern Japan within the next 130–250 years, if global climate warming continues at its present rate.

Acknowledgments

I wish to thank Y. Tanimura (National Museum of Nature and Science, Tokyo) for kind assistance of preparing the manuscript in various aspects. Thanks are also due to T. Kamiya, S. Tsukawaki, M. Kato, T. Ishii, T. Sato (Kanazawa University), T. Irizuki (Shimane University), N. Ikeya, A. Tsukagoshi (Shizuoka University), R. J. Smith (Lake Biwa Museum), H. Takata (Pusan University), K. Ikehara, H. Katayama (AIST, Japan), K. Ishida

(Shinshu University), T. Yamaguchi (Nagoya University), S. Yamada (University of Tokyo), G. Tanaka (Gunma Museum of Natural History), Y. Nakao (Nihon University), T. M. Cronin, E. M. Brouwers (U.S. Geological Survey), I. Boomer (University of Birmingham), M. Yasuhara (Smithsonian Institution), A. Nojo (Hokkaido University of Education), H. Domitsu (University of Shiga Prefecture), K. Abe (University of Tsukuba), and the late T. Matsuzaka and Y. Kuwano for many valuable suggestions for ostracode taxonomy and phylogeny, with help of access to ostracode specimens, sediment samples and literatures. I express my gratitude to captain, all crew of the *Tansei-maru* (JAMSTEC, Japan) and *Hakurei-maru* (AIST, Japan) and all onboard scientists for their help of collecting sediment samples during Cruises KT95-14, KT96-17, KT97-15, KT98-17, KT99-14, KT00-14, KT01-14, KT04-20 and GH98, and to R. J. Smith and reviewers for reading the manuscript with giving many critical comments.

REFERENCES

Abe, K. & Hasegawa, S. (2002). Distribution of the Recent benthic foraminifera from the southwestern Okhotsk Sea around Hokkaido (Preliminary report). In K. Ikehara, (Ed.), *Marine geological and geophysical studies on the collision zone of Kurile and Northeast Japan Arcs–In the area of Okhotsk Sea*, (170-179). Tsukuba, Geological Survey of Japan. (in Japanese)

Amano, K. (2004). Biogeography and the Pleistocene extinction of neogastropods in the Japan Sea. *Palaeogeography, Palaeoclimatology, Palaeoecology, 202*, 245-252.

Amano, K. (2007). The Omma-Manganji Fauna and its temporal change. *Fossils (Palaeontolological Society of Japan), 82*, 6-12. (in Japanese with English abstract)

Amano, K. & Watanabe, M. (2001). Taxonomy and distribution of Plio-Pleistocene *Buccinum* (Gastropoda) in northeast Japan. *Paleontological Research, 5*, 215-226.

Athersuch, J., Horne, D. J. & Whittaker, J. E. (1989). *Marine and brackish water ostracods, Synopsis of the British Fauna (New Series) 43*. Leiden, E. J. Brill.

Benson, R. H. (2003). The ontogeny of an ostracodologist. *The Paleontological Society Papers, 9*, 1-8.

Berger, W. H., Yasuda, M. K., Bickert, T., Wefer, G. & Takayama, T. (1994). Quaternary timescale for the Ontong Java Plateau: Milankovitch template for Ocean Drilling Program Site 806. *Geology, 22*, 463-467.

Boomer, I., Horne, D. J. & Slipper, I. J. (2003). The use of ostracodes in paleoenvironmental studies or what can you do with an ostracode shell? *The Paleontological Society Papers, 9*, 153-179.

Brouwers, E. M. (1988). Palaeobathymetry on the continental shelf based on example using ostracods from the Gulf of Alaska. In P., De Deckker, J. P. Colin, & J. P. Peypouquet, (Eds.), *Ostracoda in the earth Sciences*, (55-76). Amsterdam, Elsevier.

Cannariato, K. G., Kennett, J. P. & Behl, R. J. (1999). Biotic response to late Quaternary rapid climate switches in Santa Barbara Basin: Ecological and evolutionary implications. *Geology, 27*, 63-66.

Cronin, T. M. (1999). *Principles of Paleoclimatology*. New York, Columbia University Press.

Cronin, T. M. & Dwyer, G. S. (2003). Deep-sea ostracodes and climate change. *The Paleontological Society Papers*, *9*, 247-264.

Cronin, T. M. & Ikeya, N. (1987). The Omma Manganji ostracode fauna (Plio-Pleistocene) of Japan and the zoogeography of circumpolar species. *Journal of Micropalaeontology*, *6*, 65-88.

Cronin, T. M. & Raymo, M. E. (1997). Orbital forcing of deep-sea benthic species diversity. *Nature*, *385*, 624-627.

Cronin, T. M., Boomer, I, Dwyer, G. S. & Rodriguez-Lazaro, J. (2002). Ostracoda and paleoceanography. *The Paleontological Society Papers*, *9*, 99-119.

Cronin, T. M, Kitamura, A., Ikeya, N., Watanabe, M. & Kamiya, T. (1994). Late Pliocene climate change 3.4-2.3 Ma: paleoceanographic record from the Yabuta Formation, Sea of Japan. *Palaeogeography, Palaeoclimatology, Palaeoecology*, *108*, 437-455.

Domitsu, H. & Oda, M. (2005). Japan Sea planktic foraminifera in surface sediments: geographical distribution and relationships to surface water mass. *Paleontological Research*, *9*, 255-270.

Frenzel, P. & Boomer, I. (2005). The use of ostracods from the marginal marine, brackish waters as bioindicators of modern and Quaternary environmental change. *Palaeogeography, Palaeoclimatology, Palaeoecology*, *225*, 68-92.

Frydl, P. M. (1982). Holocene ostracods in the southern Boso Peninsula. *Bulletin, the University Museum, the University of Tokyo*, *20*, 61-140.

Hanagata, S. (2003). Miocene–Pliocene foraminifera from the Niigata oil-fields region, northeastern Japan. *Micropalaeontology*, *49*, 293-340.

Hanai, T. (1957). Studies on the Ostracoda from Japan II, Subfamily Pectocytherinae, new subfamily. *Journal of Faculty of Sciences, University of Tokyo*, *10*, 469-482.

Hanai, T. & Ikeya, N. (1991). Two new genera from the Omma-Manganji ostracode fauna (Plio–Pleistocene) of Japan–with a discussion of theoretical versus purely descriptive ostracode nomenclature. *Transactions and Proceedings of Palaeontological Society of Japan, New Series*, *163*, 861-878.

Hayashi, K. (1988). Pliocene-Pleistocene palaeoenvironment and fossil ostracod fauna from southwestern Hokkaido, Japan. In T., Hanai, N., Ikeya, & K. Ishizaki, (Eds.), *Evolutionary Biology of Ostracoda–Its fundamentals and application*, (557-568). Tokyo: Kodansha, Amsterdam: Elsevier.

Horne, D. J. (2003). Key events in the ecological radiation of the Ostracoda. *The Paleontological Society Papers*, *9*, 181-202.

Horne, D. J., Cohen, A. & Martens, K. (2002). Taxonomy, morphology and biology of Quaternary and living Ostracoda. In J. A. Holmes, & A. Chivas, (Eds.), *The Ostracoda: Applications in Quaternary Research* , *AGU Geophysical Monograph Series*, *131*(5-36). Washington DC, America Geophysical Union.

Ikeya, N. & Cronin, T. M. (1993). Quantitative analysis of Ostracoda and water masses around Japan: Application to Pliocene and Pleistocene paleoceanography. *Micropalaeontology*, *39*, 263-281.

Ikeya, N. & Hanai, T. (1982). Ecology of recent ostracods in the Hamana-ko region, the Pacific coast of Japan. *Bulletin, the University Museum, the University of Tokyo*, *20*, 15-59.

Ikeya, N. & Shiozaki, M. (1993). Characteristic of the inner-bay ostracodes around the Japanese Islands–the use of ostracodes to reconstruct paleoenvironments. *Memoirs of the Geological Society of Japan, 39*, 15-32. (in Japanese with English abstract)

Ikeya, N. & Suzuki, C. (1992). Distributional patterns of modern ostracodes off Shimane Peninsula, southwestern Japan Sea. *Report of Faculty of Sciences, Shizuoka University, 26*, 91-137.

Ikeya, N. & Yamaguchi, T. (1993). *An Introduction to Crustacean Paleobiology (UP Biology 93)*. Tokyo, University of Tokyo Press. (in Japanese)

Ikeya, N., Okubo, I., Kitazato, H. & Ueda, H. (1988). Excursion 4, Shizuoka (Pleistocene and living Ostracoda, shallow marine, brackish and fresh water). In N. Ikeya (Ed.), *Guidebook of Excursions for the 9th International Symposium of Ostracoda, Shizuoka* (1-32). Shizuoka, Organising Committee of 9th ISO.

Ikeya, N., Tanaka, G. & Tsukagoshi, A. (2003). Ostracoda. *Palaeontological Society of Japan, Special Papers, 41*, 37-131.

Ikeya, N., Zhou, B. C. & Sakamoto, J. (1992). Modern ostracode fauna from Otsuchi Bay, the Pacific coast of Northeastern Japan. In K. Ishizaki, & T. Saito, (Eds.), *Centenary of Japanese Micropaleontology*, (339-354). Tokyo, Terra Scientific Publishing Company.

Irizuki, T. (1989). Fossil ostracode assemblages from the Pliocene Sasaoka Formation, Akita City, Japan—with reference to sedimentological aspects. *Transactions and Proceedings of Palaeontological Society of Japan, New Series, 156*, 296-318.

Irizuki, T. (1993). Morphology and taxonomy of some Japanese hemicytherin Ostracoda– with particular reference to ontogenetic changes of marginal pores. *Transactions and Proceedings of Palaeontological Society of Japan, New Series, 170*, 186-211.

Irizuki, T. (1994). Late Miocene ostracods from the Fujikotogawa Formation, northern Japan–with reference to cold water species involved with trans-Arctic interchange. *Journal of Micropalaeontology, 13*, 3-15.

Irizuki, T. (1996). Ontogenetic change in valve characters in three new species of *Baffinicythere* (Ostracoda, Crustacea) from Northern Japan. *Journal of Paleontology, 70*, 450-462.

Irizuki, T. & Hosoyama, M. (2000). Fossil ostracodes (Crustacea) from the Pleistocene Noma Formation, Aichi Prefecture, central Japan. *Bulletin of Aichi University of Education (Natural Science), 49*, 9-15. (in Japanese with English abstract)

Irizuki, T. & Ishida, K. (2007). Relationship between Pliocene ostracode assemblages and marine environments along the Japan Sea side regions in Japan. *Fossils (Palaeontological Society of Japan), 82*, 13-20. (in Japanese with English abstract)

Irizuki, T. & Seto, K. (2004). Temporal and spatial variations of paleoenvironments of Paleo-Hamana bay, central Japan, during the middle Pleistocene–Analyses of fossil ostracode assemblages, and total organic carbon, total nitrogen and total sulfur contents. *Journal of the Geological Society of Japan, 110*, 309-324. (in Japanese with English abstract)

Irizuki, T., Kamiya, M. & Ueda, K. (2002). Temporal and spatial distribution of fossil ostracode assemblages and sedimentary facies in the middle Pleistocene Tahara Formation, Atsumi Peninsula, central Japan. *Geoscience Report of Shimane University, 21*, 31-39. (in Japanese with English abstract)

Irizuki, T., Kusumoto, M., Ishida, K. & Tanaka, Y. (2007). Sea-level change and water structures between 3.5 and 2.8 Ma in the central part of the Japan Sea Borderland:

Analysis of fossil Ostracoda from the Pliocene Kuwae Formation, central Japan. *Palaeogeography, Palaeoclimatology, Palaeoecology*, *245*, 421-433.

Irizuki, T., Matsubara, T. & Matsumoto, H. (2005). Middle Pleistocene Ostracoda from the Takatsukayama Member of the Meimi Formation, Hyogo Prefecture, western Japan: significance of the occurrence of *Sinocytheridea impressa*. *Paleontological Research*, *9*, 37-54.

Irizuki, T., Seto, K. & Nomura, R. (2008a). The impact of fish farming and bank construction on Ostracoda in Uranouchi Bay on the Pacific coast of southwest Japan–Faunal changes between 1954 and 2002/ 2005. *Paleontological Research*, *12*, 283-302.

Irizuki, T., Takata, Y. & Ishida, K. (2006). Recent Ostracoda from Urauchi Bay, Kamikoshiki-jima Island, Kagoshima Prefecture, southwestern Japan. *Laguna (Kisuiiki-kenkyu)*, *13*, 13-28.

Irizuki, T., Taru, H., Taguchi, K. & Matsushima, Y. (2009). Paleobiogeographical implications of inner bay ostracodes during the Late Pleistocene Shimosueyoshi transgression, central Japan, with significance of its migration and disappearance in eastern Asia. *Palaeogeography, Palaeoclimatology, Palaeoecology*, *271*, 316-328.

Irizuki, T., Yamaguchi, M. & Mizuno, K. (2008b). Middle Pleistocene ostracode faunas in paleo-Tokyo Bay–a case study of the Shobu core, Saitama Prefecture, central Japan. *Abstracts with Programs of the 2008 Annual Meeting of the Palaeontological Society of Japan*, 71. (in Japanese)

Ishii, T., Kamiya, T. & Tsukagoshi (2005). Phylogeny and evolution of *Loxoconcha* (Ostracoda, Crustacea) species around Japan. *Hydrobiologia*, *538*, 81-94.

Ishizaki, K. (1971). Ostracodes from Aomori Bay, Aomori Prefecture, Northeast Honshu, Japan. *Tohoku University, Science Report, 2nd series (Geology)*, *43*, 59-97.

Ishizaki, K. (1983). Ostracoda from the Pliocene Ananai Formation, Shikoku, Japan–Description. *Transactions and Proceedings of the Palaeontological Society of Japan, New Series*, *131*, 135-158.

Ishizaki, K. (1987). The crocodile *Tomistoma machikanense* and ostracodes from southwest Japan in the middle Pleistocene. *Fossils (Palaeontological Society of Japan)*, *43*, 35-38. (in Japanese)

Ishizaki, K. (1990a). A setback for the genus *Sinocytheridea* in the Japanese mid-Pleistocene and its implications for a vicariance event. In R. Whatley, & C. Maybury, (Eds.), *Ostracoda and Global Events*, (139-152). London, Chapman and Hall.

Ishizaki, K. (1990b). Sea level change in mid-Pleistocene time and effects on Japanese ostracode faunas. *Bulletin of Marine Science*, *47*, 213-220.

Ishizaki, K. & Kato, M. (1976). The basin development of diluvium Furuya Formation (Late Pleistocene), Shizuoka Prefecture, Japan, based on faunal analysis of fossil ostracodes. In Y. Takayanagi, & T. Saito, (Eds.), *Progress in Micropaleontology*, (118-143). New York, Micropaleontology Press.

Ishizaki, K. & Matoba, Y. (1985). Excursion 5, Akita (Early Pleistocene cold, shallow water Ostracoda). In N. Ikeya, (Ed.), *Guidebook of Excursions for 9th International Symposium of Ostracoda, Shizuoka*, (1-12). Shizuoka, Organising committee of 9th ISO.

Ishizaki, K., Irizuki, T. & Sasaki, O. (1993). Cobb Mountain spike of the Kuroshio Current detected by the Ostracoda in the lower Omma Formation (Early Pleistocene), Kanazawa City, central Japan: analysis of depositional environments. In K. G. McKenzie, & P. J.

Jones, (Eds.), *Ostracoda in the Earth and Life Sciences*, (315-334). Rotterdam, A. A. Balkema.

Itaki, T. (2007). Historical changes of deep-sea radiolarians in the Japan Sea during the last 640kyrs. *Fossils (Palaeontolological Society of Japan)*, *82*, 43-51. (in Japanese with English abstract)

Itaki, T., Funakawa, T. & Motoyama, I. (1996). Changes in the radiolarian communities in the northeast part of Japan Sea, off Shakotan, Hokkaido, since the last glacial age. In Y. Okamura, & Y. Iuchi, (Eds.), *Comprehensive study on environmental changes in the western Hokkaido coastal area, Preliminary Reports on Researches in the 1995 Fiscal Year*, (171-185). Tsukuba, Geological Survey of Japan. (in Japanese).

Iwai, M., Kondo, Y., Kikuchi, N. & Oda, M. (2006). Overview of stratigraphy and paleontology of Pliocene Tonohama Group, Kochi Prefecture, southwest Japan. *Journal of Geological Society of Japan 112 (Supplementary volume)*, 27-40. (in Japanese with English abstract).

Iwatani, H. & Irizuki, T. (2008). Geology and fossil ostracode assemblages from the Pliocene Miyazaki Group in the northern part of the Miyazaki Plain, Southwest Japan. *Fossils (Palaeontological Society of Japan*, *84*, 61-73. (in Japanese with English abstract).

Japan Meteorological Agency, (1996). *Global Warming Projection, vol. 1*. Tokyo, Printing Bureau, Ministry of Finance of Japan.

Japan Meteorological Agency, (2003). *Global Warming Projection, vol. 5*. Tokyo, Japan Meteorological Business Support Center.

Japan Meteorological Agency, (2008). *Global Warming Projection, vol. 7*. Tokyo, Japan Meteorological Business Support Center.

Kamikuru, S. & Motoyama, I. (2007). Radiolarian assemblage and environmental changes in the Japan Sea since the Late Miocene. *Fossils (Palaeontolological Society of Japan)*, *82*, 35-42. (in Japanese with English abstract)

Kamiya, T. (1988). Morphological and ethological adaptations of Ostracoda to microhabitats in *Zostera* beds. In T., Hanai, N. Ikeya, & K. Ishizaki, (Eds.), *Evolutionary Biology of Ostracoda–Its fundamentals and applications*, (303-318). Tokyo: Kodansha, Amsterdam: Elsevier.

Kamiya, T. (1997). Phylogenetics estimated from fossil information–the pore systems of Ostracoda. *Iden (Genetics)*, *51*, 28-34. (in Japanese, title translated)

Kamiya, T., Ozawa, H. & Obata, M. (2001). Quaternary and Recent marine Ostracoda in Hokuriku district, the Japan Sea coast. In N. Ikeya (Ed.), *Field Excursion Guidebook of the 14th International Symposium of Ostracoda, Shizuoka*, (73-106). Shizuoka, Organising Committee of 14th ISO.

Kanazawa, K. (1990). Early Pleistocene glacio-eustatic sea-level fluctuations as deduced from periodic changes in cold- and warm-water molluscan associations in the Shimokita Peninsula, Northeast Japan. *Palaeogeography, Palaeoclimatology, Palaeoecology*, *79*, 263-273.

Kato, M., Akada, K. Takayama, T., Goto, T. Sato, T., Kudo, T. & Kameo, K. (1995). Calcareous microfossil biostratigraphy of the uppermost Cenozoic Formation distributed in the coast of the Japan Sea–"Sawane Formation". *Annals of Science of Kanazawa University*, *32*, 21-38. (in Japanese)

Kitamura, A., Takano, O., Takata, H. & Omote, H. (2001). Late Pliocene-Pleistocene paleoceanographic evolution of the Sea of Japan. *Palaeogeography, Palaeoclimatology, Palaeoecology, 172*, 81-98.

Koizumi, I. (1992). Biostratigraphy and paleoceanography of the Japan Sea based on diatoms: ODP Leg 127. In R. Tsuchi, & J. C. Jr. Ingle, (Eds.), *Pacific Neogene: environment, evolution, and events*, (15-24). Tokyo, University of Tokyo Press.

Kuwahara, A. (1990). Chapter 34-III–Chemical features of coastal area of Sanin region. In Coastal oceanography Research committee, Oceanographical Society of Japan (Ed.), *Coastal oceanography of Japanese Islands/ Supplementary volume*, (768-771). Tokyo, Tokai University Press. (in Japanese)

Nagamine, F., Mitsuya, T., Amano, K., Takabayashi, N., Nara, Y., Shiratori, T., Yamamoto, M., Hamada, K., Fukikoshi, H. & Kindaichi, H. (1982). Report of the result for the research on the shallow sea survey lines (Mutsu Bay; Fiscal year report in 1981 of the preliminary project on the fishery and marine condition). *Report of Aomori Prefectural Fishery and Aquiculture Center, 12*, 1-24. (in Japanese with English summary)

Nagamori, H., Furukawa, R. & Hayatsu, K. (2003). *Geology of the Togakushi district. Quadrangle Series, 1: 50,000*. Tsukuba, Geological Survey of Japan, AIST. (in Japanese with English abstract)

Nakao, Y. & Tsukagoshi, A. (2002). Brackish-water Ostracoda (Crustacea) from the Obitsu River estuary, central Japan. *Species Diversity, 7*, 67-115.

Nakashima, R., Mizuno, K. & Furusawa, A. (2008). Depositional age of the Middle Pleistocene Atsumi Group in Atsumi Peninsula, central Japan, based on tephra correlation. *Journal of Geological Society of Japan, 114*, 70-79. (in Japanese with English abstract)

Nishimura, S. (1966). The zoological aspects of the Japan Sea, part III. *Publications of Seto Marine Biological Laboratory, 13*, 365-384.

Oba, T., Kato, M., Kitazato, H., Koizumi, I., Ohmura, A., Sakai, T. & Takayama, T. (1991). Paleoenvironmental changes in the Japan Sea during the last 85,000 years. *Paleoceanography, 6*, 499-518.

Okada, Y. (1979). Stratigraphy and Ostracoda from the Late Cenozoic strata of the Oga Peninsula, Akita Prefecture. *Transactions and Proceedings of the Palaeontological Society of Japan, New Series, 115*, 143-173.

Okazaki, H., Sato, H. & Nakazato, H. (1997). Two types of depositional sequences developed in Paleo-Tokyo Bay, examined from the Kamiizumi, Kiyokawa and Yokota Formations, Pleistocene Shimosa Group, Japan. *Journal of Geological Society of Japan, 103*, 1125-1143. (in Japanese with English abstract)

Ozawa, H. (1996). Ostracode fossils from the late Pliocene to early Pleistocene Omma Formation in the Hokuriku district, central Japan. *Science Reports of Kanazawa University, 41*, 77-115.

Ozawa, H. (2003a). Japan Sea ostracod assemblages in surface sediments: their distribution and relationships to water mass properties. *Paleontological Research, 7*, 257-274.

Ozawa, H. (2003b). Cold-water ostracod fossils from the southern and eastern margins of the Japan Sea. *Journal of Geological Society of Japan, 109*, 459-465.

Ozawa, H. (2004). Okhotsk Sea ostracods in surface sediments: depth distribution of cryophilic species relative to oceanic environment. *Marine Micropaleontology, 53*, 245-260.

Ozawa, H. (2006). An overview of the geographical distribution and ecological significance of species in the three families of cryophilic ostracods (Crustacea: Ostracoda) in and around the Japan Sea–with special reference to distribution of species in relation to water temperature–salinity ranges. *Taxa, Proceedings of the Japanese Society of Systematic Zoology, 20*, 26-40. (in Japanese with English abstract)

Ozawa, H. (2007). Faunal changes of cryophilic ostracods (Crustacea) in the Japan Sea, in relation to oceanographic environment: an overview. *Fossils (Palaeontolological Society of Japan), 82*, 21-28. (in Japanese with English abstract)

Ozawa, H. (2009, in press). Middle Pleistocene ostracods from the Naganuma Formation in the Sagami Group, Kanagawa Prefecture, central Japan: palaeo-biogeographical significance of the bay fauna in Northwest Pacific margin. *Paleontological Research, 13*.

Ozawa, H. & Ishii, T. (2008). Taxonomy and sexual dimorphism of a new species of *Loxoconcha* (Podocopida: Ostracoda) from the Pleistocene of the Japan Sea. *Zoological Journal of the Linnean Society, 153*, 239-251.

Ozawa, H. & Kamiya, T. (2001). Palaeoceanographic records related to glacio-eustatic sea-level fluctuations in the Pleistocene Japan Sea coast based on ostracods from the Omma Formation. *Palaeogeography, Palaeoclimatology, Palaeoecology, 170*, 27-48.

Ozawa, H. & Kamiya, T. (2005a). The effects of glacio-eustatic sea-level change on Pleistocene cold-water ostracod assemblages from the Japan Sea. *Marine Micropaleontology, 54*, 167-189.

Ozawa, H. & Kamiya, T. (2005b). Ecological analysis of benthic ostracods in the northern Japan Sea, based on water properties of modern habitats and late Cenozoic fossil records. *Marine Micropaleontology, 55*, 255-276.

Ozawa, H. & Kamiya, T. (2008). Taxonomy and palaeobiogeographical significance for four new species of *Semicytherura* (Ostracoda, Crustacea) from the Early Pleistocene Omma Formation at the Japan Sea coast. *Journal of Micropalaeontology, 27*, 135-146.

Ozawa, H. & Kamiya, T. (2009). A new species of *Aurila* (Crustacea: Ostracoda: Cytheroidea: Hemicytheridae) from the Pleistocene Omma Formation on the coast of the Sea of Japan. *Species Diversity, 14*, 27-39.

Ozawa, H. & Tsukawaki, S. (2008). Preliminary report on modern ostracods in surface sediment samples collected during R. V. *Tansei-maru* Cruise KT04-20 in the southwestern Okhotsk Sea and the northeastern Japan Sea off Hokkaido, north Japan. *Annals of the Research Institute of the Japan Sea Region, 39*, 31-48.

Ozawa, H., Ikehara, K. & Katayama, H. (1999). Recent ostracod fauna in the northeastern part of the Japan Sea, off northwestern Hokkaido. In K. Ikehara, & Y. Okamura (Eds.), *Preliminary report on researches in the 1998 Fiscal year, GSJ Interim Report no. MG/99/1* (103-117). Tsukuba, Geological Survey of Japan. (in Japanese)

Ozawa, H., Kamiya, T. & Tsukagoshi, A. (1995). Ostracode evidence for the paleoceanographic change of the middle Pleistocene Jizodo and Yabu Formations in the Boso Peninsula, central Japan. *Science Reports of Kanazawa University, 40*, 9-37.

Ozawa, H., Kamiya, T., Ito, H. & Tsukawaki, S. (2004a). Water temperature, salinity ranges and ecological significance of the three families of Recent cold-water ostracods in and around the Japan Sea. *Paleontological Research, 8*, 11-28.

Ozawa, H., Kamiya, T., Kato, M. & Tsukawaki, S. (2004b). A preliminary report on the Recent ostracodes in sediment samples from the R.V. *Tansei-maru* Cruise KT01-14 in the southwestern Okhotsk Sea and the northeastern Japan Sea off Hokkaido. *Bulletin of the Japan Sea Research Institute, 35*, 33-46.

Ozawa, H., Nagamori, H. & Tanabe, T. (2008). Pliocene ostracods (Crustacea) from the Togakushi area, central Japan; palaeobiogeography of trans-Arctic taxa and Japan Sea endemic species. *Journal of Micropalaeontology, 27*, 161-175.

Ruddiman, W. F., Raymo, M. E., Martinson, D. G., Clement, B. M. & Backman, J. (1989). Pleistocene evolution: Northern hemisphere ice sheets and North Atlantic Ocean. *Palaeoceanography, 4*, 353-412.

Sato, T. & Kamiya, T. (2007). Taxonomy and geographical distribution of recent *Xestoleberis* species (Cytheroidea, Ostracoda, Crustacea) from Japan. *Paleontological Research, 11*, 183-227.

Sato, T., Kameo, K. & Mita, I. (1999). Validity of the latest Cenozoic calcareous nannofossil datums and its implication to the tephrachronology. *Earth Science (Chikyu-Kagaku), 53*, 265-274. (in Japanese with English abstract)

Schornikov, E. I. & Tsareva, O. A. (1995). New Ostracoda of the genus *Aurila* from the N.W. Pacific. *Mitteilungen aus dem Hamburgischen Zoologischen Museum und Institut, 92*, 237-253.

Shirai, M., Tada, R. & Fujioka, K. (1997). Identification and chronostratigraphy of Middle to Upper Quaternary marker tephras occurring in the Anden coast based on comparison with ODP cores in the Japan Sea. *Quaternary Research (Daiyonki-Kenkyu), 36*, 183-196. (in Japanese with English abstract)

Sudo, H. (1986). A note on the Japan Sea Proper Water. *Progress in Oceanography, 17*, 313-336.

Sugiyama, Y., Sangawa, A., Shimokawa, K. & Mizuno, K. (1987). The Late Pliocene terrace deposits and neotectonic movement of the Omaezaki area, Shizuoka Prefecture. *Bulletin of Geological Survey of Japan, 38*, 443-472. (in Japanese with English abstract)

Tabuki, R. (1986). Plio-Pleistocene Ostracoda from the Tsugaru Basin, North Honshu, Japan. *Bulletin of College of Education, University of Ryukyus, 29*, 27-160.

Tada, K. & Nishihama, Y. (1988). Research for the water quality environment in the lake Saroma. *Fiscal year project report of the Abashiri Marine Laboratory*, 209-213. (in Japanese)

Tada, R. (1994). Paleoceanographic evolution of the Japan Sea. *Palaeogeography, Palaeoclimatology, Palaeoecology, 108*, 487-508.

Takata, H. (2000). Paleoenvironmental changes during the deposition of the Omma Formation (late Pliocene to early Pleistocene) in Oyabe area, Toyama Prefecture based on the analysis of benthic and planktonic foraminiferal assemblages. *Fossils (Palaeontological Society of Japan), 67*, 1-18. (in Japanese)

Tanaka, G. (2008). Recent benthonic ostracod assemblages as indicators of the Tsushima warm current in the southwestern Sea of Japan. *Hydrobiologia, 598*, 271-284.

Tanaka, G. & Ikeya, N. (2002). Migration and speciation of the *Loxoconcha japonica* species group (Ostracoda) in East Asia. *Paleontological Research, 6*, 265-284.

Tanaka, G., Tsukawaki, S. & Ooji, A. (2004). Preliminary report on Ostracodes from the Miocene Sunagozaka Formation, southern part of Kanazawa City, Ishikawa Prefecture, central Japan. *Bulletin of Japan Sea Research Institute, 35*, 53-63. (in Japanese with English abstract)

Tanimura, Y. (1981). Late Quaternary diatoms and paleoceanography of the Sea of Japan. *The Quaternary Research (Daiyonki-Kenkyu) 20*, 231-242. (in Japanese with English abstract)

Tsukagoshi, A. (1990). Ontogenetic change of distributional patterns of pore systems in *Cythere* species and its phylogenetic significance. *Lethaia*, *23*, 225-241.

Tsukagoshi, A. (1996). Recommendations from paleontology–fossils that demonstrate organismal phylogeny. In K. Iwatsuki, & S. Mawatari, (Eds.), *Species Diversity, Biodiversity Series 1*, (173-187). Tokyo, Shokabo. (in Japanese, title translated)

Tsukagoshi, A. (2004). Species diversity and paleontology: an example of interstitial Ostracoda, *Fossils (Palaeontolological Society of Japan)*, *75*, 18-23. (in Japanese with English abstract)

Tsukagoshi, A. & Briggs, Jr. W. M. (1998). On *Schizocythere ikeyai* Tsukagoshi & Briggs sp. nov. *Stereo-Atlas of Ostracod Shells*, *25*, 43-52.

Tsukawaki, S., Kamiya, T., Kato, M., Matsuzaka, T., Naraoka, H., Negishi, K., Ozawa, H. & Ishiwatari, R. (1997). Preliminary results from the R.V. *Tansei-maru* Cruise KT95-14 Leg 2 in the southern marginal area in the Japan Sea–Part 1: Sediments, benthic foraminifers and ostracodes. *Bulletin of Japan Sea Research Institute*, *28*, 13-43.

Tsukawaki, S., Ozawa, H., Domitsu, H., Tanaka, Y., Kamiya, T., Kato, M. & Oda, M. (1999). Preliminary results from the R.V. *Tansei-maru* Cruise KT97-15 in the eastern marginal part of the Japan Sea off Tsugaru Peninsula, Northeast Japan–Sediments, benthic and planktonic foraminifers and ostracodes. *Bulletin of Japan Sea Research Institute*, *30*, 99-140.

Yajima, M. (1978). Quaternary Ostracoda from Kisarazu near Tokyo. *Transactions and Proceedings of the Palaeontological Society of Japan, New Series*, *112*, 371-409.

Yajima, M. (1982). Late Pleistocene Ostracoda from the Boso Peninsula, Central Japan. *Bulletin, the University Museum, University of Tokyo*, *20*, 141-227.

Yajima, M. (1987). Pleistocene Ostracoda from the Atsumi Peninsula, central Japan. *Transactions and Proceedings of Palaeontological Society of Japan, New Series*, *146*, 49-76.

Yamada, K., Irizuki, T. & Tanaka, Y. (2002). Cyclic sea-level changes based on fossil ostracode faunas from the upper Pliocene Sasaoka Formation, Akita Prefecture, central Japan. *Palaeogeography, Palaeoclimatology, Palaeoecology*, *185*, 115-132.

Yamada, K., Irizuki, T. & Tanaka, Y. (2005). Paleoceanographic shifts and global events recorded in late Pliocene shallow marine deposits (2.80–2.55 Ma) of the Sea of Japan. *Palaeogeography, Palaeoclimatology, Palaeoecology*, *220*, 255-271.

Yasuhara, M. & Kumai, H. (2003). Fossil ostracodes from the Tako-Shell bed, Shimosa Group and the Somei horizontal hollow tomb floor deposits formed in its outcrop in Somei, Tako-machi, Chiba Prefecture, Japan. *Monograph of Association of Geological Collaboration of Japan*, *50*, 73-78. (in Japanese with English abstract)

Yoshikawa, S. & Mitamura, M. (1999). Quaternary stratigraphy of the Osaka Plain, central Japan and its correlation with oxygen isotope record from deep sea cores. *Journal of the Geological Society of Japan*, *105*, 332-340. (in Japanese with English abstract)

Zhou, B. C. (1995). Recent ostracode fauna in the Pacific off Southwest Japan. *Memoirs of Faculty of Science, Kyoto University, Series of Geology & Mineralogy*, *57*, 21-98.

In: Species Diversity and Extinction
Editor: Geraldine H. Tepper, pp. 111-142

ISBN: 978-1-61668-343-6
© 2010 Nova Science Publishers, Inc.

Chapter 3

NEW FRONTIERS IN GENOME RESOURCE BANKING

Joseph Saragusty[1] and Amir Arav[2]

[1]Leibniz Institute for Zoo and Wildlife Research, Berlin, Germany
[2]Institute of Animal Science, Agricultural Research Organization,
The Volcani Center, Bet Dagan, Israel

ABSTRACT

During the course of evolution species has always gone extinct; however, the rate of extinction has increased in recent decades by as much as one hundred fold. *In situ* preservation should be supported by *ex situ* efforts like captive breeding, supplemented by assisted reproductive technologies (ART) and the establishment of genome resource banks (GRB).

Semen cryopreservation protocols have been developed for many species but many others proved challenging. Apparently, specie-specific protocols for semen collection and cryopreservation should be developed. We and others (Jewgenow et al., 1997; Saragusty et al., 2006) have shown that post mortem semen collection and cryopreservation from endangered species, even hours after death, can save valuable genes. To reduce costs of liquid nitrogen storage and maintenance, we have developed large volume cryopreservation technique. Additionally, with this cryopreservation technique we showed that samples can be thawed, used and the balance refrozen (Arav et al., 2002b; Saragusty et al., 2009b). Other mid- and long-term preservation techniques are currently under exploration, including freeze-drying and electrolyte-free preservation.

Oocytes cryopreservation has proved much more challenging than sperm and, even today, success rate is very low. Vitrification is gaining the lead in this field. However, as we have recently demonstrated (Yavin and Arav, 2007), identifying the delicate balance between multiple factors is imperative for success. Alternatively, oocytes collected ante- or post mortem can be fertilized *in vitro* and cryopreserved as embryos.

Ovary freezing is a new technology developed for human fertility preservation in women that undergo cancer treatment. In recent publication we (Arav et al., 2010) documented the longest ovarian function for up to 6 years after whole organ cryopreservation in sheep. We have shown endocrine cyclisty and production of normal oocytes and embryos. This strategy could benefit endangered species if allogeneic or even xeno-transplantation after whole ovary or ovarian tissue cryopreservation could be done.

Cloning animals is currently limited to few species and has a low success rate. Nevertheless, despite this limitation, it would seem pragmatic to initiate storage of somatic cells with an eye to future improvements in nuclear transfer efficiency. However, a major obstacle to the establishment of GRB is the cost associated with the long-term maintenance of cell lines in liquid nitrogen. We have shown recently (Loi et al., 2008a; Loi et al., 2008b) the capacity of freeze-dried somatic cells, which were held at room temperature for 18 months to maintain their nuclear integrity, and subsequently be used for nuclear transfer and produce viable embryos.

In the following pages, we will review the various aspects of gametes, embryo and tissue preservation for prospective utilization in ART.

INTRODUCTION

The Species Survival Commission (SSC) of the International Union for Conservation of Nature and Natural Resources (IUCN) continuously monitors the planet's fauna and flora and launches the *IUCN Red List of Threatened Species* (http://www.iucnredlist.org). As of mid 2007, there were 5416 species of mammals described. Of these, 4863 species were evaluated and the result was that 1,094 species (22.5%) are now listed as endangered. In addition, there are 596 species listed as near threatened and another 384 species for which data is deficient. Adding all these numbers together, about 40% of the planet's mammalian species are at some level of threat for extinction. The list also reports on 70 species (1.4%) of mammals that became extinct in recent years and 4 more species that are extinct in the wild and whose survival completely depend on *ex situ* conservation programs. With each extinct species, the stability of the entire ecological system surrounding it and the food chain of which it is an integral part is shaken. Such shaking may lead to the co-extinction of dependent species (Koh et al., 2004). While species extinction is part of the evolutionary system, primarily due to human activity the rate of extinction in the past 100 years has gone up 100 to 11,000 folds according to some estimates (IUCN, 2004). Conservation can take one of two basic approaches. One approach is conservation of the habitat selected based on the biodiversity in it (Margules and Pressey, 2000). Another is the conservation efforts directed at individual species, primarily under captive conditions, through natural breeding and assisted reproductive technologies (Wildt, 1992; Holt and Pickard, 1999; Pukazhenthi and Wildt, 2004). Unfortunately, conservation often conflicts with the rapidly growing world's human population and its ever-increasing demand for land, food, water and energy. Society will obviously elect to satisfy humans' needs before any consideration of conservation, however ignoring the issue of conservation will be the wrong approach. Nature and biodiversity provide us with security, resiliency, social relations, health, and freedom of choices and actions as well as welfare and livelihood (Duraiappah et al., 2005). While conservation of the habitat is of paramount importance, this alone is not enough. *Ex situ* preservation efforts are necessary to complement it, and for many species these are the last chance for survival (IUCN, 1987). To that end, captive breeding programs, at times supplemented by assisted reproductive technologies (ART), were set into motion for a wide variety of species. Yet, detailed knowledge of the mechanisms of reproduction is available for only about 2% of the world's mammals, many of which are domestic and laboratory animals (Wildt et al., 1997;

Comizzoli et al., 2000). Regrettably, attempts to apply to wildlife species techniques developed for humans, livestock or laboratory animals, often fail to produce the expected results (Wildt et al., 1995). Such techniques are often species-specific and of little use when applied to other species. While basic evaluation techniques can usually be adapted from one species to another, species-specific methods should be devised for the following procedures: 1) semen collection and handling, 2) artificial insemination (AI), 3) oocyte collection with or without super-ovulation by ovum pick up (OPU) or post mortem, 4) *in vitro* oocyte maturation (IVM) and *in vitro* fertilization (IVF) or intracytoplasmic sperm injection (ICSI), 5) embryo culture and embryo transfer (ET) following chemical synchronization, 6) nuclear transfer and 7) the preservation of gametes, embryos, tissues and whole organs. To store and manage the preserved cells, embryos, tissues and organs, the establishment of genome resource banks (GRB) was proposed (Veprintsev and Rott, 1979; Wildt, 1992; Johnston and Lacy, 1995; Holt et al., 1996; Wildt et al., 1997) Apart from being a collection of genomes, gametes and embryos, fulfilling their function as a mean to extend the reproductive lifespan of individuals beyond their biological life and to prevent the loss of valuable individuals to the gene pool, these establishments provide multiple additional advantages. Transporting the preserved gametes or embryos is easier and cheaper than shipping the stress-susceptible live animal for breeding. The sperm collected from healthy individuals in captivity can be used to inseminate females in wild isolated small populations. Semen can also be collected from the wild to revitalize the captive population with new genetic material without the need to remove valuable individuals from the wild population. Collections in GRB act as insurance for small populations against catastrophes, epidemics etc. For example, over half of the mountain gazelle (*Gazella gazella gazella*) population in the north of Israel was wiped out following an outbreak of the foot and mouth disease in the mid 1980's (Shimshony et al., 1986; Shimshony, 1988) or the canine distemper virus epidemic of 1994 in the lion (*Panthera leo*) population of the Serengeti-Mara system lead to the death of about a third of the population (Roelke-Parker et al., 1996). In the absence of system to collect and bank samples, all these dead animals are lost for the gene pool. However, genetic diversity is lost only when animals are no longer available to reproduce (Ballou, 1992). Gametes can be collected from animals ante mortem and stored beyond their reproductive lifespan or they can be collected post mortem. GRB may also help in mating between individuals that are incompatible due to character, personal preferences, location or time. Keeping biodiversity in GRB can also eliminate the need to keep large groups of living animals to meet targeted genetic diversity and by that reducing one of the major problems faced by zoos around the world. Such collections can also be used for the purpose of research on evolving assisted reproductive technologies. Several such GRBs are already in existence. These include for example the Frozen Ark Consortium (http://www.frozenark.org), the Biological Resource Bank of Southern Africa's Wildlife (Bartels and Kotze, 2006) and a number of additional institutions listed by Andrabi and Maxwell (2007).

SPERM PRESERVATION

Probably the first step, before even starting the process of semen collection and preservation, would be to learn about the reproductive physiology of the species in question.

The knowledge on similar domestic or laboratory species of even on other members of the same family is often of little help. For example, some felids such as the lynx (*Lynx lynx*) or the Pallas' cat (*Otocolobus manul*) show clear seasonality in their reproductive behaviour and semen production (Göritz et al., 2006; Swanson, 2006), whereas other cat species such as the South American cats show weak or no seasonality at all (Morais et al., 2002; Swanson and Brown, 2004).

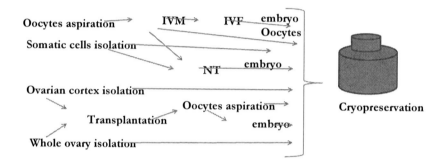

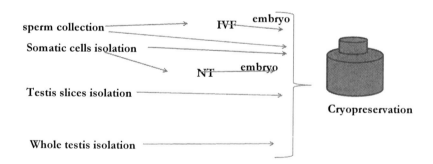

Figure 1. Reproductive biotechnology offers a variety of options for the preservation of genetic material and germplasm from both the female and the male.

Therefore, knowledge will indicate when would be the best time to collect semen samples. Evaluation of the reproductive tract and accessory glands through the use of ultrasonography may be of great help in this respect, both to identify pathologies and to understand how the system is build, what is the status of the accessory glands and what can be expected during collection (Hildebrandt et al., 2000a; Hildebrandt et al., 2000b). Monitoring of the reproductive steroid metabolites in feces, urine and saliva is a non-invasive method for the evaluation of the reproductive status of animals, suitable for animals both in zoos and in the wild (Schwarzenberger et al., 1996). While this method is primarily in use for monitoring females during the reproductive cycle and pregnancy, it was also applied in recent years to males of various species (Morais et al., 2002; Kretzschmar et al., 2004; Göritz et al., 2006).

Semen Collection and Evaluation

Numerous semen collection techniques were described in the literature for a wide variety of species. Some of the techniques involve natural mating followed by collection of the semen from the female either directly from the vagina or through the aid of intra-vaginal condom or vaginal sponge as is done, for example, in some camelids (Bravo et al., 2000). For these techniques, one should have the right setting that will lead to natural mating and the ability to access the female immediately after. Other techniques involve extensive training of the male such as the use of artificial vagina. This method is in use in some of the equids, felids, cervids and camelids species (Gastal et al., 1996; Asher et al., 2000; Bravo et al., 2000; Zambelli and Cunto, 2006) and was attempted with a lesser degree of success in other species. Other techniques that require a certain degree of training are those that involve manual stimulation of either the rectum as is routinely done in elephants (Schmitt and Hildebrandt, 1998; 2000) or through stimulation of the penis as is done for example in the flying fox (*Pteropus alecto*) or the marmoset monkey (*Callithrix jacchus*) (Schneiders et al., 2004; Melville et al., 2008). Probably by far the most popular and widely used method is the electroejaculation technique. To be successful one would need a suitable probe which often needs to be specifically designed for the animal to be collected based on preliminary knowledge of its anatomy (Hildebrandt et al., 2000a; Roth et al., 2005). However, this technique is not without drawbacks. To begin with, the animal must be immobilized, something many zoos would rather avoid when possible. It also makes it impossible to collect from the same individual on a regular, frequent basis. During collection, unintentional stimulation of the bladder often result in urinary contamination of the sample, as was recently reported, for example, in the Asiatic Black bears (*Ursus thibetanus*) (Chen et al., 2007). However, with proper anatomical understanding of the rectal region, adjustment of the stimulation level and, when possible, emptying the bladder prior to collection, this can be avoided. Pharmacologically-induced ejaculation was also described in recent years. For example in stallions, oral imipramine hydrochloride followed by intravenous xylazine hydrochloride was shown to result in ejaculation minutes after the xylazine injection in 68% of the attempts (McDonnell, 2001). Finally, there are the invasive techniques which include semen aspiration from the cauda epididymis for evaluation or salvage of semen from the epididymis and proximal portion of the vas deference following castration or post mortem (Jewgenow et al., 1997; Saragusty et al., 2006). From our and others experience, sperm survive well as long as it is in the epididymis and chilled to $4°C$, even when still attached to the testicle (Yu and Leibo, 2002) and live births were reported following artificial insemination, *in vitro* fertilization or intracytoplasmic sperm injection (Santiago-Moreno et al., 2006; Shefi et al., 2006).

Sperm evaluation also requires understanding of the species under study. In primates, semen can be as thick as paste which requires liquefaction and extraction of the cells into a diluent. In camelids, possibly due to the absence of vesicular glands, sperm is also fairly viscous and can be enzymatically liquefied (Bravo et al., 2000). We have found that similar enzymatic liquefaction can also help when attempting to separate rhinoceros sperm from the seminal plasma, something that can not be done efficiently with centrifugation alone (Behr et al., 2009b). The volume and concentration also vary several orders of magnitude among animals. In the European brown hare (*Lepus europaeus*) or the Asiatic black bear for example, volume of semen collected by electroejaculation is often in the range of 1 mL or less with concentrations exceeding 10^9 cells/mL. Low volume of up to a few mL and low concentration of few millions per mL is often the case in felids both in captivity and in the wild (Morato et al., 2001). In some other animals volumes can be very large. In the boar or the elephant semen can exceed 100 mL with concentrations of several hundred million cells per mL. Motility is also expected to be low in sperm collected from the epididymis as epididymal sperm is immotile. This is expected to change after a short incubation time in a suitable media. Some specific characteristics were also noted in certain species. For example the seminal plasma pH of the flying fox or the snow leopard (*Panthera uncial*) is high (8.2 and 8.4, respectively) (Roth et al., 1996; Melville et al., 2008) or in the elephant we found recently that the osmolarity of the seminal plasma was around 270 mOsm/kg (Saragusty et al., 2009c). Such characteristics demonstrate the need to verify multiple aspects of the semen so that suitable diluents can be made.

Many wildlife captive populations, such as the Asian elephant, small cat species, the Przewalski's horse or the Somali wild ass, originated from a small founder group. Topping this with aging of the captive population, no introduction of new genetic material through new captures from the wild and no preservation of the existing genetic diversity through GRB, and the result is genetic drift that may eventually lead such populations to extinction (Wiese, 2000; Moehlman, 2002; Swanson, 2006). A similar problem may arise in wild populations that become isolated due to habitat fragmentation or where the number of individuals left in the wild is small like in Javan rhinoceros (*Rhinoceros sondaicus*) whose wild population is estimated at under 60 or the Northern white rhinoceros (*Ceratotherum simum cottoni*) whose wild population is believed to be extinct (http://www.rhinos-irf.org/). Inbreeding is a problem that affects semen quality as well. An inverse correlation was found between the level of inbreeding and the quality of the semen collected from three endangered gazelle species (*Gazella dorcas*, *Gazella dama* and *Gazella cuvieri*) populations (Gomendio et al., 2000). Poor semen quality was also found in the clouded leopard (*Neofelis nebulosa*) in North American zoos whose population there originated from a founding group of 7 individuals (Pukazhenthi et al., 2006). In this population close to 90% of the spermatozoa were with abnormal, damaged acrosomes. A similar low quality sperm in association with poor gene diversity was also reported in the Florida panther (*Puma concolor coryi*) (Roelke et al., 1993; Barone et al., 1994; Holt and Pickard, 1999).

Sperm Cryopreservation

To start with, sperm cryopreservation in endangered species is not an easy undertaking. For most species, there is little or no available information about the various factors associated with the process. Does the seminal plasma need to be removed by centrifugation? Which diluents and cryoprotectants will be most suitable and at what concentrations, osmolarity and pH? Are the cells chilling-sensitive? Which freezing method should be used? And how to thaw the samples? Should the cryoprotectants be removed or diluted after thawing? To determine all these, basic research is needed for each species under evaluation. Since the first report on semen cryopreservation 61 years ago which resulted from a chance observation (Polge et al., 1949), the field of sperm cryopreservation made progress primarily through trial and error. Even today, after six decades of intensive studies propelled primarily by research related to human infertility and livestock and laboratory animal production, our knowledge about the exact mechanisms that eventually lead to success, failure or anywhere in-between is still very limited. Describing the advances made in each endangered species is beyond the scope of this report. We will concentrate here on recent advances related to the freezing and thawing process and point out some unique aspects of it.

The first step towards cryopreservation would be to determine the composition of the freezing extender. The four basic components of the freezing extender are a buffer, a cryoprotectant, a source of energy and often a source of lipids. Alterations in the buffer composition can be made to change the osmolarity and pH of the solution. Cryoprotectant are divided into cell permeating and non-permeating materials. Among the permeating ones, glycerol is probably the most widely used. While the spermatozoa of some species tolerate high glycerol concentrations, those of others can survive only very low concentrations. Spermatozoa of marsupials such as the koala (*Phascolarctos cinereus*) or the kanguru (*Macropus giganteus*) freezes in concentrations as high as 14% or more (Johnston et al., 1993; Holt and Pickard, 1999), whereas spermatozoa of other species such as the mouse or pig can tolerate only very low glycerol concentrations. In some cases, like we have found recently in the Asian elephant, gradual addition of the glycerol with the bulk of it added shortly before freezing, can minimize its damage while utilizing its protective affects (Saragusty et al., 2009c). Other permeating cryoprotectants include dimethyl sulfoxide (Me_2SO; a.k.a. DMSO), ethylene glycol and propylene glycol. Non-permeating cryoprotectants are usually oligosaccharides (lactose, raffinose, trehalose, sucrose) or polymers such as polyvinylpyrrolidone (PVP). The concentration to be adopted for any of these to achieve optimal results is a kind of a delicate balance between the toxicity of the cryoprotectant and its protective affect. To avoid the risk of toxicity, attempts were done to cryopreserve mouse sperm with no cryoprotectant at all. Sperm frozen in EGTA-Tris-HCl buffered solution produced similar fertilizing ability to fresh semen when used in ICSI, and developed to viable fetuses when transferred at the 2-cell stage (Ward et al., 2003). The presence of the cryoprotectant may hinder fertilization in some animals so it should be removed after thawing and before AI. In the Poitou donkey, pregnancies were achieved only if the 4% glycerol was removed from the thawed samples before the jennies were inseminated (Trimeche et al., 1998).

The phospholipids of mammalian spermatozoa possess a distinctive and highly unusual fatty acid composition, the most unique feature of which is a very high proportion of long chain (C20-22) highly polyunsaturated fatty acyl groups (Rooke et al., 2001). In most mammals, docosahexaenoic acid (DHA; 22:6,n-3) is the dominant polyunsaturated fatty acid (PUFA) although, in several species, docosapentaenoic acid (22:5,n-6) is also a major component (Rooke et al., 2001). The lipid composition is a major determinant of the membrane flexibility required for the characteristic flagella movement of spermatozoa and for the fusogenic properties of the membranes, associated with the acrosome reaction and fertilization (Papahadjopoulos et al., 1973; Langlais and Roberts, 1985). During the process of cryopreservation, the cells go through two major stages - they are chilled from body temperature down to about 4°C and they are then frozen to cryogenic temperatures. In the first stage the chilling process causes changes in the consistency of the cellular plasma membrane, called lipid phase transition. These changes result in the transition of the membranes' lipids from the liquid state to the gel-like state. The longer the lipid chain, and the higher the degree of saturation, the higher the temperature of transition. This means that during chilling, the long, saturated fatty acids form raft-like structures that drift in the still fluid long, unsaturated lipids, a process known as lipid phase separation. These membrane changes are damaging, resulting in chilling injury and increased membrane permeability (Arav et al., 1996; Arav et al., 2000a). The ability of the spermatozoal plasma membrane to resist structural damage during cryopreservation may therefore be related to its fatty acids composition and the strength of the bonds between membrane components (Hammerstedt et al., 1990). Not all animals' sperm are sensitive to chilling. It was shown, for example, that spermatozoa of Asian elephants, which are more chilling sensitive than African elephant sperm, contain lower concentration of PUFA in their membranes and are especially poor in DHA (Swain and Miller, 2000). Analysis of membrane lipid composition (Drobnis et al., 1993) or the use of Fourier transform infrared spectroscopy can help in identifying the temperature range at which species-specific spermatozoa are most sensitive, as we have showed in the Asian elephant (Saragusty et al., 2005).

There are several ways to influence the lipid composition of the sperm membrane. Most sperm freezing extenders contain egg yolk or other lipid source. We have showed that when Asian elephant spermatozoa were incubated in extender containing egg yolk or even only with egg-phosphatidylcholine liposomes, the lipid phase transition temperature was down-shifted by more than 12 degrees, making it much less sensitive to chilling injury (Saragusty et al., 2005). Others showed that the addition of cholesterol also increases membrane fluidity and provide similar protection (Purdy and Graham, 2004; Moore et al., 2005). Various reports on studies done by us and others show that enrichment of the diet with PUFA can improve motility and concentration, alter the biophysical properties and enhance the reproductive performance of mammalian and avian sperm (Blesbois et al., 1997; Culver, 2001; Gacitua et al., 2002; Brinsko et al., 2003; Mitre et al., 2004; Strzezek et al., 2004).

While long chain unsaturated fatty acids seems to be beneficial to the spermatozoal motility, such fatty acids are also the subject for oxidation, resulting in reactive oxygen species (ROS) which are known to be damaging to both the sperm plasma membrane and the DNA (Baumber et al., 2000; Chen et al., 2002). In the attempt to prevent these damaging effects, numerous studies were conducted using various antioxidants (vitamin C, vitamin E, glutathione, β-carotene just to name a few), showing the beneficial effect of these compounds on sperm characteristics (Comhaire et al., 2000; Ball et al., 2001). These antioxidants are often added to the freezing extender to protect the cells during the processing and after thawing.

One of the questions that should be answered is whether it is necessary to remove the seminal plasma before suspending the sperm in the freezing extender. In some mammals such as the goat, the seminal plasma contains enzymes that react with components in the egg yolk, leading to severe damage to the cells in the sample. Yet, in other mammals, such as the camelids or the kuala, the seminal plasma contain ovulation-inducing factor, rendering it essential during AI (Pan et al., 2001; Johnston et al., 2004; Adams et al., 2005). If the seminal plasma needs to be removed, we found that cushioned centrifugation was beneficial in terms of total sperm yield and centrifugation force damage reduction (Saragusty et al., 2006; Saragusty et al., 2007). In this process, the semen sample is under-laid with an inert, non-ionic, high-density fluid such as 60% aqueous iodixanol solution. At the end of centrifugation, the sperm pellet is at the interface between the iodixanol and the seminal plasma. The iodixanol and the seminal plasma can then be aspirated. With iodixanol, a higher centrifugation force can be used to increase the yield.

Most freezing protocols call for slow and gradual chilling of the extended sperm sample down to the pre-freezing temperature of 4°C. This slow chilling provide the sperm with ample time for equilibration and thus to a reduced chilling injury. Recently a report on human spermatozoa suggested that diluting the semen with cold (4°C) extender, and thus facilitating fast chilling, resulted in improvement of 15% in post-thaw sperm motility (Clarke et al., 2003). Despite previous reports that fast chilling rates are damaging to sperm cells (Fiser and Fairfull, 1986) and the fact that it is known the Asian elephant spermatozoa is sensitive to chilling (Saragusty et al., 2005), we tested this method of cold extender addition on elephant sperm. Our results showed that the use of cold extender was clearly damaging to the elephant sperm cells (Saragusty et al., 2009c). Thus, it can only be assumed that such an improvement can be achieved in some mammals but not in others.

Several freezing methods are in use for the cryopreservation of spermatozoa. Probably the most widely used one is the vapor freezing method during which the extended semen, packaged into 0.25 or 0.5 mL plastic straws is held at a predetermined distance above liquid nitrogen for several minutes before being plunged into the liquid nitrogen for storage. Another method is known as the pellet method in which a small volume (usually around 200 µL) of extended semen is placed directly on carbon dioxide ice ("dry ice") and then stored in liquid nitrogen. In more recent years, controlled rate freezing machines started penetrating the market. In these machines, suitable for freezing extended semen packaged in straws, the rate of both the chilling and the freezing can be programmed and precisely controlled. The drawback of the various conventional freezing methods (vapor, pellet, controlled rate) is that ice crystal growth is uncontrolled in terms of both velocity and morphology, and the crystals may therefore disrupt and kill cells in the sample (Watson, 2000). The alternative technique, which was described in a recent review as "the only recent significant advancement in semen preservation technology" (Gillan et al., 2004) is the directional freezing technique. This technique uses the Multi- Thermal Gradient device (MTG®; IMT Ltd, Ness Ziona, Israel) (Figure 2). After the initial seeding stage, the semen sample is advanced at a constant velocity through a linear temperature gradient. The ice crystal propagation can thus be controlled, to optimize crystal morphology, and to achieve continual seeding and a homogenous cooling rate throughout the entire freezing process. Damage to the cells, even when freezing in large volumes of 8 mL, is thereby minimized (Arav, 1999; Arav et al., 2002a; Arav et al., 2002b; Saragusty et al., 2009a). Several studies, conducted by us and others, that used this technology, demonstrated the viability and fertilizing ability of frozen–thawed semen from a variety of species (Arav et al., 2002b; Hermes et al., 2003; Arav et al., 2005; Gacitua and Arav, 2005; O'Brien and Robeck, 2006; Saragusty et al., 2006; Si et al., 2006; Saragusty et al., 2007; Reid et al., 2009; Saragusty et al., 2009c). The advantages of large volume freezing are apparent when one has to consider the storage space and costs of a large number of samples under liquid nitrogen over extended period of time. In studies conducted on bovine bull semen, we have demonstrated that semen frozen in this way can be thawed, packaged in insemination-dose straws, re-frozen and later thawed and used for AI with acceptable fertility results, similar to those achieved in the conventional single-freezing method (Arav et al., 2002b; Saragusty et al., 2009b) Several studies on human sperm showed that some cells preserve their functional capacity even after several freeze-thaw cycles and can be used to fertilize oocytes through ICSI (Polcz et al., 1998; Rofeim et al., 2001; Bandularatne and Bongso, 2002).

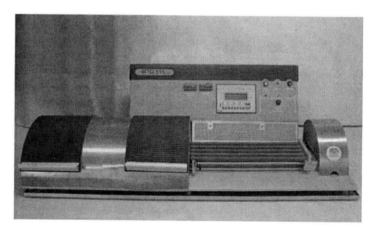

Figure 2. Multi-Gradient Thermal device in which large-volume test tubes with samples are advanced at a constant velocity through a pre-defined temperature gradient to achieve optimal heat transfer and extra-cellular ice crystallization.

Another method, which is similar to freezing but very different from it, is vitrification. Vitrification, also known as ice-free cryopreservation, is a process in which liquid is transformed into an amorphous, glass-like solid, free of any crystalline structures. To achieve this state, either very high concentrations of cryoprotectants or very high cooling rates are needed. Since the high cryoprotectant concentrations needed are beyond what the spermatozoa can tolerate, the frozen volume is being considerably reduced to achieve high cooling rate with adequate heat transfer through the sample. In a series of three studies conducted on human sperm, cryoprotectant-free (but with some sugars and 10% bovine serum albumin) vitrification was achieved by loading the sperm suspension into a copper loop to form a thin film. The copper loop was then plunged into liquid nitrogen to achieve cooling rates as high as 7.2×10^5 °C per minute. To match the ultra-fast cooling rate, thawing was also done in a very fast warming rate. Under these conditions, the researchers were able to achieve with vitrification similar post-thaw motility, normal morphology and fertilizing ability as compared to the standard liquid nitrogen vapor freezing with glycerol (Nawroth et al., 2002; Isachenko et al., 2003; Isachenko et al., 2004). In a very recent study, these authors showed that even vitrification of 30 µL of sperm, suspended in human tubal fluid containing 1% human serum albumin and 0.25M sucrose, can be done by dropping it into liquid nitrogen. After such treatment, post thaw forward motility was 57.1% (Isachenko et al., 2008).

In many cases, sperm can be collected in the field, away from any fully equipped andrology and cryobiology laboratory, and transferring the samples to a facility for processing may take time. For such cases, or when the sample is destined to be used but not immediately, mid-term non-freezing preservation techniques may help. In some species, such as the pig, chilled storage is the most widespread method of conservation as thus far sperm cryopreservation has provided only mediocre post-thaw results. Preliminary experiments conducted in our laboratory showed that using the large volume directional freezing technique, boar semen can be frozen with as high as 75% post-thaw motility, suggesting that this technique might be successful where the conventional methods has thus far failed (unpublished data). Boar semen can usually be stored at 15-17°C for several days. When planning on extended chilled storage, several sperm energy-metabolism aspects should be taken into consideration. Both the glycolysis and the Krebs cycle play an important role in sperm energy metabolism. Besides monosaccharides, sperm use other substrates such as citrate and lactate to produce energy as well. A fine balance between these metabolites is required to produce an optimal preservation solution (Rodriguez-Gil, 2006). For mid-term non-freezing preservation, two methods were recently proposed. After evaluating various solutions, osmolarities and storage temperatures, Van Thuan and colleagues found that the optimal conditions for mouse spermatozoa preservation were in 800 mOsm potassium simplex optimized medium, containing amino acids and 4 mg/mL bovine serum albumin, and a holding temperature of 4°C (Van Thuan et al., 2005). More than 40% of oocytes injected with sperm heads stored under these conditions for 2 months developed to the morula/blastocyst stage *in vitro* and 39% of the embryos developed to term after being transfer to recipient mice. Using the same medium, these researchers also showed that mouse spermatozoa can retain competence for ICSI if they are held for one week at room temperature and then for up to 3 months at -20°C. In another study, conducted on human and mouse spermatozoa, the cells were stored at 4°C in electrolyte-free solution containing glucose and bovine serum albumin. Under these conditions, cells preserved their motility for 3 weeks (mice) or 4 weeks (human) and their viability for up to 6 weeks. Both mouse and human sperm stored under these conditions were capable of fertilizing oocytes when used in ICSI (Riel et al., 2007).

Storage of cryopreserved samples under liquid nitrogen is very demanding in terms of maintenance, storage space, storage equipment and costs. An alternative that would minimize costs, storage and maintenance has been gaining a foothold in the field of sperm preservation in recent years - the dry storage. Drying of cells can be achieved by either freeze-drying or by convective-drying. Freeze-drying was known for hundreds of years as a method of meat and vegetable preservation among the people who lives in very high altitudes, like in the Andes mountain dwellers of South America. In more recent times, freeze-drying is used for preparation of pharmaceuticals, viral, bacterial, fungal or yeast in a dry and convenient form for transporting and storage. Freeze-drying is achieved by sublimation of the ice after freezing the sample to subzero temperatures. Convective drying, on the other hand, is achieved by placing the sample in a vacuum oven at ambient temperatures. In nature, many plants and animals can enter the state of anhydrobiosis by accumulating disccharides such as trehalose in their cells to as much as 50% of their dry weight (Crowe et al., 1984; Womersley and Ching, 1989; Westh and Ramløv, 1991). During the dehydration process, water molecules leave the interface between the lipid bilayer, resulting in damage to the membrane and an increase in the lipid phase transition temperature. If trehalose is present in the system, it is able to replace the water molecules and keep the lipid phase transition down (Leslie et al., 1994). When the water content is very low, trehalose can reach its glass transition at ambient temperatures, a state of vitrification (Sun et al., 1996; Buitink et al., 1998). The process is damaging to the cellular membrane and the rehydrated spermatozoa are devoid of motility and viability. Some degree of chromosomal damage also take place due to endogenous nucleases. This damage can be greatly reduced by the use of calcium chelating agents such as EGTA (Martins et al., 2007). Attempts to freeze-dry spermatozoa were first reported more than 5 decades ago (Polge et al., 1949; Sherman, 1957). Experiments in fertility of freeze-dried bull spermatozoa were reported in 1976 (Larson and Graham, 1976) and since then reports on a growing number of species are accumulating in the literature. To date, embryonic development after ICSI with freeze-dried sperm heads has been reported in humans and hamster (Katayose et al., 1992), cattle (Keskintepe et al., 2002; Martins et al., 2007), pigs (Kwon et al., 2004), rhesus macaque (Sanchez-Partida et al., 2008) and cats (Moisan et al., 2005), and live offspring were reported in mice (Wakayama and Yanagimachi, 1998; Kaneko et al., 2003; Ward et al., 2003), rabbits (Yushchenko, 1957; Liu et al., 2004), rat (Hirabayashi et al., 2005; Hochi et al., 2008) and fish (Poleo et al., 2005). Although storage at room temperature would have been the ideal solution, it seems that the storage temperature is affecting the DNA integrity. Studies on both mouse and rat spermatozoa showed that at room temperature there is progressive damage to the cells whereas at $+4^\circ C$, and even more so at $-196^\circ C$, these damages are reduced (Kaneko and Nakagata, 2005; Hochi et al., 2008). Based on extrapolation of Arrhenius plots, there will be on going deterioration of sperm stored at $+4^\circ C$ but there would be no decline in the fertilizing ability (by ICSI) and the rate of development to blastocyst stage of freeze-dried spermatozoa stored at $-80^\circ C$ (Kawase et al., 2005). Convective drying was tested on rhesus macaque spermatogonial stem cells. After loading the cells with 50 mM trehalose, as many as 80% of the cells maintained viability at water content as low as $0.5g\ H_2O^-g$ (Meyers, 2006).

While spermatogonial stem cells preserved in the dry form can only be used for ICSI or nuclear transfer, viable cells, for example freshly isolated or following cryopreservation, can be transplanted into the testes of another male of the same or of closely related species. This technique was first reported in mice in 1994, when it was demonstrated that such transplantation can lead to spermatogenesis (Brinster and Zimmermann, 1994). This study was followed by the transplantation of spermatogonial stem cells of several other mammalian species (rat, hamster, rabbit, dog, non-human primate and human) into sterilized (chemically or irradiated) mice. The next stage was the demonstration that this technique can work also when the transplantation is done between larger mammals, such as boars or bovine bulls (Honaramooz et al., 2002; Izadyar et al., 2003). Finally, the ultimate demonstration came recently, showing that such xenotransplanted spermatogonial stem cells can actually produce normal, functioning spermatozoa. Spermatogonial stem cells collected from immature rats were transplanted into chemically-sterilized mice and the spermatozoa or spermatids collected from the recipient mice produce normal, fertile rat offspring both when freshly used and following cryopreservation (Shinohara et al., 2006). The demonstration that such stem cells can be collected, transplanted and produce viable spermatozoa indicate that this is yet another useful way for the conservation of genetic material, for example from immature deceased animals.

Testicular tissue preservation can basically be done in two forms. Either the tissue is cryopreserved for future use or it is transplanted. When preserved in the frozen form, both spermatozoa and spermatogonial stem cells can be harvested from the tissue after thawing. Recently it was demonstrated that spermatozoa or spermatids retrieved from reproductive tissues (whole testes or epididymes) frozen for up to one year at -80°C or from whole mice frozen at -20°C for up to 15 years can produce normal offspring following ICSI (Ogonuki et al., 2006). The technique of testicular tissue cryopreservation is widely used today in both adult and pediatric human medicine as a mean to preserve spermatozoa from patients undergoing cancer treatments. From these tissues, spermatozoa and round and elongated spermatids can all be retrieved and used in ICSI (Hovatta et al., 1996; Gianaroli et al., 1999). Testicular tissue xenografting involves the transplantation of miniscule testicular parenchymal tissue from one species into orchidectomized immunodeficient recipient (usually mouse). The graft is supported by the recipient system and, after some time, starts producing spermatozoa, which can be harvested by surgical excision of all or part of the graft (Schlatt et al., 2002). Although dependent on the recipient system for support, the spermatogenesis cycle length seems to be inherent to the spermatogonial stem cells and they preserve the length of the donating species (Zeng et al., 2006). The sperm produced this way does not go through the epididymal maturation process so the only way it can be utilized is by ICSI (Shinohara et al., 2002).

One final procedure that holds great potential for wildlife conservation is the pre-selection of the offspring gender through the use of spermatozoa that were sorted according to the sex chromosome they carry. This technique is based on the small, but significant, difference in DNA content between the X- and Y-chromosome bearing spermatozoa that exist in many (but not all) species (Johnson et al., 1987; Johnson, 1988). A cell sorting machine, specifically modified for this purpose, can separate the sample into three populations - X-chromosome bearing spermatozoa, Y-chromosome bearing spermatozoa and unsortable cells (dead or maloriented in relation to the laser ray). The sorting purity can exceed 90%, although this comes with a price of reduced sorting rate. Tens of thousands of viable, normal offspring were produced thus far, primarily in the livestock industry. Development of sex sorting technique as part of the ART tool box of wildlife management will improve our ability to better control the sex ratio within social groups, simulating that in the wild and, in endangered species, where the production of a large number of offspring is crucial, produce more females. As a first step, the DNA content difference between the X- and Y-chromosome bearing spermatozoa should be determined. This was done already for a large number of species (for review see Garner, 2006) and attempts at using this technique to sort the spermatozoa of wildlife species were done by us and others in a variety of non-human primates, alpaca, rhino and elephant (O'Brien et al., 2004; O'Brien et al., 2005a; O'Brien et al., 2005b; Morton et al., 2008; Behr et al., 2009a; Hermes et al., 2009), with offspring born in elk and dolphins (Schenk and DeGrofft, 2003; O'Brien and Robeck, 2006). Some difficulties are still associated with this technique, making it difficult to utilize its full potential. One problem is the location of such sorting machine in relation to both the sperm donor and the female to be inseminated with the sorted sample. There are only a handful of such sorting machines around the world so the chances they will be in proximity to where they are needed are small. The double freezing technique mentioned earlier can come handy in this respect. The semen can be cryopreserved, shipped to the sorting machine, thawed, sorted, refrozen in insemination doses and shipped to where it will be used for insemination.

This idea was recently demonstrated in bull and ram (de Graaf et al., 2007; Underwood et al., 2007). Another problem is the sorting rate. Under optimal conditions, only about 20 million cells of each sex can be sorted by one machine every hour. Usually the rate is much lower so the number of sorted cells that can be collected within reasonable time is small. While this is less of a problem in animals where small insemination doses are used and when intrauterine insemination is possible, it certainly is a problem in animals, such as the elephant, where large sperm numbers in large volumes are used. The cells, after going through all the stresses involved in the sorting and post-sorting handling, also suffer from reduced viability over time. This field is still young and progress is fast. The future will probably bring the needed technologies that will make things work better for the species that need it so much.

While the ultimate goal is to obtain viable, motile, intact cells after cryopreservation so that they can be used for AI or IVF, since the development of ICSI (Goto et al., 1991), sperm motility became obsolete and in many cases, sonication is used to remove the tails altogether prior to their injection into the oocyte. In some species, such as the clouded leopard (*Neofelis nebulosa*), sperm quality is very poor, it hardly produce any embryos during IVF and it does not survive well the cryopreservation process (Pukazhenthi et al., 2006). Such sperm, however, can still be preserved and used in ICSI as part of the conservation efforts of such species. The eventual use of the preserved spermatozoa, be it for AI, IVF or ICSI, requires both the ability to handle the gametes *in vitro* and in the case of IVF or ICSI, also the ability to obtain and handle oocytes, fertilize them and eventually transfer the resulting embryos to a recipient females. And, of course, in all these cases, a thorough understanding of the female reproductive biology is a pre-requisite.

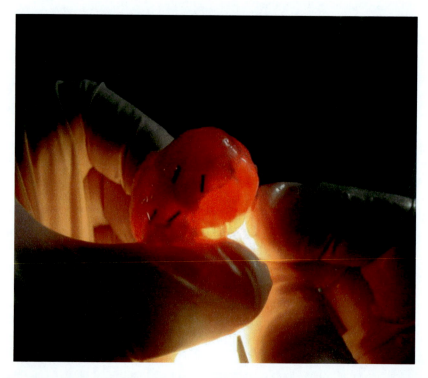

Figure 3. Trans-illumination of sheep ovary shows many antral follicles. The ovary was isolated from a sheep six years after transplantation of a frozen-thawed ovary (Arav, 2001; Arav et al., 2007).

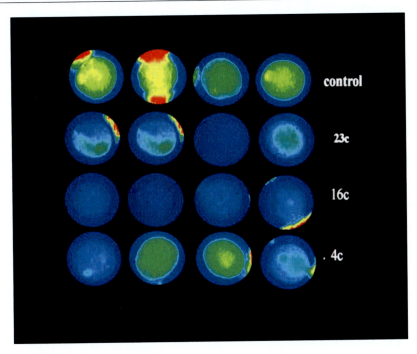

Figure 4. shows bovine oocytes stained with vital dye (cFDA) following rapid cooling (>1000C/min) and short exposure to different temperature. Green indicates membrane integrity and blue is for membrane's damage. It is possible to note that 16°C is the most damaging temperature while exposure to lower temperature is less damaging.

OOCYTES ASPIRATION AND VITRIFICATION

Collecting cells from dead animals is a valuable technique for cryopreservation of endangered species. We have shown that it is possible to collect sperm from the cauda epididymis up to 24h after death and to maintain cellular viability after freezing and thawing (Saragusty et al., 2006). For female preservation, the parallel procedure should be the collection of oocytes at the germinal vesicle (GV) stage, followed by in vitro maturation and oocytes vitrification. It was shown by us that oocytes aspiration, using trans-illumination technique (Figure 3), increase the number of oocytes collected from isolated ovaries by 50% (Arav, 2001). In cows, an average of 7.3 oocytes was collected from each ovary (Arav, 2001) and from pigs and sheep the number of GV oocytes that can be collected is even higher.

Oocytes are cells that present a high sensitivity to low temperature upon cooling and before freezing (chilling injury). Chilling injury depend on the biochemical composition and the thermobehavior of the membranes' lipids (Arav et al., 1996; Arav et al., 2000a; Zeron et al., 2002a; Zeron et al., 2002b). Recently we have shown that human oocytes are highly sensitive to cryopreservation while fertilized oocytes are very resistant. The reason is the temperature at which lipid phase transition (LPT) occur - between 20 and 16°C for mature oocytes and 10°C lower for fertilized oocytes (Ghetler et al., 2005). Chilling injury, which is a kinetic process, is reduced by temperature and exposure time (Zeron et al., 1999). Therefore,

LPT that occur at high temperature and for longer time will be more damaging than LPT that occur at lower temperature and for a very short time.

We and others have shown that rapid cooling rate will overcome (outrace) the problems associated with chilling injury (Figure 4) in a variety of cells models (Arav, 1992; Mazur et al., 1992; Arav et al., 2000b). It is proposed that vitrification, which is a very rapid cooling method, will allow cooling to low temperatures without chilling injury.

In the year 2008 we celebrated 70 years since the first publication of Luyet on the vitrification of frog sperm (Luyet and Hoddap, 1938). Vitrification is a thermodynamic process by which liquid solution will go through solidification without the formation of ice crystals. Vitrification is depending on three factors which are cooling rate, viscosity and volume. Increasing the cooling rate and the viscosity of the solution and decreasing the volume will increase the probability for vitrification (Arav, 1992; Yavin and Arav, 2007). In 1992 we developed the minimum drop size (MDS) technique, the reduction of the sample volume which allow reducing the chance for crystallization and hence increasing the probability of vitrification (Arav, 1992; Arav and Zeron, 1997). The use of cryoprotectants (CPs) with high glass formation properties and high permeability also contribute to the progress in oocytes vitrification (Vajta et al., 1998; Kasai, 2002; Yavin and Arav, 2007). Finally, high cooling rate (and warming rate) is expected to decrease crystallization, which is kinetic process, and take time to grow. We developed a device, the VitMaster® (Figure 5) which cools the liquid nitrogen (LN) close to its freezing temperature (between -205 and -210°C). LN slush, defined as nitrogen in a liquid state close to its freezing point (-210°C), increase by 2 to 5 fold the cooling rate over LN (Yavin and Arav, 2007) leading to the decrease in both the probability of crystallization and chilling injury during vitrification. Specific papers that evaluated LN in comparison to LN slush include our reports on bovine oocytes (Arav and Zeron, 1997; Arav et al., 2000b), and more recently by Santos et al. (2006), on sheep oocytes (Isachenko et al., 2001) and recent publication on manipulated mouse embryos (Lee et al., 2007). These studies showed 20-50% survival improvement when using LN slush. The LN slush was also successfully applied to the vitrification of human oocytes (Yoon et al., 2007).

We believe that aspirating immature oocytes from collected ovaries of dead animal or by ovum pick up will be an optional for oocytes cryobanking by vitrification. These oocytes can later be warmed and fertilized by ICSI or IVF and transfered to recipients by embryo transfer technique. To that end, there are already reports on vitrification of wild animals oocytes (Pereira and Marques, 2008) and frozen-thawed embryo transfer has been reported in a wide variety of species including cow, eland, baboon, buffalo, cat, goat, horse, human, cynomolgus monkey, marmoset monkey, rhesus monkey, mouse, rabbit, red deer and sheep (Leibo and Songsasen, 2002). To enhance the propagation of endangered species with the help of more common species, several attempts to conduct interspecies or intergeneric embryo transfer were reported. In this technique, embryos from endangered species are gathered or oocytes are collected, matured and fertilized *in vitro* and then transferred into the uterus of a common species, where they are gestated. Using this technique, offspring of gaur (*Bos gaurus*) were born after transferring the embryos to domestic cow, bongo (*Tragelaphus euryceros*) offspring were following embryo transfer to eland (*Taurotragus oryx*) and offspring of both zebra (*Equus bruchelli*) and Przewlaski's horse (*Equus przewalskii*) were born following embryo transfer to domestic mares (Dresser et al., 1985; Summers et al., 1987; Wildt et al., 1997).

Following Is the Protocol That We Use for Oocytes Vitrification

Mature oocytes at the MII stage or embryos can be vitrified with the same protocol. Oocytes are exposed to 10% vitrification solution (VS) for 1 minute, transferred into 50% VS and immediately thereafter into a final 100% VS (containing 38% v/v ethylene glycol (EG), 0.5M Trehalose and 6% BSA in PBS). Oocytes are then loaded into super open pulled straw (SOPS) that is sealed, and vitrified at a rapid cooling rate (CR) of 15,500°C/minute, using the VitMaster® apparatus. Straws are then stored under LN until used. Warming of the oocytes is performed by plunging the SOPS into the warming chamber of the VitMaster, at 38°C. Oocytes are then immersed in 0.6M Trehalose solution for 2.5 minutes at 37°C and transferred though a series of solutions containing decreasing concentrations of trehalose: 0.5M, 0.4M, 0.3M, 0.2M and 0.1M for 2 minutes each.

Figure 5. Vitmaster® is a device which cools liquid nitrogen to temperature close to its freezing point (-210°C) by applying negative pressure.

WHOLE OVARY FREEZING

Since oocytes freezing is problematic as been discussed earlier, the solutions offered are preservation of embryos or, albeit still investigational, preservation of ovarian tissue. There are two options for fertility preservation involving ovarian tissue. The first is represented by cryopreservation of ovarian cortical slices and the second is the cryopreservation of the whole ovary. One of the problems associated with the first technique is the ischemic damage that leads to reduced follicular survival and thus shortened ovarian function upon transplantation. In fact, although in sheep autotransplantation of frozen-thawed ovarian cortex (Gosden et al., 1994) and of hemi-ovaries (Salle et al., 2003) has resulted in deliveries and hormone production (Baird et al., 1999; Salle et al., 2003), the duration of the function was transient due to ischemia which caused severe reduction of the total number of follicles (Liu et al., 2002). The experience in humans with ovarian cortex has also confirmed the short lifespan and thus the suboptimal quality of this approach (Kim et al., 2005). The results of the second option (freezing the whole ovary), albeit preliminary and gathered in one animal model, seem to confirm prolonged ovarian survival when the organ is cryopreserved as a whole and then utilized for retransplantation.

Before this report, our group (Arav and Elami, 2005) documented the longest ovarian function for up to 3 years after whole organ cryopreservation in sheep. Recently we extended that work by reporting, in the same animal model, the longest documented ovarian functional survival 6 years after whole organ cryopreservation, thawing and orthotopic transplantation (Arav et al., 2010).

The cryopreservation of whole organ was very difficult for several mass transfer and heat transfer problems. Following the introduction of directional freezing, we were able to overcome most of these problems. We will describe here in brief the multi-thermal gradient (MTG) device and the protocol used for ovary preparation and slow freezing. The device is built of four temperature domains within 270 mm copper blocks (Fig 2). The test tube is advanced at a constant velocity (V) through the predetermined temperature gradient (G). The value of G is based on $G= \Delta T/d$, where ΔT is temperature differences and d is the distance between temperatures. The result is a cooling rate (B) according to the equation $B=G*V$. The cooling rate set to 0.3°C/min by adjusting the speed at which the tube passes through the temperature gradient. Seeding is performed at the tip of the test tube and ice interface is propagated according to the freezing point of the solution. The University of Wisconsin solution (UW) (Madison, WI, USA), used in organ transplantation, supplemented with the cryoprotectant DMSO, is employed for vascular perfusion.

The ovarian artery can be perfused under a microscope with cold (4°C) UW supplemented with 10% DMSO for 3 minutes and then inserted into a freezing tube containing the same cryoprotectant. Slow freezing is performed as follows: slow cooling to -6°C when seeding is performed. Then directional freezing commence to -14°C or to -30°C at 0.03 mm/sec or 0.01mm/sec, respectively, resulting in a cooling rate of 0.3°C/min. When reaching the target temperature, the tubes are plunged into liquid nitrogen (LN). Thawing is performed by plunging the test tube into a 68°C water bath for 20 seconds and then into a 37°C water bath for 2 minutes. Careful temperature measurements are taken to avoid heating the ovaries above 20°C during thawing.

Tissue preservation could be applied to endangered species is several ways. Xenografting tissue (ovaries or testis) from endangered species can be done into nude mice, as was shown, for example, in the common wombat and elephant (Gunasena et al., 1998; Cleary et al., 2003). Another option is to transfer a whole ovary, together with the preparation for anti rejection, to a recipient female (Chen et al., 2006; Silber et al., 2007). Isolation of primordial follicles from the ovary can be done for use in *in vitro* culture and maturation (Eppig and O'Brien, 1996).

FREEZE DRYING TECHNIQUE

Lyophilisation has been used for the preservation of fowl spermatozoa already in the late forties (Polge et al., 1949). The original protocol was applied later to other species, including humans (Sherman, 1954), but the results in terms of offspring production were contradictory (Saacke and Almquist, 1961). The definitive proof that dry spermatozoa retain genetic integrity was established only when microsurgical procedures for bypassing the lack of mobility of freeze dried spermatozoa were developed, and normal mice were produced from the intracytoplasmic sperm injection (ICSI) of freeze-dried sperm (Wakayama and Yanagimachi, 1998). A following paper from the same group demonstrated the preservation of genomic integrity in freeze-dried spermatozoa (Kusakabe et al., 2001), and more recently these results have been demonstrated in other species (Keskintepe et al., 2002; Liu et al., 2004). The possibility to store male gametes in a dry state represents a major breakthrough for storing and shipping male gametes of laboratory, farm and wild animal species as well as humans, and be used through ICSI, which has become a routine procedure in assisted reproduction.

The progressive decline of large mammals worldwide (Hilton-Taylor, 2000; Margules and Pressey, 2000), the establishment of genetic banks from species threatened by extinction was suggested (Myers et al., 2000), with the aim to use the cells for somatic cell nuclear transfer (SCNT, Wilmut et al., 1997). SCNT has indeed an obvious potential for the multiplication of rare genotypes (Corley-Smith and Brandhorst, 1999; Loi et al., 2008a; Loi et al., 2008b), but its wide application is prevented by the low efficiency in terms of offspring outcomes. To date, successful cloning was reported in sheep (Campbell et al., 1996; Wilmut et al., 1997; Loi et al., 2008a; Loi et al., 2008b), cow (Cibelli et al., 1998), mice (Wakayama and Yanagimachi, 1998), goat (Baguisi et al., 1999), pigs (Polejaeva et al., 2000), cats (Shin et al., 2002), dogs (Jang et al., 2007), rabbits (Chesne et al., 2002), ferrets (Li et al., 2006), mule (Woods et al., 2003), horse (Galli et al., 2003), gaur (*Bos gaurus*) (Lanza et al., 2000), water buffalo (*Bubalus bubalis*) (Lu et al., 2005; Shi et al., 2007), European mouflon (*Ovis orientalis musimon*) (Loi et al., 2001), African wildcat (*Felis silvestris*) (Gómez et al., 2003), wolves (Kim et al., 2007), mountain bongo antelope (*Tragelphus euryceros*) (Lee et al., 2003) and African eland antelope (*Taurotragus oryx*) (Nel-Themaat et al., 2008). In addition to the low success rate, the majority of the endangered mammals are practically unknown from a reproductive biology point of view. Therefore, the storage of somatic cells for future use, once the procedure of SCNT will be reliable, is certainly a wise step to be undertaken.

However, the establishment of genetic banks in the form of cell lines encounters several problems, represented by the costs of liquid nitrogen. Recently, our group has demonstrated that somatic cells rendered unviable by heat treatment retained the potential to generate a normal lamb after nuclear transplantation (Loi et al., 2002). We also demonstrated the feasibility of using freeze-dried sheep somatic cells preserved at room temperature, and we tested their ability to direct early embryonic development of enucleated oocytes (Loi et al., 2008a; Loi et al., 2008b). The ensuing results demonstrate for the first time the production of normal embryos from nuclear transfer of somatic cells stored freeze dried at room temperature for more than three years.

REFERENCES

Adams, G.P., Ratto, M.H., Huanca, W., Singh, J., 2005. Ovulation-inducing factor in the seminal plasma of alpacas and llamas. Biology of Reproduction 73, 452-457.

Andrabi, S.M.H., Maxwell, W.M.C., 2007. A review on reproductive biotechnologies for conservation of endangered mammalian species. Animal Reproduction Science 99, 223-243.

Arav, A., 1992. Vitrification of oocytes and embryos, In: A., L., Gandolfi, F. (Eds.), New Trends in Embryo Transfer, Portland Press, Cambridge, UK, pp. 255-264.

Arav, A., 1999. Device and methods for multigradient directional cooling and warming of biological samples., Arav, Amir, US Patent.

Arav, A., 2001. Transillumination increases oocyte recovery from ovaries collected at slaughter. A new technique report. Theriogenology 55, 1561-1565.

Arav, A., Elami, A., 2005. Cryopreservation of intact ovaries--size is a variable? Reply of the Authors. Fertility and Sterility 83, 1588.

Arav, A., Gavish, Z., Elami, A., Natan, Y., Revel, A., Silber, S., Gosden, R.G., Patrizio, P., 2010. Ovarian function 6 yers after cryopreservation and transplantation of whole sheep ovaries. Reproductive Biomedicine Online In press.

Arav, A., Gavish, Z., Elami, A., Silber, S., Patrizio, P., 2007. Ovarian survival 6 years after whole organ cryopreservation and transplantation. Fertility and Sterility 88, S352 (abstract).

Arav, A., Pearl, M., Zeron, Y., 2000a. Does lipid profile explains chilling sensitivity and membrane lipid phase transition of spermatozoa and oocytes? CryoLetters 21, 179-186.

Arav, A., Revel, A., Nathan, Y., Bor, A., Gacitua, H., Yavin, S., Gavish, Z., Uri, M., Elami, A., 2005. Oocyte recovery, embryo development and ovarian function after cryopreservation and transplantation of whole sheep ovary. Hum. Reprod. 20, 3554-3559.

Arav, A., Yavin, S., Zeron, Y., Natan, D., Dekel, I., Gacitua, H., 2002a. New trends in gamete's cryopreservation. Mol Cell Endocrinol 187, 77-81.

Arav, A., Zeron, Y., 1997. Vitrification of bovine oocytes using modified minimum drop size technique (MDS) is effected by the composition and the concentration of the vitrification solution and by the cooling conditions. Theriogenology 47, 341 (abstract).

Arav, A., Zeron, Y., Leslie, S.B., Behboodi, E., Anderson, G.B., Crowe, J.H., 1996. Phase transition temperature and chilling sensitivity of bovine oocytes. Cryobiology 33, 589-599.

Arav, A., Zeron, Y., Ocheretny, A., 2000b. A new device and method for vitrification increases the cooling rate and allows successful cryopreservation of bovine oocytes. Theriogenology 53, 248 (abstract).

Arav, A., Zeron, Y., Shturman, H., Gacitua, H., 2002b. Successful pregnancies in cows following double freezing of a large volume of semen. Reprod Nutr Dev 42, 583-586.

Asher, G.W., Berg, D.K., Evans, G., 2000. Storage of semen and artificial insemination in deer. Animal Reproduction Science 62, 195-211.

Baguisi, A., Behboodi, E., Melican, D.T., Pollock, J.S., Destrempes, M.M., Cammuso, C., Williams, J.L., Nims, S.D., Porter, C.A., Midura, P., Palacios, M.J., Ayres, S.L., Denniston, R.S., Hayes, M.L., Ziomek, C.A., Meade, H.M., Godke, R.A., Gavin, W.G., Overstrom, E.W., Echelard, Y., 1999. Production of goats by somatic cell nuclear transfer. Nature Biotechnology 17, 456-461.

Baird, D.T., Webb, R., Campbell, B.K., Harkness, L.M., Gosden, R.G., 1999. Long-term ovarian function in sheep after ovariectomy and transplantation of autografts stored at -196 C. Endocrinology 140, 462-471.

Ball, B.A., Medina, V., Gravance, C.G., Baumber, J., 2001. Effect of antioxidants on preservation of motility, viability and acrosomal integrity of equine spermatozoa during storage at 5°C. Theriogenology 56, 577-589.

Ballou, J.D., 1992. Potential contribution of cryopreserved germ plasm to the preservation of genetic diversity and conservation of endangered species in captivity. Cryobiology 29, 19-25.

Bandularatne, E., Bongso, A., 2002. Evaluation of human sperm function after repeated freezing and thawing. Journal of Andrology 23, 242-249.

Barone, M.A., Roelke, M.E., Howard, J., Brown, J.L., Anderson, A.E., Wildt, D.E., 1994. Reproductive characteristics of male Florida panthers: Comparative studies from Florida, Texas, Colorado, Latin America, and North American zoos. Journal of Mammalogy 75, 150-162.

Bartels, P., Kotze, A., 2006. Wildlife biomaterial banking in Africa for now and the future. Journal of Environmental Monitoring 8, 779-781.

Baumber, J., Ball, B.A., Gravance, C.G., Medina, V., Davies-Morel, M.C., 2000. The effect of reactive oxygen species on equine sperm motility, viability, acrosomal integrity, mitochondrial membrane potential, and membrane lipid peroxidation. Journal of Andrology 21, 895-902.

Behr, B., Rath, D., Hildebrandt, T.B., Goeritz, F., Blottner, S., Portas, T.J., Bryant, B.R., Sieg, B., Knieriem, A., de Graaf, S.P., Maxwell, W.M.C., Hermes, R., 2009a. Germany/Australia index of sperm sex sortability in elephants and rhinoceros. Reprod Dom Anim 44, 273-277.

Behr, B., Rath, D., Mueller, P., Hildebrandt, T.B., Goeritz, F., Braun, B.C., Leahy, T., de Graaf, S.P., Maxwell, W.M.C., Hermes, R., 2009b. Feasibility of sex-sorting sperm from the white and the black rhinoceros (*Ceratotherium simum*, *Diceros bicornis*). Theriogenology 72, 353-364.

Blesbois, E., Lessire, M., Grasseau, I., Hallouis, J., Hermier, D., 1997. Effect of dietary fat on the fatty acid composition and fertilizing ability of fowl semen. Biology of Reproduction 56, 1216-1220.

Bravo, P.W., Skidmore, J.A., Zhao, X.X., 2000. Reproductive aspects and storage of semen in Camelidae. Animal Reproduction Science 62, 173-193.

Brinsko, S.P., Varner, D.D., Love, C.C., Blanchard, T.L., Day, B.C., Wilson, M.E., 2003. Effect of feeding a DHA-enriched nutriceutical on motion characteristics of cooled and frozen stallion semen, 49th Annual Convention of the American Association of Equine Practitioners, www.ivis.org Document No. P0653.1103, New Orleans, Louisiana.

Brinster, R.L., Zimmermann, J.W., 1994. Spermatogenesis following male germ-cell transplantation. Proceedings of the National Academy of Sciences of the United States of America 91, 11298-11302.

Buitink, J., Claessens, M.M., Hemminga, M.A., Hoekstra, F.A., 1998. Influence of water content and temperature on molecular mobility and intracellular glasses in seeds and pollen. Plant Physiol 118, 531-541.

Campbell, K.H., McWhir, J., Ritchie, W.A., Wilmut, I., 1996. Sheep cloned by nuclear transfer from a cultured cell line. Nature 380, 64-66.

Chen, C.-H., Chen, S.-G., Wu, G.-J., Wang, J., Yu, C.-P., Liu, J.-Y., 2006. Autologous heterotopic transplantation of intact rabbit ovary after frozen banking at -196°C. Fertility and Sterility 86, 1059-1066.

Chen, H., Cheung, M.P., Chow, P.H., Cheung, A.L.M., Liu, W., O, W.-S., 2002. Protection of sperm DNA against oxidative stress in vivo by accessory sex gland secretions in male hamsters. Reproduction 124, 491-499.

Chen, L.M., Hou, R., Zhang, Z.H., Wang, J.S., An, X.R., Chen, Y.F., Zheng, H.P., Xia, G.L., Zhang, M.J., 2007. Electroejaculation and semen characteristics of Asiatic Black bears (*Ursus thibetanus*). Animal Reproduction Science 101, 358-364.

Chesne, P., Adenot, P.G., Viglietta, C., Baratte, M., Boulanger, L., Renard, J.P., 2002. Cloned rabbits produced by nuclear transfer from adult somatic cells. Nature Biotechnology 20, 366-369.

Cibelli, J.B., Stice, S.L., Golueke, P.J., Kane, J.J., Jerry, J., Blackwell, C., Ponce de Leon, F.A., Robl, J.M., 1998. Cloned transgenic calves produced from nonquiescent fetal fibroblasts. Science 280, 1256-1258.

Clarke, G.N., Liu, D.Y., Baker, H.W.G., 2003. Improved sperm cryopreservation using cold cryoprotectant. Reproduction, Fertility and Development 15, 377-381.

Cleary, M., Paris, M.C., Shaw, J., Jenkin, G., Trounson, A., 2003. Effect of ovariectomy and graft position on cryopreserved common wombat (*Vombatus ursinus*) ovarian tissue following xenografting to nude mice. Reproduction Fertility and Development 15, 333-342.

Comhaire, F.H., Christophe, A.B., Zalata, A.A., Dhooge, W.S., Mahmoud, A.M.A., Depuydt, C.E., 2000. The effects of combined conventional treatment, oral antioxidants and essential fatty acids on sperm biology in subfertile men. Prostaglandins, Leukotrienes and Essential Fatty Acids 63, 159-165.

Comizzoli, P., Mermillod, P., Mauget, R., 2000. Reproductive biotechnologies for endangered mammalian species. Reproduction Nutrition Development 40, 493-504.

Corley-Smith, G.E., Brandhorst, B.P., 1999. Preservation of endangered species and populations: a role for genome banking, somatic cell cloning, and androgenesis? Mol Reprod Dev 53, 363-367.

Crowe, J.H., Crowe, L.M., Chapman, D., 1984. Preservation of membranes in anhydrobiotic organisms: The role of trehalose. Science 223, 701-703.

Culver, J.N., 2001. Evaluation of tom fertility as affected by dietary fatty acid composition, Animal and Poultry Sciences, Virginia Polytechnic Institute and State University, Blacksburg, p. 60.

de Graaf, S.P., Evans, G., Maxwell, W.M.C., Cran, D.G., O'Brien, J.K., 2007. Birth of offspring of pre-determined sex after artificial insemination of frozen-thawed, sex-sorted and re-frozen-thawed ram spermatozoa. Theriogenology 67, 391-398.

Dresser, B.L., Pope, C.E., Kramer, L., Kuehn, G., Dahlhausen, R.D., Maruska, E.J., Reece, B., Thomas, W.D., 1985. Birth of bongo antelope (*Tragelaphus euryceros*) to eland antelope (*Tragelaphus oryx*) and cryopreservation of bongo embryos. Theriogenology 23, 190.

Drobnis, E.Z., Crowe, L.M., Berger, T., Anchordoguy, T.J., Overstreet, J.W., Crowe, J.H., 1993. Cold shock damage is due to lipid phase transitions in cell membranes: A demonstration using sperm as a model. Journal of Experimental Zoology 265, 432-437.

Duraiappah, A.K., Naeem, S., Agardy, T., Ash, N.J., Cooper, H.D., Díaz, S., Faith, D.P., Mace, G., McNeely, J.A., Mooney, H.A., Oteng-Yeboah, A.A., Miguel Pereira, H., Polasky, S., Prip, C., Reid, W.V., Samper, C., Johan Schei, P., Scholes, R., Schutyser, F., van Jaarsveld, A., 2005. Ecosystems and Human Well-being: Biodiversity Synthesis., In: Sarukhán, J., Whyte, A. (Eds.), Millennium Ecosystem Assessment, World Resources Institute, Washington, DC, pp. VI, 86.

Eppig, J.J., O'Brien, M.J., 1996. Development in vitro of mouse oocytes from primordial follicles. Biology of Reproduction 54, 197-207.

Fiser, P.S., Fairfull, R.W., 1986. The effects of rapid cooling (cold shock) of ram semen, photoperiod, and egg yolk in diluents on the survival of spermatozoa before and after freezing. Cryobiology 23, 518-524.

Gacitua, H., Arav, A., 2005. Successful pregnancies with directional freezing of large volume buck semen. Theriogenology 63, 931-938.

Gacitua, H., Zeron, Y., Arav, A., 2002. Dietary supplementation with polyunsaturated fatty acid modifies sperm quality and fatty acid composition in the bull. Theriogenology 57, 376 (abstract).

Galli, C., Lagutina, I., Crotti, G., Colleoni, S., Turini, P., Ponderato, N., Duchi, R., Lazzari, G., 2003. A cloned horse born to its dam twin. Nature 424, 635.

Garner, D.L., 2006. Flow cytometric sexing of mammalian sperm. Theriogenology 65, 943-957.

Gastal, M.O., Henry, M., Beker, A.R., Gastal, E.L., Goncalves, A., 1996. Sexual behavior of donkey jacks: Influence of ejaculatory frequency and season. Theriogenology 46, 593-603.

Ghetler, Y., Yavin, S., Shalgi, R., Arav, A., 2005. The effect of chilling on membrane lipid phase transition in human oocytes and zygotes. Human Reproduction 20, 3385-3389.

Gianaroli, L., Magli, M.C., Selman, H.A., Colpi, G., Belgrano, E., Trombetta, C., Vitali, G., Ferraretti, A.P., 1999. Diagnostic testicular biopsy and cryopreservation of testicular tissue as an alternative to repeated surgical openings in the treatment of azoospermic men. Human Reproduction 14, 1034-1038.

Gillan, L., Maxwell, W.M.C., Evans, G., 2004. Preservation and evaluation of semen for artificial insemination. Reproduction, Fertility and Development 16, 447-454.

Gomendio, M., Cassinello, J., Roldan, E.R.S., 2000. A comparative study of ejaculate traits in three endangered ungulates with different levels of inbreeding: fluctuating asymmetry as an indicator of reproductive and genetic stress. Proceedings of the Royal Society of London. Series B: Biological Sciences 267, 875-882.

Gómez, M.C., Jenkins, J.A., Giraldo, A., Harris, R.F., King, A., Dresser, B.L., Pope, C.E., 2003. Nuclear transfer of synchronized African wild cat somatic cells into enucleated domestic cat oocytes. Biology of Reproduction 69, 1032-1041.

Göritz, F., Neubauer, K., Naidenko, S.V., Fickel, J., Jewgenow, K., 2006. Investigations on reproductive physiology in the male Eurasian lynx (*Lynx lynx*). Theriogenology 66, 1751-1754.

Gosden, R.G., Baird, D.T., Wade, J.C., Webb, R., 1994. Restoration of fertility to oophorectomized sheep by ovarian autografts stored at -196 degrees C. Human Reproduction 9, 597-603.

Goto, K., Kinoshita, A., Takuma, Y., Ogawa, K., 1991. Birth of calves after the transfers of oocytes fertilized by sperm injection. Theriogenology 35, 205 (abstract).

Gunasena, K.T., Lakey, J.R.T., Villines, P.M., Bush, M., Raath, C., Critser, E.S., McGann, L.E., Critser, J.K., 1998. Antral follicles develop in xenografted cryopreserved african elephant (*Loxodonta africana*) ovarian tissue. Animal Reproduction Science 53, 265-275.

Hammerstedt, R.H., Graham, J.K., Nolan, J.P., 1990. Cryopreservation of mammalian sperm: what we ask them to survive. Journal of Andrology 11, 73-88.

Hermes, R., Arav, A., Saragusty, J., Göritz, F., Pettit, M., Blottner, S., Flach, E., Eshkar, G., Boardman, W., Hildebrandt, T.B., 2003. Cryopreservation of Asian elephant spermatozoa using directional freezing, Annual meeting of the American Association of Zoo Veterinarians, Minneapolis, MN, USA, p. 264 (abstract).

Hermes, R., Behr, B., Hildebrandt, T.B., Blottner, S., Sieg, B., Frenzel, A., Knieriem, A., Saragusty, J., Rath, D., 2009. Sperm sex-sorting in the Asian elephant (*Elephas maximus*). Anim Reprod Sci 112, 390-396.

Hildebrandt, T.B., Hermes, R., Jewgenow, K., Göritz, F., 2000a. Ultrasonography as an important tool for the development and application of reproductive technologies in non-domestic species. Theriogenology 53, 73-84.

Hildebrandt, T.B., Hermes, R., Pratt, N.C., Fritsch, G., Blottner, S., Schmitt, D.L., Ratanakorn, P., Brown, J.L., Rietschel, W., Göritz, F., 2000b. Ultrasonography of the urogenital tract in elephants (*Loxodonta africana* and *Elephas maximus*): an important tool for assessing male reproductive function. Zoo Biology 19, 333-345.

Hilton-Taylor, C., 2000. IUCN Red List of Threatened Species, IUCN species survival commission, Gland, Switzerland.

Hirabayashi, M., Kato, M., Ito, J., Hochi, S., 2005. Viable rat offspring derived from oocytes intracytoplasmically injected with freeze-dried sperm heads. Zygote 13, 79-85.

Hochi, S., Watanabe, K., Kato, M., Hirabayashi, M., 2008. Live rats resulting from injection of oocytes with spermatozoa freeze-dried and stored for one year. Molecular Reproduction and Development 75, 890-894.

Holt, W.V., Bennett, P.M., Volobouev, V., Watwon, P.F., 1996. Genetic resource banks in wildlife conservation. Journal of Zoology 238, 531-544.

Holt, W.V., Pickard, A.R., 1999. Role of reproductive technology and genetic resource banks in animal conservation. Reviews of Reproduction 4, 143-150.

Honaramooz, A., Megee, S.O., Dobrinski, I., 2002. Germ cell transplantation in pigs. Biology of Reproduction 66, 21-28.

Hovatta, O., Foudila, T., Siegberg, R., Johansson, K., von Smitten, K., Reima, I., 1996. Pregnancy resulting from intracytoplasmic injection of spermatozoa from a frozen-thawed testicular biopsy specimen. Human Reproduction 11, 2472-2473.

Isachenko, E., Isachenko, V., Katkov, I.I., Dessole, S., Nawroth, F., 2003. Vitrification of mammalian spermatozoa in the absence of cryoprotectants: from past practical difficulties to present success. Reproductive BioMedicine Online 6, 191-200.

Isachenko, E., Isachenko, V., Weiss, J., Kreienberg, R., Katkov, I., Schulz, M., Lulat, A., Risopatron, J., Sanchez, R., 2008. Acrosomal status and mitochondrial activity of human spermatozoa vitrified with sucrose. Reproduction 136, 167-173.

Isachenko, V., Alabart, J.L., Nawroth, F., Isachenko, E., Vajta, G., Folch, J., 2001. The open pulled straw vitrification of ovine GV-oocytes: positive effect of rapid cooling or rapid thawing or both? Cryo Letters 22, 157-162.

Isachenko, V., Isachenko, E., Katkov, I.I., Montag, M., Dessole, S., Nawroth, F., van der Ven, H., 2004. Cryoprotectant-free cryopreservation of human spermatozoa by vitrification and freezing in vapor: Effect on motility, DNA integrity, and fertilization ability. Biology of Reproduction 71, 1167-1173.

IUCN, 1987. IUCN Policy Statement on Captive Breeding, Gland, Switzerland.

IUCN, 2004. IUCN Red List of Threatened Species, International Union for the Conservation of Nature, Gland, Switzerland.

Izadyar, F., Den Ouden, K., Stout, T.A., Stout, J., Coret, J., Lankveld, D.P., Spoormakers, T.J., Colenbrander, B., Oldenbroek, J.K., Van der Ploeg, K.D., Woelders, H., Kal, H.B., De Rooij, D.G., 2003. Autologous and homologous transplantation of bovine spermatogonial stem cells. Reproduction 126, 765-774.

Jang, G., Kim, M.K., Oh, H.J., Hossein, M.S., Fibrianto, Y.H., Hong, S.G., Park, J.E., Kim, J.J., Kim, H.J., Kang, S.K., Kim, D.Y., Lee, B.C., 2007. Birth of viable female dogs produced by somatic cell nuclear transfer. Theriogenology 67, 941-947.

Jewgenow, K., Blottner, S., Lengwinat, T., Meyer, H.H., 1997. New methods for gamete rescue from gonads of nondomestic felids. Journal of Reproduction and Fertility, supplement 51, 33-39.

Johnson, L.A., 1988. Flow cytometric determination of sperm sex ratio in semen purportedly enriched for X- or Y-bearing sperm. Theriogenology 29, 265.

Johnson, L.A., Flook, J.P., Look, M.V., Pinkel, D., 1987. Flow sorting of X and Y chromosome-bearing spermatozoa into two populations. Gamete Research 16, 1-9.

Johnston, L.A., Lacy, R.C., 1995. Genome resource banking for species conservation: Selection of sperm donors. Cryobiology 32, 68-77.

Johnston, S.D., McGowan, M.R., Carrick, F.N., Tribe, A., 1993. Preliminary investigations into the feasibility of freezing koala (*Phascolarctos cinereus*) semen. Australian Veterinary Journal 70, 424-425.

Johnston, S.D., O'Callaghan, P., Nilsson, K., Tzipori, G., Curlewis, J.D., 2004. Semen-induced luteal phase and identification of a LH surge in the koala (*Phascolarctos cinereus*). Reproduction 128, 629-634.

Kaneko, T., Nakagata, N., 2005. Relation between storage temperature and fertilizing ability of freeze-dried mouse spermatozoa. Comparative Medicine 55, 140-144.

Kaneko, T., Whittingham, D.G., Yanagimachi, R., 2003. Effect of pH value of freeze-drying solution on the chromosome integrity and developmental ability of mouse spermatozoa. Biology of Reproduction 68, 136-139.

Kasai, M., 2002. Advances in the cryopreservation of mammalian oocytes and embryos: Development of ultrarapid vitrification. Reproductive Medicine and Biology 1, 1-9.

Katayose, H., Matsuda, J., Yanagimachi, R., 1992. The ability of dehydrated hamster and human sperm nuclei to develop into pronuclei. Biology of Reproduction 47, 277-284.

Kawase, Y., Araya, H., Kamada, N., Jishage, K., Suzuki, H., 2005. Possibility of long-term preservation of freeze-dried mouse spermatozoa. Biology of Reproduction 72, 568-573.

Keskintepe, L., Pacholczyk, G., Machnicka, A., Norris, K., Curuk, M.A., Khan, I., Brackett, B.G., 2002. Bovine blastocyst development from oocytes injected with freeze-dried spermatozoa. Biology of Reproduction 67, 409-415.

Kim, M.K., Jang, G., Oh, H.J., Yuda, F., Kim, H.J., Hwang, W.S., Hossein, M.S., Kim, J.J., Shin, N.S., Kang, S.K., Lee, B.C., 2007. Endangered wolves cloned from adult somatic cells. Cloning and Stem Cells 9, 130-137.

Kim, S.S., Kang, H.G., Kim, N.H., Lee, H.C., Lee, H.H., 2005. Assessment of the integrity of human oocytes retrieved from cryopreserved ovarian tissue after xenotransplantation. Human Reproduction 20, 2502-2508.

Koh, L.P., Dunn, R.R., Sodhi, N.S., Colwell, R.K., Proctor, H.C., Smith, V.S., 2004. Species coextinctions and the biodiversity crisis. Science 305, 1632-1634.

Kretzschmar, P., Ganslosser, U., Dehnhard, M., 2004. Relationship between androgens, environmental factors and reproductive behavior in male white rhinoceros (*Ceratotherium simum simum*). Hormones and Behavior 45, 1-9.

Kusakabe, H., Szczygiel, M.A., Whittingham, D.G., Yanagimachi, R., 2001. Maintenance of genetic integrity in frozen and freeze-dried mouse spermatozoa. Proceedings of the National Academy of Sciences of the United States of America 98, 13501-13506.

Kwon, I.K., Park, K.E., Niwa, K., 2004. Activation, pronuclear formation, and development in vitro of pig oocytes following intracytoplasmic injection of freeze-dried spermatozoa. Biology of Reproduction 71, 1430-1436.

Langlais, J., Roberts, K.D., 1985. A molecular membrane model of sperm capacitation and the acrosome reaction of mammalian spermatozoa. Gamete Research 12, 183-224.

Lanza, R.P., Cibelli, J.B., Diaz, F., Moraes, C.T., Farin, P.W., Farin, C.E., Hammer, C.J., West, M.D., Damiani, P., 2000. Cloning of an endangered species (*Bos gaurus*) using interspecies nuclear transfer. Cloning 2, 79-90.

Larson, E.V., Graham, E.F., 1976. Freeze-drying of spermatozoa. Developments in Biological Standartization 36, 343-348.

Lee, B., Wirtu, G.G., Damiani, P., Pope, E., Dresser, B.L., Hwang, W., Bavister, B.D., 2003. Blastocyst development after intergeneric nuclear transfer of mountain bongo antelope somatic cells into bovine oocytes. Cloning and Stem Cells 5, 25-33.

Lee, D.R., Yang, Y.H., Eum, J.H., Seo, J.S., Ko, J.J., Chung, H.M., Yoon, T.K., 2007. Effect of using slush nitrogen (SN2) on development of microsurgically manipulated vitrified/warmed mouse embryos. Human Reproduction 22, 2509-2514.

Leibo, S.P., Songsasen, N., 2002. Cryopreservation of gametes and embryos of non-domestic species. Theriogenology 57, 303-326.

Leslie, S.B., Teter, S.A., Crowe, L.M., Crowe, J.H., 1994. Trehalose lowers membrane phase transitions in dry yeast cells. Biochimica et Biophysica Acta (BBA) - Biomembranes 1192, 7-13.

Li, Z., Sun, X., Chen, J., Liu, X., Wisely, S.M., Zhou, Q., Renard, J.P., Leno, G.H., Engelhardt, J.F., 2006. Cloned ferrets produced by somatic cell nuclear transfer. Developmental Biology 293, 439-448.

Liu, J., Van der Elst, J., Van den Broecke, R., Dhont, M., 2002. Early massive follicle loss and apoptosis in heterotopically grafted newborn mouse ovaries. Human Reproduction 17, 605-611.

Liu, J.L., Kusakabe, H., Chang, C.C., Suzuki, H., Schmidt, D.W., Julian, M., Pfeffer, R., Bormann, C.L., Tian, X.C., Yanagimachi, R., Yang, X., 2004. Freeze-dried sperm fertilization leads to full-term development in rabbits. Biology of Reproduction 70, 1776-1781.

Loi, P., Clinton, M., Barboni, B., Fulka, J., Jr., Cappai, P., Feil, R., Moor, R.M., Ptak, G., 2002. Nuclei of nonviable ovine somatic cells develop into lambs after nuclear transplantation. Biology of Reproduction 67, 126-132.

Loi, P., Matsukawa, K., Ptak, G., Clinton, M., Fulka Jr., J., Natan, Y., Arav, A., 2008a. Freeze-dried somatic cells direct embryonic development after nuclear transfer. PLoS ONE 3, e2978.

Loi, P., Matsukawa, K., Ptak, G., Nathan, Y., Fulka, J., Jr., Arav, A., 2008b. Nuclear transfer of freeze dried somatic cells into enucleated sheep oocytes. Reprod Dom Anim 43, 417-422.

Loi, P., Ptak, G., Barboni, B., Fulka, J., Jr., Cappai, P., Clinton, M., 2001. Genetic rescue of an endangered mammal by cross-species nuclear transfer using post-mortem somatic cells. Nature Biotechnology 19, 962-964.

Lu, F., Shi, D., Wei, J., Yang, S., Wei, Y., 2005. Development of embryos reconstructed by interspecies nuclear transfer of adult fibroblasts between buffalo (*Bubalus bubalis*) and cattle (*Bos indicus*). Theriogenology 64, 1309-1319.

Luyet, B.J., Hoddap, A., 1938. Revival of frog's sprmatozoa vitrified in liquid air, Proceedings of the Meeting of Society for Experimental Biology, pp. 433-434.

Margules, C.R., Pressey, R.L., 2000. Systematic conservation planning. Nature 405, 243-253.

Martins, C.F., Bao, S.N., Dode, M.N., Correa, G.A., Rumpf, R., 2007. Effects of freeze-drying on cytology, ultrastructure, DNA fragmentation, and fertilizing ability of bovine sperm. Theriogenology 67, 1307-1315.

Mazur, P., Cole, K.W., Hall, J.W., Schreuders, P.D., Mahowald, A.P., 1992. Cryobiological preservation of Drosophila embryos. Science 258, 1932-1935.

McDonnell, S.M., 2001. Oral imipramine and intravenous xylazine for pharmacologically-induced ex copula ejaculation in stallions. Animal Reproduction Science 68, 153-159.

Melville, D.F., Crichton, E.G., Paterson-Wimberley, T., Johnston, S.D., 2008. Collection of semen by manual stimulation and ejaculate characteristics of the black flying-fox (*Pteropus alecto*). Zoo Biology 27, 159-164.

Meyers, S.A., 2006. Dry storage of sperm: applications in primates and domestic animals. Reproduction, Fertility and Development 18, 1-5.

Mitre, R., Cheminade, C., Allaume, P., Legrand, P., Legrand, A.B., 2004. Oral intake of shark liver oil modifies lipid composition and improves motility and velocity of boar sperm. Theriogenology 62, 1557-1566.

Moehlman, P.D., 2002. Status and action plan for the African Wild Ass (*Equus africanus*), In: Moehlman, P.D. (Ed.), Equids: Zebras, Asses and Horses. Status survey and conservation plan, IUCN/SSC Equid Specialist Group, Gland, pp. 2-10.

Moisan, A.E., Leibo, S.P., Lynn, J.W., Gómez, M.C., Pope, C.E., Dresser, B.L., Godke, R.A., 2005. Embryonic development of felid oocytes injected with freeze-dried or air-dried spermatozoa. Cryobiology 51, 373 (abstract).

Moore, A.I., Squires, E.L., Graham, J.K., 2005. Adding cholesterol to the stallion sperm plasma membrane improves cryosurvival. Cryobiology 51, 241-249.

Morais, R.N., Mucciolo, R.G., Gomes, M.L.F., Lacerda, O., Moraes, W., Moreira, N., Graham, L.H., Swanson, W.F., Brown, J.L., 2002. Seasonal analysis of semen characteristics, serum testosterone and fecal androgens in the ocelot (*Leopardus pardalis*), margay (*L. wiedii*) and tigrina (*L. tigrinus*). Theriogenology 57, 2027-2041.

Morato, R.G., Conforti, V.A., Azevedo, F.C., Jacomo, A.T., Silveira, L., Sana, D., Nunes, A.L., Guimaraes, M.A., Barnabe, R.C., 2001. Comparative analyses of semen and endocrine characteristics of free-living versus captive jaguars (*Panthera onca*). Reproduction 122, 745-751.

Morton, K.M., Ruckholdt, M., Evans, G., Maxwell, W.M.C., 2008. Quantification of the DNA difference, and separation of X- and Y-bearing sperm in alpacas (*Vicugna pacos*). Reprod Dom Anim 43, 638-642.

Myers, N., Mittermeier, R.A., Mittermeier, C.G., da Fonseca, G.A.B., Kent, J., 2000. Biodiversity hotspots for conservation priorities. Nature 403, 853-858.

Nawroth, F., Isachenko, V., Dessole, S., Rahimi, G., Farina, M., Vargiu, N., Mallmann, P., Dattena, M., Capobianco, G., Peters, D., Orth, I., Isachenko, E., 2002. Vitrification of human spermatozoa without cryoprotectants. Cryo Letters 23, 93-102.

Nel-Themaat, L., Gomez, M.C., Pope, C.E., Lopez, M., Wirtu, G., Jenkins, J.A., Cole, A., Dresser, B.L., Bondioli, K.R., Godke, R.A., 2008. Cloned embryos from semen. Part 2: Intergeneric nuclear transfer of semen-derived eland (*Taurotragus oryx*) epithelial cells into bovine oocytes. Cloning and Stem Cells 10, 161-172.

O'Brien, J.K., Hollinshead, F.K., Evans, K.M., Evans, G., Maxwell, W.M., 2004. Flow cytometric sorting of frozen-thawed spermatozoa in sheep and non-human primates. Reproduction, Fertility and Development 15, 367-375.

O'Brien, J.K., Robeck, T.R., 2006. Development of sperm sexing and associated assisted reproductive technology for sex preselection of captive bottlenose dolphins (*Tursiops truncatus*). Reprod Fertil Dev 18, 319-329.

O'Brien, J.K., Stojanov, T., Crichton, E.G., Evans, K.M., Leigh, D., Maxwell, W.M., Evans, G., Loskutoff, N.M., 2005a. Flow cytometric sorting of fresh and frozen-thawed spermatozoa in the western lowland gorilla (*Gorilla gorilla gorilla*). American Journal of Primatology 66, 297-315.

O'Brien, J.K., Stojanov, T., Heffernan, S.J., Hollinshead, F.K., Vogelnest, L., Chis Maxwell, W.M., Evans, G., 2005b. Flow cytometric sorting of non-human primate sperm nuclei. Theriogenology 63, 246-259.

Ogonuki, N., Mochida, K., Miki, H., Inoue, K., Fray, M., Iwaki, T., Moriwaki, K., Obata, Y., Morozumi, K., Yanagimachi, R., Ogura, A., 2006. Spermatozoa and spermatids retrieved from frozen reproductive organs or frozen whole bodies of male mice can produce normal offspring. Proceedings of the National Academy of Sciences of the United States of America 103, 13098-13103.

Pan, G., Chen, Z., Liu, X., Li, D., Xie, Q., Ling, F., Fang, L., 2001. Isolation and purification of the ovulation-inducing factor from seminal plasma in the bactrian camel (*Camelus bactrianus*). Theriogenology 55, 1863-1879.

Papahadjopoulos, D., Poste, G., Schaeffer, B.E., 1973. Fusion of mammalian cells by unilamellar lipid vesicles: Influence of lipid surface charge, fluidity and cholesterol. Biochimica et Biophysica Acta (BBA) - Biomembranes 323, 23-42.

Pereira, R., Marques, C., 2008. Animal oocyte and embryo cryopreservation. Cell Tissue Bank. 9, 267-277.

Polcz, T.E., Stronk, J., Xiong, C., Jones, E.E., Olive, D.L., Huszar, G., 1998. Optimal utilization of cryopreserved human semen for assisted reproduction: recovery and maintenance of sperm motility and viability. Journal of Assisted Reproduction and Genetics 15, 504-512.

Polejaeva, I.A., Chen, S.H., Vaught, T.D., Page, R.L., Mullins, J., Ball, S., Dai, Y., Boone, J., Walker, S., Ayares, D.L., Colman, A., Campbell, K.H., 2000. Cloned pigs produced by nuclear transfer from adult somatic cells. Nature 407, 86-90.

Poleo, G.A., Godke, R.R., Tiersch, T.R., 2005. Intracytoplasmic sperm injection using cryopreserved, fixed, and freeze-dried sperm in eggs of Nile tilapia. Marine Biotechnology (New York, NY) 7, 104-111.

Polge, C., Smith, A.U., Parkes, A.S., 1949. Revival of spermatozoa after vitrification and dehydration at low temperatures. Nature 164, 666.

Pukazhenthi, B., Laroe, D., Crosier, A., Bush, L.M., Spindler, R., Pelican, K.M., Bush, M., Howard, J.G., Wildt, D.E., 2006. Challenges in cryopreservation of clouded leopard (*Neofelis nebulosa*) spermatozoa. Theriogenology 66, 1790-1796.

Pukazhenthi, B.S., Wildt, D.E., 2004. Which reproductive technologies are most relevant to studying, managing and conserving wildlife? Reproduction, Fertility and Development 16, 14.

Purdy, P.H., Graham, J.K., 2004. Effect of cholesterol-loaded cyclodextrin on the cryosurvival of bull sperm. Cryobiology 48, 36-45.

Reid, C.E., Hermes, R., Blottner, S., Goeritz, F., Wibbelt, G., Walzer, C., Bryant, B.R., Portas, T.J., Streich, W.J., Hildebrandt, T.B., 2009. Split-sample comparison of directional and liquid nitrogen vapour freezing method on post-thaw semen quality in white rhinoceroses (*Ceratotherium simum simum* and *Ceratotherium simum cottoni*). Theriogenology 71, 275-291.

Riel, J.M., Huang, T.T., Ward, M.A., 2007. Freezing-free preservation of human spermatozoa--a pilot study. Archives of Andrology 53, 275-284.

Rodriguez-Gil, J.E., 2006. Mammalian sperm energy resources management and survival during conservation in refrigeration. Reproduction in Domestic Animals 41, 11-20.

Roelke, M.E., Martenson, J.S., O'Brien, S.J., 1993. The consequences of demographic reduction and genetic depletion in the endangered Florida panther. Current Biology 3, 340-350.

Roelke-Parker, M.E., Munson, L., Packer, C., Kock, R., Cleaveland, S., Carpenter, M., O'Brien, S.J., Pospischil, A., Hofmann-Lehmann, R., Lutz, H., et al., 1996. A canine distemper virus epidemic in Serengeti lions (*Panthera leo*). Nature 379, 441-445.

Rofeim, O., Brown, T.A., Gilbert, B.R., 2001. Effects of serial thaw-refreeze cycles on human sperm motility and viability. Fertility and Sterility 75, 1242-1243.

Rooke, J.A., Shao, C.C., Speake, B.K., 2001. Effects of feeding tuna oil on the lipid composition of pig spermatozoa and in vitro characteristics of semen. Reproduction 121, 315-322.

Roth, T.L., Stoops, M.A., Atkinson, M.W., Blumer, E.S., Campbell, M.K., Cameron, K.N., Citino, S.B., Maas, A.K., 2005. Semen collection in rhinoceroses (*Rhinoceros unicornis, Diceros bicornis, Ceratotherium simum*) by electroejaculation with a uniquely designed probe. Journal of Zoo and Wildlife Medicine 36, 617-627.

Roth, T.L., Swanson, W.F., Collins, D., Burton, M., Garell, D.M., Wildt, D.E., 1996. Snow leopard (*Panthera uncia*) spermatozoa are sensitive to alkaline pH, but motility in vitro is not influenced by protein or energy supplements. Journal of Andrology 17, 558-566.

Saacke, R.G., Almquist, J.O., 1961. Freeze-drying of Bovine Spermatozoa. Nature 192, 995-996.

Salle, B., Demirci, B., Franck, M., Berthollet, C., Lornage, J., 2003. Long-term follow-up of cryopreserved hemi-ovary autografts in ewes: pregnancies, births, and histologic assessment. Fertility and Sterility 80, 172-177.

Sanchez-Partida, L.G., Simerly, C.R., Ramalho-Santos, J., 2008. Freeze-dried primate sperm retains early reproductive potential after intracytoplasmic sperm injection. Fertility and Sterility 89, 742-745.

Santiago-Moreno, J., Toledano-Diaz, A., Pulido-Pastor, A., Gomez-Brunet, A., Lopez-Sebastian, A., 2006. Birth of live Spanish ibex (*Capra pyrenaica hispanica*) derived from artificial insemination with epididymal spermatozoa retrieved after death. Theriogenology 66, 283-291.

Santos, R.M.d., Barreta, M.H., Frajblat, M., al., e., 2006. Vacuum-cooled liquid nitrogen increases the developmental ability of vitrified-warmed bovine oocytes. Ciencia Rural 36, 1501-1506.

Saragusty, J., Gacitua, H., King, R., Arav, A., 2006. Post-mortem semen cryopreservation and characterization in two different endangered gazelle species (*Gazella gazella and Gazella dorcas*) and one subspecies (*Gazella gazelle acaiae*). Theriogenology 66, 775-784.

Saragusty, J., Gacitua, H., Pettit, M.T., Arav, A., 2007. Directional freezing of equine semen in large volumes. Reproduction in Domestic Animals 42, 610-615.

Saragusty, J., Gacitua, H., Rozenboim, I., Arav, A., 2009a. Do physical forces contribute to cryodamage? Biotech Bioeng 104, 719-728.

Saragusty, J., Gacitua, H., Zeron, Y., Rozenboim, I., Arav, A., 2009b. Double freezing of bovine semen. Anim Reprod Sci 115, 10-17.

Saragusty, J., Hildebrandt, T.B., Behr, B., Knieriem, A., Kruse, J., Hermes, R., 2009c. Successful cryopreservation of Asian elephant (*Elephas maximus*) spermatozoa. Anim Reprod Sci 115, 255-266.

Saragusty, J., Hildebrandt, T.B., Natan, Y., Hermes, R., Yavin, S., Göritz, F., Arav, A., 2005. Effect of egg-phosphatidylcholine on the chilling sensitivity and lipid phase transition of Asian elephant (*Elephas maximus*) spermatozoa. Zoo Biology 24, 233-245.

Schenk, J.L., DeGrofft, D.L., 2003. Insemination of cow elk with sexed frozen semen. Theriogenology 59, 514 (abstract).

Schlatt, S., Kim, S.S., Gosden, R., 2002. Spermatogenesis and steroidogenesis in mouse, hamster and monkey testicular tissue after cryopreservation and heterotopic grafting to castrated hosts. Reproduction 124, 339-346.

Schmitt, D.L., Hildebrandt, T.B., 1998. Manual collection and characterization of semen from Asian elephants (*Elephas maximus*). Animal Reproduction Science 53, 309-314.

Schmitt, D.L., Hildebrandt, T.B., 2000. Corrigendum to "Manual collection and characterization of semen from asian elephants" [Anim. Reprod. Sci. 53 (1998) 309-314]. Animal Reproduction Science 59, 119.

Schneiders, A., Sonksen, J., Hodges, J.K., 2004. Penile vibratory stimulation in the marmoset monkey: a practical alternative to electro-ejaculation, yielding ejaculates of enhanced quality. Journal of Medical Primatology 33, 98-104.

Schwarzenberger, F., Mostl, F., Palme, R., Bamberg, E., 1996. Fecal steroid analysis for non-invasive monitoring of reproductive status in farm, wild and zoo animals. Animal Reproduction Science 42, 515-526.

Shefi, S., Raviv, G., Eisenberg, M.L., Weissenberg, R., Jalalian, L., Levron, J., Band, G., Turek, P.J., Madgar, I., 2006. Posthumous sperm retrieval: analysis of time interval to harvest sperm. Human Reproduction 21, 2890-2893.

Sherman, J.K., 1954. Freezing and freeze-drying of human spermatozoa. Fertility and Sterility 5, 357-371.

Sherman, J.K., 1957. Freezing and freeze-drying of bull spermatozoa. American Journal of Physiology 190, 281-286.

Shi, D., Lu, F., Wei, Y., Cui, K., Yang, S., Wei, J., Liu, Q., 2007. Buffalos (*Bubalus bubalis*) cloned by nuclear transfer of somatic cells. Biology of Reproduction 77, 285-291.

Shimshony, A., 1988. Foot and mouth disease in the Mountain gazelle in Israel. Revue Scientifique et Technique 7, 917-923.

Shimshony, A., Orgad, U., Baharav, D., Prudovsky, S., Yakobson, B., Bar Moshe, B., Dagan, D., 1986. Malignant foot-and-mouth disease in Mountain gazelles. Veterinary Record 119, 175-176.

Shin, T., Kraemer, D., Pryor, J., Liu, L., Rugila, J., Howe, L., Buck, S., Murphy, K., Lyons, L., Westhusin, M., 2002. A cat cloned by nuclear transplantation. Nature 415, 859.

Shinohara, T., Inoue, K., Ogonuki, N., Kanatsu-Shinohara, M., Miki, H., Nakata, K., Kurome, M., Nagashima, H., Toyokuni, S., Kogishi, K., Honjo, T., Ogura, A., 2002. Birth of offspring following transplantation of cryopreserved immature testicular pieces and in-vitro microinsemination. Human Reproduction 17, 3039-3045.

Shinohara, T., Kato, M., Takehashi, M., Lee, J., Chuma, S., Nakatsuji, N., Kanatsu-Shinohara, M., Hirabayashi, M., 2006. Rats produced by interspecies spermatogonial transplantation in mice and in vitro microinsemination. Proceedings of the National Academy of Sciences of the United States of America 103, 13624-13628.

Si, W., Hildebrandt, T.B., Reid, C., Krieg, R., Ji, W., Fassbender, M., Hermes, R., 2006. The successful double cryopreservation of rabbit (*Oryctolagus cuniculus*) semen in large volume using the directional freezing technique with reduced concentration of cryoprotectant. Theriogenology 65, 788-798.

Silber, S.J., Pineda, J.A., DeRosa, M., Gorman, K.S., Patrizio, P., Gosden, R.G., 2007. Comparison of ovarian cortical tissue grafting vs. intact whole ovary microvascular homotransplantation and allotransplantation for patients with premature ovarian failure. Fertility and Sterility 88, S45 (abstract).

Strzezek, J., Fraser, L., Kuklinska, M., Dziekonska, A., Lecewicz, M., 2004. Effects of dietary supplementation with polyunsaturated fatty acids and antioxidants on biochemical characteristics of boar semen. Reproductive Biology 4, 271-287.

Summers, P.M., Shephard, A.M., Hodges, J.K., Kydd, J., Boyle, M.S., Allen, W.R., 1987. Successful transfer of the embryos of Przewalski's horses (*Equus przewalskii*) and Grant's zebra (*E. burchelli*) to domestic mares (*E. caballus*). Journal of Reproduction and Fertility 80, 13-20.

Sun, W.Q., Leopold, A.C., Crowe, L.M., Crowe, J.H., 1996. Stability of dry liposomes in sugar glasses. Biophysical Journal 70, 1769-1776.

Swain, J.E., Miller, R.R., Jr, 2000. A postcryogenic comparison of membrane fatty acids of elephant spermatozoa. Zoo Biology 19, 461-473.

Swanson, W.F., 2006. Application of assisted reproduction for population management in felids: The potential and reality for conservation of small cats. Theriogenology 66, 49-58.

Swanson, W.F., Brown, J.L., 2004. International training programs in reproductive sciences for conservation of Latin American felids. Animal Reproduction Science 82-83, 21-34.

Trimeche, A., Renard, P., Tainturier, D., 1998. A procedure for poitou jackass sperm cryopreservation. Theriogenology 50, 793-806.

Underwood, S.L., Bathgate, R., Maxwell, W.M.C., O'Donnell, M., Evans, G., 2007. Pregnancies after artificial insemination of frozen-thawed, sex-sorted, re-frozen-thawed dairy bull sperm. Reproduction in Domestic Animals 42, 78 (abstract).

Vajta, G., Holm, P., Kuwayama, M., Booth, P.J., Jacobsen, H., Greve, T., Callesen, H., 1998. Open pulled straw (OPS) vitrification: A new way to reduce cryoinjuries of bovine ova and embryos. Molecular Reproduction and Development 51, 53-58.

Van Thuan, N., Wakayama, S., Kishigami, S., Wakayama, T., 2005. New preservation method for mouse spermatozoa without freezing. Biology of Reproduction 72, 444-450.

Veprintsev, B.N., Rott, N.N., 1979. Conserving genetic resources of animal species. Nature 280, 633-634.

Wakayama, T., Yanagimachi, R., 1998. Development of normal mice from oocytes injected with freeze-dried spermatozoa. Nature Biotechnology 16, 639-641.

Ward, M.A., Kaneko, T., Kusakabe, H., Biggers, J.D., Whittingham, D.G., Yanagimachi, R., 2003. Long-term preservation of mouse spermatozoa after freeze-drying and freezing without cryoprotection. Biology of Reproduction 69, 2100-2108.

Watson, P.F., 2000. The causes of reduced fertility with cryopreserved semen. Animal Reproduction Science 60-61, 481-492.

Westh, P., Ramløv, H., 1991. Trehalose accumulation in the tardigrade *Adorybiotus coronifer* during anhydrobiosis. Journal of Experimental Zoology 258, 303-311.

Wiese, R.J., 2000. Asian elephants are not self-sustaining in North America. Zoo Biol 19, 299-309.

Wildt, D., Pukazhenthi, B., Brown, J., Monfort, S., Howard, J., Roth, T., 1995. Spermatology for understanding, managing and conserving rare species. Reproduction, Fertility and Development 7, 811-824.

Wildt, D.E., 1992. Genetic resource banks for conserving wildlife species: justification, examples and becoming organized on a global basis. Animal Reproduction Science 28, 247-257.

Wildt, D.E., Rall, W.F., Critser, J.K., Monfort, S.L., Seal, U.S., 1997. Genome resource banks: living collections for biodiversity conservation. Bioscience 47, 689-698.

Wilmut, I., Schnieke, A.E., McWhir, J., Kind, A.J., Campbell, K.H., 1997. Viable offspring derived from fetal and adult mammalian cells. Nature 385, 810-813.

Womersley, C., Ching, C., 1989. Natural dehydration regimes as a prerequisite for the successful induction of anhydrobiosis in the nematode Rotylenchulus reniformis. Journal of Experimental Biology 143, 359-372.

Woods, G.L., White, K.L., Vanderwall, D.K., Li, G.P., Aston, K.I., Bunch, T.D., Meerdo, L.N., Pate, B.J., 2003. A mule cloned from fetal cells by nuclear transfer. Science 301, 1063.

Yavin, S., Arav, A., 2007. Measurement of essential physical properties of vitrification solutions. Theriogenology 67, 81-89.

Yoon, T.K., Lee, D.R., Cha, S.K., Chung, H.M., Lee, W.S., Cha, K.Y., 2007. Survival rate of human oocytes and pregnancy outcome after vitrification using slush nitrogen in assisted reproductive technologies. Fertility and Sterility 88, 952-956.

Yu, I., Leibo, S.P., 2002. Recovery of motile, membrane-intact spermatozoa from canine epididymides stored for 8 days at 4°C. Theriogenology 57, 1179-1190.

Yushchenko, N.P., 1957. Proof of the possibility of preserving mammalian spermatozoa in a dried state. Proceedings of the Lenin Academy of Agricultural Sciences of the USSR 22, 37-40.

Zambelli, D., Cunto, M., 2006. Semen collection in cats: Techniques and analysis. Theriogenology 66, 159-165.

Zeng, W., Avelar, G.F., Rathi, R., Franca, L.R., Dobrinski, I., 2006. The length of the spermatogenic cycle is conserved in porcine and ovine testis xenografts. Journal of Andrology 27, 527-533.

Zeron, Y., Pearl, M., Borochov, A., Arav, A., 1999. Kinetic and temporal factors influence chilling injury to germinal vesicle and mature bovine oocytes. Cryobiology 38, 35-42.

Zeron, Y., Sklan, D., Arav, A., 2002a. Effect of polyunsaturated fatty acid supplementation on biophysical parameters and chilling sensitivity of ewe oocytes. Molecular Reproduction and Development 61, 271-278.

Zeron, Y., Tomezak, M., Crowe, J.H., Arav, A., 2002b. The effect of liposomes on thermotropic membrane phase transitions of bovine spermatozoa and oocytes: implications for reducing chilling sensitivity. Cryobiology 45, 143-152.

Chapter 4

THE DIVERSITY OF CYPRINIFORMS THROUGHOUT BANGLADESH: PRESENT STATUS AND CONSERVATION CHALLENGES

Mostafa A. R. Hossain[*] *and Md. Abdul Wahab*
Faculty of Fisheries, Bangladesh Agricultural University
Mymensingh-2202, Bangladesh

ABSTRACT

Bangladesh is endowed with a vast expanse of inland openwaters characterised by rivers, canals, natural and man-made lakes, freshwater marshes, estuaries, brackish water impoundments and floodplains. The potential fish resources resulting from these are among the richest in the world; in production, only China and India outrank Bangladesh. The inland openwater finfish fauna is an assemblage of ~267 species, the diversity of which is attributed to the habitats created by the Bengal Delta wetlands and the confluence of the Brahmaputra, Ganges and Jamuna rivers that flow from the Himalayan Mountains into the Bay of Bengal.

The indigenous fish fauna of Bangladesh's inland openwaters, however, are dominated by the cypriniforms - 87 species under 35 genera. Although representatives are rarely encountered in brackish waters, certain species have adapted to some of the country's most extreme freshwater environments.

There are, however, serious concerns surrounding the slow decline in the condition of openwater fish stocks which have been negatively impacted upon through a series of natural and anthropogenic induced changes. These include disturbances resulting from water management programmes including the large scale abstraction of water for irrigation and the construction of water barrages and dams, human activity resulting in the overexploitation of stocks, the unregulated introduction of exotic stocks and pollution from industry. Also, natural phenomena, regular flooding etc cause rivers to continually change course creating complications of soil erosion or oversiltation of waterways. As a consequence, many Bangladeshi species are either critically endangered or extinct. The

---

[*] Corresponding author: marhossain@yahoo.com

biodiversity status of many of these have now changed from that listed in the IUCN Red Book almost a decade ago.

Assessment is based primarily on the study of specimens maintained in the Fish Museum and Biodiversity Center (FMBC) of Bangladesh Agricultural University and through surveys conducted over the last ten years. The threat to inland openwater biodiversity is countrywide, but that facing cypriniform species is severe. More than 15% of cypriniforms appear to have disappeared; only one or two individuals of a further 20% of species have been found in the last ten years.

The needs of Bangladesh's poor fisher community to eat what they catch and lack of a legal legislative framework means the situation can only worsen. Hope, however, is offered through several new conservation initiatives including the establishment of fish sanctuaries at strategic points in rivers and floodplains, concerted breeding programmes and the maintenance of captive stocks and cryogenically stored materials.

1. INTRODUCTION

Bangladesh is situated in the northeastern part of the South Asia and lies between $20°34'$ and $26°38'$ North longitudes and $88°01'$ and $92°41'$ East latitudes. The country is bordered by India on the West, North and North-East (2,400 kilometer land frontier) and Myanmar on the Southeastern tip (193 km land and water frontier). On the south is a highly irregular deltaic coastline of about 710 kilometers, fissured by many rivers and streams flowing into the Bay of Bengal. The territorial waters of Bangladesh extend 22 km, and the exclusive economic zone of the country is 370 km. The total landmass of the country is about 144,400 km^2 and extends 820 kilometers north to south and 600 kilometers east to west. The country stretches out at the junction of the Indian and Malayan sub-regions of the Indo-Malayan zoogeographic realm.

Formed by a deltaic plain, Bangladesh is virtually the only drainage outlet for a vast complex river basin made up of the Ganges (local name the Padma), the Brahmaputra and the Meghna rivers and their network of tributaries. The Padma unites with the Jamuna (main channel of the Brahmaputra) and later joins the Meghna to eventually empty into the Bay of Bengal. The alluvial soil deposited by these rivers every year has created some of the most fertile plains in the world. Most parts of the delta are less than 12 metres above the sea level, and it is believed that about 50% of the land would be flooded if the sea level rise by a metre. Straddling the Tropic of Cancer, Bangladesh has a tropical monsoon climate characterised by heavy seasonal rainfall, high temperatures, and high humidity. There are three broad physiographic regions in the country. The floodplains occupy about 80%, terrace about 8% and hills about 12% of the land area (Table 1). Moreover, it is a country dominated by wetland having more than 50% of its territory under true wetlands that is freshwater marshes, swamps, rivers estuaries and the world's largest contiguous mangrove forest - the Sundarbans.

Bangladesh has a total inland water area of 6.7 million ha of which 94% is used for open water capture fishery and 6% for closed water culture fishery (Table 2). The inland open water fishery resources have been playing a significant role in the economy, culture, tradition and food habit of the people of Bangladesh. Rivers and their ramified branches cover about 479,735 ha area of land. Seasonal floodplain expands over a massive 5.5 million ha for 4-6 months of the year. Inland open water also contains estuarine areas with semi-saline waters

(0-10 ppt), huge number of *beels* (natural depressions often with permanent area of water) and *haors* (bowl-shaped deeply flooded depressions) in the north and east and the manmade Kaptai lake – the largest lake of the country in the south. The country is blessed with 0.26 million of closed waters in the form of ponds, ditches, oxbow lakes (channel of dead rivers) and brackish water farms. More than 2 million people directly or indirectly depend on inland capture fisheries for their livelihood.

Fish have been an integral part of life of the people of Bangladesh from time immemorial. Many aspects of the Bangladeshi culture, economy and tradition are based around fishing and fish culture activities. The sector plays a vital role in the country's economy, employment generation, animal protein supply, foreign currency earning and poverty alleviation. Fish is a natural complement to rice in the national diet, giving rise to the adage "Maache-Bhate Bangali", literally meaning – 'fish and rice make a Bangladeshi'. Fisheries, second only to agriculture in the overall economy of Bangladesh, contribute nearly 5.0% to the gross domestic product (GDP), 23% of gross agriculture products and 5.71% to the total export earnings (DoF 2008). It accounts for about 63% of animal protein intake in the diet of the people of Bangladesh (DoF 2005). The fisheries sector provides full-time employment to an estimated 1.2 million fishermen and an estimated 10 million households or about 64% of all households are partly dependent on fishing, e.g. part time fishing for family subsistence in flooded areas. Another 10% poor and middle class people are engaged in part-time fishing, aquaculture, fish seed production and collection of shrimp and prawn seed, fish handling, processing and marketing, net making, input supply etc.

The people of Bangladesh largely depend on fish to meet their protein needs in both the rural and urban areas. Until 70s, there was an abundance of fish in the natural waters of the country to well-satisfy the demand. In recent years, however, capture fish production has declined to about 50%, with a negative trend of 1.24 % per year (Ahmed 1995). Despite the constant depletion of the natural waterbodies for years, Bangladesh still holds one of the most diverse inland fisheries in the world. However, the availability of many fish species has been drastically declined, and many are either critically endangered or extinct. On the migration journey to the floodplains, and the return to rivers, fish now face many blockage and danger, which seriously interfere with the breeding and resulting recruitment.

Table 1. Major physiographic areas of Bangladesh

Description	Area (km^2)	% of total area
Rivers, canals, streams	8,300	5.76
Estuarine, brackishwater water-bodies	1,828	1.27
Floodplains	112,010	77.76
Wetlands	2,930	2.03
Freshwater ponds and tanks	794	0.55
Artificial lakes	906	0.63
Hill areas	17,286	12.00
Total Bangladesh	144,054	100

Source: Hoq (2009)

Table 2. Extent of different type of water areas

Types of water areas	Area (ha)
a) Inland open waters	
1. Rivers (during dry season)	
The Ganges	27,165
The Padma	42,325
The Jamuna	73,666
The Meghna (upper)	33,592
The Meghna (lower	40,407
Other rivers and canals	262,580
2. Estuarine area	551,828
3. *Beels* and *haors*	114,161
4. Kaptai Lake	68,800
5. Inundated flood plains (seasonal)	5,486,609
Total	6,701,133
b) Closed waters	
1. Ponds and ditches	146,890
2. *Baors* (oxbow lakes)	5,488
3. Brackish water farms	108,000
Total	260,378

Source: FRSS-DoF 2008

2. THE KEY WATERBODIES AND THE STATUS

The inland open water fishery of Bangladesh is composed of highly diverse and unique aquatic systems. It has an extensive network comprising of floodplains, large and small rivers, *beels*, *haors* and *baors* offering tremendous scope and potential for fish production. It has also large impounded water areas in manmade ponds, ditches, borrow pits, lakes and enclosures (DoF 2005).

2.1. The Floodplain

Floodplains are relatively low laying land area, bordering rivers and seasonally over-flooded by overspill from the main river channel. There are two distinct flooding patterns, one resulting in flow direction from the floodplains to the rivers (from flush flood due to local rainfall) and the other from the rivers to the floodplains (river overspill due to the heavy rainfall in the upstream). The ecodynamics of floodplain are influenced by the river water incursion and retreat and the timing and intensity of monsoon. There are great differences in the area flooded from year to year, and this greatly influences the population dynamics of many fish species.

The seasonally flooded area is highly productive for growth of fish and other aquatic animals. During the dry season, as pasture land, the floodplain receives nutrients in the form of animal dropping and rotting vegetation. As the monsoon approaches the accumulated nutrients rapidly enters into the solution combined with river-borne silt, led to an upsurge of productivity resulting in rapid growth of plants and other forms of aquatic biota. This productivity phase offers an ideal condition for feeding and breeding of many riverine fishes and other aquatic animals which migrate to floodplain with the rising waters. Floodplains inundated during monsoons are nutrient rich and play a significant role as nurseries for larvae and juvenile of many fish species (Junk *et al.* 1989). A large number of fresh water fish species migrate from rivers and *beels* to floodplains for breeding and grazing and are harvested by the rural professional and amateur fishers. The floodplains are essential for most of the rural people of Bangladesh for their livelihood.

The floodplains are very rich in both floral and faunal diversity and harbour a large number of finfish, crustaceans, molluscs amphibians, reptiles and a large number of aquatic vegetation. FAP-6 (1992) recorded 154 finfish and prawn in the floodplain of the northeastern region of the country. FAP-17 (1994) documented 89 finfish and prawn species in the floodplain of Tangail. Major species under the two studies reported were Indian major carps, several species of minor carps and loaches belonging to the Order Cypriniformes.

Fish production in 2006-07 was 819,446 mt in the floodplains, which was 77.29% of total inland capture fish production and 51.32% of the total inland fish production. The rate of present floodplain fish production in Bangladesh is 289 kg ha^{-1} year^{-1} which is higher than the production in the floodplains of many neighbouring countries (DoF 2008).

Since 1970, the annual flooding of approximately 2-3 million ha of floodplain has been either controlled or prevented altogether by means of sluice gates or pumps positioned along earth embankments or levees (ESCAP 1998). This reduction in area is believed to be one of the major reasons for declining floodplain fisheries in Bangladesh (FAP 17 1994). Overexploitation of inland fish stocks has also been reported (Graaf *et al.* 2001).

2.2. The Rivers

Bangldesh is a riverine country. It has numerous rivers and their tributaries (Figure 1). The Ganges, the Brahmaputra and the Meghna are the mighty rivers. The three rivers along with their innumerable tributaries form one of the richest habitats of fishes in the Indian Subcontinent. In addition to three main rivers, the other main rivers are the Karnafuli, Matamuhuri, Halda and Sangu in the southern Chittagong sub-region. The major rivers are the Surma, Kushiara, Kangsha and Someshwari in the north-east region and the Tista, Korotoa, Atrai, Bangalee, Mohananda in the north-west. The total length of the network of about 310 rivers in Bangladesh together covers more than 24,000 km with a catchment area of 1,031,563 ha. Annual flooding of the rivers inundates about 70% of the total land surface. The total annual discharge passing through the rivers system into the Bay of Bengal reaches up to 1,174 billion m^3 (Banglapedia 2004).

The rivers are not evenly distributed in the country. For instance, the numbers and size of the rivers gradually increase from the northwest of the northern region to the southeast of the southern region. All the rivers, except those of Chittagong hilly sub-region, belong to three major river systems, the Ganges, the Bhramaputra and Meghna. In the global context, the Brahmaputra is the 22nd longest (2,850 km) and the Ganges is the 30th longest (2,510 km) river in the world. Rivers and canals roughly cover 5.8 % of the total area of the country. According to (BWDB 2005) 57 of the rivers are trans-boundary – 54 originate from India and 3 from Myanmar. The river system of Bangladesh is divided into 6 hydrological regions as shown in Table 3.

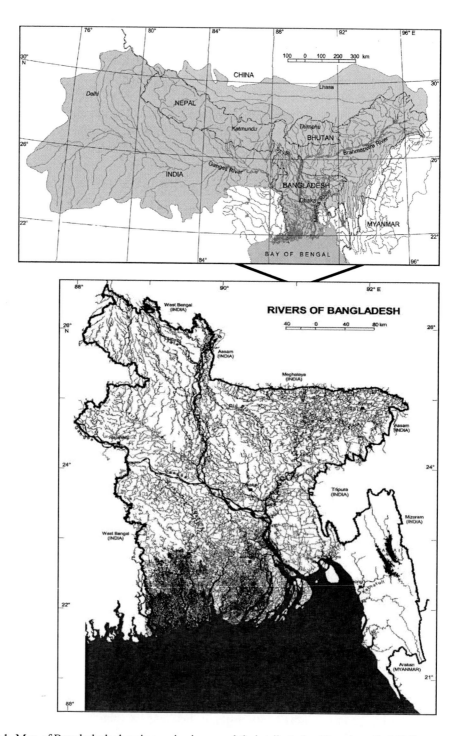

Figure 1. Map of Bangladesh showing main rivers and their tributaries (Banglapedia 2004).

Table 3. Hydrological regions of river system in Bangladesh

Hydrological region	Number of rivers	Length (km)	Catchments area (km^2)
North West Region	96	4,908	63,718
North Central Region	20	1,311	18,404
North East Region	55	3,250	47,616
South East Region	24	1,320	10,068
Eastern Hilly Region	17	1,131	6,253
South West Region	98	4,969	35,576
Total	310	16,889	1,81,635

Source: BWDB (2005)

During rainy season, the rivers carry high amount of silt which makes the water turbid. In winter, the water level decreases and water becomes clear. The depth of the coastal rivers usually ranges from 2 m to 5.5 m and reaches up to 36.5 m near the Bay of Bengal. Salinity of about 1 ppt extends nearly 56 km upstream in these rivers.

The rivers of Bangladesh have a great importance in respect of fisheries and other hydrological and navigation benefits. Rivers are the migratory routes of fishes with adjacent floodplains and vice-versa. Among the riverine fishes, Cypriniform - carps, barbs and minnows and loaches are very important. Many of the Cypriniforms migrate upstream (floodplain) in order to spawn in nutrient rich water, where they feed on plankton and grow (Rahman 2005). At the end of monsoon the adult and young fishes escape to the rivers and most likely to the adjacent deeper *beels* to avoid harsh condition of the floodplain during dry season.

The hill-streams in the north and north-east are swift flowing with clear water. Many of the Cypriniform mainly loaches inhabiting the hill-streams usually have modified paired fins and grooved thoracic disk, which acts as adhesive apparatus. The fishes with compressed head and horizontally placed pectoral and pelvic fins can easily stick to the bottom and do not swept away in swift flowing water. Among the riverine Cypriniform that inhabit the hill-streams are the members of the genera – *Nemachilus*, *Schistura*, *Balitora* and *Psilorhynchus* under the Families of Balitoridae and Psilorhynchidae. The fish species of the Genera - *Garra*, *Barilius* and *Raiamas* under the Family Cyprinidae are also available in these streams.

2.3. The Beels

The *beel* is a Bengali term used for relatively large surface, static waterbody that accumulates surface run-off water through an internal drainage channel (Banglapedia 2004). This type of shallow, seasonal waterbody is common in low-lying floodplain areas throughout Bangladesh. The total area of *beels* in Bangladesh was estimated to be 114,161 ha, occupying 27.0% of the inland freshwater (Ahmed *et al.* 2007). The number of *beels* in the north-eastern part of the country recorded was 6,034 having an area of 69,870 ha (Bernaesek *et al.* 1992).

The most famous *beel* in the country known as the Chalan *beel* is located in the north-west. The other major *beels* in this region are Hilna, Kosba, Uthrail, Manda, Sobna and *Beel* Mansur. In central region, Arial *beel* and Balai *beel* now lost their importance as natural fish habitat. Other important *beels* in this region are Chanda, Boro, Mollar and Tungipara *beels*. There are many *beels* in the south and south west and the notable are Chapaigachi, Garalia, Panjiapatra, Chenchuri and Dakatia *beels*.

The *beels* are parts of riverine floodplain formed due to changes in the river course or strengthening of river embankments for controlling flood (Saha *et al.* 1990). The *beel* water is very productive in terms of fertility and nutrient, full of organic debris and organic vegetation and provide food and shelter to many larvae and juvenile as well as adult fishes and other aquatic organisms (Graff 2003).

Beels are mainly of two types: a) perennial *beel* (retain water throughout the year) and b) seasonal *beel* (dried up in the dry season but during the rains expand into broad and shallow sheets of water). The perennial *beels* are the dry season refuges (over wintering ground) of many fish species. These waterbodies are very favourable natural habitat of small and large indigenous fishes to feed, grow and breed during monsoon as well. Among the various parameters that influence the *beel* ecosystem are depth, nature of catchments area or river basin, precipitation and duration of connection to river (Sugunan and Bhattacharjya 2000).

Beels are generally rich in fisheries resources in Bangladesh and provide considerable fish production of the country. Some important *beels* in greater Sylhet and Mymensingh region are known as "mother fishery". *Beels* play an important role in the fish production as well as shelter of brood fish in the country. The *beels* of Bangladesh comprise about 2.82% of total aquatic resources which supplied 77,524 mt of fish in 2007-08. *Beel* fishery of Bangladesh is being deteriorating day by day due to over fishing, uncontrolled use of chemical fertilizer and insecticide, destruction of natural breeding and feeding grounds, harvesting of wild brood fishes and for many other causes (Azher *et al.* 2007).

2.3.1. Chalan beel

Chalan *beel* is the largest and most important watershed in the North Central Bangladesh. It comprises of a series of depressions interconnected by numerous channels to form more or less one continuous sheet of water during monsoon covering an area of about 375 km^2. The watershed serves about five million people predominantly through fisheries and agriculture. Though far from its past glory, Chalan *beel* is still an abode of large variety of ichthyofauna with a huge importance in local economy and people's livelihood.

During the dry season, the water area decreases down to 52-78 km^2 and looks like a cluster of small *beels* of different sizes. Besides being a giant junction of a number of water ways, the *beel* also served a springboard where many rivers flowed further south and east to meet finally with the river Padma and the Brahmaputra (Iqbal 2006).

The Chalan *beel* comprised of 21 rivers and 93 small *beels* and their floodplains, 12,817 ponds and 214 borrow pits. There are 21 rivers streaming into the Chalan *beel* which cover a total area of about 709 ha and 3,300 ha in dry and monsoon, respectively. Most of the rivers and *beels* are at the risk of partial or total degradation due to manifold reasons like agricultural encroachment, siltation along with other anthropogenic activities. The critical dry out condition (0-5% of the monsoon size) was observed in 83% of the rivers and 68% of the *beels* in the lean season (Hossain *et al.* 2009)

Hossain *et al.* (2009) documented 114 finfish species from 29 families in the largest *beel* of Bangladesh – the Chalan *beel*. Among the thirty nine Cyprinifoms in the Chalan *beel* - thirty one species were under Cyprinidae one under Balitoridae, six under Cobitidae and one under Psilorhynchidae. The most abundant fish species were two Cypriniform - *Puntius sophore* and *Puntius ticto*.

The number of fishers in the Chalan *beel* has been changed over time with a 58% reduction from 1982 to 2006 (Table 4). They either left their profession or migrated elsewhere in Bangladesh. Presently there are 75,000 professional and subsistence fishers in the Chalan *beel* area maintaining their livelihoods from this resource directly or indirectly.

Most of the fishermen families have been leading a sub-human life, many have been catching eggs, spawns, fry, undersized fish and broods indiscriminately which have resulted in scarcity of fish in the *beel* area. In this situation, almost all the fishermen, who used to earn their livelihood by fishing, are now facing hardships as there is virtually no fish in the waters in most of the time of the year.

It has been reported that the fish production in Chalan *beel* reduced by 31% and 52% in 1992 and 2002, respectively, as compared to the production in 1982 (Figure 2). Fish species availability and production are tightly bound to the pattern of the flooding which takes place during the monsoon in the Chalan *beel* (Ahmed 1991).

Human interferences including some development interventions such as construction of roads, dams and embankments and human settlement adversely affected the aquatic ecosystem and habitat of fish population in the Chalan *beel* by obstructing their migratory routes. Therefore, the breeding of migratory species has been interrupted which hampered natural recruitment of fish in the *beel*. Though, this is not the case for non-migratory resident fish spawners, over-exploitation and drying up of the waterbody perhaps responsible for overall low abundance of many species.

Table 4. Changes in number of fishermen over the years in Chalan *beel*

Fishermen category	Year			
	1982	1992	2002	2006
Professional	53,446	46,534	33,445	22,316
Subsistence	1,23,615	1,06,335	73,612	52,684
Total	177,061	152,869	107,057	75,000

Source: Hossain *et al.* (2009)

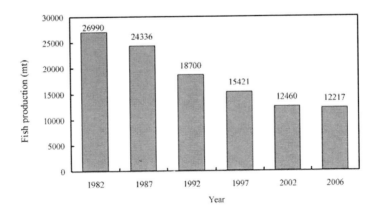

Figure 2. Fish production tends in Chalan *beel* (Hossain *et al.* 2009).

2.4. The Haors

The *haors* are back swamps or bowl-shaped depressions between the natural levees of rivers, or in some cases, much larger areas incorporating a succession of these depressions. The Bengali word *haor* basically derived from the word *sagor* (literally meaning sea) and dialectically *sagor - saior - haor* has been evolved (Khan *et al.*1990). In terms of morphology and hydrology, a *haor* can be subdivided into three major areas, the piedmont area around the hill foot, the floodplains and the deeply flooded area (Hossain and Nishat 1989). The *haors* vary in size from as little as a few hectares to thousands of hectares. The *haors* flood to a depth of as much as 6 m during the rainy season and in many cases two or more neighboring *haors* unite to form a much larger water body.

Greater part of the north east region of Bangladesh is characterized by the presence of numerous large, deeply flooded depressions, known as *haors*, between the rivers. There are altogether 411 *haors* (47 major and large sized) comprising an area of about 8,000 km^2 dispersed in the north-eastern Sylhet and Mymensingh districts. The *haor* basin is bounded by the several Indian states - hills of Meghalaya on the north, hills of Tripura and Mizoram on the south, and the highlands of Manipur on the east.

The two big rivers in the region – Surma and Kushiyara in association with several smaller hill-streams - Manu, Khowai, Jadhukata, Piyangang, Mogra and Mahadao form the dense network and supply the massive water to the *haors*. The rivers are primarily responsible for providing inputs - rainwater and sediment load to the haors. The *haors* remain flooded for about 7 to 8 months. During the rainy season, the *haors* look just like vast inland sea and the villages within appear as islands.

In greater Sylhet the most prominent *haors* are Saneer, Hail, Hakaluki, Dekar, Maker, Chayer, Tangua and Kawadighi. In consideration of the environmental importance and heritage, the government has decided to save the Tangua *haor* (9,500 ha) by symbolizing it as an internationally critical environment area under the Environmental Protection Law of 1995 and registered as a wetland of international importance (Ramsar site, site no. 1031, declared in 10.07.2000) under Ramsar Convention. The location-wise principal *haor* systems are given in Table 5.

The *haors* are considered to be the most productive wetland resources of Bangladesh. The basin supports a large variety of aquatic biodiversity and works as natural reservoir. In past, most of the *haor* basin was covered with swamp forests and reed lands had a characteristic feature of flooded forests made the entire *haor* a suitable habitats for small and large fish and other aquatic animals. Livelihoods of the villagers living in and around the *haor* were largely dependent on the *haor* resources for food, nutrition, grazing, boating, housing, income and other forms of livelihood security.

The *haors* serve as the natural brood stocks of many indigenous fishes including large carps and catfishes. With the advent of the dry season water recedes, the relatively elevated parts of *haor* area begin to dry when paddy is raised on the dried upland areas. The relatively depressed areas, however, remain under water where fishes take shelter. The *haors* also act as important breeding and spawning ground of many fishes. There are a total of 141 finfish species found in the *haors*. The important Cypriniform fishes available in the *haors* are - *Labeo rohita, L. gonius, L. boga, L. calbasu, Catla catla, Cirrhinus mrigala, C. reba, Crossocheilus latius, Bengala elanga, Rasbora rasbora, Osteobrama cotio cotio* and *Tor tor*.

Table 5. The principal *haor* systems of Bangladesh

Location	Name of the *haors*
Eastern and lowest part of the basin, Mymensingh	Baram, Banka, Habibpur, Maka, Makalkandi and Ghulduba
Foot of the Meghalaya Hills	Tangua, Shanir, and Matian
East of the Tangua system	Dekhar, Pathar Chanli, and Jhilkar
Eastern rim of the basin	Jamaikata, Mahai, Nalua, and Parua
Central Sylhet lowlands	Hakaluki, Chatal Bar, Haila, Kawadighi, Pagla and many smaller
Tarap and Banugach hill ranges in the southeast	Hail
South of the basin	Dingapota, Ganesher, Tolar, Anganer, Bara, and Humaipur
Kishorganj	Etna and Sania
East Mymensingh	Khaliajhuri

Source: NERP (1995)

2.5. The Baors

In the southwest region of Bangladesh there are a number of meandering rivers changed their courses, part of the old course got silted up and cut-off from the main course. As a result horseshoe shaped oxbow lakes known as *baor* was created. A *baor* apparently looks like a lake, but unlike lakes, it remains connected with original river through channels during monsoon. This way, the *baors* annually receive fresh supply of riverine water carrying fry, fingerlings and adult fishes and other aquatic animals. *Baors* are very important wetlands of Bangladesh and support a wide range of aquatic flora and fauna.

There are more than 87 *baors* in Bangladesh covering an area of 5,488 ha (DoF 2008). Most of the larger *baors* are in southwestern Jessore region. The *baors* range in size from about 25 ha to a maximum of 500 ha (Bhuiyan and Choudhury 1997). The important *baors* of the country are Arial, Bahadurpur Baluhar, Bookbhara, Harina, Habullah, Rustampur, Ichhamati, Jaleshwar, Jogini Bhagini, Joydia, Kannadah, Kathgara, Khedapara, Marjat, Pathanpara, Rampur, Sagarkhali, Sirisdia and Sonadia.

The *baors* are an abode of small and large indigenous fish species. Most of the fish species breed and thrive in the *baors* throughout the year. The waterbodies are shelters and breeding grounds for fishes, amphibians, reptiles and a gamut of aquatic invertebrates. The *baors* support a continuous daily harvest of subsistence and commercial fishing round the year. About 50 species of indigenous fishes belonging to 31 genera and 20 families could be found in a *baor*. All the *baors* are now under a heavy fishing pressure. The construction of dams and other flood control structures have reduced the natural recruitment and contributed to stock depletion. The total catch area in the baors is about 5,488 ha and the annual production is about 2,460 mt (DoF 2008).

2.6. The Ponds and Ditches

There are more than 1.3 million ponds having a water surface of 0.3 million ha in the country (DoF 2008). Though in past ponds were constructed for washing, bathing and irrigation purposes, recently, many ponds are being constructed absolutely for fish culture purposes. There are two types of ponds on the basis of water retention capacity – the perennial ponds - contain water round the year and the seasonal ponds - contain water at a certain times or seasons (mainly in monsoon).

The pond culture fisheries have always been considered as being crucial for the livelihoods of the most vulnerable people of the country. In addition, it is also good for the fish diversity as it encourages the domestication of wild fishes through artificial breeding and rearing in the captivity. Selective aquaculture, however, could be detrimental for fish biodiversity as the culture technologies advice farmers to remove all small indigenous fishes from the ponds before releasing the fry of target fish. Farmers often use piscicide and insecticide to clean their ponds. The practice has been going on in Bangladesh since the carp polyculture being introduced in late 70s. As a result, though harvests from fish culture are rapidly increasing, the catches of small fishes are declining at an alarming rate.

The fish production from ponds and ditches in Bangladesh in the year 2007-08 reached to 866,049 mt with an average production of 2,839 kg ha^{-1} (DoF 2008). The major culture species are the Cypriniforms – *Labeo rohita, L. bata, Cirrhinus mrigala* and *Catla catla*. *Labeo calbasu* and *Cirrhinus reba* are also being cultured in smaller scale. The small indigenous Cypriniforms used to be abundantly available in the past in most of the homestead ponds, now-a-days seldom found are *Amblypharyngodon mola, Salmostoma phulo, Esomus danricus, Osteobrama cotio cotio, Puntius sophore* and *Puntius ticto*.

2.7. The Kaptai Lake

There are only three true natural lakes in the country. Rainkhyongkine lake and Bogakine lake are located in the Chittagong Hill Tracts, and a lake named Ashuhila *beel* at the northern end of the Barind Tract. The largest man-made lake in South Asia is Kaptai lake of 68,800 ha (surface area – 58,300 ha). The H-shaped Kaptai lake, the only major reservoir in Bangladesh was created from the construction of dam across the river Karnafuli near Kaptai town in 1961. It has drowned almost the whole of the middle-Karnafuli valley and the lower reaches of the Chengi, Kasalong and Rinkhyong rivers. Shoreline and the basin of Kaptai Lake are very irregular. The volume of the lake is 524,700 m^3 with a mean depth 9 m (maximum depth – 32 m and mean water level fluctuation - 8.14 m). Though the lake was created primarily with a vision to generate hydroelectric power, it substantially contributes to the national economy through freshwater fish production, navigation, flood control and agriculture. The lake is confined within the hill district of Rangamati and embraces sub-districts of Rangamati Sadar, Kaptai, Nannerchar, Langadu, Baghaichhari, Barkal, Juraichhari and Belaichhari

In 2007-08 fish production in Kaptai lake was 8,248 mt with an average of 120 kg ha^{-1}. At the early stage of the creation of lake, Indian major carps were the dominant species of about 60% of total catch, which is reduced to 5.69% in 2007-08. Presently the major catch in the Kaptai lake is kachki (*Corica soborna*) - 29.90% followed by chapila (*Gudusia chapra*) -

29.81%, respectively. Halder *et al.* (2002) recorded 66 species of indigenous fish in the lake. The major Cypriniforms available in the lake are *Catla catla, Cirrhinus mrigala, Lebeo rohita, L. calbasu and L. goinus* and *Puntius sarana*.

There are about 10,000 people directly or indirectly involved in fishing and fishery related activities at the reservoir. The reservoir has also provided income and employment opportunities to people, particularly in the areas of drying and retail marketing of the fish. As the local poor people remove the protective vegetation around the lake, the rocks are exposed to the monsoon rains and thus eroded easily. This results in regular landslides, and the loose rocks is washed down the slopes and carried by rivers into the lake. As a result, the lake is silting up rapidly. By early 1990s, in its 30-year existence, it had already lost 25% of its volume due to siltation.

3. AQUATIC FAUNA OF BANGLADESH

Bangladesh is a transitional zone of flora and fauna, because of its geographical settings and climatic characteristics. It is natural that the water resources of the existing extent and magnitude should harbour and support populations of a large variety of vertebrate and invertebrate aquatic living organisms. This country is rich in fish and aquatic resources, and other biodiversity (Table 6). Bangladesh's water bodies are known to be the habitat of 267 freshwater fishes, 475 marine fishes, 23 exotic fishes and a number of other vertebrates and invertebrates. Among the documented aquatic fauna, finfish tops the list, followed by the crustaceans and molluscs (Figure 3).

Table 6. Diversity of aquatic animals in Bangladesh water

Animal group	Number of Species	
	Freshwater	Marine
Finfish	267	475
Shrimp	-	41
Prawn	20	-
Mollusc	26	336
Crab	4	11
Lobster	-	6
Frog	10	-
Turtle & tortoise	24	7
Crocodiles	2	1
Snakes	18	6
Otters	3	-
Dolphin	1	8
Whale	-	3
Total	375	894

Source: Ahmed and Ali (1996), Ali (1997) and Banglapedia (2004).

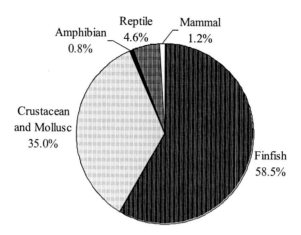

Figure.3. Percentages of major aquatic animal groups of Bangladesh.

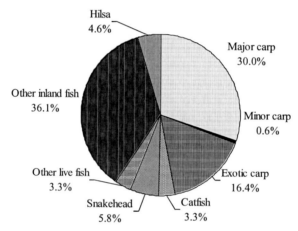

Source: FRSS-DoF 2008.

Figure 4. Group-wise catch of freshwater fishes (2006-07).
Major carp - Rui, Catla, Mrigal and Kalibaus; Exotic carp - Silver carp, Common carp, Mirror carp and Grass carp; Minor carp – Gonia, Reba and Bata; Catfish - Rita, Boal, Pangas, Silong and Air; Snakehead - Shol, Gazar and Taki; Other live fish - Koi, Singh and Magur; and Other Fish - all other fishes.

The total inland open water fish production of Bangladesh in the year of 2006-07 (July-June) was 1.783 million tons. The catch was dominated by major carps (30%) followed by exotic carps (16.4%) and snakeheads (5.8%) (Figure 4).

4. THE CYPRINIFOM FISHES

The Oriental region (i.e. monsoonal Asia, south of 30°N) hosts over 3,500 species under 105 families of freshwater finfishes. Cypriniforms are the most diverse group of fishes in most of the Asian countries like Bangladesh, India, Nepal, Pakistan, Myanmar, Thailand, Vietnam, Cambodia, Indonesia and Malaysia. In terms of generic richness, the top two fish families in Asia are Cyprinidae (147 genera) and Balitoridae (38 genera) under the Order

Cypriniformes (Dudgeon 2002). Globally the Order contains 6 families (Balitoridae, Catostomidae, Cobitidae and Gyrinocheilidae under Super Family Cobitoidea and Cyprinidae and Psilorhynchidae under Super Family Cyprinoidea), 320 genera and more than 3,250 species.

In Bangladesh, there are 267 species of freshwater bony fishes under 156 genera and 52 families (Table 7). Among the freshwater fishes, the three Orders that dominate are Cypriniformes, Siluriformes and Perciformes. More than one third of the total freshwater fish belong to the three Orders – Cypriniformes (33%), Siluriformes (22%) and Perciformes (24%). The major fish groups available in the country's freshwater are major carps, large catfishes, minor carps, small catfishes, river shads, snakeheads, freshwater eels, feather backs, parches, loaches, anchovies, gobies, glass fishes, mullets, minnows, barbs and flounders.

The diversity in size, shape, colors, habitat preference, feeding and breeding of the freshwater fishes of the country is high. If one considers only the size, there are very small fish (maximum length - only a few cm. *Danio rerio, Oryzias dancena* etc.) and also there are large fish (maximum length – more than two meters, *Wallogo attu, Bagarius bagarius* etc).

The Cypriniformes is the largest order of freshwater fishes in Bangladesh and includes carps, barbs, loaches and minnows. This is usually the dominant group of freshwater fishes, very rarely entering brackish water, and adapted to the most extreme freshwater environment. Members of this order have compressed head and body and most species lack fin spines. They do not have any adipose fin. The mouth is usually more or less protractile. Cyprinifoms also have a weberian apparatus that connects the swim bladder to the auditory system (the inner air) through a chain of small bones to facilitate an acute sense of hearing. The presence of the weberian apparatus is one of the most important and phylogenetically important distinguishing characteristics of the Cypriniformes and some other fishes. Body of the Cypriniform fishes are either with cycloid scale or naked and heads are scaleless.

Table 7. Number of freshwater fish species, genera, families and orders present in Bangladesh

Order	Family	Genus	Species
Anguiliformes	5	6	8
Osteoglossiformes	1	2	2
Elopiformes	1	1	1
Clupeiformes	3	12	18
Cypriniformes	4	35	87
Siluriformes	13	36	59
Cyprinodontiformes	1	1	1
Syngnathiformes	1	2	3
Synbranchiformes	2	4	6
Perciformes	15	46	65
Beloniformes	3	5	7
Pleuronectiformes	3	4	7
Tetraodontiformes	1	2	3
Total = 13	52	156	267

Table 8. The key distinguishing characters of the four Cypriniform families available in the freshwaters of Bangladesh

Characters	Psilorhynchidae	Balitoridae	Cobitidae	Cyprinidae
Pectoral fin	Inserted horizontally		Pectoral and pelvic inserted laterally	
Pectoral fin ray	At least two unbranched rays		All rays branched or only outermost anterior ray unbranched	
Body shape	Greatly arched dorsally and flattened ventrally	Depressed	Worm-like to fusiform	Usually laterally compressed
Barbel	Absent	3 or more pairs	3 - 4 pairs	1 - 2 pairs or absent
Air bladder	Greatly reduced; free in the abdominal cavity	Large; partly enclosed in a bony capsule	-	-
Spine	-	-	Erectile spine near eyes	No suborbital or preorbital spine
Pharyngeal teeth	-	-	1 row	1-3 rows

Modified from Talwar and Jhingran (1991)

Among the six families of Cypriniformes available worldwide, fishes of four families (Balitoridae, Cobitidae, Cyprinidae and Psilorhynchidae) are available in Bangladesh. Under the four families, there are 35 genera and 87 species reported from the freshwaters of Bangladesh. The four families can be separated by a number of key characters (Table 8).

4.1. Family Psilorhynchidae

The family only has one genus – Psilorhynchus and the members are found in primarily in the Gangetic river system of Southeast Asia - Bangladesh, Nepal, Myanmar and Indian state of Assam. These fishes are found in the rapidly flowing permanent hill-streams without tidal influence.

There are about 12 species under the Genus worldwide. In Bangladesh, only three species have been documented (Box 1.) They have been found only in a few hill-streams near the Bangladesh-India border in Dinajpur, Mymensingh, Sylhet and Chittagong region.

In the Red Book published in 2000, all three fishes, - *Psilorhynchus balitora*, *P. gracilis* and *P. sucatio* have been described as data deficient (IUCN Bangladesh 2000). Through surveys conducted over the last ten years by the research team of Bangladesh Agricultural University (BAU), all three species were found in a few occasions and are maintained in the Fish Museum and Biodiversity Center (FMBC) of the BAU. They, however, were found in a very little number (one or two individuals) together with many other small fish in fishermen's net and never found in the fish markets. The biodiversity status of all three *Psilorhynchus* should be considered as critically endangered (Table 9).

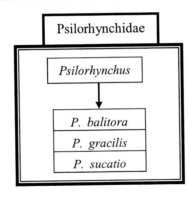

Box 1. The available species under the family - Psilorhynchidae in Bangladesh.

Table 9. Biodiversity status of the fishes of the family Psilorhynchidae

Fish	English name	Local name	IUCN Red Book 2000 status	Present status
Psilorhynchus balitora	Balitora minnow	Balitora	DD	CR
Psilorhynchus gracilis	Rainbow minnow	Balitora	DD	CR
Psilorhynchus sucatio	River stone carp	Titari	DD	CR

DD- Data deficient; CR – Critically endangered

4.2. Family Balitoridae

Members of the family are known as river loaches or hillstream loaches. The rheophilic fishes are found in the well-oxygenated torrential and swift streams of the south and east Asian countries, and Indo-Australian archipelago except in New Guinea. Pectoral and pelvic fins of Balitorids are usually horizontally inserted. The fishes use the modified paired fins for cleaning the rocks. The family includes more than 60 genera and over 600 species. In the waters of Bangladesh, however, only 4 genera containing 9 species have been reported (Box 2).

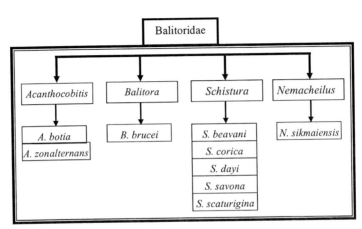

Box 2. The available species under the family - Balitoridae in Bangladesh.

Table 10. Biodiversity status of the fish of the family Balitoridae

Fish	English name	Local name	IUCN Red Book 2000 status	Present status
Acanthocobitis botia	Mottled loach	Bilturi	DD	NO
Acanthocobitis zonalternans	-	-	DD	NO
Balitora brucei	Stone loach	-	NE	NF
Nemacheilus sikmaiensis	-	-	DD	CR
Schistura beavani	Creek loach	-	DD	NF
Schistura corica	-	Korica	DD	CR
Schistura dayi	-	-	NE	NF
Schistura savona	-	Savon korika	NO	NO
Schistura scaturigina	-	Dari	NO	EN

NO – Not threatened; VU – Vulnerable; EN – Endangered; CR – Critically endangered; DD- Data deficient; NE – Not Evaluated; and NF – Not found.

As the distribution of Balitorids goes in Bangladesh, the fishes are only available in a few hillstreams of Dinajpur, Mymensingh and Sylhet region. Presently, all the fishes under the family Balitoridae are very rare in Bangladesh. IUCN Bangladesh (2000) described five fishes as data deficient and two as not threatened. The biodiversity of the *Balitora brucei* and *Schistura dayi* were not determined. In the BAU survey, six species have been found in varying numbers, however, *Balitora brucei*, *Schistura beavani* and *S. dayi* were not found. According to the number of occurrences and number of individuals present in each occurrence, the present biodiversity status of the Balitorids is given in Table 10.

4.3. Family Cobitidae

Members of the family are known as true loaches. They are distributed in Eurasia and northern part of Africa, but the diversity is nowhere greater than in Asia. Bottom dwelling Cobitids are mostly of small size (less than 30 cm). Presently more than 200 species and nearly 30 genera are idntified under this family. New species under the family are being described regularly. Family Cobitidae in Bangladesh is represented by 5 genera and 12 species. Most of the fishes are inhibiting the hillstreams of Dinajpur, Mymensingh and Sylhet. A few of the members are also available in rivers, swamps and ditches throughout the country.

Among the 12 Cobitids, the biodiversity status of 3 were not evaluated by the IUCN Bangladesh (2000), 3 fishes described as data deficient, 2 as endangered and remaining four as not threatened. Among the twelve fishes, the BAU survey found 11 except *Neoeucirrhichthys maydelli* (Table 11). The BAU survey agrees with the biodiversity status of four Cobitids - *Lepidocephalichthys annandalei*, *L. guntea*, *Pangio oblonga* and *Somileptus gongota* as described by the IUCN Bangladesh (2000) as not threatened (NO). Although Bengal loach *Botia dario* was described as endangered by IUCN, the BAU survey found the fishes regularly in large quantities without noticing any decreasing trend in its biodiversity from most of the areas of Bangladesh. Therefore, the biodiversity status of the fish should be changed to the category of not threatened (NO). Remaining three *Botia sp.* and *Pangio pangia* are described as critically endangered (CR), and *Lepidocephalichthys berdmorei, L. irrorata* as endangered (EN) based on the survey findings.

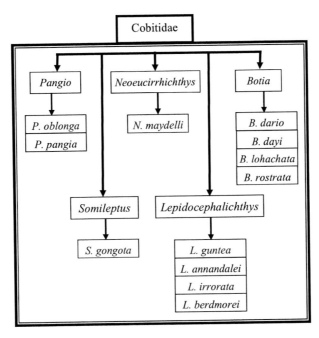

Box 3. The available species under the family - Cobitidae in Bangladesh.

Table 11. Biodiversity status of the fish of the family Cobitidae

Fish	English name	Local name	IUCN Red Book 2000 status	Present status
Botia dario	Bengal loach	Bou	EN	NO
Botia dayi	Hora loach	Rani	DD	CR
Botia lohachata	Reticulate loach	Rani	EN	CR
Botia rostrata	Gangetic loach	Rani	NE	CR
Lepidocephalichthys annandalei	Annandale loach	Gutum	NO	NO
Lepidocephalichthys berdmorei	Burmese loach	Puiya	DD	EN
Lepidocephalichthys guntea	Guntea loach	Gutum	NO	NO
Lepidocephalichthys irrorata	Loktak loach	Puiya	NE	EN
Neoeucirrhichthys maydelli	Goalpara loach	-	DD	NF
Pangio oblonga	Java loach	Panga	NO	NO
Pangio pangia	-	Panga	NE	CR
Someileptus gongota	Gongota loach	Cheng gutum	NO	NO

NO – Not threatened; VU – Vulnerable; EN – Endangered; CR – Critically endangered; DD- Data deficient; NE – Not Evaluated; and NF – Not found.

4.4. Family Cyprinidae

Cyprinidae is the largest family of freshwater fish in the world with about 220 genera containing over 2,400 species. The members are known as carps, barbs, loaches and minnows. Goldfish, rasboras, danios are also included in the family. These fishes have soft finrays without true spines. In some members, however, the last unbranched ray is hardened and forms a stiff spine-like structure, and in a few species, the last anal fin ray is also hardened. Fishes have toothless jaw but strong pharyngeal teeth. The Cyprinids are egg-layers and most fishes do not guard their eggs, only a very few species build nests and/or guard the eggs.

This important family of primarily freshwater fishes is widely distributed in Asia, Africa (excluding Madagasker), Europe and North America. The family is very dominant in most areas within its distribution and is of considerable economic importance in many Asian countries. This large family has been divided into various subfamilies. Sixty three cyprinid species under four subfamilies Leuciscinae, Rasborinae, Cyprininae and Garrinae are found in Bangladesh waters (Box 4).

There are five species available in Bangladesh belong to the subfamily Leuciscinae. The leuciscine minnows could be differentiated from the other Cyprinids by the possessions of a sharply keeled abdominal rim and presence of highly irregular scale pattern on the dorsal side. The 5 species are placed under 2 genera *Securicula* (1 species) and *Salmostoma* (4 species). According to the status given by the IUCN, 3 species are not threatened, one data deficient and one not evaluated. In the BAU survey, all five species were found. Based on the survey findings, the biodiversity status of thee leuciscine - *Securicula gora*, *Salmostoma bacaila* and *S. phulo* remains unchanged, ie., not threatened as described by IUCN Bangladesh (2000), *S. acinaces* is vulnerable and *S. sardinella* is critically endangered (Table 12).

Subfamily Rasborinae has 9 genera containing 21 species in Bangladesh. In the BAU survey all except two rasborine - *Raiamas guttatus* and *Devario aequipinnatus* were found. The genus *Esomus* has 2 species- *E. danricus* and *E. lineatus*. *E. danricus* is abundantly available throughout the country (not threatened). *E. lineatus*, however, is rarely found and the diversity is rapidly decreasing (critically endangered). The two *Chela* species - *Chela cachius* and *C. laubuca* should now be considered as endangered. The 2 species of *Aspidoparia* was described as data deficient by the IUCN. Both the species were found in the BAU survey and the present status of two fish is endangered. According to the BAU survey, the biodiversity status of *Begala elonga*, four *Barilius sp. Raiamas bola* and *Danio dangila* are critically endangered. The biodiversity status of two *Rasbora sp.*, two *Barilius sp.*, and *Devario devario* should be considered as endangered and *Danio rerio* as vulnerable.

There are 11 genera containing 34 species under the subfamily Cyprninae found in Bangladesh. Among the 34 species, the BAU survey did not find 6 species - *Osteochilus hasseltii*, *Labeo angra*, *L. fimbriatus*, *L. dero*, *L. nandina* and *Neolissochilus hexagonolepis*. The survey found 14 fishes regularly in large quantities without any decreasing trend in their biodiversity over the years from most of the areas of their distribution (not threatened). Although *Puntius sarana* was described by IUCN as critically endangered, the stable stocks of the species in a number of rivers and floodplain were found and the fish has been successfully bred in the laboratory. *P. sarana* should be described as vulnerable. According to the findings of the survey the number of endangered and critically endangered fishes under the subfamily Cyprninae is 3 and 10, respectively (Table 12).

The outstanding identifying character of the Subfamily Garrinae is the absence of a groove between the upper lip and snout, the upper lip is being coalescent with the skin of snout. The pectoral and pelvic fins are horizontally placed. It has two genera containing three species in Bangladesh. In our survey we did not find *Garra annandalei*. The present biodiversity status of the two Garrinae is endangered.

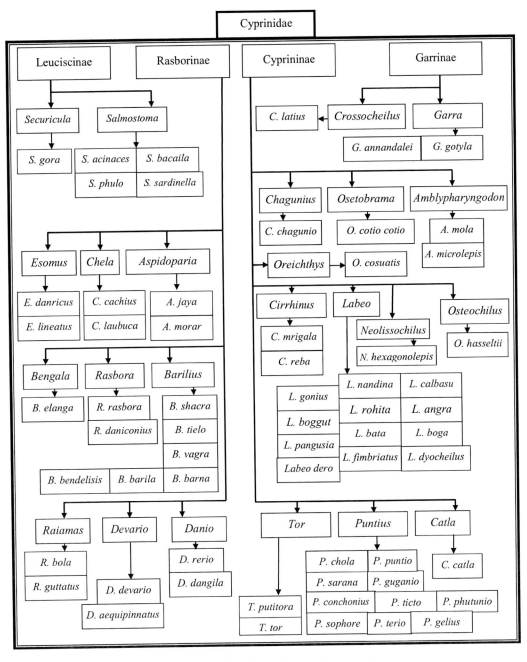

Box 4. The available species under the family - Cyprinidae in Bangladesh.

Table 12. Biodiversity status of the fish of the family Cyprinidae

Fish	English name	Local name	IUCN Red Book 2000 status	Present status
Securicula gora		Ghora chela	NO	NO
Salmostoma acinaces	Silver minnow	Chela	DD	VU
Salmostoma bacaila	Large minnow	Katari	NO	NO
Salmostoma phulo	Finescale minnow	Ful chela	NO	NO
Salmostoma sardinella	Sardinella minnow	Chela	NE	CR
Esomus danricus	Flying barb	Darkina	DD	NO
Esomus lineatus	Striped barb	Darkina	NE	CR
Chela cachius	Glass barb	Chhep chela	DD	EN
Chela laubuca	Glass barb	Chhep chela	EN	EN
Aspidoparia jaya	Joya	Joya	DD	EN
Aspidoparia morar	-	Morari	DD	EN
Begala elonga	Megarasbora	Elong	EN	CR
Rasbora daniconius	Slender rasbora	Darkina	DD	EN
Rasbora rasbora	Gangetic rasbora	Darkina	EN	EN
Barilius barila	-	Barali	DD	EN
Barilius barna	-	Koksa	DD	CR
Barilius bendelisis	-	Joiya	EN	EN
Barilius shacra	-	Koksa	DD	CR
Barilius tileo	-	Pathorchata	DD	CR
Barilius vagra	-	Koksa	EN	CR
Raiamas bola	Trout barb	Bhol	EN	CR
Raiamas guttatus	Burmese trout	Bhol	NE	NF
Devario aequipinnatus	Giant danio	-	EN	NF
Devario devario	Sind danio	Debari	NO	EN
Danio dangila	-	Nipati	DD	CR
Danio rerio	Zebra danio	Anju	NO	VU
Amblypharyngodon microlepis	Indian carplet	Mola	NO	NO
Amblypharyngodon mola	Mola carplet	Mola	NO	NO
Osteobrama cotio cotio	-	Dhela	EN	EN
Chagunius chagunio	Chaguni	Chaguni	DD	CR
Osteochilus hasseltii	Silver sharkminnow	-	DD	NF
Labeo angra	-	Angrot	NO	NF
Labeo bata	Bata	Bata	EN	NO
Labeo boga	-	Bhangon	CR	CR
Labeo boggut	Labeo boggut	-	DD	CR

Table 12. (Continued)

Fish	English name	Local name	IUCN Red Book 2000 status	Present status
Labeo calbasu	Orange-fin labeo	Kalibaus	EN	NO
Labeo dyocheilus	-	Ghora machh	DD	CR
Labeo fimbriatus	Fringed-lipped carp	-	NE	NF
Labeo gonius	Kuria labeo	Sada gonia	EN	NO
Labeo nandina	-	Nandil	CR	NF
Labeo pangusia	-	Baitka	CR	CR
Labeo rohita	Rohu	Rui	NO	NO
Labeo dero	Kalabans	Kursa	NO	NF
Neolissochilus hexagonolepis	Copper mahseer	-	DD	NF
Cirrhinus mrigala	Mrigal	Mrigal	NO	NO
Cirrhinus reba	Reba	Raek	VU	NO
Catla catla	Catla	Katla	NO	NO
Puntius chola	Swamp barb	Chalapunti	NO	NO
Puntius conchonius	Rosy barb	Kanchonpunti	NO	NO
Puntius gelius	Dwarf barb	Gilipunti	DD	EN
Puntius guganio	Glass barb	Molapunti	NO	CR
Puntius phutunio	Spottedsail barb	Phutanipunti	NO	EN
Puntius puntio	Puntio barb	Punti	DD	CR
Puntius sarana	Olive barb	Sarpunti	CR	VU
Puntius sophore	Pool barb	Bhadipunti	NO	NO
Puntius terio	Onespot barb	Teripunti	NO	NO
Puntius ticto	Ticto barb	Titpunti	VU	NO
Oreichthys cosuatis	-	Kosuati	NO	CR
Tor putitora	Putitor mahseer	Mohashol	NE	CR
Tor tor	Mahseer	Mohashol	CR	CR
Crossocheilus latius	-	Kalabata	EN	EN
Garra annandalei	Sucker head	Ghorpoiya	DD	NF
Garra gotyla gotyla	Sucker head	Ghorpoiya	DD	EN

NO – Not threatened; VU – Vulnerable; EN – Endangered; CR – Critically endangered; DD- Data deficient; NE – Not Evaluated; and NF – Not found.

Among the 87 Cyprinifoms reported in Bangladesh, the BAU survey observed 74 over the last ten years. The biodiversity status of many of these have now changed from that listed in the IUCN Red Book published in 2000. The changed biodiversity status proposed here is based primarily on the study of specimens maintained in the Fish Museum and Biodiversity Center (FMBC) of the BAU and through surveys conducted over the last ten years. Figure 5 shows that although the percentage of not threatened and vulnerable Cypriniform remained mostly unchanged over the last ten years, the percentage of critically endangered fish increased almost five times. Many Cypriniforms described as data deficient by IUCN were found in the BAU survey. The biodiversity of these fishes, however, remains under severe threat and most are described as either endangered or critically endangered.

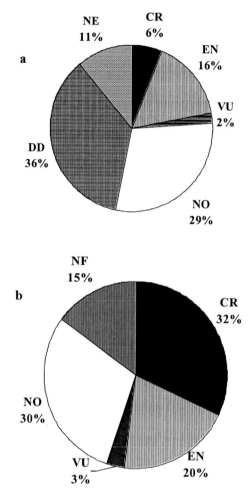

Figure 5. Comparison of percent of Cypriniform fishes under different categories – a. IUCN Bangladesh 2000 and b. Present status NO – Not threatened; VU – Vulnerable; EN – Endangered; CR – Critically endangered; DD- Data deficient; NE – Not Evaluated; and NF – Not found.

5. THE MAJOR CAUSES OF LOSS OF FISH BIODIVERSITY

In the past the major source of fish production in Bangladesh was the inland open water capture fisheries. During 1960s, it contributed about 90% of the country's total fish production. Rapid growth of population coupled with lack of proper management policy, however, created increasing pressure on fish resources and aquatic environment. Due to over exploitation of fish including use of harmful fishing gears and system (fishing by dewatering), degradation and loss of fish habitats, obstruction of fish migration routes by construction of embankment and water control structures mainly to increase agriculture production and road communication, siltation of water bodies by natural process, introduction of a number of alien invasive fish species and water pollution by industry, and agrochemicals, the natural inland fish stocks have declined significantly and fish biodiversity and poor fishers' livelihood have been affected seriously (Ali 1997).

Fish stocks in the rivers and floodplains are declining for a variety of reasons. Most of the indigenous fish are migratory and rely on seasonal flooding for spawning cues and access to larval rearing habitat (floodplain). Almost all dams and embankment interfere directly with the successful completion of the fish migration (breeding and feeding). Agriculture (excessive removal of surface water and extraction of groundwater for irrigation), pollution (domestic and industrial), and unregulated discharge of untreated industrial and farm effluents, habitat destruction also have significant impact, as does the regular overflooding and lack of flooding rain in the last few decades. Introduced species (primarily tilapia, Chinese carp and Thai pangas) are significant contributors to aquaculture production, but also threaten the biodiversity of indigenous fishes. In past, stocking of rivers and floodplain is carried out with both indigenous and introduced species by government and through different projects. The effectiveness of stocking activities has generally not been well assessed. Furthermore, the impacts of aquaculture (both commercial and small scale) have not been accurately assessed in this country.

Most of the literal and floodplains areas are cultivated with rice and other crops, providing multiple annual harvests. Thus, government policy has always prioritized cereal food production. Consequently, most development initiatives in the country have focused on crop cultivation, rather than biological management of the rich floodplain system for fish production, ignoring the needs of poorer people for access to renewable protein sources. Capture fisheries in inland waters which are based on natural productivity generally have reached the level of overexploitation. The inland open water fisheries, where the floodplains assume an important position in the livelihoods and nutrition of the rural poor have now been under serious threat of resource depletion due to various man-made and natural causes. The majority of the waters of this type have been depleted to an alarming state and warrant urgent interventions for conservation and sustenance. Ecosystem integrity has often been destabilized and aquatic systems now fail to support decent levels of aquatic life. As a result the livelihoods of fishers and rural Bangladeshis, previously supported by the inland open waters, are seriously compromised (Coates 1995). Some rivers and floodplains have been modified to a level where they are only recognized as narrow ditches and paddy fields.

During 1960s, the inland capture fisheries contributed about 90% of the country's total fish production. Production from inland capture fisheries has declined significantly over the years and in 2005-06 it accounted only about 40% (Figure 6a). During 1960s, production from inland capture fisheries was almost 20 times higher compared to the then aquaculture production of the country (Figure 6b). However, aquaculture production both in fresh water and brackish water has significantly increased during the last two and a half decades with development of technology. Due to the rapid increase of aquaculture production and sharp decrease of capture fishery production, in 2007-08, the aquaculture contributed almost equally (about 40 %) as inland capture fisheries in total fish production of the country (FRSS-DoF 2008). There has been a qualitative degradation of fish catch in terms of valued species which included cypriniforms like Indian major carps and olive barb. The Indian major carps contributed 67% of the total stock in 1967 in Sylhet-Mymensingh *haor* basin that rapidly declined to 50% in 1973 and only 4% in 1984 (Tsai and Ali 1987).

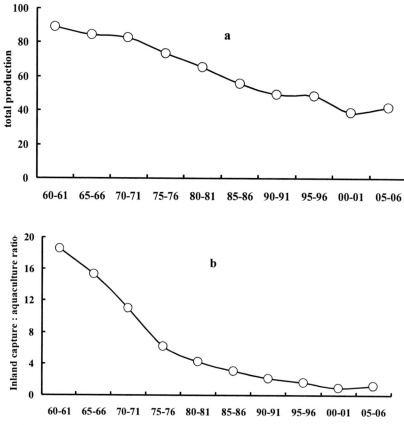

Source: Ali *et al.* (2009).

Figure 6. Trend of fish production in Bangladesh 1960-2006, a. Contribution of inland capture (%) in total fish production, and b. Inland capture to aquaculture ratios.

5.1. Effects of Usage of Pesticides and Chemicals

Every year, there are thousands of tons of different pesticides (insecticides, herbicides, piscicide, miticides, fungicides, weedicides etc.) used around the globe that enter into aquatic systems from direct application and indirectly through terrestrial runoff or wind-borne drift. Pesticide affects the aquatic ecosystem by interrupting the aquatic food chain of open water fish species resulting loss of natural diversity (Parveen and Faisal 2002).

The Bangladesh Pesticides Rule clearly states that "no person shall import, manufacture, formulate, repack, sale, hold in stock, or in any other manner advertise any brand of pesticides which has not been registered." The naive and illiterate farmers are, however, convinced by glib sales talk at promotional camps, and through incentive schemes, to buy new unregistered formulations that promise to protect crops against pest attacks and disease. Suppliers continue to sell many chemicals banned by the government. The increased reliance on pesticides in rice and other crop production has, in some areas, proved to be unsuitable and unsustainable due to pesticide-induced outbreaks of insect pests, development of pesticide resistant pests, rising cost of pesticide use and the negative effects of pesticide use on human health and the environment (Pingali and Gerpacio 1997).

The inundated floodplains of Bangladesh during monsoon are the seasonal habitat of the many indigenous fish. The residual effects of pesticides applied to these floodplains for agricultural purpose before monsoon lead to the fish mass mortality. Besides fish killing, there are also many other chronic effects of pesticides on fish including changes in their reproductive system, metabolism, growth patterns, food availability and population size and numbers (Rohar and Crumrine 2005). Lower abundance of phytoplankton and, consequently, lower abundance of zooplankton are observed as a result of pesticide use in the waterbodies. The application of a pesticide might kill all individuals and it can be substantial perturbation to the ecosystem.

The pesticides affect the aquatic biodiversity in two ways depending on the intensity: sublethal (chronic) effect and lethal (acute) effect. The sublethal concentrations of pesticides can alter a wide range of individual traits including changes in neurotransmitters, hormones, immune response, reproduction, physiology, morphology and behaviour including reduced foraging and changes in swimming ability, predator detection, learning and social interactions (Weis *et al.* 2001). At relatively high concentrations, pesticides become lethal and kill the organisms immediately. However, pesticides that are sublethal for short exposure can also be lethal to aquatic organisms when they are exposed for longer durations (chronic exposure).

The indiscriminate use of insecticides and pesticides in the crop fields by the farmers is one of the major causes of disappearance of many fish from the natural waters in Bangladesh High yielding varieties (HYV) of rice have replaced the indigenous ones resulting in substantial increase in insecticides and pesticides use and causing total disappearance of fish from many monsoon fed water bodies (Mazid 2002). Prolonged misuse of pesticides and fertilizers over the years has also halted the development of inland fisheries and aquaculture (Abdullah *et al.* 1997).

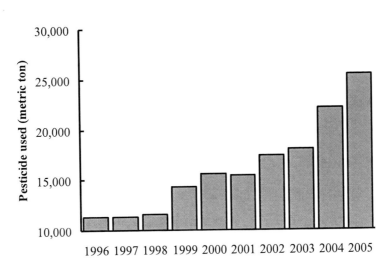

Source: Bangladesh Crop Protection Association, Aziz (2005) and www.moa.gov.bd/statistics/Table4.15CP.htm.

Figure 7. Trend in pesticide use in Bangladesh during 1996-2005.

Pesticide use in Bangladesh got started from mid 1950s and gained momentum in late 1960s with the introduction of green revolution through the use of HYV rice in the country (Rahman 2004). A total of 94 pesticides, with 299 trade names of different groups and formulations have been registered for use in the crop fields. In 1999, the total use of pesticides was about 14,340 mt (active ingredient 2,462 mt) (Banglapedia 2004). In pesticide sector, farmers have been receiving extension services and considerable subsidies from the government over the years (Hossain 1988). As a result of the expansive policy and to minimize the increasing demand of staple crops, pesticide use in Bangladesh has been more than double since 1996, rising from 11,700 mt to 25,466 mt in 2005 (Figure 6). Among the different pesticides, more than 60% are insecticides and used mainly in the paddy field.

Table 13. The exotic fishes introduced into the freshwaters of Bangladesh and the countries they imported from

Common name	Scientific Name	Source	Year of introduction
Siamese gourami	*Trichogaster pectoralis*	Singapore	1952
Goldfish	*Carassius auratus*	Pakistan	1953
Tilapia	*Oreochromis mossambicus*	Thailand	1954
Guppy	*Poecilia reticulata*	Thailand	1957
Common carp	*Cyprinus carpio*	India, Nepal	1960
Mirror carp	*Cyprinus carpio var specularis*	India, Nepal	1979
Scale carp	*Cyprinus carpio var communis*	India, Nepal	1965
Leather carp	*Cyprinus carpio var nudus*	India, Nepal	-
Grass carp	*Ctenopharyngodon idella*	Hong Kong,	1966
Silver carp	*Hypophthalmichthys molitrix*	Hong Kong	1969
Nilotica	*Oreochromis niloticus*	Thailand	1974
Thai sarpunti	*Barbonymus gonionotus*	Thailand	1977
Bighead carp	*Hypophthalmichthys nobilis*	Nepal	1981
Black carp	*Mylopharyngodon piceus*	China	1983
African magur	*Clarias gariepinus*	Thailand	1990
GIFT (genetically improved farmed tilapia)	*Oreochromis niloticus*	Philippines	1994
Genetically improved scale carp	*Cyprinus carpio var communis*	Vietnam	1995
Thai pangas	*Pangasius hypophthalmus*	Thailand	1990
Giant pangas	*Pangasius gigus*	Thailand	-
Mosquito fish	*Gambusia affinis*	India	-
Sucker mouth catfish	*Hypostomus plecostomus*	Hong Kong, Singapore	-
Red piranha	*Pygocentrus nattereri*	Hong Kong, Singapore	2003
Pirapatinga	*Piaractus brachypomus*	Hong Kong, Singapore	2003

Modified from Rahman (2005).

Table 14. The negative impacts of exotic fishes on the cyprniform fishes

Exotic fish	Impact
Tilapia	Their prolific breeding surpasses the carrying capacity of the waterbody leading to stunting of tilapia and a number of cypriniform SIS – mola, dhela, anju, darkina, chela, punti etc.
Common carp	Destroy pond embankments, make water turbid by stirring up mud. Reduce the water transparency and dissolved O_2 in water. Destroy the habitat of SIS living closed to the pond dyke and loaches in the bottom.
Grass carp	High feeding competition with many herbivorous cypriniforms.
Silver carp	Strong feeding and habitat competition with cypriniform – catla in both captive condition and in the wild
Thai sarpunti	Compete with local sarpunti for foods and space
African magur	Predation and voracity of this catfish is legendary, predate on almost all small fishes including most cypriniforms
Thai pangas	Natural diet is finfish, crustacean and insects, periphyton and benthos. This predatory fish is the major cause of disappearance of SIS from the pond system
Mosquito fish	They live in the littoral zone of the waterbody and compete with small cypriniforms for food and habitat
Suckermouth catfish	One of the dangerous catfish, now found in the floodplain allover the country, feeds on small crustaceans and cypriniforms like small loaches and freshwater eel
Red piranha	One of the most dangerous and aggressive species of piranha, feeds on insects, worms and small and large fish. The cultured fish in the pond system and escapees in the wild actively predate on the indigenous fishes including many cyprinifomrs
Pirapatinga	The natural diet is terrestrial plants, fruits, insects and crustaceans, however, in captivity where the natural food is scarce the pirapatinga predate on small cypriniforms. The fish has strong, human like teeth used to crush food items.

In the inland open waters of Bangladesh, mass mortality of fish by pesticides mainly occurs due to the use of pesticides in improper doses and use of banned chemicals. The most commonly used pesticide in the crop field is organochlorine which is highly toxic to fish and other aquatic organism. In sub-lethal doses, organochlorine affects the reproductive physiology of fish and other aquatic fauna. A few drops of endrin can kill all fish in a pond. Hossain and Halder (1996) reported that the main cause of disappearance of the fish from the inland open water of Bangladesh was the use of excessive and banned pesticide and 100% fish mortality occurred within 96 hours of the application of a number of pesticides following even recommended dose for the crop. Lethal dose and even at sub lethal dosage of chemical residues of pesticides largely attributed to cropland runoff contaminants killed fish as well as other aquatic organisms (Parveen and Faisal 2002).

5.2. Introduction of Exotic Fish

Allover the world the exotic species have been recognized as an agent of the loss of indigenous biodiversity. Alteration of species and ecosystem caused by exotic invasive animals and plants influence the functioning and overall health of the affected ecosystems (Ameen 1999).

As a country of rivers and wetlands, Bangladesh is very rich in fish diversity. Even then, over the last six decades a total of 23 fishes have been introduced (Table 13). The invasive species rapidly spread over the wetlands as biological explosives during the rainy seasons. Most of the introduced species were meant only for captive cultivation in closed pond systems but nobody succeeded to maintain the fish in captivity. During monsoon and/or flood the escapees easily found their ways to the rivers and floodplains throughout the country. This posed one of the major threats to the biodiversity of many indigenous fishes in this country.

Several introduced species are highly carnivorous and predatory and eat almost everything including the small indigenous species of fish (SIS - which grow to a maximum length of 5- 25 - Felts *et al.* 1996). Sixty fishes out of 87 cyprinids are considered as SIS. Several exotic species also compete with the SIS and gradually occupy their niches. The ecological, economic and biodiversity consequences of the introductions of exotic fish species have never been taken into consideration. It is very unfortunate that the long-term, and even short-term adverse effects were not considered while introducing the invasive species in Bangladesh. The excessive fecundity and growth rate of these species created pressure on the carrying capacity of the habitat, and the ecosystem balance itself by reducing the indigenous species diversity and population. Some of the negative impacts of exotic species on Cypriniforms are given in Table 14.

6. THE CONSERVATION MEASURES

The government of Bangladesh and a number of non government organizations (NGOs) have taken a number of regulatory and development interventions for sustainable management of the natural fisheries. In order to reverse the loosing trend and ensure sustainability of fish biodiversity and production from inland open waters various measures for protection, conservation and management of fisheries resources have been adopted time to time. Among the measures are the implementation of Fish Protection and Conservation Act 1950 and related rules including new fisheries management policy (licensing the fishing rights directly to the true fishers), community based fisheries management (CBFM), establishment of fish sanctuary in the strategic points of the rivers and floodplains, fish stock enhancement through releasing fish seed in seasonal floodplains, and fish habitat improvement through excavation of link canals (between rivers and floodplains) and *beels*.

The Fish Act 1950 provides regulations for: (i) restriction on capture size of some fish for a specific period, (ii) restriction on catch of any species for specific time or season, (iii) closure of fishing in any water body for any stipulated time period, (iv) restriction of fishing by dewatering or any other destructive method, (v) restriction on the use of any kind of gear and mesh size of net, and (vi) restriction on placing fixed engine in a water course, which may restrict fish migration.

Implementation of fisheries regulations has proved to be very difficult in this country due to institutional weakness of implementing authorities and the socio-economic conditions of the fishers. However, the Fish Act 1950 element – 'closure of fishing in specific area for specific period' as may be termed as 'fish sanctuary' is easier than applying other regulations of the Fish Act. Sanctuary has been tested and found as a powerful tool for protection and conservation of fish stock in Bangladesh.

The dry season is the critical time for the fishes, when water levels in the rivers, canals, *beels* etc. recede drastically leaving a very few refuge for the inland fishes. Fish are exposed to greater predation and increased susceptibility to fishing pressure as the water level drops due to water extraction for irrigation and evaporation due to persistent heat of the dry season. Loss of surface water in the dry season results in the reduction in the brood fish stock. The fishes become increasingly vulnerable to intensive fishing and thereby the fish stock particularly the brood stock depletes to such a level that cannot sustain the fisheries and gradually fish diversity and production decline. Therefore, the major issue for biodiversity conservation is to provide sufficient dry season refuges to maintain the population at sustainable level.

Among all measures, fish sanctuary has been apparently found most effective for fish biodiversity conservation, when other measures are difficult to implement in the present administrative and social contexts. With this notion, Bangladesh government has established fish sanctuaries under different development projects following a number of management approaches since 1960 and more intensively in last decade. The NGOs like BRAC, CARITAS, CNRS, PROSHIKA and WorldFish Centre (CBFM project) have also been involved in fish stock development by establishing traditional sanctuaries in *beels* and rivers of Bangladesh.

In addition, a number of silted up *beels, baors,* dead rivers and link canals have been re-excavated by the government under the food for work programs over the years. By 2000 a total of about 8,300 ha water area of borrow pit, *baors*, dead rivers, canals and *beels* had been excavated (DoF 2005). In the late 1990s the government approved a series of sectoral policies including National Fisheries Policy (1998), National Environment Policy (1995), and National Land Use Policy (2001) with a new emphasis on maintaining and protecting the moribund inland waterbodies. Under the National Fisheries Policy, government has formulated strategies for inland capture fisheries and emphasized on fisher community participation in fisheries management, along with fish sanctuaries as a key management tool (DoF 2005).

6.1. Fish Sanctuaries

The massive siltation has threatened the existence of most of the inland waterbodies – rivers, floodplains, *beels, haors* and *baors*. Many waterbodies once the blessings for Bangladesh providing fishing, communication and irrigation facilities are now drying up at an alarming rate. Most of the waterbodies are becoming empty of fish. The causes of reduced abundance of fish are over-fishing, reduced flooding, siltation, agricultural and industrial pollution etc. These activities have severely affected the indigenous fish diversity of the country. The complete drying up in many parts of the river and other waterbodies is a common scenario during lean season, which is detrimental to fish populations. Where perennial waterbodies have been transformed to seasonal waters due to several manmade and natural factors, establishing a fish sanctuary (refuges where fish are protected during the dry season) can help to restore the fish habitat and fish diversity. The establishment of fish sanctuaries in the deeper parts of the waterbodies where fish reside during dry season and grow and attain maturity for spawning in the next monsoon– is particularly very important. At the onset of early monsoon rains, these fish disperse on the rivers and adjacent floodplains for breeding and feeding.

Following the provision of the Fish Act 1950, Govt. declared closed season for fishing of certain species or for all species of fish in specified water bodies under normal fisheries management programme and under different development schemes/projects of DoF. In 1952 Govt. prohibited catching of cypriniforms - rui, catla, mrigal, kalibaus and gonia of any size in rivers and canals for different time period between mid March to 31 July every year except for pisciculture purposes. Under the Development and Management Scheme of Department of Fisheries (DoF), 23 sanctuaries were established in different floodplains during 1960-1965. Upon having good results of the established sanctuaries, 25 more sanctuaries were established under the same scheme of DoF during 1965-70. Afterwards 10 more sanctuaries were established in 1987 under the Integrated Fisheries Development Project of DoF.

Most of the fish sanctuaries now focus on the need of involvement of fisher community and local govt. in the management system, long tenure of lease period and also strong monitoring and supervision. Besides, to safeguard of the fishers interest, the Govt. policy now is to establish sanctuary in part of the floodplain and the remaining part is open for fishing by the local fishers. Based on this idea, under different development projects, government has established a number of sanctuaries involving the fisher communities with support of NGOs. In a government declared fish sanctuary, catching/killing of fishes is prohibited by law and order of competent authority for all the times to come or for a specified period mainly with objective of protecting/conserving the fish.

A total of 463 permanent fish sanctuaries covering an area of 1,745 ha have been established in 98,405 ha water bodies by 2007 (Table 15). A number of the sanctuaries have been closed after the projects ended. Management has deteriorated in many sanctuaries due to the conflict of interests among the stakeholders, lack of funding and lack of coordination among the organizations.

Fish sanctuary in Bangladesh was proved to be an important and efficient tool for management in protection and conservation of fishes and other aquatic organisms (Ali *et al.* 2009). Since mid 80s, concept of the involvement/participation of local fisher community in setting up and managing sanctuaries has been the government policy. However, a major problem in managing sanctuaries in public water bodies is the policy conflict between the government ministries. Although the national fisheries policy envisages establishing fish sanctuaries, there is no clear guideline for establishment and management of fish sanctuaries. To make the fish sanctuaries more effective, the following stages should be followed - mitigation of all the conflicts among the stakeholders involved, formulation of clear guidelines of sanctuary management, selecting the strategic place and size of the sanctuary, proper awareness building among the stakeholders, ensuring proper community organization and full participation and continuous monitoring and impact assessment.

Table 15. Fish sanctuaries established in Bangladesh by 2007

Project/ Programme	Area of water body ha	Area of Sanctuary ha	Number of Sanctuary
Fourth Fisheries Project	12,233	1,022	63
Community Based Fisheries Project (CBFM-2)	9,602	93	182
Management of Aquatic Ecosystems through Community Husbandry Project (MACH)	785	76	65
New Fisheries Management Policy	1,698	77	21
Fisheries/ Fish Culture Development in *Beel* and Chharas project	1,294	18	29
Aqua Development Project (Faridpur)	454	11	14
Patuakhali Barguna Aquaculture Extension Project (PBAEP)	307	26	19
Fish Habitat Restoration Project	3,890	73	45
Fisheries Development in *Jabai Beel* project	75	4	4
Sustainable Environment Management Programme (SEMP-17)	50	17	12
Community Based Wetland Management Project(CBWM- 4)	17	4	7
Kaptai Lake	68,000	324	2
Total	98,405	1,745	463

Modified from Ali *et al.* (2009).

6.2. Fish Breeding, Domestication and Gene Banking

As more fish species of Bangladesh become threatened, there is tremendous need to preserve the disappearing genetic material as well as to conserve the existing gene pools. The ideal strategy for conservation of threatened and endangered fish species is through restoration of the native habitat of the species (*in situ* approach). Unfortunately, most habitat damages are irrevocable and where remediation is possible it is costly and requires a great deal of time, as the restoration process is slow. One alternative is to maintain *ex situ* conservation (outside the natural environment) as live populations or in a cryopreserved sperm bank (Pullin *et al.* 1991).

Domestication of wild fishes in most cases benefits both the farmer and the environment. Investments in domestication have to pay off; therefore, researches should take into account the biodiversity and production scenario and overall socioeconomic and environmental outcome at a broader scale. In Bangladesh, to date about 22 fish species have been domesticated and their breeding and rearing protocols have been developed. Around 50% of the domesticated fishes are cypriniforms and now under nation-wide aquaculture (Table 16). Though there is high possibility of working with reduced gene pool, it is optimistically believed that the biodiversity of the domesticated fish are well-preserved.

Table 16. The domesticated indigenous fishes of Bangladesh

Order	Fish	Culture status
Cypriniformes	*Catla catla*	Country-wide commercial
	Labeo rohita	Country-wide commercial
	Labeo gonius	Country-wide commercial
	Labeo bata	Country-wide commercial
	Labeo calbasu	Small scale, sporadic
	Cirrhinus mrigala	Country-wide commercial
	Cirrhinus reba	Small scale, sporadic
	Tor putitora	Breeding protocol developed
	Puntius sarana	Small scale, sporadic
	Lepidocephalichthys guntea	Breeding protocol developed
	Botia dario	Breeding protocol developed
Osteoglossiformes	*Chitala chitala*	Small scale, sporadic
Siluriformes	*Ompok bimaculatus*	Small scale, sporadic
	Ompok pabda	Small scale, sporadic
	Mystus vittatus	Small scale, sporadic
	Mystus gulio	Breeding protocol developed
	Clarias batrachus	Small scale, sporadic
	Heteropneustes fossilis	Small scale, sporadic
Synbranchiformes	*Mastacembelus armatus*	Breeding protocol developed
	Macrognathus aculeatus	Breeding protocol developed
Perciformes	*Anabas testudineus*	Breeding protocol developed
	Colisa fasciata	Breeding protocol developed

Recently there has been expanded development of cryogenic sperm banks (preservation of fish sperm in liquid N_2 at -196 °C) for fish in Europe and North America. These sperm banks are more cost effective than maintaining live gene banks which require wide space, maintenance and high costs. Cryogenic gene banking avoids the risk of genetic contamination and requires little space and minimal facilities.

Fish sperm cryopreservation assists conservation of fish biodiversity through gene banks of endangered species, and assists aquaculture by providing flexibility in spawning of females and selective breeding through synchronizing artificial reproduction, efficient utilization of semen, and maintaining the genetic variability of broodstocks (Lahnsteiner 2004). The technique also ensures preservation of genetic materials of the genetically superior wild fish populations and the gene transfer between wild and hatchery stocks (Tiersch *et al.* 1998).

The sperm cryopreservation protocol for different fish species seems variable and species-specific. Although fish are the main protein source in Bangladesh and other countries in the sub-continent, and the fish biodiversity and production from open water are declining, little attention has been paid to cryopreservation of fish sperm. In Bangladesh, research on fish sperm cryopreservtion was started in early 2004. The studies have focused on aquacultured or commercial species and so far none of the threatened species have been considered (Table 17).

Table 17. Cryopreservation of sperm of some fish species in Bangladesh

Fish group	Fish
Indigenous - cypriniform	*Catla catla*
	Cirrhinus mrigala
	Labeo rohita
	Labeo calbasu
	Puntius sarana
Indigenous - others	*Ompok bimaculatus*
	Mastacembelus armatus
	Channa striatus
	Rita rita
Exotic fishes	*Cyprinus carpio*
	Hypophthalmichthys molitrix
	Hypophthalmichthys nobilis
	Barbonymus gonionotus
	Oreochromis niloticus

Genetic stock conservation for wild and domesticated fishes is very important, as the genetic diversity of every species develops through a long evolutionary process over millions of years. Cryogenic techniques can assist in the conservation of biodiversity, to bring back the threatened species to natural environment with restocking programmes, as well as in improving aquaculture production. Cryogenic sperm banks for more fish need to be established as means of germplasm conservation in Bangladesh.

7. CONCLUSION

There is a crying need to adjust the existing laws and legislation of the country for integrated resource management to save the fisheries resources. Although much of the damage to the habitat and biodiversity of the inland water of Bangladesh over recent decades is likely to be irreversible, there is still time to act. From now on, Bangladesh government, the NGOs and national and international bodies should foster a social and technical environment in which the enormous richness of the fisheries resources can stabilize and eventually rebuild so as to continue to feed people of today and tomorrow. Poverty in fishing communities should be reduced in part by ensuring a stable supply of fish, something can only be achieved through improved knowledge, integration of fisheries and freshwater management, and greater public involvement. In case of fishing closure in areas or for certain time, the fishers should be provided with alternative income generating activities, credit with low interest and other sustainable means. Creating public awareness of the importance of maintenance of fish diversity in Bangladesh is extremely necessary and should be the first priority for a lasting change. Sustenance of fish diversity can only be achieved with public support. Bangladeshi fishers, fish farmers, traders, processors, and general people as a whole need - to understand the issues, to be involved in the formulation of management plans and to benefit from the whole process. A key step in building fisheries co-management and fish biodiversity conservation with community participation is to bring all the various stakeholders in a common front with a view to sharing resource and knowledge, creating an environment for meaningful discussion on cross-cutting themes and valuing each other.

A renewable resource like fish, when under intense exploitation, needs a management regime as it is not inexhaustible. Therefore, management measures should be applied in such a way that young fish are protected to grow before capture and enough are left as breeding stock for future generations. The management measures should include – regulate fishing intensity at sustainable level, control gear selectivity, gear type and size of fish, closed season, prohibition of destructive fishing, closed fish sanctuary, and allocation of resources to different types of fisheries.

For sustainable and well-protected fish diversity for present and for future, the country should go for -

- Rational use of inorganic fertilizers and pesticides, and proper management of industrial effluents,
- Maintenance of minimum water depth (at least 1 m) during water extractions from critical waterbodies,
- Regulation of selective fishing gears, mesh sizes, and fishing by dewatering,
- Establishment of more fish sanctuaries and natural *beel* nurseries in strategic points,
- Stock enhancement programmes,
- Establishment of community-based organizations (CBO) among the fishers,
- Zero tolerance to new exotic fish introduction, and
- Strict application of existing fisheries rules and regulations.

This is the high time to care for the biodiversity of the most valuable Cypriniform and other indigenous fishes – the pride, heritage and livelihood of Bangladesh before they are lost forever. The researchers, policy makers, GOs and NGOs and national and international bodies should come forward to conserve the fish species using both *in situ* and *ex situ* approaches.

ACKNOWLEDGMENTS

We wish to acknowledge the research fellows and students of the Laboratory of Fish Biodiversity and Conservation, Bangladesh Agricultural University for their sincere cooperation and assistance in preparing this manuscript. Thanks also are given to Dr. Andy P. Shinn, Institute of Aquaculture, University of Stirling, UK and Mr. Md. Nahiduzzaman, Professor Md. Samsul Alam and Professor A.K.M. Nowsad Alam, Bangladesh Agricultural University for improving the English, and helping to rewrite some portion of this manuscript.

REFERENCES

Abdullah, A. R., Bajet, C. M., Matin, M. A., Nhan, D. D. & Sulaiman, A. H. (1997). Ecotoxicology of pesticides in the tropical paddy field ecosystem. *Environmental Toxicological Chemistry, 16*, 59-70.

Ahmed, M. (1991). A model to determine benefits obtainable from the management of riverine fisheries of Bangladesh. *ICLARM Tech. Rep, 28*, 133.

Ahmed, A. T. A. (1995). Impact of Other Sectoral Development on the Inland Capture Fisheries of Bangladesh. Proceeding of the Fourth Asian Fisheries Forum. *The Fourth Asian Fisheries Forum*. China Ocean Press. Beijing, China.

Ahmed, A. T. A. & Ali, M. L. (1996). Fisheries Resources Potential in Bangladesh and Manpower Development for Fisheries Activities. *In: Population Dimension of Fisheries Development and Management Policies of Bangladesh*. Dhaka, Bangladesh.

Ahmed, K. K. U., Ahamed, S. U., Hasan, K. R. & Mustafa, M. G. (2007). Option for formulating community based fish sanctuary and its management in beel ecosystem in Brahmanbaria. *Bangladesh J. Fish. (Special Issue), 30*, 1-10.

Ali, M. Y. (1997). Fish, water and people, Reflections on inland openwater fisheries resources in Bangladesh. *The University Press Limited*, Dhaka, 154.

Ali, M. L., Hossain, M. A. R. & Ahmed, M. (2009). Impact of Sanctuary on Fish Production and Biodiversity in Bangladesh. Final Project Report. *Bangladesh Fisheries Research Forum (BFRF)*, Dhaka, Bangladesh. 80.

Ameen, M. (1999). Development of guiding principles for the prevention of impacts of alien species. *Paper presented at a consultative workshop in advance of the 4th meeting of SBSTTA to the CBD*, organized by IUCN Bangladesh at Dhaka on 25 May 1999.

Azher, S. A., Dewan, S., Wahab, M. A., Habib, M. A. B. & Mustafa, M. G. (2007). Impacts of fish sanctuaries on production and biodiversity of fish and prawn in Dopi beel, Joanshahi haor, Kishoregonj. *Bangladesh J. Fish. (Special Issue), 30*, 23-36.

Aziz, M. A. (2005). Country Reports, Bangladesh. In the Proceedings of the Asia regional workshop on the implementation, monitoring and observance of the international code of conduct on the distribution and use of pesticides. *FAO Regional Office for Asia and the Pacific*, Bangkok, Thailand. 26-28 July 2005 RAP Publication - 2005/29. 258.

Banglapedia. (2004). National Encyclopedia of Bangladesh, Asiatic Society of Bangladesh, 1st edn. February 2004. Dhaka, Bangladesh. Available from URL: www.banglapedia.org.

Bernacsek, G. M. S., Nandi, S. & Paul, N. C. (1992). Draft Thematic Study: Fisheries in the North East Region of Bangladesh. North East Regional Water Managemenet Project (FAP-6). *Bangladesh Engineering and Technological Services*, Dhaka, Bangladesh, 122.

Bhuiyan, A. K. M. A. & Choudhury, S. N. (1997). Freshwater aquaculture: potentials, constraints and management needs for sustainable development. *National Workshop on Fisheries Development and Management in Bangladesh*. BOBP/REP/74. Madras, India. 77-113.

BWDB (Bangladesh Water Development Board). (2005). Bangladesher Nad Nadi, Pani biggyan, June 2005, *Bangladesh Water Development Board*, Dhaka. 621.

Coates, D. (1995). Inland capture fisheries enhancement: status, constraints and prospects for food security. *Report of the International Conference on the Sustainable Contribution of*

Fisheries to Food Security, 4-9 December, 1995. Kyoto, Japan. Government of Japan. Document # KC/ FI/95/TECH/3. 85.

DoF (Department of Fisheries) (2005). Matshya Pakkhya Saranika-2005. Department of Fisheries, Ministry of Fisheries and Livestock. *The Government of the Peoples Republic of Bangladesh*, Ramna, Dhaka. Bangladesh. 109.

DoF (Department of Fisheries). (2008). Matshya Pakkhya Saranika-2008. Department of Fisheries, Ministry of Fisheries and Livestock. *The Government of the Peoples Republic of Bangladesh*, Ramna, Dhaka. Bangladesh. 96.

Dudgeon, D. (2002). An inventory of riverine biodiversity in monsoonal Asia: present status and conservation challenges. *Water Science and Technology*, 45, 11-19.

ESCAP (Economic and Social Commission for Asia and the Pacific). (1998). Integrating Environmental Considerations into the Economic Decision Making Process.*ESCAP*, Bangkok, Thailand. www.unescap.org/drpad/pub3/integra/ modalities/ bangladesh /4bl02b01.htm.

FAP (Flood Action Plan)-6. (1992). Northeast Regional Water Management Project, Fisheries Specialist Study, Main Report (Volume-1), *Engineering and Planning Consultants Ltd*, CIDA. 67.

FAP (Flood Action Plan)-17. (1994). Fisheries Studies and Pilot Project- Final Report. ODA, UK. 185.

Felts, R. A., Rajts, F. & Akhteruzzaman, M. (1996). Small Indigenous Fish Species Culture in Bangladesh. Technical Brief, EC & DoF, *Integrated Food Assisted Development Project*, Gulshan, Dhaka, Bangladesh. 41.

FRSS-DoF. (2008). Fisheries Statistical Year Book 2006-07. Department of Fisheries, *Ministry of Fisheries and Livestock*, Bangladesh. 42 p.

Graaf, G. J., Born, A. F., Uddin, A. M. K. & Marttin, F. (2001). Floods, Fish and Fishermen. *Eight Years Experience with Floodplain Fisheries in Bangladesh*. The University Press Limited, Bangladesh. 174.

Graaf, G. J. (2003). The flood pulse and growth of floodplain fish in Bangladesh. *Fisheries Management Ecology, 10(4)*, 241-247.

Haldar, G. C., Ahmed, K. K., Alamgir, M., Akhter, J. N. & Rahman, M. K. (2002). Fish and fisheries of Kaptai Reservoir, Bangladesh. *In*: Management and Ecology of Lake and Reservoir Fisheries (ed. I.G. Cowx), 144-160. Blackwell Science Publishers, Oxford, UK.

Hoq, M. E. (2009). Fisheries Resources and Management Perspectives of a World Heritage Site- Sundarbans Mangrove, Bangladesh. In: Fisheries: Management, Economics and Perspectives (N. F. McManus, & D. S. Bellinghouse, Eds.)., *Published by Nova Science Publishers Inc.*, New York, USA. (ISBN 978-1-60692-303-0). 199-227.

Hossain, M. (1988). Nature and Impact of the Green Revolution in Bangladesh. Research Report No. 67. *International Food Policy Research Institute*, Washington DC, USA. 146.

Hossain, G. M. & Nishat, A. (1989). Planning considerations for water resources development in the *haor* areas. *AIT-BUET Workshops on Development and Technology*, Bangladesh University of Engineering and Technology, Dhaka, Bangladesh. 12.

Hossain, Z. & Halder, G. C. (1996). Impact of pesticide on environment and fisheries. In: Fisheries Fortnight publication: Technologies and management for fisheries development. (M. A. Mazid, Ed.), *Bangladesh Fisheries Research Institute*, Mymensingh, Bangladesh. 160.

Hossain, M. A. R., Nahiduzzaman, M., Sayeed, M. A., Azim, M. E., Wahab, M. A. & Olin, P. G. (2009). The Chalan beel in Bangladesh: habitat and biodiversity degradation, and implications for future management. *Lakes & Reservoirs: Research and Management, 14*, 3-19.

Iqbal, I. (2006). The railway in colonial India: Between ideas and impacts. In: Our Indian Railway: Themes in Indian Railway History. (R., Srinivasan, M. Tiwari, & S. Silas, Eds.), 173-186. *Foundation Books Pvt. Ltd.*, Darayganj, New Delhi, India

IUCN Bangladesh. (2000). Red Book of Threatened Fishes of Bangladesh. *IUCN- The World Conservation Union*. 116.

Junk, W. B., Bayley, P. B. & Sparks, R. E. (1989). The flood pulse in river floodplain systems. In: Proc. International Large River Symposium (D. P. Dodge, ed.). *Canadian Special Publication Fisheries Aquatic Sciences*, 106, 110-127.

Khan, M. A., Srivastava, K. P., Dwivedi, R. K., Singh, D. N. Tyagi, R. K & Mehrotra, S. N. (1990). Significance of ecological parameters in fisheries management of a newly impounded reservoir-Bachhara reservoir. In: Contribution to the Fisheries of Inland Open Water Systems in India (Part I) (eds. A.G. Jhingran, V.K. Unnithan and A. Gosh). IFSI, *Barrackpore*, India. 100-108.

Lahnsteiner, F. (2004). Cryopreservation of semen of the Salmonidae with special reference to large-scale fertilization. *Methods in Molecular Biology, 253*, 1-12.

Mazid, M. A. (2002). Development of Fisheries in Bangladesh: Plans and strategies for income generation and poverty alleviation. *Nasima Mazid Publications*, Dhaka, Bangladesh. *176*.

NERP. (1995). Northeast Regional Water Management Project; Wetland Resources Specialist Study, Northeast Regional Water Management Plan, Bangladesh Flood Action Plan 6 (IEE NERP FAP 6). Flood Plan Coordination Organisation, Water Development Board, *Government of the People's Republic of Bangladesh*, Bangladesh.

Parveen, S. & Faisal, M. I. (2002). Open Water Fisheries in Bangladesh: A critical Review. *Environmental Studies*, North South University, Dhaka, Bangladesh. 20.

Pingali, P. L. & Gerpacio, R. V. (1997). Living with reduced insecticide use for tropical rice in Asia. *Food policy, 22*, 107-118.

Pullin, R. S. V., Eknath, A. E., Gjedrem, T., Tayamen, M., Julie, M. & Tereso, A. (1991). The genetic improvement of farmed tilapias (GIFT) Project. *Naga, the ICLARM Quarterly, 14(2)*, 210-315.

Rahman, M. M. (2004). Uses of persistent organic pollutants (POPs) in Bangladesh. Paper presented at the Inception Workshop of the Project Bangladesh: Preparation of POPs National Implementation Plan under Stockholm Convention (POP NIP), Department of Environment, *held at Hotel Sonargaon*, Bangladesh.

Rahman, A. K. A. (2005). Freshwater Fishes of Bangladesh. Second edition, *Zoological Society of Bangladesh*, Dhaka, Bangladesh. 394.

Rohar, J. R. & Crumrine, P. W. (2005). Effects of an herbicide and an insecticide on pond community structure and processes. *Ecological Applications, 15(4)*, 1135-1147.

Saha, S. B., Bhagat, M. J. & Pathak, V. (1990). Ecological changes and its impact on fish yield of Kulia *beel* in Ganga Basin. *J. Inland Fish. Soc. India, 22(1-2)*, 7-11.

Sugunan, V. V. & Bhattacharjya, B. K. (2000). Ecology and fisheries of *beels* in Assam. Central Inland Capture Fisheries Research Institute. Bull. No. 104. *Barrackpore*, India. 65.

Talwar, P. K. & Jhingran, A. G. (1991). *Inland Fishes of India and Adjacent Countries.* Oxford & IBH Publishing Co., Calcutta, *1158*.

Tiersch, T. R., Figiel, C. R., Wayman, W. R., Williamson, J. H., Carmivhael, G. J. & Gorman, O. T. (1998). Cryopreservation of sperm of the endangered razorback sucker. *Transactions of the American Fisheries Society*, *127*, 95-104.

Tsai, C. & Ali, L. (1987). The Changes in Fish Community and Major Carp Population in Beels in the Sylhet-Mymensingh Basin, Bangladesh. *Indian Journal of Fisheries*, *34(1)*, 78-88.

Weis, J. S., Smith G., Zhou T., Santiago-Bass, C. & Weis, P. (2001). Effects of contaminants on behavior: biochemical mechanisms and ecological consequences. *Bioscience*, *51*, 209-217.

Chapter 5

BASIC ECOLOGICAL INFORMATION ABOUT THE THREATENED ANT, *DINOPONERA LUCIDA* EMERY (HYMENOPTERA: FORMICIDAE: PONERINAE), AIMING ITS EFFECTIVE LONG-TERM CONSERVATION

Amanda Vieira Peixoto[1], *Sofia Campiolo*[1,2,3] *and Jacques Hubert Charles Delabie*[1,3,4,*]

[1]Graduate program in Zoology, Universidade Estadual de Santa Cruz (Santa Cruz State University), 45650-000 Ilhéus, Bahia, Brazil
[2]Biological Science Department, Universidade Estadual de Santa Cruz (Santa Cruz State University), 45650-000 Ilhéus, Bahia, Brazil
[3]Myrmecology Laboratory, Cocoa Research Center, CEPLAC, Mailbox 7, 45600-000 Itabuna, Bahia, Brazil
[4]Department of Agricultural and Environmental Sciences, Universidade Estadual de Santa Cruz (Santa Cruz State University), 45650-000 Ilhéus, Bahia, Brazil

ABSTRACT

The giant ants of the *Dinoponera* genus belong to a convergent group of ants in which there is no morphologically specialized caste of reproducing females and reproduction is done by fertilized workers known as gamergates. The *Dinoponera* genus is endemic of South America. The Brazilian Atlantic rain forest native species, *Dinoponera lucida* Emery, is included on the Brazilian official red list, because of its habitat loss and fragmentation and peculiarities of its biology. Population ecology and biological cycle studies were carried out from August, 2004 to July, 2005, in five forest areas in the states of Bahia and Espirito Santo, Brazil. A range of information was accumulated, with the purpose of providing strong arguments for further implantation of

---

[*] Corresponding author: Email: jacques.delabie@gmail.com

an effective conservation plan for the species: i) ant nest architecture: dynamics of ant nest populations, suggesting colony division per fission, as already seen in other species, was observed; ii) aggregate distribution of ant nests was defined and is explained by the particular reproductive biology of the ant; iii) ant nest colonization by other terrestrial arthropods was studied; and iv) the ant foraging behavior (activity time, prey categories) was studied. Such information is necessary to justify the long-term effort necessary to implant an effective conservation plan for this ant.

INTRODUCTION

Between other factors, the Ponerinae subfamily is characterized by its retention of ancestral morphological features (Peeters & Ito, 2001). One of such features is the large size it can reach, as in the case of the *Dinoponera* genus. The large size of such species is very noticeable and can reach 4 cm in length, therefore being considered the biggest ants on the planet (Kempf, 1971; Paiva & Brandão, 1995). The *Dinoponera* genus is composed of six valid species (Bolton, 1995), all of which are endemic to South America and most of which are allopatric, except for the superposition of some areas for some species distributed in the Amazon (Kempf, 1971; Paiva & Brandão, 1995). All *Dinoponera* species are black and large (2.5 to 4 cm); and morphological differences are very discreet, but clearly defined (Kempf, 1971). *Dinoponera lucida* Emery is exclusively distributed in a part of the Atlantic Rain Forest that grossierly follows the eastern coast of Brazil, called the *Mata Atlântica* (Figure 1). In 2003, it was included on the Brazilian Red List (Fundação Biodiversitas, 2003; Campiolo & Delabie, 2008). Its main threat is the human destruction of its natural habitat, which is the Atlantic Coastal Forest. This habitat is becoming smaller and increasingly isolated.

Little is known about the ants of the genus *Dinoponera*, with the exception of *Dinoponera quadriceps* Santschi, which is basically studied for its reproductive behavior (Araújo *et al.*, 1988, 1990a&b; Araújo & Jaisson, 1994; Araújo, 1995; Monnin & Araújo, 1994, 1995; Monnin & Peeters, 1997, 1998; 1999; Monnin & Ratnieks, 1999, 2001). However, aspects of the biology and ecology of this genus are sparcely known (Mariano et al., 2004, Fowler, 1985; Morgan, 1993; Paiva & Brandão, 1995; Fourcassié & Oliveira, 2002, Vasconcellos *et al.,* 2004). Topics such as nest spatial distribution, the place of nidification, nest architecture, size and composition of colonies, foraging habits and food sources were yet poorly studied (Fowler, 1985; Morgan, 1993; Paiva & Brandão, 1995; Fourcassié & Oliveira, 2002, Mariano *et al.*, 2004; Vasconcellos *et al.,* 2004). Patterns of spatial distribution may explain the existence or non-existence of intra-specific competition for nidification place and food resources (Ludwing & Reynolds, 1988). Patterns of spatial distribution varied from random to uniform in the species *Dinoponera australis* Emery and *D. quadriceps* (Paiva & Brandão, 1995; Vasconcellos *et al.* 2004, respectively).

Ants of the *Dinoponera* genus belong to a convergent group of queenless ant species, which have a sub-caste of workers called gamergates (Peeters, 1993, 1997; Monnin & Peeters, 1998) able to mate and reproduce. The sexual reproduction of workers occurs in at least an hundred ant species in the subfamily Ponerinae (Peeters, 1991) and in some ants of the subfamily Myrmicinae (Heinze *et al.* 1999; Hölldober *et al.* 2002). The process through which new *Dinoponera* colonies are formed seems to be exclusively the fission of mature colonies which have large populations (Fowler, 1985; Araújo *et al.*, 1990a; Araújo & Jaisson,

1994). Hence, formation of new aggregates of populations occurs nearby the original colony, resulting in aggregates which are closely related, at least through female gene flow. This is due to the fact that the reproductive function is performed by a worker (the queen caste totally inexists), the gamergate, which dominates the colony and mates in the entrance of the nest with only one male (males, all haploid, are the only individuals to have wings in this species) from the other colony. In such species, reproduction is made by the workers who have spermatheca and mature ovaries, making them able to mate and, for this reason, to produce diploid offspring (females [=other workers]). The gamergate will play the same role as a queen of a normal colony, once it is mated. As the reproducing female is apterous, there is no dissemination at distance on the part of the females; and that kind of dissemination remains the male rule. Since all female workers of a queenless colony have ovaries that can mature, the gamergate must maintain its superiority according a dominance hierarchy in order to continue being the reproducer of the colony (Monnin & Peeters, 1999). For this reason, it is common to observe agonistic interactions among workers in such colonies (Medeiros et al., 1992; Cuvillier-Hot et al., 2002) as well as egg cannibalism (Monnin & Peeters, 1997).

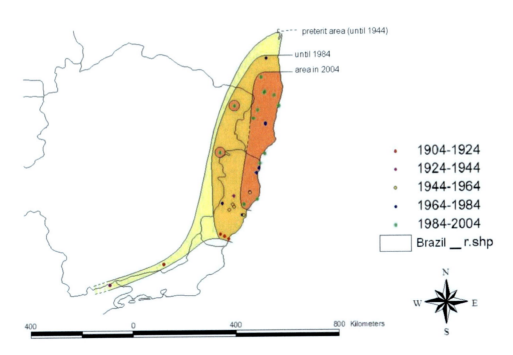

Figure 1. Preterit and current distribution of *Dinoponera lucida* in the Brazilian States of Bahia, Espirito Santo, Minas Gerais and São Paulo, along the Atlantic Ocean coast.

All species of the *Dinoponera* genus build subterranean nests (Fowler, 1985; Morgan, 1993; Paiva & Brandão, 1995; Fourcassié & Oliveira, 2002; Vasconcellos et al. 2004), varing from 40 cm to 200 cm deep. The number of workers in colonies of different species varies from 20 to 200 workers. In all species in which it was studied, foraging was always solitary (Figure 2), without recruitment, and food habits were always generalists (Morgan, 1993; Fourcassié & Oliveira, 2002). Studies concerning *Dinoponera lucida* are still incipient and most of them were initiated after this specie was included in the Brazilian Red List (Fundação Biodiversitas, 2003). Such studies include general elements which characterize nests and colonies (Paiva & Brandão, 1995), behavior (Peixoto *et al.*, 2008), and cytogenetic studies (Mariano *et al.*; 2004, 2008). This study aimed at accurately describing aspects of *D. lucida* biology focusing on patterns of the spatial distribution of ant nests, time and space dynamics, characterization of ant nests and ant populations, foraging methods and the food habits.

MATERIAL AND METHODS

The experiments were carried out between August 2004 and June 2005 in five forest fragments located in the Central Corridor of the Atlantic Coastal Forest (Brasil, 2006) in the following localities: a) Belmonte, Barrolândia district, Bahia (Experimental Station Gregório Bondar [EGREB], *Comissão Executiva do Plano da Lavoura Cacaueira* (Executive Commission of the Plan of Cocoa Agriculture) [CEPLAC], 15°40'S 38°57'W, with 500 hectares of native forest reserves); b) Teixeira de Freitas, Bahia (Regional Agricultural High School of CEPLAC [EMARC], CEPLAC, 17°32'S 39°44W, with 25 hectares of secondary vegetation); c) Mucuri, Bahia (Settlement Paulo Freire, 18°03'S 39°32'W, with 1,100 hectares of native forest); d) Linhares, Espírito Santo (National Forest [FLONA] of Goyatacazes, 19°23'S 40°04'W, with 1,610 hectares of native forest); e) Santa Teresa, Espírito Santo (Biological Station Santa Lúcia, 19°56'S 40°36'W, with 400 hectares of native forest). In all areas studied, the Lowland Rain Forest prevails except in Linhares where the vegetation belongs to the type the Submontane Semideciduous Forest (Oliveira Filho & Fontes, 2000; SEAMA, 2002). In Teixeira de Freitas, due to the small population of *D. lucida* in the area, only studies concerning density, temporal and spatial distribution and patterns of spatial distributions were performed, avoiding causing a greater disturbance in the existing population. Information concerning ant fauna in Teixeira de Freitas fragment, as well as concerning environmental problems experienced by such reserve, may be found in Conceição *et al.* (2006).

A. In order to verify the standard of distribution of *D. lucida* nests, an experiment was conducted in Belmonte fragment, presuming that the population was distributed in spots due to the peculiar reproductive biology of these ants. In this model, we suggest that nests have a clumped dispersion. During the experiment, two parallel transects were located (each one measuring 1,850 m) along the fragment. In these transects, pitfall traps without fixing liquid were placed each 50 meters (see Bestelmeyer et al., 2000). They remained in the field for 24 hours. That allowed us to capture foraging workers alive and release when traps were checked. Then, it was possible to map the dispersion of colony patches, since another experiment evaluated the average distance foraged by workers.

B. Ant nest densities were verified in ant nest concentration areas in the five study areas. In order to calculate *D. lucida* ant nest density in five forest fragments, the nests were checked for locating, mapping and marking of sites. An area with 2,500 m^2 was delimited, in which all ant nests found were marked. In order to locate easily ant nests, sardine baits were offered to foraging workers and these were followed while returning to their nest. Each nest was marked with colored tape on the closest tree or bush. The marked areas were revisited several times and the procedure of nest search was repeated in order to register the variation of ant nest density and mobility, allowing inference concerning the temporal and spatial dynamics of the populations in the areas. Density data were compared with the aid of a Quisquare test. The nest spatial pattern in patch spots was calculated based on the distance of the closest neighbor (Clark & Evans, 1954).

Figure 2. Worker of *Dinoponera lucida* foraging (Photo: Fábio Falcão).

C. Architecture and population of three *D. lucida* ant nests were analyzed in each of the four areas: Belmonte (Barrolândia), Mucuri, Linhares and Santa Teresa. Normality and homoscedasticity (Kolmogorov-Smirnov) of all data analyzed were verified concerning the graphic record and number of individuals in the colonies. The relation between nest characteristics and population size was verified. In case of normal distribution, data were tested with the Pearson Correlation Coefficient. If not, the Spearman Correlation Coefficient was used.

D. Nest architecture was studied in the following manner: a 50 cm deep and 70 cm long hole was dug 30 cm from the entrance. After that, successive 5 cm vertical cuts were made in the soil in order to locate chambers and underground passages until the entire ant nest was dug out. Each chamber found was measured in order to graphically reconstruct the ant nests later. The population found was recorded chamber by chamber according to gender and stage (eggs, larva, pupa and imagos). Insects were released after information record. Data were analyzed by ANOVA.

E. Others organisms found in the nest were collected and identified when possible. Arachnids were forwarded to the Arachnida and Myriapoda Collection of the Butantan Institute, São Paulo (IBSP) for deposit and identification by Dr. Antonio D. Brescovit. Formicidae were identified at the Myrmecology Laboratory of the Cocoa Research Center (CPDC) where they were deposited. The relation between depth, number of chambers in the ant nests and diversity of related fauna was tested through the Pearson Correlation Coefficient test.

F. In order to verify the influence of ant nests on surrounding vegetation and/or the preference for microhabitat related to vegetation structure, the height of all plants located at most 2 m from the entrance of the ant nests was recorded and the profile of vegetation on each ant nest was verified. The method consists of counting the number of contacts of leaves with an imaginary 10 cm wide vertical cylinder (Hubbel & Foster, 1986).

G. *Dinoponera lucida* activity was observed during an entire day in 10 ant nests of each forested area studied, except at Santa Teresa, where only eight ant nests were observed. Each ant nest was observed for five minutes, every hour, for 24 hours. Each individual that entered or left the ant nest was counted. Activity, time and air temperature were recorded. Activities observed were classified into: i) foraging, ii) nest maintenance (which includes excavation and refuse remotion) and, iii) guard behavior of individuals in the ant nest (which may be induced by any disturbance in the habitat produced by the coming and going of people, or by the incidence of light, even red light, at night).

H. The foraging activity of *D. lucida* was studied on ten ant nests in each of the four experimental areas in which ant nest architecture and population were studied. The following aspects were specifically determined: a) the maximum distance reached by foraging workers: three workers were followed in each ant nest. The distance between the maximum point reached and ant nest entrance was determined; b) the number of workers who were foraging at the same time: all workers which entered or left the ant nests for at least one hour were marked, and the times during which they were foraging were recorded; c) the number of successive returns to the same food source: sardine bait was offered at 50 cm from the entrance of each ant nest and the number of workers which visited the bait, the number of workers which returned to the bait and the frequency of successive returns to the bait by the same worker were recorded for each ant nest. Workers were marked with non-toxic ink.

I. Identification of the items gathered by foraging ants for ant nest feeding was done in three ways: a) the items were gathered during the observations described above, fixed in alcohol and taken to the laboratory for further identification; b) in certain cases (evaluation of the frequency of visits to the same food source and number of foraging ants) they were not gathered, so that such workers were not disturbed, avoiding any interference to the behavior of the individual. In such cases, the visualized items were simply recorded; c) the refuse excavated during the ant nest architectonic characterization was sorted at the laboratory where food itens were identified.

RESULTS AND DISCUSSION

Organization and Density of Nests in Patches

In both transects, *D. lucida* foragers were captured in dry pitfall traps at the initial portion and throughout the larger transect extensions, but the ants were totally absent at the final part of the sample area, although the forested area seems perfectly homogenous and continuous. This is consistent with the presumption that *D. lucida* is organized in colony aggregates on a large scale, distributed in areas which may be over one hectare and where nests are frequently observed, giving the wrong impression that population distribution is continuous if the observation scale is short. According to the theory, the aggregate pattern is principally related to heterogeneous habitats, to the gregarious behavior of the specie or to reproductive ecology, among other factors (Ludwing & Reynolds, 1988). For *D. lucida,* it is probable that the aggregate pattern is a reflex of its reproduction mode, which is by colony fission and fecundation of an apterous female (gamergate) at the nest entrance.

Ant nest distribution was mapped in each studied area (Figure 3). Mean densities of ant nests in each area during the observation period were initially determined as 2,500 m^2 and projections of colony densities for each area were estimated: a) Belmonte (Barrolândia): 48

nests per hectare; b) Teixeira de Freitas: 36 nests per hectare; c) Mucuri: 52 nests per hectare; c) Linhares: 36 nests per hectare; e) Santa Teresa: 20 nests per hectare. There was no significant difference among the densities of the five areas ($X^2=7,167$; $p=0,067$).

As observations were performed on a limited superficial area within the ant nest aggregate area (patch), the impression is that nest quantity is always high. However, that is not the case in Teixeira de Freitas, where the nine nests observed in the same sample block are the only ones remaining in a secondary vegetation area whose degradation (Conceição et al., 2006) gives little hope of a future recovery of the population. The colony aggregation pattern was confirmed in each area: Barrolândia: R=0.144; Teixeira de Freitas: R= 0.166; Mucuri: R=0.138; Linhares R=0.166 and Santa Teresa R=0.223.

For *D. quadriceps*, Vasconcellos *et al.* (2004) found 15 to 40 ant nests per hectare, with a tendence to regular distribution. Paiva & Brandão (1995) found 80 ant nests of *D. australis* per hectare randomly distributed, but they assume a possibility of error in their estimations since they have recorded evidence of the migration of *D. australis* colonies in the field. However, it is possible that the spatial distribution patterns found by such authors reflect the small scale of extrapolated samples for a 1-hectare area and that the standards observed correspond to the situation of populations within the patches.

At Belmonte (Barrolândia), there are ant nests evidently recently built near old ones. Furthermore, some of the colonies are sometimes not found on successive visits. Stochastic events, such as the falling of a tree or the digging up of an ant nest by a mammal predator may have caused colony migration. The observations of such new ant nests near old ones and their posterior disappearance suggest there was a provisory change of colony caused by some temporary disturbance, although nothing evidently uncommon was observed in the environment. Although nests are frequently considered fixed structures, ants abandon them when they face environmental changes or when they are invaded by other ants or predators (Cerdá *et al.*, 1998). Ecitoninae is the main group of litter predator ants in the Neotropical Region (Hölldobler & Wilson, 1990) and, on one occasion, the effect of an Ecitoninae trail on a *D. lucida* nest was observed. *Dinoponera lucida* left the nest very quickly, carrying its larvae, eggs and pupas, running away to a distant place and, after that, they would return to the nest. It is also possible that new ant nests were provisory, serving as a stopover during the colony migration or fission.

In Barrolândia, workers were observed going from one ant nest to another, at a 0.6 m distance from each other. The same fact was observed in two ant nests which were 6.8 m away from each other in Linhares. This kind of observation of ants entering different ant nests could be interpreted as the fact that such ants are polydomic, that is, one colony occupies more than one ant nest (Delabie *et al.*, 1991; Jaffé, 1993). However, repeated observations of the same *D. lucida* ant nests showed that after several weeks, the workers stopped visiting both nests and started visiting only one, which indicated that the colonies had recently been separated through fission. Overal (1980) states that when colonies of *Dinoponera gigantea* Perty migrate, the workers leave in pairs, one following the other, to a new ant nest. This kind of recruiting, which was also observed for *D. lucida* at Barrolândia, is known as "tandem-running" and is common in less derived ants (Hölldobler & Wilson, 1990; Fowler *et al.*, 1991).

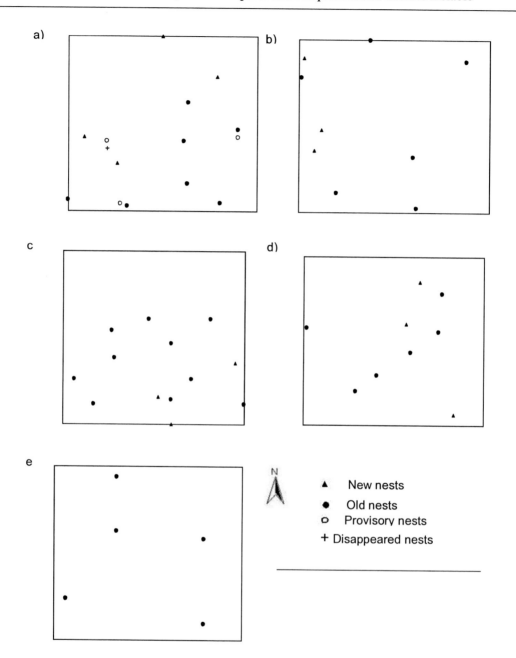

Figure 3. Distribution of nests of *Dinoponera lucida* in five fragments of Atlantic Rain Forest, in Barrolândia (a), Teixeira de Freitas (b), Mucuri (c), Linhares (d), Santa Teresa (e).

Nest Characterization: Architecture and Microhabitat

Information concerning architecture and microhabitat of *D. lucida* ant nests studied were summarized in Table 1.

Table 1. Characterization of nests of *Dinoponera lucida* of Barrolândia (B), Mucuri (M), Linhares (L) and Santa Teresa (S). * Number of contacts, method of Hubbel & Foster (1986); - data not available

code	date	number of chambers	number of entrances	depth of the nest (cm)	proximity to tree (m)	height of tree (m)	profile of foliage *
B1	13/04/05	3	1	30	0,39	15	74
B2	04/08/04	6	4	27	0,27	6	12
B3	05/08/04	3	3	46	0	2	7
M1	24/10/04	3	3	31	>3,00	>3	4
M2	25/10/04	7	2	27	0,90	8	8
M3	26/10/04	5	1	65	0	8	9
L1	18/12/04	7	1	16	0,60	1	17
L2	19/12/04	4	1	21	0,40	1,45	34
L3	23/12/04	3	1	33	0,20	1,1	12
S1	11/05/05	7	1	56	1,30	4	25
S2	14/10/04	5	2	26	0,30	0,8	-
S3	18/10/04	4	1	43	2,60	2,5	11
Mean		4,75	1,75	35,1	0,63		

Seventy-five per cent of *D. lucida* ant nests studied were less than a meter away from a tree (mean = 0.63 m, Table 1). According to data concerning foliage, all ant nests were located under tree canopies (Table 1). Around the ant nests, there are sticks, leaves and soil balls resulting from digging out of underground galeries of the ant nest itself, which were deposited by workers. Sticks were also observed abandoned around the entrance of ant nests of *D. quadriceps* and *D. lucida* by Paiva & Brandão (1995), as well as of *D. gigantea* by Fourcassié & Oliveira (2002). It is speculated that such behavior confer a kind of camouflage upon the ant nests. It is not impossible that small rocks are also used to collect dew in order to supply water to the ants (see Moffet, 1985, *apud* Karpakakunjaram *et al.*, 2003). *Dinoponera lucida* workers were observed collecting water at a drinking fountain in the laboratory, but such behavior was not observed in the field.

Dinoponera lucida seems not have preference to nidificate near trees but, generally, ant nests are constructed in shady areas in the middle of the day, when the air and soil temperatures are higher. Paiva & Brandão (1995) suggests that *D. quadriceps* prefers the areas near trees to establish their ant nests. However, this observation is invalidated by Vasconcellos *et al.* (2004) who observed that 60% of the ant nests were at least 3 meters away from trees and only the other ones were under the branches and near the trunks.

The *D. lucida* ant nest is underground and has one to four evident entrances. Half of the nests had more than one entrance (mean+1.75; standard deviation = 1.05), and only one ant nest with four evident entrances was observed. The entrances were generally elliptical and are at least 2.6 cm high and 4.7 cm wide. On the other hand, the number of entrances in *Dinoponera longipes* Emery ant nests is extremely variable (1 to 30), with an average of 11 orifices per ant nest (Morgan, 1993). Fowler (1985) found *D. australis* ant nests with only one 2 cm opening. In *D. gigantea*, the number of entrances varies from one to eight (Fourcassié & Oliveira, 2002). In the case of *D. quadriceps*, there is no consensus: according to Paiva & Brandão (1995), all *D. quadriceps* ant nests have two 4 cm cylindrical openings, while Vasconcellos *et al.* (2004) found only 20% of the ant nests with two openings, and all the other ant nests had only one opening. This difference may be related to the different habitats in which the observations were made: caatinga and cerrado for Paiva & Brandão (1995) and Atlantic Forest for Vasconcellos *et al* (2004).

The minimum ant nest depth found was 16 cm, while maximum depth was 65 cm (mean = 35.1, standard deviation = 14.6). The number of chambers varied from three to seven, and seven (n=3) and four (n=3) chambers were more frequent. The chambers had sizes varying from 4 to 39 cm in diameter (mean = 12.0 cm) and from 2 to 7 cm in height (mean = 4.2 cm). One atypical chamber was observed: it was 17 cm high in Mucuri, which suggests that the ants may have re-used the refuge cavity built by another animal. Differences between areas were not observed concerning the number of chambers, entrances and depth of the anthills (Levene: p= 0,934; ANOVA: F=0.285; p=0.835, Levene: p= 0.094; ANOVA: F=1.788; p=0.227, Levene: p=0.285; ANOVA: F=1.030, p=0.430, respectively). Ant nests, as big as the ones observed in this study, might be considered small if compared to the ant nests of *D. quadriceps* or *D. australis* which, according to Paiva & Brandão (1995), may reach, respectively, 2 m and 1.43 m in depth. Vasconcellos *et al.* (2004) found ant nests of D. *quadriceps* which were 1.2 m deep, and ant nests of *D. longipes* which can be 2.1 meters in depth (Morgan, 1993), while ant nests of *D. australis* can be at most 0.4 meters deep, with two or three chambers (Fowler, 1985). The depth of *D. gigantea* nests was around 0.4 m and the number of chambers varied from one to eight (Fourcassié & Oliveira, 2002). Morgan (1993) found underground galeries in nests of *D. longipes* which were 10 to 15 cm wide, and two of them reached 25 cm.

Excavations and laboratory observations revealed that, the foragers stay in the upper part (20 cm wide) of the ant nest, while the gamergates and the ants in charge of taking care of offspring, eggs, larvae and pupas stay in the deeper part (10 cm wide and 3 cm high). In the case of *D. lucida*, most of the offspring were found in the deepest chambers (72.7% of the ant nests). However, it is very probable that the workers took offspring deeper inside the ant nest, due to the disturbance caused by the excavation, although observations of artificial ant nests show that offspring are placed in chambers which are farther from the entrance, but not necessarily deeper.

According to Paiva & Brandão (1995), the distances between passages became larger with depth in *D. quadriceps*. It is possible that the number and depth of the passages indicate the age of the colony: the older the ant nest, the deeper and the higher in number are the chambers, because of having been dug out by several generations (Paiva & Brandão, 1995). However, our observations do not allow such a statement for *D. lucida*, since no correlation between the number of passages and their depth was found in relation to the number of individuals in the ant nests. In addition to this, it is known that there may be a colonization of empty nests by ants of the same or of another species (Hölldobler & Wilson, 1990), that is, the age of a structure does not always reveal the age of an insect society.

When the ant's waste matter was found, the study noticed that it was not deposited in only one place, but was distributed in more than one. Observations of *D. lucida* colonies in artificial plaster dwellings show that workers place food waste and stools in the holes between the plaster dwelling and the glass lid, or even on the walls of the nest. However, the most frequent destination of the waste is its deposition in the foraging area. In the field, it was observed that there were workers leaving the nest carrying pieces of arthropods or bits of dry leaves, which were abandoned shortly thereafter. The occurrence of associated organisms, of commensals or others in the dwelling may also explain the absence of a trash repository area, since it is probable that such organisms eat the leftover food of the *D. lucida*.

Characterization of Colony Population

Information concerning the population found in *D. lucida* nests was reported in Table 2. The number of imagos varied from 22 to 106 per colony (Table 2), and considering all of the individuals, adult and young ones, 27 to 189 per colony. However, no differences between areas were verified considering the total of individuals in the colonies, or just the workers (Levene: p=0.374; ANOVA: F=2.714; p=0.115 and Levene: p=0.246; ANOVA: F=1.488; p=0.290, respectively). Male ants were found in the ant nests of Barrolândia (data gathering from August/2004 to March/2005), of Mucuri (October/2004), and of Santa Teresa (October/2004 and May/2005). No male was found in Linhares during the excavations. However, in other field works in this locality, we observed males near the entrance of the ant nest (December/2004). The size of colony populations of *D. lucida* is not essentially different from information available for *D. quadriceps*: average population: 55.8 workers (Vasconcellos *et al.*, 2004), 141 workers in an ant nest which was studied isolatedly (Monnin & Peeters, 1998); *D. gigantea*: around 84 workers on average (Fourcassié & Oliveira, 2002) and this seems to be much higher than in the colonies of just *D. australis*: average population: 13.5 ± 6 individuals. The greatest population observed was made of 31 workers (Paiva & Brandão, 1995) and 25 workers in an ant nest which was studied in isolation (Fowler, 1985).

Table 2. Population of *Dinoponera lucida* found in nests in Barrolândia (B), Mucuri (M), Linhares (L) and Santa Teresa (S)

code	date of collection	workers	males	eggs	larvae	pupae	total
B1	13/04/05	22	0	0	3	2	27
B2	04/08/04	35	2	0	0	13	50
B3	05/08/04	33	0	0	1	10	44
Mean		30	0,7	0	1,3	8,3	40,3
M1	24/10/04	61	0	53	34	22	170
M2	25/10/04	45	1	3	22	11	82
M3	26/10/04	54	0	29	35	17	135
Mean		53,3	0,3	28,3	30,3	16,7	129
L1	18/12/04	69	0	2	17	37	125
L2	19/12/04	106	0	0	17	66	189
L3	23/12/04	29	0	2	18	9	58
Mean		68		1,3	17,3	37,3	124
S1	11/05/05	37	3	12	0	4	56
S2	14/10/04	91	0	6	21	2	120
S3	18/10/04	63	0	4	21	0	88
Mean		63,6	1	7,3	14	2,0	88

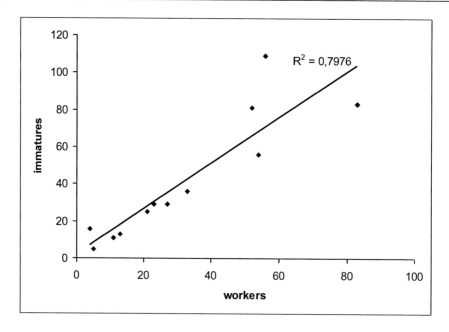

Figure 4. Relationship between the number of workers and immature (R^2=0,797, p=0,00).

The number of workers per ant nest is positively related to the number of offspring (Figure 4) ($R^2 = 0.7976$, p=0.00). The total number of workers was always higher than the number of offspring, except in Mucuri, where it was lower in two colonies. In the M1 ant nest, for example, the number of offspring was higher than the double of workers, which suggests that colonies M1 and M3 in Mucuri were preparing themselves for colony fission.

Reproductive cycle of *Dinoponera lucida*

The great number of offspring observed in certain ant nests in relation to the number of workers in the beginning of the summer suggests that the colonies were in a phase of growth before fission. Based on such information and on other observations about *D. lucida* societies made before or after this study, in addition to the data available in literature about other species of the same genus, we were led to see the ant reproductive cycle as it is presented in Figure 5. The gamergate egg production period seems to be limited to the austral winter. The following months correspond to larva breeding, and the population reaches its apex during austral summer, when fission occurs (the colony is divided into two populations). According to Peeters & Ito (2001), fission is the division of a colony and it occurs in monogynic species. It is associated with the production of new queens (Bourke & Franks, 1995) (gamergates, in this case) which cannot live in the same colony that the other queens because the colonies are monogynic. In fission, a reproductive female abandons the original colony to found a new one, followed by sterile workers which collaborate in such process (Peeters & Ito, 2001).

In the case of *D. lucida*, the number of individuals leaving the initial colony was not quantified, but it is estimated that this number is relatively small, if data originating from other species of the same genus are considered. For example, based on their study about maturation phases of colonies of D. *quadriceps,* Araújo & Jaisson (1994) considered recently-founded colonies the ones which had a small number of workers (1-3) and some eggs; young colonies the ones which had 10 to 40 imagos and offspring in all development phases and mature colonies the ones which had more than 40 imagos, also accompanied by offspring in several development phases. Morgan (1993) found a colony of *D. longipes* with a total population of only seven workers, which suggests recent fission.

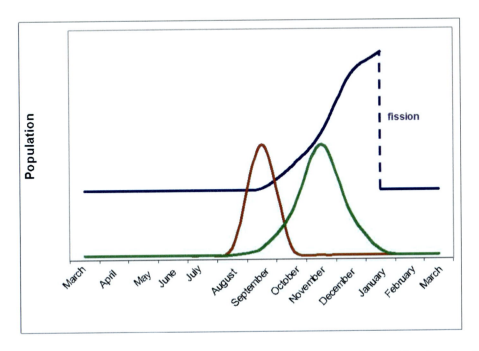

Figure 5. Dynamics of *Dinoponera lucida* population (red: eggs; green: larvae; blue: immature).

On the other hand, sexual offspring production (in this case, only males) also characterizes a mature colony (Hölldobler & Wilson, 1990). As males were found in four out of five months of data gathering, it is probable that they do not have any specific time for reproduction, like *D. quadriceps* (Vasconcellos *et al.*, 2004) and *Thaumatomyrmex* spp. (Delabie *et al.*, 2000). The fact that the production of males of *D. lucida* is probably continuous (this study, and other unpublished information found in the Myrmecology Laboratory Collection) is certainly related to the possibility of a gamergate being replaced by a highly-hierarchical worker, called alpha virgin, in case the gamergate is dead or missing (Monnin & Peeters, 1999). As such events take place in a random manner; mature males must be able to be released at any time of the year in order to fecundate one of these females in one of neighboring ant nests.

Fauna Associated with *Dinoponera lucida* Nests

Eight groups of animals living in *D. lucida* ant nests were identified. These were identified by order or at the family level (Table 3). Some spiders were identified at the genus level and the ants, at species level. Some elements of such fauna could not be identified, principally the immatures. The groups found were registered for their presence or absence only, since it was not possible to count the number of individuals in each group. The relation between the diversity of inquilines and the depth of the ant nest and number of chambers was not observed (Pearson: R=-0.035; p=0.915 and R=-0.278; p=0.381, respectively). According to Schultz & McGlynn (2002), most of these animals are frequently found in ant nests. Such animals may steal food brought there by ants, pick their trash or even continually attack the ants, progeny and inquilines (Delabie & Jahyny, 2007). Thysanura may be found in the ant nests of several kinds of ants, where they steal food or even eat pupae of the host ant (Wojcik, 1990). Millipedes are usually ignored by ants. However, they may eat them under stress (Wojcik, 1990). A case of associations of millipedes and isopoda has already been reported with a desert ant, *Messor bouvieri* Bondroit (Sanchéz-Piñero & Gómez, 1995). Gastropods have also been observed being carried to the ant nest by workers. The gastropod *Allopeas myrmekophilos* Janssen & Witte produces a substance which attracts the ponerine *Leptogenys distinguenda* (Emery), incitating it to carry the gastropod to its nest. It is eventually carried together with the ant offspring when there is a migration of a colony (Witte et al., 2002).

It is possible that the observed Chilopodae and Orthoptera are not really associated with the *D. lucida* breeding ground, but may have accidentally fallen into the ant nests during excavations from leaf litter.

Spiders found in the ant nests in Santa Teresa are female *Corinna* sp. (IBSP 57326, Corinnidae) and the ones found in the ant nests in Linhares belong to the family of Ctenidae (one offspring) and Nemesiidae (*Pselligmus* sp.: one female – IBSP 12602 – and one offspring). The following was found at Barrolândia: *Corinna* sp. (one female – IBSP 57327 – and one offspring) and one spider belonging to the family Ctenidae (one offspring). The following were found in Mucuri: *Isoctenus* sp. females (IBSP 57325, Ctenidae). Vasconcellos et al. (2004) also found Corinnidae species (*Abapeba* sp.) in nests of *D. quadriceps*. Ctenidae e Corinnidae are usually errant and hunters and have already been found in leaf litter in the Atlantic rain forest (Dias, 2001). Nemesiidae are cosmopolitan and construct underground refuges, many of which have a door at the entryway (Dippenaar-Shoeman & Jocqué, 1997 apud Dias, 2001). Thus, it is possible that the Ctenidae and Corinnidae observed settle in the ant nests of *D. lucida*. However, the Nemesiidae must have been accidentally gathered, in the leaf litters which surrounded the ant nest or during the actual excavations, if their refuge was near the ant nest.

A new species of *Pheidole* was found in *D. lucida* nests (Table 4): *Pheidole* sp.1. Initially, such ants were found only in ant nests in Mucuri and Linhares, but they were also detected in further observations gatherings at the Barrolândia site. Paiva & Brandão (1995) also found *Pheidole* spp. in ant nests of *D. quadriceps, D. lucida* and *D. australis*. Eventually, the *Pheidole* gathered in *D. australis* ant nests was described by Wilson (2003) as *Pheidole dinophila* Wilson, while the same author identified as *Pheidole rudigenis* Emery the species found in an ant nest of *D. lucida* in Itaúnas (ES). Paiva & Brandão (1995) found larvae of Geometridae in ant nests of *D. quadriceps,* Thysanura, Coleoptera, larvae of Phoridae (Diptera), Isopoda and Araneae living in the ant nests of *D. australis*. Vasconcellos et al. (2004) also found *Pheidole* sp. living in the ant nests of *D. quadriceps*, while *Pheidole* sp. and some Coleoptera were found in those of *D. longipes* by Morgan (1993). Paiva & Brandão (1995) found trash repositories in ant nests of *D. australis*, but only when they did not contain *Pheidole*, and the authors suggest that such ants fed themselves from the refuse of *D. australis*.

Table 3. Fauna associated with nests of *Dinoponera lucida* in Barrolândia, Mucuri, Linhares and Santa Teresa. (Gastropoda (GA), Isopoda (IS), Araneae (AR), Chilopoda (CH), Diplopoda (DI), Thysanura (TH), Formicidae (FO), Orthoptera (OR), unidentified immatures (INI))

	Barrolândia			Mucuri			Linhares			Santa Teresa			Occurrence of taxa in nests (%)
	B1	B2	B3	M1	M2	M3	L1	L2	L3	S1	S2	S3	
Gastropoda			+	+			+	+					11
Isopoda				+	+	+							8
Araneae	+			+		+	+	+		+	+		19
Chilopoda	+												3
Diplopoda	+			+	+								8
Thysanura	+	+	+	+	+	+				+			16
Formicidae	+		+	+	+	+	+	+	+	+		+	14
Orthoptera	+										+		8
Immature unidentified				+	+	+	+	+					14

Table 4. Formicidae found in the nests of *Dinoponera lucida* of Barrolândia (B), Mucuri (M), Linhares (L) and Santa Teresa (S)

Nest	Occurrence of *Pheidole* sp.1	Other ants
B 1		*Dolichoderus imitator* Emery, *Pachycondyla unidentata* (Mayr)
B 2		*Pachycondyla unidentata* (Mayr)
B 3		
M 1	observed	*Strumigenys elongata* Roger, *Gnamptogenys acuminata* Emery, *Pachycondyla constricta* (Mayr)
M 2	observed	*Pachycondyla arhuaca* (Forel)
M 3	observed	*Pachycondyla venusta* (Forel)
L 1	observed	*Sericomyrmex* sp.
L 2	observed	
L 3	observed	
S 1		*Solenopsis virulens* (Smith)
S 2		
S 3		*Solenopsis virulens* (Smith)

Nycthemeral Activity of *Dinoponera lucida*

Dinoponera lucida activity was observed during one entire day in all the sites, except at Teixeira de Freitas.

In order to verify if the replication of observations would increase data accuracy, observation days in Mucuri were compared hour per hour with those in Santa Teresa. An analysis of variance (ANOVA) was applied for each hour of observation in both places, separately, comparing the activity of the species at each hour of the observation day. In Santa Teresa, activity difference was not verified (p>0.05), while in Mucuri there are differences in two of the time periods (12:00 pm, p=0.008, e 6:00 pm, p=0.034). However, the difference in the activity observed may be attributed to pluviometric precipitation which occurred once, not at the time during which the data were gathered. Thusly, there being no significant differences between observation days in the same period of the year, the curves of activity in Mucuri and Santa Teresa were plotted according to the hourly averages. Similar trials were done in Linhares and Barrolândia.

The cycles of daily activities were not different between areas. Foraging activities are diurnal (Figure 6), guarding is diurnal and nocturnal, but it is more important at night

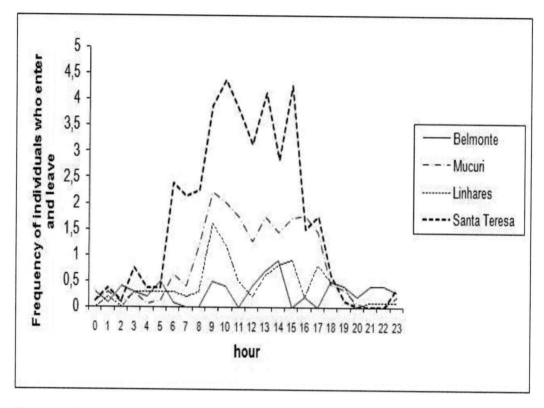

Figure 6. Entrances and exits of nests of *Dinoponera lucida* (numbers of individuals). Mean observations of all nests (five minutes per hour): Belmonte (Barrolândia) (n=10, one day of observation), Mucuri (n=10, three days of observation), Linhares (n=10, one day of observation) and Santa Teresa (n=8, two days of observation).

(Figure 7), when there is no worker flow. On the other hand, maintenance of nests occurs at any time of the day (Figure 8). Foraging activity of *Dinoponera lucida* starts at the beginning of the morning when the sun rises at around 6:00 am (Figure 6). However, the flow of workers leaving and entering the ant nests becomes more important from 9:00 am on, reduces during the day, increases at 2:00 pm and finally finishes at 6:00 pm. Therefore, there are around 12 foraging hours, coincident with the daylight hours. Like *D. lucida*, *D. quadriceps* forages during the day, between 6:00 am and 5:00 pm (Araújo & Rodrigues,

2006), while *D. gigantea* (Fourcassié & Oliveira, 2002) and *D. longipes* (Morgan, 1993) show a higher rate of activity at sunrise and sunset, or even at night, in the latter case.

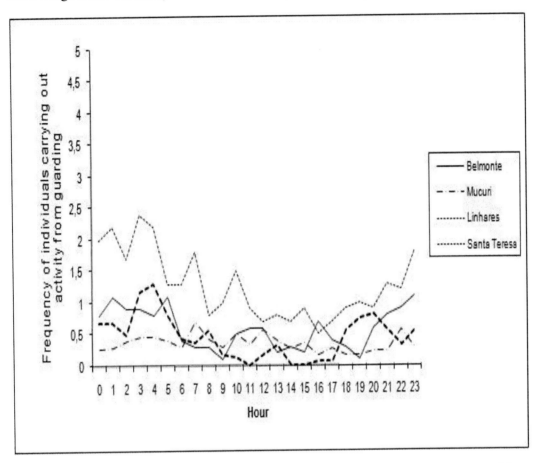

Figure 7. Activity guarding (number of individuals observed) *Dinoponera lucida* in the area of focus for five minutes per hour. Barrolândia (Belmonte) (n = 10, one day of observation), Mucuri (n = 10, three days of observation), Linhares (n = 10, one day of observation) and Santa Teresa (n = 8, two days of observation).

Figure 7 shows that workers stay at nest entrances during the day and during the night. Such behavior is more frequent at night. That might be the consequence of the disturbance caused by the flow of observers during observation, and is not being ruled out.

In all the localities, workers were observed removing dirt, as well as leaves and sticks, from their nests, as a result of the excavation of galleries and chambers. This aspect of the behavior was called "nest maintenance". Such activity was verified during the whole observation period (Figure 8), but it was more frequently seen after rains and in nests not protected from rain. This activity was most frequently observed in Linhares, since constant rains occurred there during the observation period. Differences among those activities observed in Mucuri, Barrolândia and Santa Teresa were explained by great variations in the regional weather conditions during the observation period.

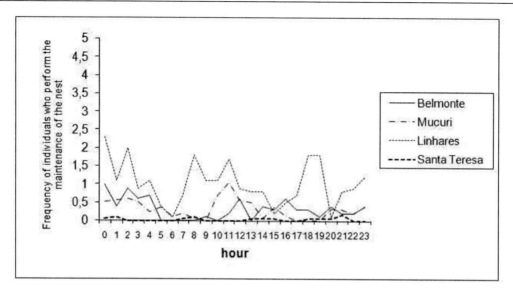

Figure 8. Activity maintenance of nests of *Dinoponera lucida* (number of individuals). Mean observations of all nests (five minutes per hour): Barrolândia (Belmonte) (n = 10, one day of observation), Mucuri (n = 10, three days of observation), Linhares (n = 10, one day of observation) and Santa Teresa (n = 8, two days of observation).

Dinoponera lucida Foraging Ecology

In all cases, worker activities outside the nest were observed during the day, as in Mucuri, where the temperature varied from 16°C to 31.5°C. Paiva & Brandão (1995) also observed *D. lucida* foraging at the warmest hours of the day. In Santa Teresa, foraging activity was higher between 20 and 23°C. In Linhares, it was higher between 25 and 27°C. Temperature conditions for the regions in which observations were performed, varying from 16 to 31.5°C during our study, did not seem to be limiting factors for ant activities in the external area of the ant nest. As has already been observed, rain certainly inhibits such activity, while daylight activates it. The period in which ants are active is strongly related to the physiological characteristics of species, particularly their tolerance in respect to the oscillation of temperature and humidity in the environment (Hölldobler & Wilson, 1990). Ectothermic animals need a certain level of temperature for their physiological comfort, and too high or too low a temperature may be harmful or even lethal to them (Cerdá *et al.*, 1998). Thusly, and certainly for this reason, *D. lucida* activity presents a slight activity reduction around 12:00 pm, exactly when the temperature is higher. Although bigger ants are in theory more resistant to higher temperatures, they do not necessarily forage at the warmest hours of the day (Cerdá *et al.*, 1998). Such a behavior pattern in relation to temperature is shared with Mexican populations of *Pachycondyla villosa* (Fabricius), a Neotropical ponerine (Valenzuela-Gonzalez *et al.*, 1994). The *Dinoponera* species originating from the rain forest react differently to temperature changes: foraging of *D. longipes* ceases when the temperature falls from 24 to 19°C (Morgan, 1993), while the activity rhythm of *D. gigantea* is negatively associated with temperature (Fourcassié & Oliveira, 2002).

Like other poneromorph (*sensu* Bolton, 2003) ants, such as *Amblyopone pallipes* (Haldeman), *Pachycondyla berthoudi* (Forel), *Pachycondyla apicalis* (Latreille), *P. villosa*, *Diacamma ceylonense* Emery (Tranielo, 1982; Peeters & Crewe, 1987; Goss *et al.* 1989;

Valenzuela-Gonzalez *et al*; 1994; Karpakakujaram *et al.*, 2003) and like other species of *Dinoponera* (Fowler, 1985; Morgan, 1993; Fourcassié & Oliveira, 2002; Araújo & Rodrigues, 2006), the mode of foraging verified for *D. lucida* is solitary and always on the surface of the pile of leafy debris. Foraging workers, when returning to the nest, would frequently deliver the hunted or gathered item to a worker who was waiting at the entrance of the ant nest, and would then return to a new food search. This fact may be related to the polyethism of age, in which activities performed by ants are related to physiological changes in accordance with individual age (Hölldobler & Wilson, 1990). Thus, it is probable that the worker which stays at the entrance of the ant nest is still inexperienced, while older workers are those who leave the ant nest in search for food (Hölldobler & Wilson, 1990). Age polyethism related to foraging was similarly observed in *P. villosa* (Valenzuela-Gonzalez, 1994) and in *P. apicalis* (Fresneau, 1994), for example.

The distances at which ants forage vary little among the study areas, except in Santa Teresa. These distances are reported in Table 5. Santa Teresa is the area where the longest foraging distance was registered, but it is also the area with the lowest ant nest density: maximum foraging distance is inversely related to ant nest density, although the relation has little significance (Pearson, r=-0.748, p=0.252). Competition for foraging territory is probably lower in areas with lower colony density, allowing proportionally more opportunities to find food. Paiva & Brandão (1995) observed workers of *D. lucida* foraging more than 20 m far from the Itaúnas ant nests, in the State of Espírito Santo, while *D. gigantea* forages at least 10 m far from the nest (Fourcassié & Oliveira, 2002).

It was observed that one to eight workers per colony foraged simultaneously in the highest period of external activity of ants. Fidelity of ants to attractive bait was tested and such fidelity varies from 79.3% to 48.3% (Linhares) (Figure 9). This fidelity was observed both for a short-distance baiting (50 cm from the ant nest) and for long-distance baiting (10 meters from the ant nest). The number of successive returns may be very high regardless of the area (Figure 10). Factors which regulate the number of returns to the same food source are unknown, but the return of workers to the same food source, even at great distances, may be interpreted as a certain fidelity to trails, which was verified in other ponerines, such as *P. apicalis* (Fresneau, 1985). It was verified that when bait supply was suspended and, if light conditions were favorable to foraging, workers would continue foraging in other places. Once it was observed that some workers would stop only when it rained or when light conditions were not favorable to foraging, it is suggested that it is the weather or light conditions which determine the number of successive returns to the food source, certainly reinforced by the quality of the nutritional substratum found.

Table 5: Maximum and mean distance of foraging for *Dinoponera lucida* in Barrolândia, Mucuri, Linhares and Santa Teresa

Areas	Distance maximum (m)	Mean (m)	Total nest observed
Barrolândia	15,3	8,9	10
Mucuri	18,6	9,5	10
Linhares	15,4	8,5	10
Santa Teresa	26,4	13	10

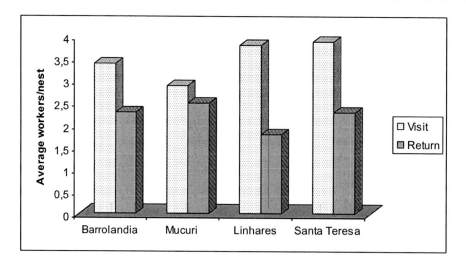

Figure 9. Fidelity of workers to an attractive bait. Mean per nest.

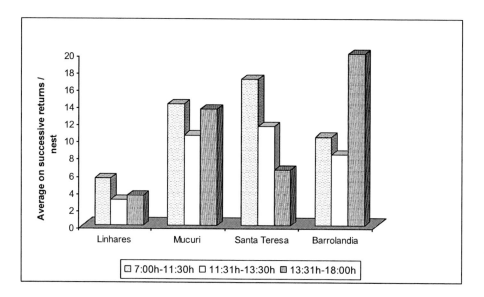

Figure 10. Maximum number of successive returns to the same food source per worker/nest.

All the food items observed and gathered during foraging and ant nest openings are summarized in Table 6. *Dinoponera lucida* appears as an omnivore whose diet is mostly made of invertebrates (70.6%). Although workers have been observed in the field transporting live, dead or moribund organisms to the ant nest, it is not evident, however, if *D. lucida* hunts its food actively, or if it just gathers dead or weaker animals. Regardless of the answer, slow animals are hunted, since it has been observed that *D. lucida* was carrying larvae of insects or annelids to the ant nests. That suggests that *D. lucida* is an opportunist predator which gathers any kind of live or dead food which casually appears during foraging activity. Morgan's hypothesis (1993), according to which offspring of *D. longipes* depend on a food source based on vertebrate protein to survive, was not confirmed for *D. lucida*, due to the absence of vertebrate remains (skeletons, scales) in the ant nests and due to their rarity as seen with *D. lucida* during foraging.

Table 6. Frequency of food items that comprise the diet of *Dinoponera lucida*, observed in all areas during excavations of their nests and foraging. ni = unidentified

Food item	Freqüência (n)
Angiospermae	**20**
Fragments (ni)	3
Fruits (ni)	2
Seeds Sapotacea	1
Seeds Aracacea	2
Seeds (ni)	9
Annelida	**3**
Arthropoda	**88**
Araneae	4
Insecta (ni)	28
Blattodea	4
Coleoptera	1
Diptera	1
Hemiptera	2
Hymenoptera Apoidea	2
Hymenoptera Formicidae	2
Lepidoptera	2
Orthoptera	12
Other Insecta	2
Other fragments (ni)	14
Immature (ni)	13
Exuvia (ni)	1
Anura	**1**
Squamata	**2**
Lizard tail	1
Scale of snake	1
Fecal pellets (ni)	**12**

Seeds were also observed among food items abandoned in ant wastes; they seemed intact, without evidence of granivory. In one of the studied nests, a germinating seed was found, which suggests that, through the gathering of other fruits or seeds, ants may occasionally act as secondary sowers of seeds. There are several myrmecochoric plants (whose diaspores are exclusively or partially dispersed by ants) (Westoby *et al.*, 1991), but few cases were studied in Brazil (Peternelli *et al.*, 2001; Passos & Oliveira, 2002). Ants which disperse seeds belong to one of the following subfamilies: Fomicinae, Myrmicinae, Ponerinae or Dolichoderinae (Beattie, 1985). The seeds of such plants may present an elaiosome which is rich in lipids that induces a great variety of ants to gather diaspores. The removal of elaiosome by ants and the fissure originating from such process increases the germination chances of seeds (Gómez &

Espadaler, 1998). However, it is not known if the seeds found with *D. lucida* had elaisome or other attracting agent. The seeds which were lost or discarded by ants may germinate around the ant nest or in the nest itself (Handel & Beattie, 1990; Westoby *et al.*, 1991; Gómez & Espadaler, 1998; Gorb et al., 2000; Passos & Oliveira, 2002), sometimes inside the garbage of the colony, *a priori* rich in nutrients favorable to the germination or growth of the seedling (Handel & Beattie, 1990; Passos & Oliveira, 2002).

Omnivorous characteristics of *D. lucida* are similar to the ones described for *D. longipes* (Morgan, 1993), *D. gigantea* (Fourcassié & Oliveira, 2002) or even *D. quadriceps* (Araújo & Rodrigues, 2006) and are similar to other ponerines (*P. villosa*: Valenzuela-Gonzalez *et al.*, 1994). Although workers carrying live, moribund or dead animals or seeds and fruits to the nest were seen, the role of *D. lucida* does not seem to be that of an active predator. The ants probably collect any live food item which appears by chance in the foraged territory. The high frequency of vertebrates on their food list, however, places such ants at the top of the food chain of mesofauna associated with the soil where they are found.

CONCLUSION

The population distribution of *D. lucida* follows an aggregate pattern. The ant nests are not deep and the size of the population varies from two dozen to less to two hundreds individuals per colony, depending on the study area and the fission cycle of colonies. They shelter a complex associated fauna, mainly Araneae, Thysanura and Formicidae, which seem, in part, to be related to the absence of a garbage deposit area in the ant nests. Foraging is diurnal and solitary. Food habits are generalist with arthropod predominance. The presence of seeds in the ant nests generates speculations concerning the ant ability to sow seeds.

ACKNOWLEDGMENTS

The authors are grateful for the financing by PROBIO/MMA/BIRD of the Sub-Project "Elaboração do Plano de Manejo de Dinoponera lucida Emery, a Formiga Gigante do Corredor Central da Mata Atlântica" (IBAMA license n°118/2004), to Fundação de Amparo à Pesquisa do Estado da Bahia (FAPESB), to CNPq, to the PRONEX Program (CNPq-FAPESB) PNX0011/2009, as well as to the Zoological Society of Chicago, for the initial support which motivated more research on the ant. They would like to express their gratitude to Eduardo Mariano Neto, José Crispim Soares do Carmo and José Raimundo dos Santos Maia for field and laboratory help, to Antônio Brescovit for identifying the spiders, to Braulio Dias for encouragements, as well as to Instituto Dríades and CEPLAC for logistics. JHCD acknowledges his research grant by CNPq.

EXPERT COMMENT

The Causes of the Threat

The principal causes which may be related to the extinction process of *D. lucida* populations are related, at first sight, to the intense deforestation suffered by the Brazilian Mata Atlântica after the Second World War at the proximities of the east Atlantic coast. Such process caused an extreme fragmentation of native forests. Nowadays, the remaining forested areas are very isolated, and are frequently small. Such forested areas are isolated in the middle of environments of agricultural production or in silvicultural ones, principally pastures or eucalyptus thickets, which inhibit (biogeographical barriers) any genetic flow (flights allowing male fecundation of gamergate at the nest entry) between distinct populations. In addition to this, as populations are isolated, once local extinction occurs, there is no chance of recolonization.

Observations indicate that the occurrence of the species is not related only to the quality of forest habitat. We may affirm this, since the ant is not exclusively found in forests in optimal state of conservation, but also in areas at initial and medium stages of regeneration within its distribution area.

Isolation, associated with the probability of extinction by stochastic events, explains the absence of the ant in several forest fragments which are relatively equivalent in terms of forest quality. The preservation of populations as a whole is thus blocked when a metapopulation pattern, with its consequent dynamics of extinction/recolonization, would be expected.

The available genetic information has already proved the occurrence of an important genetic drift which contributes to the differentiation of one population from another, and it is certainly responsible for variations which have been observed in their karyotypes. Beyond this, those populations surely suffer from a strong endogamy which reinforces the accelerated differentiation process of the genomes.

Soil Coverage

The analyses of soil coverage point to at least one other factor that limits the installation of this species. Most nests were found in areas which had little or no herbaceous vegetation. Changes which cause a higher herbaceous coverage, regardless of the origin of the perturbation, certainly restrict the effective area for ant nest allocation.

Dense herbaceous vegetation is associated with the penetration of light during the day to the forest soil, a condition which only occurs when there is a large clearing in the forest, an opening in the branchy and leafy canopy, owing to the falling of trees. These may be caused by natural dynamic processes, but also may be accelerated in cases where there is intense extraction of timber and firewood.

Fragmentation

The extreme habitat loss makes remaining populations small, increasing the possibility of local extinction.

Isolation harms the movement of individuals between forested areas, which precludes not only genetic changes, but also the recolonization of forest fragments where the ant is locally extinct.

The study revealed that entire distribution region of *D. lucida* is extremely fragmented. Most of the analyzed fragments had small areas, isolated one from the other and from other areas by inhospitable conditions for *D. lucida*, such as pastures and eucalyptus plantations. Isolation associated with the probability of extinction by stochastic events explains the absence of the species in several forest areas which are relatively equivalent in terms of forest quality. As populations are isolated, there is no chance of recolonization after events of local extinction.

In some regions of the distribution area, the greatest problem is linked to a homogeneous matrix - eucalyptus. Such culture is to-day the principal economic activity in the region, acting as a matrix which is nearly or effectively impermeable for the ant in the landscape context.

The presence of landless workers movement settlements in the region represents a threat and, at the same time, an opportunity for the few forest remanescent, depending on the measures that come to be implanted in the region. The ant may also survive and increase its population as well the occupied area in a region, living in groups, as was already observed. Such information is important, since it indicates that actions taken to involve the recovery of areas where human beings are living may hold, in a short time, populations of *D. lucida*, since the species is able to occupy areas of forests in initial stages of regeneration.

The problem of the reduction and isolation of populations, in any forests of the regions of *D. lucida* occurrence, is that, in the first case, the risk of endogamy and genetic drift increases, reducing drastically the diversity of such populations and thus reducing their ability of survival and adaptation to changes. Concerning the isolation of forested areas surrounded by inhospitable matrixes, such as pastures or eucalyptus plantations, there is an impediment to recolonization. Therefore, several forested areas which were surveyed and can at first sight support populations of the ant, showed to lack it totally.

It is important to point out that there are forest fragments under excellent conservation status in mountain areas. However, studies have been concluding that one of the factors that limit the colonization of *D. lucida* is altitudes higher than 800 m. As the Santa Teresa area of study is located between valleys, there are places in which altitudes surpass 800 m and this seems to have restricted the occurrence of this species. The same thing may occur in other localities of the State of Espírito Santo. The connection of forested areas by lands which are located near the hillsides could provide a propitious environment for *D. lucida* colonization. However, forests on the hillsides are rare, which makes the connection with the forested areas on the hilltops difficult.

Implications Related to the Eventual Extinction of *Dinoponera lucida*

The eventual extinction of *D. lucida* may cause the simultaneous extinction of species which depend on such ants are: a) one specie of the type *Pheidole* (Formicidae, Myrmicinae) which has not been described yet and which lives in the *D. lucida* nests; b) one still non-identified specie of Phoridea (Diptera), which is a specific parasite of foraging workers of *D. lucida*.

Other species living in nests and belonging to Araneae, Mollusca (Gastropoda) e Insecta (mainly Thysanura) Classes would probably be in the same situation, but we do not have enough information concerning its degree of dependence on the ant for such a statement.

There is no evidence concerning an eventual close association between vegetal species and *D. lucida*: its role as an agent for seed dispersion is probably a casual one, although it is relatively frequent, as well as the plant near which the ant nest is lodged must be casual. However, we do not have enough information to affirm any degree of dependence between these ants and vegetation.

What to Do

The threat of extinction of the species cannot be unlinked from that of the extinction of its natural habitat and of all fauna and flora elements which constitute it, with no exception. Significant efforts to conserve *D. lucida* mean conserving consistent parts of the Atlantic Forest in Brazil. In Bahia and in the north of the State of Espirito Santo, a particular effort to recover and preserve different kinds of vegetation in the remaining areas of Atlantic rain forest in which the ant occurs must be developed. If such effort does not happen, such populations will not be recovered, and they will suffer inexorable extinction, which will be related at the long term to a complete discharacterization of the landscape. In the south of Espirito Santo, a greater flexibility is possible, due to the characteristics of the existing forested areas, which are generally larger and more preserved in better protected areas than in the first case. Also in such case, preservation of such ants naturally implies the preservation of the regional landscape.

Such statements point to the urgency of taking measures which promote forest connectivity on a local scale, increasing the effective area of the forest and reconnecting populations. Actions may be taken through the planning of the recovery of areas of permanent protection, or even of establishment of areas of legal preservation (many properties within such present distribution area do not preserve any of those areas which are protected by Brazilian legislation). On the other hand, the occurrence of *D. lucida* in modified forest areas and initial regeneration stages, together with more preserved forest nuclei, indicates that it may be possible to restore the connectedness of forested areas, for such species, in a relatively short time, with the simple existence of a forest with bushy ground cover, even a small one.

The extreme reduction and fragmentation to which the forests of the species occurrence are submitted demand concrete actions of environmental recovery. In most visited regions, the attitude of preserving areas of permanent protection such as river shores and steep hills, together with the preservation of legal reserve areas on the properties, would represent a significant increase in the current size of forests. In such cases, it would be possible to work with the local prosecutor's office, and use awareness actions for landowners, in order to make effective the recovery of forested areas and their maintenance. It is also necessary to invest in the production of seedlings which are typical of each region where the ant occurs, based on the great turnover present between different places. It would also be interesting to try to coordinate the allocation of areas of legal reserve, in order to create areas of continuous forests made of the legal reserves and permanent protection areas of neighbor farms, instead of the current situation of isolated areas on each property. Such coordinated action would also have in view the joining of the existing forest fragments.

As we have already seen, such measures have a short term impact, as we observe *D. lucida* invading forested areas in their initial regeneration stage, which may occupy or at least make use of recently recovered land areas. Such recovery and recolonization measures provide unique opportunities for a follow-up and monitoring project of the spatial and temporal dynamics of ants.

However, planning which areas must be joined, and how they should be connected, in order to increase the population of the ant, must be realized taking into account specially the genetic characteristics of the different remaining populations.

The characteristics of *D. lucida* observed concerning habitat and microhabitat, distribution and bioecology indicate that actions taken for reforestation and recovery of forested areas may help populations of this ant, provided that such recovery process is already established in areas of shaded understorey. This is applicable in both ombrophilous forests in Espirito Santo and Bahia and for semidecidual forests in Espirito Santo, Bahia and Minas Gerais and in other flat forested regions in Espirito Santo and Bahia.

Such results suggest that measures which involve the recovery of areas where human beings live may hold, in a short time, populations of *D. lucida,* once the species is able to occupy areas of forests in initial stages of regeneration, provided that there had already shaded understory.

Another important factor is the tolerance of the ant to areas near watercourses, which indicates a tolerance to humid soils which are subject to floods. Consequently, recovery of ciliary forests would reinforce the preservation of the specie both through connectivity and through the increase of areas of the environment adequate to its establishment.

As these actions, even if started quickly, would present only medium-term effects, direct interventions are suggested through the reintroduction of the ant in the case of areas with small populations. This reintroduction should be done in order to reestablish populations in suitable habitat areas which may have suffered local extinction processes due to stochastic events.

Eventual translocations would aim the reduction of endogamy in the areas where there are still existent populations. In this case, due to the great variation found in Bahia, each measure should be carefully studied in accordance with the knowledge of genetic diversity of each studied microregion.

REFERENCES

Araújo, C. Z. D. (1995). Duração dos estágios e longevidade de operarias e machos de *Dinoponera quadriceps* Santschi Hymenoptera: Formicidae no campo e em cativeiro. *An. Soc. Entomol. Brasil, 24*, 33-38.

Araújo, C. Z. D., Fresneau D. & Lachaud. J. P. (1988). Premiers résultats sur l'éthologie d'une fourmi sans reine : *Dinoponera quadriceps*. *Actes Coll. Insect. Soc., 4*, 149-155.

Araújo, C. Z. D., Fresneau, D. & Lachaud, J. P. (1990a). Données biologiques sur la fondation des colonies de *Dinoponera quadriceps* (Hymenoptera, Formicidae). *Actes Coll. Insect. Soc., 6*, 281-286.

Araújo, C. Z. D., Fresneau, D. & Lachaud, J. P. (1990b) . Le système reproductif chez une ponérine sans reine: *Dinoponera quadriceps* Santschi. *Behav. Process, 22*, 101-111.

Araújo, C. Z. D. & Jaisson, P. (1994). Modes de fondation des colonies chez la fourmi sans reine *Dinoponera quadriceps* Santschi (Hymenoptera, Formicidae, Ponerinae). *Actes Coll. Insect. Soc., 9*, 79-88.

Araújo, A. & Rodrigues, Z. (2006). Foraging Behavior of the Queenless Ant *Dinoponera quadriceps* Santschi (Hymenoptera: Formicidae). *Neotropical Entomology, 35(2)*, 159-164.

Beattie, A. J. (1985). *The evolutionary ecology of ant-plant mutualisms.* Cambridge. Cambridge University Press.

Bestelmeyer, B. T., Agosti, D; Alonso, L. E., Brandão, C. R. F., Brown, W. L., Jr; Delabie, J. H. C. & Silvestre, R. (2000). Field techniques for the study of ground-living ants: an overview, description, and evaluation. In: (D. Agosti, J.D. Majer, L. Tennant de Alonso & T. Schultz, Eds), *Ants: Standart Methods for Measuring and Monitoring Biodiversity* (122-144), Washington: Smithsonian Institution.

Bolton, B. (1995). *A new general catalogue of the ants of the World.* Cambridge, Massachusetts. Harvard University Press.

Bolton, B. (2003). Synopsis and classification of Formicidae. *Memoirs of the American Entomological Institute, 71*, 1-370.

Bourke, A. F. G & Franks, N. R. (1995). *Social evolution in ants.* Princeton, NJ. Princeton Univ. Press.

Brasil. (2006). Ministério do Meio Ambiente. *O Corredor Central da Mata Atlântica : uma Nova Escala de Conservação da Biodiversidade.* Brasília, Ministério do Meio Ambiente , Conservação Internacional.

Campiolo, S. & Delabie, J. H. C. (2008). *Dinoponera lucida* Emery 1901. In: A. B. M., Machado, G. M. Drummond, & A. P. Paglia, (Eds.). *Livro vermelho da fauna brasileira ameaçada de extinção, Volume I*, Ministério do Meio Ambiente, *Biodiversidade, 19*, 388-389.

Cerda, X., Retana, J. & Cros, S. (1998). Critical thermal limits in Mediterranean ant species: trade-off between mortality risk and foraging performance. *Functional Ecology, 12*, 45-55.

Clark, P. J. & Evans, F. C. (1954). Distance to nearest neighbor as a measure of spatial relationships in populations. *Ecology, 35(4)*, 445-452.

Conceição, E. S., Costa-Neto, A. O., Andrade, F. P., Nascimento, I. C., Martins, L. C. B., Brito, B. N., Mendes, L. F. & Delabie, J. H. C. (2006). Assembléias de Formicidae da serapilheira como bioindicadores da conservação de remanescentes de Mata Atlântica no extremo sul do estado da Bahia. *Sitientibus série Ciências Biológicas, 6(4)*, 296-305.

Cuvillier-Hot, V., Gadagkar, R., Peeters, C. & Cobb, M. (2002). Regulation of reproduction in a queenless ant: aggression, pheromones and reduction in conflict. *Proc. R. Soc. Lond. B, 269*, 1295:1300.

Delabie, J. H. C., Benton, F. P. & Medeiros, M. A. de. (1991). La polydomie de Formicidae arboricoles dans les cacaoyères du Brésil: optimisation de l'occupation de l'espace ou stratégie défensive? *Actes Coll. Insectes Sociaux, 7*, 173-178.

Delabie, J. H. C., Fresneau, D. & Pezon, A. (2000). Notes on the ecology of *Thaumatomyrmex* spp. (Hymenoptera: Formicidae: Ponerinae) in Southeast Bahia, Brasil. *Sociobiology, 36(3)*, 571-584.

Delabie, J. H. C. & Jahyny, B. (2007). A mirmecosfera animal: relações de dependência entre formigas e outros animais. *O Biológico*, São Paulo, *69(2)*, 7-12.

Dias, M. F. R. (2001). Efeito da fragmentação da Mata Atlântica sobre a comunidade de aranhas de solo (Arachnida: Araneae) na região de Una, Bahia. *Dissertação*, UESC/PRODEMA. 61.

Fourcassié, V. & Oliveira, P. S. (2002). Foraging ecology of the giant Amazonian ant *Dinoponera gigantea* (Hymenoptera, Formicidae, Ponerinae): activity schedule, diet and spatial foraging patterns. *Journal of Natural History, 36*, 2211-2227.

Fowler, H. G. (1985). Populations, foraging and territoriality in *Dinoponera australis* Hymenoptera, Formicidae. *Rev. Brasil. Entomol, 29*, 443-447.

Fowler, H. G, Forti, L. C, Brandão, C. R. F., Delabie, J. H. C. & Vasconcelos, H. L. (1991). Ecologia Nutricional de Formigas. In: (A. R., Panizzi, & J. R. P. Parra, Eds.). *Ecologia nutricional de insetos e suas implicações no manejo de pragas*, (131-209) São Paulo: Manole.

Fresneau, D. (1985). Individual foraging and path fidelity in a ponerine ant. *Insectes Sociaux, 32(2)*, 109-116.

Fresneau, D. (1994). *Biologie et comportament social d'une fourmi ponérine néotropicale (Pachycondyla apicalis)*. Thèse de Doctorat d'Etat ès-Sciences, Université Paris XIII, Villetaneuse, France.

Fundação Biodiversitas. (2003). Lista das Espécies Ameaçadas da Fauna Brasileira. http://www.biodiversitas.org.br

Gómez, C. & Espalader, X. (1998). Myrmecochorous dispersal distances: a world survey. *Journal Biogeography, 25*, 573-580.

Gorb, S. N., Gorb, E. V. & Puntilla, P. (2000). Effects of redispersal of seeds by ants on the vegetation pattern in a deciduous forest: A case study. *Acta Oecologica, 21(4-5)*, 293-301.

Goss, S., Fresneau, D., Deneubourg, J. L., Lachaud, J. P. & Valenzuela-Gonzalez, J. (1989). Individual foraging in the ant *Pachycondyla apicalis*. *Oecologia, 80*, 65-69.

Handel, S. & Beattie, A. (1990). La dispersion des graines par les fourmis. *Pour la Science. 156*, 54-61.

Heinze, J., Hölldobler, B. & Alpert, G. (1999). Reproductive conflict and division of labor in *Eutetramorium mocquerysi*, a myrmicine ant without morphologically distinct female reproductives. *Ethology, 94*, 690-606.

Hölldobler B. & Wilson, E. O. (1990). *The ants*. Cambridge. Harvard University Press.

Hölldobler, B., Liebig, J. & Alpert, G. D. (2002). Gamergates in the myrmicine genus *Metapone* (Hymenoptera: Formicidae). *Naturwissenschaften, 89*, 305-307.

Hubbel, S. P. & Foster, R. B. (1986). Commonness and rarity in a neotropical forest: implications for tropical tree conservation. In: (M. E. Soulé, Ed.), *Conservation Biology: Science of Scarcity and Diversity* (205-231), Sunderland, Massachusetts: Sinauer Associates, Inc.

Jaffé, K. C. (1993). *El mundo de las hormigas*. Baruta: Universidade Simon Bolivar.

Karpakakunjaram, V., Nair, P., Varguese, T., Royappa, G., Kolatkar, M. & Gadagkar, R. (2003). Contribuitions to the biology of the queenless ponerine ant *Diacamma ceylonense* Emery (Formicidae). *Journal Bombay Natural History Society, 100(2-3)*, 533-543.

Kempf, W. W. (1971). A preliminary review of the ponerine ant genus *Dinoponera* Roger Hymenoptera: Formicidae. *Stud. Entomol, 14*, 369-394.

Ludwing, J. A. & Reynolds, J. F. (1988). *Statistical Ecology: a primer on methods and computing*, New York: John Wiley & Sons.

Mariano, C. S. F., Delabie, J. H. C, Ramos, L. S., Lacau, S. & Pompolo, S. G. (2004). *Dinoponera lucida* Emery (Formicidae: Ponerinae): largest number of chromosomes known in Hymenoptera. *Naturwissenschaften, 91*, 182-185.

Mariano, C. S. F., Pompolo, S. G., Barros, L. A. C., Mariano-Neto, E, Campiolo, S. & Delabie, J. H. C. (2008). A biogeographical study of the threatened ant *Dinoponera lucida* Emery (Hymenoptera: Formicidae: Ponerinae) using a cytogenetic approach. *Insect Conservation & Diversity, 1,* 161-168.

Medeiros, F. N. S., Lopes, L. E., Moutinho, P. R. S., Oliveira, P. S. & Hölldobler, B. (1992). Functional polygyny, agonistic interactions and reproductive dominance in the neotropical ant *Odontomachus chelifer* (Hymenoptera, Formicidae, Ponerinae). *Ethology, 91(2)*, 134-146.

Monnin, T. & Araújo, C. Z. D. (1994). Formation and maintenance of the hierarchy in the queenless ponerine ant *Dinoponera quadriceps*. *Les Insectes Sociaux*. Proc. 12th I.U.S.S.I. Congress, Paris. A. Lenoir, G. Arnold & M. Lepage (Eds.). Paris: Université Paris Nord: 284.

Monnin, T. & Araújo, C. Z. D. (1995). Dominance hierarchy in the queenless ant *Dinoponera quadriceps* Hymenoptera, Formicidae, Ponerinae. *Rev. Bras. Entomol, 39*, 911-920.

Monnin, T. & Peeters, C. (1997). Cannibalism of subordinates' eggs in the monogynous queenless ant *Dinoponera quadriceps. Naturwissenschaften, 84*, 499-502.

Monnin, T .& Peeters. C. (1998). Monogyny and regulation of worker mating in the queenless ant *Dinoponera quadriceps. Anim. Behav, 55*, 299-306.

Monnin, T. & Peeters, C. (1999). Dominance hierarchy and reproductive conflicts among subordinates in a monogynous queenless ant. *Behav. Ecol., 10(3)*, 323-332.

Monnin, T. & Ratnieks, F. L. W. (1999). Reproduction versus work in queenless ants: when to join a hierarchy of hopeful reproductives? *Behav. Ecol. Sociobiol, 46*, 413-422.

Monnin, T. & Ratnieks, F. L. W. (2001). Policing in queenless ponerine ants. *Behav. Ecol. Sociobiol, 50*, 97-108.

Morgan, R. C. (1993). Natural history notes and husbandry of the Peruvian giant ant *Dinoponera longipes* Hymenoptera: Formicidae. *SASI/ITAG 1993 Invertebrates in Captivity Conference Proceedings*, 140-151. Tucson, Arizona.

Oliveira-Filho, A. T. & Fontes, M. A. L. (2000). Patterns of floristic differentiation among Atlantic forests in southeastern Brazil and the influence of climate. *Biotropica, 32*, 793-810.

Overal, W. L. (1980). Obsevations of colony founding and migration of *Dinoponera gigantea*. *J. Georgia Entomol. Soc., 15*, 466-469.

Paiva, R. V. S. & Brandão, C. R. F. (1995). Nests, worker population, and reproductive status of workers, in the giant queenless ponerine ant *Dinoponera* Roger Hymenoptera Formicidae. *Ethol. Ecol. Evol., 7*, 297-312.

Passos, L. & Oliveira, P. S. (2002). Ants affect the distribution and performance of seedlings of *Clusia criuva*, a primarily bird-dispersed rain forest tree. *Journal of Ecology, 90*, 517-528.

Peeters, C. (1991). The occurrence of sexual reproduction among ant workers. *Biol. J. Linn. Soc., 44*, 141-152.

Peeters, C. (1993). Monogyny and polygyny in ponerine ants with or without queens. In: (L. Keller, Ed.) *Queen Number and Sociality in Insects*, (235-261). Oxford: Oxford University Press.

Peeters, C. (1997). Morphologically 'primitive' ants: comparative review of social characters, and the importance of queen-worker dimorphism. In: (By J. Choe, & B. Crespi, Ed.), *The Evolution of Social Behaviour in Insects and Arachnids*. (372-391). Cambridge, Cambridge University Press.

Peeters, C. & Crewe, R. (1987). Foraging and recruitment in Ponerine ants: solitary hunting in the queenless *Ophthalmopone berthoudi* (Hymenoptera: Formicidae). *Psyche, 94*, 201-214.

Peeters, C. & Ito, F. (2001). Colony dispersal and the evolution of queen morphology in social Hymenoptera. *Annu. Rev. Entomol, 46*, 601-30.

Peixoto, A. V., Campiolo, S., Lemes, T. N., Delabie, J. H. C. & Hora, R. R. (2008). Comportamento e estrutura reprodutiva da formiga *Dinoponera lucida* Emery (Hymenoptera, Formicidae). *Rev. Bras. Entomol, 52(1)*, 88-94.

Peternelli, E. F., de, O., Della Lucia, T. M. C. & Borges, E. E. (2001). Mirmecocoria – dispersão de sementes por formigas. *Folha Florestal, 100*, 17-18.

Sánchez-Piñero, F. & Gómez, J. M. (1995). Use of ant-nest debris by darkling beetles and other arthropod species in an arid system in south Europe. *Journal of Arid Environments. 31*, 91-104.

Schultz, T. R. & McGlynn, T. P. (2000). The interactions of ants with other organiusms. In: (D. Agosti, J. D., Majer, L. E. Alonso & T. R. Schultz, Eds.) *Ants: Standard methods for measuring and monitoring biodiversity*, (35-44), Washington: Smithsonian Institution Press.

SEAMA: Secretaria de Estado de Meio Ambiente e de Recursos Hídricos. Espírito Santo: (2002). Disponível em: http://www.seama.es.gov.br/Scripts/sea1805.asp

Traniello, J. F. A. (1982). Population structure and social organization in the primitive ant *Amblyopone pallipes* (Hymenoptera: Formicidae). *Psyche, 89*, 65-80.

Valenzuela-Gonzalez, J., Lopez-Mendez, A. & Garcia-Ballinas, A. (1994). Ciclo de actividad y aprovisionamiento de *Pachycondyla villosa* (Hymenoptera, Formicidae) em agroecosistemas cacaoteros del Soconusco, Chiapas, Mexico. *Folia Entomol. Mex, 91*, 9-21.

Vasconcellos, A., Santana, G. G. & Souza, A. K. (2004). Nest spacing and architecture, and swarming of males of *Dinoponera quadriceps*, (Hymenoptera, Formicidae), in a remnant of the Atlantic Forest in northeast Brazil. *Brazilian Journal of Biology, 64(2)*, 357-362.

Westoby, M., French, K., Hughes, L, Rice, B. & Rodgerson, L. (1991). Why do more plant species use ants for dispersal on infertile compared with fertile soils? *Australian Journal of Ecology, 16*, 445-455.

Wojcik, D. P. (1990). Behavioral Interacions of fire ants and their parasites, predadors and inquilines. In: (R. K. Vander Meer, K. Jaffe & A. Cedeno, Eds.), *Applied Myrmecology: a World Perspective*, (329-344), Boulder: Westiew Press.

Wilson, E. O. (2003). *Pheidole in the New World; a Dominant, Hyperdiverse Ant Genus.* Cambridge: Harvard University Press.

Witte, V., Janssen, R., Eppenstein, A. & Maschwitz, U. (2002). *Allopeas myrmekophilos* (Gastropoda, Pulmonata), the first myrmecophilous mollusc living in colonies of the ponerine army ant *Leptogenys distinguenda* (Formicidae, Ponerinae). *Insectes Sociaux, 49*, 301-305.

In: Species Diversity and Extinction
Editor: Geraldine H. Tepper, pp. 215-238

ISBN: 978-1-61668-343-6
© 2010 Nova Science Publishers, Inc.

Chapter 6

EXPLORING MICROBIAL DIVERSITY – METHODS AND MEANING

Lesley A. Ogilvie[*1] *and Penny R. Hirsch*[2,3]

[1]Department of Molecular Biology, Max Planck Institute for Infection Biology, Charitéplatz 1, 10117 Berlin, Germany
[2]Plant Pathology & Microbiology Department, Rothamsted Research, Harpenden, Herts AL5 2JQ, UK
[3]Cross Institute Programme for Sustainable Soil Function, Rothamsted Research, Harpenden, Herts AL5, 2JQ, UK

ABSTRACT

Since species diversity is central to a large amount of ecological theory, its accurate measurement is key to understanding community structure and dynamics. Techniques to assess the diversity of macro-organisms are well established, but when it comes to micro-organisms many challenges still remain. The rapidly developing suite of molecular microbial community analysis methods, from high throughput sequencing to DGGE, has provided unprecedented insights into microbial community structure, revealing unappreciated levels of diversity. However, the analysis of microbial community profiling data is still in its infancy. Diversity indices and analysis methodology used for macro-organisms are often adopted for microbial community profiling data *en masse*. But no comprehensive analysis of their suitability for micro-organisms has been carried out. Here we review the currently available profiling techniques and posit an analysis framework that will facilitate the translation of this data into credible assumptions about microbial diversity and community structure.

---

[*] Corresponding author: Email: ogilvie@mpiib-berlin.mpg.de, Tel : +49 30 28 460 461.

1. INTRODUCTION

Although small (<10^{-6}m), micro-organisms represent the most diverse and abundant (10^{30} individuals globally) group of organisms known (Curtis et al., 2006; Dykhuizen, 1998). These 'small giants', representing around 60% of the Earth's biomass (Whitman et al., 1998), play crucial roles in all major biogeochemical cycles, are key sources and catalysts of biotechnological products and are of practical importance for a range of activities, including agriculture, waste water and pollution treatment, and most recently, fuel production. The importance of complex microbial communities that are essential to the normal "healthy" functioning of plants and animals (including humans) is already well established, less is known concerning biogeochemical processes. Thus, in these times of climate change, emerging diseases and an ever-increasing human population, understanding microbial community structure, diversity and function within diverse habitats could be a prerequisite for our continuing sustainable existence on Earth.

The enduring intractability of the majority (>99%) of micro-organisms to laboratory culture has driven the development of culture-independent methods to assess the vast unknown 'out there'. A suite of molecular techniques has arisen over the last 20 years largely based on small subunit rRNA genes (mostly 16S rRNA) heterogeneities. These molecular methods, including PCR clone libraries, community profiling methods such as denaturing gradient gel electrophoresis (DGGE) and terminal restriction fragment length polymorphism (T-RFLP) and, most recently, metagenomics coupled with high-throughput sequencing (pyroseqeuncing), have provided unprecedented insights into microbial ensembles, revealing unappreciated and somewhat unfathomable levels of diversity. Estimates per gram of soil range from a conservative 4000 (Schloss and Handlesman, 2006a) 'species' (the definition of species in bacteria remains controversial – see below) to a mind-blowing 10^6 (Gans et al., 2005). The accurate assessment of microbial diversity may tell us if indeed 'everything is everywhere' (Bass-Becking 1934), elucidate the diversity/ecosystem function relationship (Mayr, 1948), plus we may be able to truly investigate our footprint on the environment, from the effects of agricultural practices and land use to pollution. And as evidence further increases for microbial links to a range of diseases including Cystic fibrosis (Tümmler et al., 1999; Willner et al., 2009), Crohn's disease (Peterson et al., 2008; Dicksved et al., 2008) and cancer (Mbulaiteye et al., 2009; Dorer et al., 2009), comparative assessments of microbial community diversity and dynamics in healthy and diseased people may provide some novel perspectives for therapy.

The ongoing development of theory to facilitate the understanding of relationships between organisms with the environment, i.e. ecology and specifically macroecology[1], has resulted in the recognition of diversity as its linchpin (see Macarthur 1955, Loreau et al., 2001). Since complete inventories of species ensembles in any environment are fraught with difficulties, a statistical framework for a more accurate estimation of diversity has arisen. Numerous species abundance models and indices of diversity have been developed each with their pros and cons (see Margurran 2004 for a comprehensive assessment of diversity statistics). Quantification of animal and plant diversity using this theoretical framework has a rich history in facilitating the evaluation of the role of ecological, evolutionary and

---

[1] Understanding statistical patterns of abundance, distribution and diversity through the study of relationships between organisms and their environment on large spatial scales.

anthropogenic factors on ecosystem processes, helping to predict extinctions, guiding conservation strategies and determining functionally important species. Hoping to mirror these insights, microbial ecologists have adopted *en masse* the macroecological statistical tools of the trade despite the unknown and newly emerging role of ecological theory in microbial ecology (see Prosser et al., 2007 for an insightful commentary).

With a focus on prokaryotes, here we review the ability of the currently available molecular profiling techniques and their associated statistical analysis to provide credible insights into microbial diversity and community structure, highlighting their strengths, weaknesses and underlying assumptions. The aim is not to provide a comprehensive review of the technical aspects of each of the molecular and statistical methodologies but to show how they have been used to provide insights into microbial diversity.

2. THE CURRENCY OF DIVERSITY

At the heart of ecological predictions and assessments is the concept of species diversity. The concept originated from Mayr (1948), who envisaged a species as a group of interbreeding individuals that are ecologically distinct due to barriers to recombination. But an asexual mode of reproduction plus a propensity to incorporate foreign DNA into their genome via horizontal gene transfer makes this definition problematic when applied to bacteria. Cohan (2002) posited the ecological species concept, in which species are dictated by niche partitioning. Genetic divergence between groups of microbes will occur as a function of their discrete ecological niches. However, there is growing recognition that bacterial genomes consist of a core and accessory genome (see e.g. Young et al., 2006); the former comprises genes essential for everyday functioning, the latter are nomadic genes, readily lost and gained, usually conferring specialist functions and/adaptations from increased pathogenicity (Groisman and Ochman, 1996) to enhanced catabolic abilities (Todd et al., 2009). For example, strains of *E. coli* can differ by as much as 25% in gene number and content (Welch et al., 2002). Both Mayer's and Cohan's concept of species fails to truly assimilate the accessory nature of the genome.

Typically, 16S rRNA sequences with greater than 97% identity are assigned to the same species (Devereux et al., 1990; Hagström et al., 2002; Schloss and Handelsman, 2005), and those with greater than 93% identity are designated the same genus (Devereux et al., 1990). However, much discussion continues over these somewhat arbitrary cut-offs (Keswani et al., 2001; Morales et al., 2009) and indeed what level is most appropriate for diversity measurement in general – genetic, species, at higher taxon levels such as genus or family, or perhaps even at the indicator groups or organism level, if they exist. For instance, a recent meta-analysis of 81 microbial communities (each containing 100-664 16S rRNA gene sequences) highlighted the importance of genetic diversity in determining the consequences of biodiversity loss due to anthropogenic influence (Parnell et al., 2009). Due to a lack of consensus, microbial ecologists group phylogenetically similar (based on DNA sequence) bacteria into operational taxonomic units (OTUs), or ribotypes, or may refer to guilds or functional groups. As Noel Fierer points out (Fierer, 2007) 'there is no reason to assume that species level is the most appropriate level of taxonomic resolution for comparing levels of microbial diversity'.

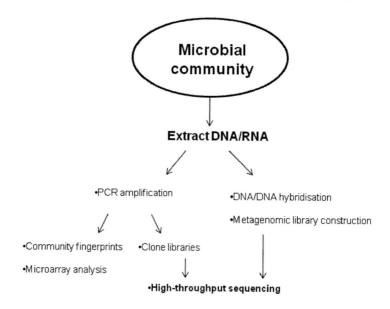

Figure 1. Flow chart showing different community DNA/RNA handling routes.

3. COUNTING MICROBES

An overview of the many molecular methods used to analyse extracted microbial community DNA/RNA is shown in Figure 1. Some of the most influential and/or frequently used methods are discussed below.

3.1. DNA-DNA Hybridization and Reassociation Kinetics

One of the earliest and perhaps most ground breaking techniques for the investigation of microbial diversity within environmental samples. Based on the principle that reassociation of denatured DNA (i.e. single stranded) can be used to determine the size of a bacterial genome (Britten and Kohne, 1968), Torsvik and co-workers (1990a) inferred genetic diversity from the DNA reassociation kinetics of microbial community DNA. Thus, the time taken for reassociation of microbial community DNA represents the amount of heterologous DNA within a mixture. The authors calculated the $C_0t_{1/2}$, where C_0 is the molar concentration of nucleotides in single stranded DNA when reassociation starts and $C_0t_{1/2}$ is the time taken in seconds for 50% reassociation, and used this as a proxy for diversity, analogous to a univariate species diversity indices (Torsvik et al., 1990b). This seminal work led to the much quoted estimate of soil microbial diversity of approximately 10,000 species per gram of forest soil (Torsvik et al., 1990b). However, this estimate was based on the assumption that all microbial species are equally distributed within the sample. Diversity estimates derived from a reworking of this approach incorporating different species-abundance models dwarfed these original calculations by three orders of magnitude. Perhaps more importantly, DNA reassociation kinetics were used to identify different abundance patterns highlighting the large-scale losses of rare members of the soil community due to the toxic effects of metal pollution (Gans et al., 2005). This broad-scale approach provides an over-view of the whole community – not what is there of course, but how many and how they are distributed. Although many studies have demonstrated that community composition and species richness have a role to play in ecosystem functioning (e.g. Naeem et al., 2000; Griffiths et al., 2004), the contribution each plays is still under debate (e.g. see Bell et al., 2005). Nevertheless, if more detailed phylogenetic information is required, investigation of DNA reassociation kinetics could be an important determinant of these finer-scale sampling strategies.

3.2. The 16S rRNA Phylogenetic Framework

The desire to know more than just how many has driven the development of techniques utilizing the phylogenetic framework of highly conserved molecular marker genes, in particular the small subunit ribosomal (SSU) 16S rRNA gene. The 16S rRNA gene has been favoured in microbial diversity studies due to its ubiquitous distribution, i.e. it is found in all Bacteria and Archaea, its constant and constrained functions that have been established at early stages of evolution, plus its relative lack of response to environmental pressures (Woese 1987). It was Pace et al., (1986) who realized the potential use of the 16S rRNA genes (and other SSUrRNA) within phylogenetic studies via the development of probes and primers to facilitate these investigations. Burgeoning sequence databanks pay testament to the overwhelming adaptation of this approach; there are now (October 2009) more than 1 million 16S rRNA sequences in the Ribosomal Database Project (RDP; http://rdp.cme.msu.edu/) – equivalent to the number of bacterial species in gram of soil (albeit that many entries are related isolates of well-known bacteria of medical, veterinary or biotechnological importance). There are other housekeeping genes that are universally present with single copies but the databases are much less extensive, for example the NCBI database has $< 2 \times 10^4$ entries for *gyrB* compared with $>10^6$ for 16S rRNA genes.

Although such molecular techniques avoid the need for cultivation, they are of course subject to the usual biases associated with DNA/RNA extraction and handling, PCR amplification (see Suzuki et al., 1996; Osborn et al., 2000; Ranjard et al., 2003) and the fact that any bacterium or archaeon can harbour multiple copies of the 16S rRNA gene resulting in erroneous estimates of diversity. Despite these caveats, a census of 16S rRNA genes within microbial communities remains the gold standard for most microbial diversity studies.

3.2.1. Community fingerprints

Community fingerprint techniques aim to simplify community structure by providing a representation of the community as a whole or a functional part of it, as defined by primers used. The two main players, Denaturing Gradient Gel Electrophoresis (DGGE) and Terminal Restriction Fragment Length Heterogeneity (T-RFLP) have been used extensively to investigate microbial diversity and community dynamics. DGGE is based on the electrophorectic separation of PCR amplicons according to sequence composition on a denaturing gradient gel (Muyzer 1993). Bands differing in only one base pair composition can be separated; however, in practice multiple amplicons often co-migrate. For T-RFLP, primers (one or both) are labeled with a fluorescently dye (5' or 3' end) and PCR products are digested with restriction enzymes; terminal fragments are detected by capillary or gel electrophoresis (Avaniss-Aghajani et al., 1994; Liu et al., 1997). Less well used techniques such as length heterogeneity PCR (LH-PCR) (Suzuki et al., 1998; Mills et al., 2003), automated ribosomal intergenic spacer analysis (ARISA) (Fisher and Triplett, 1999; Danovaro et al., 2006) and single strand conformation polymorphism (SSCP) (see Schweiger and Tebbe, 1998) also use PCR amplicons heterogeneities to discriminate types or 'species'.

These techniques have facilitated the comparison of microbial diversity within numerous habitats over space and time, providing insights into the microbial inhabitants of terrestrial (Osborn et al., 2000) and agricultural soils (Enwall et al., 2005; Ogilvie et al., 2008), freshwater and estuarine sediments (Konstantinidis et al., 2003; Ben Said et al., 2009), marine (Moeseneder et al., 1999) and freshwater (Danavoro et al., 2006) aquatic environments and the human gut (Dicksved et al., 2008). These methods have also been used to test the functional diversity of habitats; for instance, nitrogen fixers, nitrifiers and denitrifiers (e.g. Patra et al, 2006; Gamble et al., 2009; Zhang et al., 2008), mercury resistance (e.g. Bruce et al., 1997) and ammonia oxidisers (e.g. Nicol et al., 2008).

There is a growing awareness, however, of the limited ability of these techniques to provide an accurate reflection of species diversity, as they characterise only the dominant members of the community and therefore miss taxa that exist in low-abundance (i.e. less than 1% abundance) (Dunbar et al., 2000; Blackwood et al., 2003; Engebretson and Moyer, 2003; Bent et al., 2007; Forney et al., 2004). Therefore, the rare species that often make up the vast majority of diversity in microbial communities (Sogin et al., 2006; Bent and Forney, 2008) will never be detected.

Blackwood et al., (2007) compared empirical and theoretical datasets to find T-RFLP data inaccurately portrays diversity within samples, with all tested diversity indices, leading to underestimates of species diversity. Likewise, Loisel et al., (2006) found DGGE bands were a poor indicator of species number within soil samples. Comparative studies have further highlighted the reduced sensitivity these barcoding methods possess in comparison to the clone and sequence approach; for instance, a recent comparison of microbial diversity of seafloor basalts using small clone libraries (71-246 sequences) and T-RFLP (Orcutt et al., 2009) revealed T-RFLP grossly underestimated the number of species in high diversity samples, missing at least 50% of species identified by the clone library approach. In lower diversity samples, however, T-RFLP provided similar richness estimates and relative diversity as the clone library analysis. Similarly, T-RFLP failed to detect seasonal changes in microbial communities from the Western English Channel, whereas high-throughput sequencing revealed evidence for seasonally structured communities and succession, which was correlated to prevailing environmental conditions (Gilbert et al., 2009).

Fingerprint methodologies, however, are not obsolete when it comes to diversity assessments. Although the age of deep sequencing is here (see below), it is still not a routine part of laboratory life. Until then, the relative ease, financially and time-wise, of producing community profiling data has one main advantage over its more powerful cousin, in that one can easily compare microbial diversity and community structure over space and time. Comparative diversity of a large number of samples at a coarser taxonomic resolution will suffice for many research purposes, but perhaps more importantly these techniques can guide future more detailed investigations. Indeed, such an approach is currently being undertaken by the National Soil Inventory of Scotland (NSIS) to explore structural and functional diversity. T-RFLP analysis of a broad range of soil samples in tandem with functional analysis using GeoChip (see 3.2.3) will identify areas of interest for further finer-scale investigations (Singh et al., 2009).

3.2.2. Sequencing

One of the most enduring approaches facilitating the characterisation of microbial diversity is the construction of clone libraries, in which individual PCR products (normally 16S rRNA gene fragments) are cloned into a vector and analysed via sequencing. Species number is inferred from the number of variants cloned. Over the last ten years small-scale libraries (10-1000 clones) have been routinely reported, for example enabling comparison of microbial communities in grassland soil with different treatments (McCaig et al., 1999). Despite revealing invaluable insights into the microbial domain and uncovering the extent of diversity and advancing our understanding of microbial community structure, these studies also highlighted the inadequacy of such sampling and sequencing effort to reflect in situ diversity. Dunbar and co-workers were the first to quantify the short-fall, predicting, via a comparison of empirical and theoretical datasets (Dunbar et al., 2002), that a sequencing effort of up to two orders of magnitude more was required to obtain a more accurate portrayal of diversity within soils.

The advent of advanced sequencing technologies has ushered a renaissance for the sequence approach (see Maclean et al., 2009 for review of sequencing technologies). Pyrosequencing is based on the 'sequencing by synthesis' technique (Nyrén et al., 1997), i.e. the enzymatic synthesis of a complimentary strand of DNA from a single strand PCR template. As each base pair is added, a chemiluminescent enzyme, luciferase, detects DNA polymerase activity. This method of directly sequencing pooled PCR amplification products presents the opportunity to make more accurate assessments of diversity a step closer to reality by decreasing the time requirements and costs. In the last three years reports from large-scale sequencing projects are started to refine our idea of how much sequencing effort is really required to reflect diversity. Roesch et al., (2007) sequenced over 149,000 16S rRNA sequences for the comparison of diversity within agricultural and forest soils, revealing 52, 000 different OTUs per gram of soil. Bacterial diversity of the forest soils was phylum rich but species poor, whereas agricultural soil was species rich but phylum poor. The authors predict the maximum number of sequences required to identify 90% of the 52,000 OTUs is less than 713,000, which they estimate requires 1 day of operation with the Roche Genome 454 Sequencer FLX system. But these heavy-duty approaches are still not feasible for most laboratories. Even using the most conservative estimates of species diversity (e.g. Schloss and Handlesman, 2006a; see Table 1 for overview of diversity estimates), ca. 36,000 sequences will be required to estimate the total number of 'species' in a gram of soil. And again no method is free of bias; recent work by Kunin et al., (2009) revealed serious over-estimation of diversity could result through sequence errors such as the formation of homopolymer length errors and urged the use of stringent quality control criteria for pyrosequencing efforts (Huse et al., 2007). Despite these remaining barriers to large-scale adaptation of these techniques, unprecedented insight into microbial diversity and function is being made. Recent work by Elshahed et al., (2008) provided a rare glimpse of the 'rare biosphere' by sequencing only 13,001 near-full-length 16S rRNA gene clones derived from an undisturbed tall grass prairie soil in central Oklahoma, revealing many members of the 'rare biosphere' belonged to novel taxonomic lineages or were related to bacteria usually found in non-soil environments. In addition, methodological and analytical improvements are occurring rapidly, facilitating the accurate determination of microbial diversity determination, e.g. the algorithm Pyronoise (Quince et al., 2009).

3.2.3. Microarrays

The development of high-density microarrays for the massive parallel detection of 16S rRNA genes (or functional genes) in recent years is providing a potential rival approach to the clone and sequence standard (DeSantis et al., 2007). One of the most comprehensive, the Phylochip, designed at the Lawrence Berkeley National Lab (LBNL), USA, arrays 500,000 16S rRNA gene DNA probes for the detection of over 9000 bacterial species (or taxa). For functional diversity analysis, the GeoChip, developed at the University of Oklahoma, USA, detects over 10,000 genes covering more than 150 functional groups. The major drawback of the microarray technique is its reliance on predefined sequences, meaning it is less likely to detect novel species or functional groups than, e.g. the sequence approach. This is especially true of environments that have poor coverage, e.g. Antarctic soils (Yergeau et al., 2009). In addition, low abundance microorganisms may be missed as detection sensitivities are thought to be 100-10,000-fold lower than PCR (Zhou et al., 2002). On the other hand, one must balance the possibility for replication and speedy assessment of data that this method provides. Accordingly, microarray methodology has been demonstrated to be well suited to the accurate and reproducible study of bacterial population dynamics, e.g. during metal remediation (Brodie et al., 2006), during development of White Plague Disease type II in reef building corals (Sunagawa et al., 2009), and functional gene diversity in response to organic carbon decomposition (Zhang et al., 2007).

3.3. Metagenomic Analyses

A more recent method based on the clone and sequence approach is the metagenomic analysis of microbial communities. Rather than sequencing of conserved genes, randomly cloned whole community DNA is sequenced using whole genome shotgun cloning. Thus, both the core genome, and the accessory genome are accessed using this approach, providing the full complement of genetic information. Metagenomic libraries have been used to provide a spatial and temporal 'snapshot' of diversity and metabolic potential of diverse habitats such as the Sargasso Sea (Venter et al., 2004), the human gut (Gill et al., 2006), the termite hindgut (Warnecke et al., 2007), honey-bee colonies (Cox-foster et al., 2007), Acid mine drainage (Tyson et al., 2004), agricultural soils (Rondon et al., 2000), and glacial ice (Simon et al., 2009). These investigations are revealing new microbial species and functional genes (e.g. the new class of genes of the rhodopsin family, named proteorhodopsin were discovered in marine picoplankton (Beja et al., 2000), which were later found to function as a proton light pump (Beja et al., 2000b).

As with all techniques a number of biases are implicit; for instance, a surprising lack of the cosmopolitan Sar11 group in a marine metagenomic data set (Beja et al., 2000a) prompted a reminder that new applications of old technologies still have their inherent biases. Bias in this particular data set is attributed to toxicity of cloned genes to *E.coli* (Sorek et al., 2007) and low GC content (Temperton et al., 2009). Most recently the library curators themselves suggested the Sar11 group is underrepresented due to the presence of toxic effects of some of their proteins (Feingersch and Beja, 2009). The use of advanced sequencing technologies, such as pyrosequencing, that bypass the need to clone can now overcome this problem. One major drawback that still needs to be solved is establishing how representative the metagenome sampled really is; Johnston et al., (2005) highlighted this problem in their trawl for *nif* genes in a number of different metagenomic datasets, revealing a paucity of *nif* genes in the Sargasso Sea metagenome (Venter et al., 2004) and in two ocean whale-fall sites, but an over-representation in a third whale-fall site (Tringe et al., 2005)– these observations vividly exemplify the patchy nature of gene distribution within natural environments and highlight the need for careful interpretation of 'diversity' data gained in such studies.

Table 1. Current estimates of species diversity

Habitat	Method	Statistics	Wt/Vol DNA source	Estimate of species richness (OTUs)	Reference
Pristine and metal-contaminated soil	DNA-DNA reassociation kinetics	*Power law model* *Zipf*	10 g	10^7 (pristine) 10^4 (contaminated)	Gans et al., 2005
Alaskan soil	16S rRNA gene clone library, 1033 sequences	*Parametric* Rarefaction	0.5 g	5000	Schloss and Handlesman 2006a
Four geographically separated soils	Pyrosequencing (149,000 16S rRNA sequences, 3% dissimilarity)	*Non-parametric* Chao 1, ACE *Parametric* Rarefaction	1 g	1700 (Chao1) 1725 (ACE) 902 (Rarefaction)	Roesch et al., 2007
Subsurface uranium-contaminated soil; Subsurface water; Air	Clone libraries (485, 253, 417 16s rRNA soil, water, air sequences respectively); Microarrays (OTUs 94% identity: subfamily)	*Non-parametric* clone libraries Chao 1 *Directly observed* microarray	0.5 g(soil)	119-276 (279 on microarray)	DeSantis et al., 2007
			150 ml (water)	4-17 (99 on microarray)	
			14400 L (air)	155-396 (238 on microarray)	
Marine Sediment	556 16S rRNA sequences	Comparison of *parametric* and *non-parametric* models	5 g	2000-3000	Hong et al., 2006
Eight deep water marine sites	Pyrosequencing :118,000 16S rRNA sequences	*Parametric* Chao1, ACE (DOTUR)	1 L	5482 – 23315	Sogin et al., 2006

4. ELUCIDATING DIVERSITY

The techniques outlined above are providing vast datasets that now need to be understood in an ecological context. The adaptation of the macroecological quantitative framework gives microbial ecologists a convenient starting point for the exploration and understanding of factors shaping the microbiome.

Statistical methods for describing diversity generally fall into one of three categories, (i) Diversity indices, (ii) Distribution analysis and (ii) Multivariate methodologies (see Clarke and Warwick 1994). For all categories, the first step is the conversion of sequence or community fingerprint data into a presence/absence or taxon abundance matrix (see Figure 2 for analyses 'pipeline'). At this stage critical decisions have to be made that will define the subsequent analyses. The issue of defining a species or OTU – what % cut-off will be used for sequence data?; what baseline threshold used for community fingerprint data?; if raw data is used, variations in the abundance of a small number of common OTUs will affect the outcome strongly; or will you log/square root transform data to reduce the influence of common OTUs? - an approach that is *de rigueur* in eukaryotic studies. If one is interested in common or rare species then raw or binary data, respectively, may be more appropriate (Grant and Ogilvie, 2003). The analysis route must be driven by sample size, habitat and scientific question asked. Here we consider the suitability of commonly used approaches to microbial diversity.

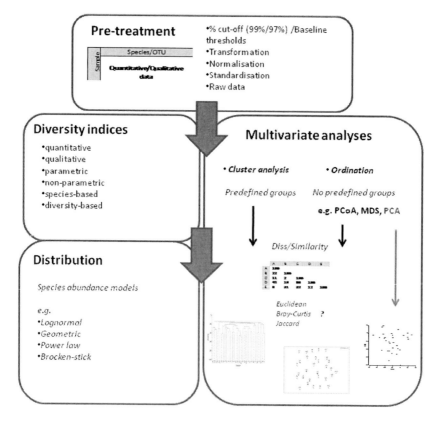

Figure 2. Analyses pipeline showing key stages affecting analyses outcome.

4.1. Parametric and Non-Parametric Diversity Indices

One of the mainstays of diversity measurement is the diversity index. These single indices of community diversity providing easily digestible and functional numerical units are commonly used in microbial ecology (see Hughes et al., 2001, Bohannan and Hughes 2003, Hill et al., 2003). Frequently used indices (see Table 1) for the measurement of within community diversity (alpha diversity) or between community diversity (beta diversity)[2] usually take into account (i) species richness (the total number of species) and/or (ii) evenness, which signifies how evenly individuals are distributed among species. Species number more often changes in relation to sample size than the distribution of abundances (Huston, 1996). Each index describes (i) or (ii) or emphasizes these two facets to a different degree (see Table 1; and Magurran (2004) for a comprehensive overview). If diversity comparisons are required confidence intervals and significance of differences can be calculated across replicate samples using analysis of variance (ANOVA), assuming the univariate indices are normally distributed and variance is consistent across sample groups (see Clarke and Warwick 1994). Although these indices generally make no assumptions about the underlying species abundance distribution, they do require representative samples– still a major stumbling block for microbial ecologists. One major drawback with the approach for microbial communities is that the distillation of data into a single number does not exploit the phylogenetic information inherent within molecular data sets, i.e. samples can have exactly the same diversity but have not one single species in common. Hill et al., (2003) were the first to make a comparison of ecological diversity measurements using amplified rDNA restriction analysis (ARDRA) of relatively small 16S rRNA gene clone libraries (236 clones from metal contaminated and control soil). They found the number of clones and the weight given to the rare versus abundant species were the most important consideration and recommended the use of the log series index alpha, Q statistic (when coverage is >50%), Berger-Parker and Simpson indices and the ubiquitous Shannon H' and evenness indices; however, the latter two led to inaccurate values when coverage was low.

Most recent publications have tended to use non-parametric indices, such as Chao 1 and ACE, based on the mark-recapture model (Krebs, 1989) that determine the heterogeneity of capture probability (Chao, 1984); therefore, the precision and significance of the estimate between two samples can be derived. Both use the frequency of detection of rarer members of the communities to calculate richness. Hughes et al., (2001) indicated Chao 1 may be a particularly useful measure of richness as it is relatively unaffected by sample size.

Roesch et al., (2007) tested the ability of Chao1 and ACE as well as rarefaction to accurately estimate diversity using 2702 known 16S rRNA gene sequences from 2410 species/OTUs and 685 genera of bacteria from the Ribosomal Database Project II. At each of A range of dissimilarity percentage cut-offs were used (0,3,5,10%) to reflect different phylogenetic levels from species to phyla. ACE faired best at 3% dissimilarity with an estimate of 1775 OTUs, but when a 5% cut-off was used, the estimate based on rarefaction made an accurate prediction of 689 genera. Another recent study found these non-parametric approaches unsuitable for estimating microbial diversity within soils as estimates failed to

---

[2] Whittaker (1972) proposed the distinction between α and β diversity as well as γ diversity, which is species richness over a range of habitats.

reach an asymptote (Fierer et al., 2007); however, this study used a considerably lower amount of sequences for analyses (between 86 and 408 for each habitat sampled).

Divergence-based measures of diversity estimate the degree to which pairs of organisms differ, i.e. they do not assume all species are equally related to one another, thereby exploiting the rich phylogenetic information inherent in sequence data. For example, the Phylogenetic Diversity (PD) measure (Faith, 1992) is a divergence-based measure of richness that determines which communities are most diverse; however this index is very sensitive to sample size. Another divergence-based index, theta, provides a measure of evenness and abundance by determining how distinct individuals within a community are (Martin, 2002). To date, however, these indices have been rarely used in microbial diversity studies (e.g. Eckburg et al., 2005; Lozupone et al. 2007; Costello et al., 2009). See Lozupone and Knight (2008) for a comprehensive review on divergence-based diversity measures.

4.2. Species Abundance Models[3]

The observation that ecological communities are generally compiled of numerous low abundance species (Preston, 1948), that cannot be exhaustively sampled particularly in complex communities, has led to the development of a range of species abundance models for the accurate estimation of number and abundance of species within an area or habitat. The number of unobserved species (or OTUs) in a community is estimated by fitting obtained sample data to a species abundance model. This framework is a central tenet of ecology that has facilitated the understanding of plant and animal community dynamics (Fisher et al., 1943). Intuitively these models seem to be a good starting point for microbial communities – lots of rare species and unfathomable diversity that cannot be sampled sufficiently. Moreover, understanding of species distributions will lay the foundations for further more in-depth investigations.

The most often used models are based on lognormal (Preston, 1948), geometric (Motomura, 1932), power-law (Tokeshi, 1993), Brocken-stick (MacArthur, 1957) and Fisher's log-series (Fisher et al., 1943) distributions (see Magurran 2004 for in-depth description of these models). However, the appropriateness of these models for microbial diversity has not as yet been determined empirically. Attempts to do so using T-RFLP /GC-TRFLP and clone library data sets have often corroborated the validity of using the lognormal distribution as a best fit (Schloss and Handlesmann 2006a; Doroghazi et al., 2008; Dunbar et al., 2002), although sometimes only in certain circumstances, i.e. low diversity environments (Hill et al 2003). Fierer et al., (2007) also found the lognormal model useful, but the power law model provided the most accurate estimates of microbial OTU richness. In all these studies, the authors acknowledge sampling effort fell short of that required, i.e. greater than 80% of the total number of species (Preston, 1962). Thus, if there are really up to 10,000 species per gram of soil (Torsvik et al., 1990a,b) then approximately 1 million 16S rRNA sequences would be required to accurately predict microbial diversity (Gans et al., 2005). The lognormal distribution is commonly used as a convenient starting point for many ecological studies, but theoretically testing of the abundance models has cautioned against the whole-

---

[3] For information on other distributional analyses techniques such as k-dominance and ABC curves please see Magurran (2004), Clarke and Warwick (1994), Lambshead (1983) and (Warwick, 1986).

scale adaptation of this approach. Both Narang et al., (2004) and Sloan et al., (2007) modeled different distributions to exemplify how misleading small-scale datasets are in inferring large-scale distribution. The latter study recommends the development of neutral community models (Bell, 2000; Hubbell, 2001) incorporating features of the theory of Island Biogeography (MacArthur and Wilson, 1967) such as evolutionary and ecological processes of speciation, environmental selection, dispersal/immigration and local competition to encompass sampling bias for the accurate estimation of species distribution.

4.3. Multivariate Analyses

For a more in-depth exploration of shifts in microbial diversity over temporal, spatial or physicochemical gradients or episodes, multivariate analyses may be more appropriate. This approach, although just finding its feet in microbial diversity studies, has strong foundations in macroecology (see Clarke and Warwick, 1994). Multivariate analysis allows the investigator to visualise interrelationships between biological samples (and their corresponding environmental characteristics) over space and time, potentially allowing statistically valid discrimination of diss/similarities between microbial communities.

The most frequently used method of identifying similarities is cluster analysis (Ramette, 2007), a technique that groups datasets into well-defined categories based on dissimilarities (Legendre and Legendre, 1998). However, when exploring microbial diversity and structure over temporal and spatial gradients there is often no *a priori* defined structure or groupings (see Chatfield and Collins, 1980), therefore samples are often explored by ordination techniques such as principle components analysis (PCA) and its derivatives, e.g. principle coordinates analysis (PCoA), or multidimensional scaling (MDS). PCA is one of the longest-established ordination methods, but has inherent limitations that make it unsuited to diversity analyses: although based on the quantitative or qualitative species matrix, PCA uses Euclidean distance as an implicit measure of dissimilarity; Euclidean distance can be unsuited to species diversity data as it is often non-continuous in nature, i.e. lots of zeros. In addition, PCA excludes rare species to provide comparable numbers of species and samples (see Chatfield and Collins, 1980; Clarke and Warwick, 1994). PCA is thought to be more suited to environmental variables, which often form a continuum and are of limited number. Nevertheless, many publications have used this method to determine differences in microbial community structure and diversity (e.g. see Miletto et al., 2008; Talbot et al., 2009).

Most other ordination analyses of diversity start with the construction of a similarity/dissimilarity matrix using transformed species abundance data. Blackwood et al., (2003) recommended the use of Euclidean distance, one of the most commonly used measures, on square root transformed T-RFLP peak height data for exploring their data; however, in addition to above-mentioned drawbacks, Euclidean distance measures similarities based on shared absences, meaning two sites could have a high similarity but not have one species in common! For this reason, the Bray-Curtis (Czekanowski) co-efficient (Bray and Curtis, 1957), which is unaffected by shared absences (Clarke and Warwick, 1994), could be a more pertinent choice. For qualitative data, Jaccard distance could be an unbiased choice as it gives equal weighting to common and rare OTUs.

Table 2. Diversity indices

Index	Features	Reference
Species/OTU richness		
Richness (S)	Number of species, OTUs. Non-parametric, qualitative, species-based	
Margalef's index (D_{Mg})	Non-parametric, qualitative. The number of species present for a given number of individuals. Species-based.	Margalef, 1958
Choa1	Non-parametric, species-based. No underlying assumptions. Abundance data required. Not affected by size of sample.	Chao, 1984
ACE	Non-parametric, species-based. No underlying assumptions. Abundance data required. Not affected by size of sample.	Chao and Lee, 1992
Shannon diversity (H')	Measure of the difficulty of predicting the next species. Incorporates species richness and abundance. Non-parametric, quantitative, species-based. More weight attached to abundant species. Assumes all species are represented in a sample. Sensitive to sample size.	Shannon and Weaver, 1963
Phylogenetic diversity (PD)	Non-parametric, qualitative, divergence-based. Calculates which communities are the most diverse.	Faith 1992
Theta	Measure of the total genetic variation. Non-parametric, quantitative, divergence based. Measures how phylogenetically distinct individuals are within a community.	Martin 2002
Log series alpha (α)/ Fishers α	Parametric, quantitative, species-based. Describes the number of species and no. of individuals in those species. Underlying assumption that species distribution is log series. Biased by the abundance of dominant species.	Fisher et al., 1943
The Q statistic (Q)	Parametric, quantitative, species-based. Very abundant or very rare species do not bias the outcome. Best used when >50% of species are sampled.	Kempton and Taylor, 1978
Evenness/Dominance		
Simpson's dominance (D)	Probability that two species are the same. Range: 0 (taxa are present in equal abundance)-1 (one taxa dominates the community). More weight attached to abundant species. Sensitive to sample size. Species-based.	Simpson, 1949
Pielou's evenness	Relative abundance of each species. Sample size sensitive. Species-based.	Pielou, 1966
Berger-Parker index (*d*)	Numerical importance of the most abundant species. Sensitive to sample size. Species-based.	Berger and Parker, 1970
Diss/Similarity		
Euclidean distance	Simple measure of distance between two co-ordinates. Similarity determined by shared absences.	
Sørensen Index /Bray-curtis similarity	Comparison of communities, beta diversity. Unaffected by shared absences	Sørensen, 1948
Jaccard co-efficient	Comparison of communities beta diversity. Equal weighting to common and rare OTUs	Jaccard, 1901

An ordination is then used to rank the samples in dimensional space. For example, one of the frequently used ordination techniques, MDS, attempts to preserve the ranked order of the similarity of communities as an inverse function of the distance between the points representing, e.g., DGGE or T-RFLP profiles or sequence data, on a graphical plot (Kruskal, 1964). The greater the similarity of the profiles from each sample the closer they are plotted together, profiles with the lowest similarity are plotted furthest apart. Significance and confidence in the ordination can then be calculated by analysis of similarities (ANOSIM) (Clarke, 1993) and Kruskal values, respectively. The ANOSIM R value is a particularly useful measure as it indicates whether replicates within one habitat are more similar to each other than to any samples from the other habitats. MDS and ANOSIM analysis have a generality of application due to a lack of assumptions and flexibility to use unbalanced data sets (Clarke and Warwick, 1994). Readers are referred to Chatfield and Collins (1980), Clarke and Warwick (1994) and Ramette (2007) for further information.

Certain aspects of both univariate and multivariate analyses are now being incorporated into software interfaces facilitating the robust statistical analyses of the ever growing datasets resulting from high-throughput sequencing projects. For instance, UniFrac a software that compares sequence libraries based on phylogenetic difference, and tests if significantly different, visualizing these differences using principle components analysis (PCA) (see Lozupone and Knight, 2005 for further details of software); DOTUR (Distance-based OTU richness determination; Schloss and Handlesman, 2005) constructs rarefaction and collector's curves and estimates OTU and richness using Chao and ACE1 indices; SONS, also developed by Schloss and Handlesmann, implements these nonparametric estimators for the fraction and richness of OTUs shared between two communities (Schloss and Handlesman, 2006b).

4.4. Case Study

To exemplify the alternate diversity outcomes using different statistical treatments, we re-analysed T-RFLP 16S rRNA gene fingerprints derived from estuarine sediment microbial communities located in a gradient of metal pollution spanning two orders of magnitude (Ogilvie and Grant, 2008). All analysis was carried out using the Primer (v5) statistical package. Raw data was used to provide a diversity estimate biased by common OTUs, binary data to highlight rare OTUs, and transformed data to provide equal weighting to both. Diversity estimates varied according to type of data and index applied (see Table 3): Maximum and minimum OTU richness corresponding to the binary and raw data, respectively, was calculated using the Margalef's richness index. In addition, Fisher's alpha index varied widely upon transformation of data.

Table 3. Diversity index comparison

Sample	S	N	d	J'	Fisher α	H'	Transformation
K	51	163	9.816	0.8183	25.49	3.217	Raw data
K	51	77	11.52	0.9587	66.61	3.769	sq. root
K	51	56	12.45	0.955	302.4	3.755	log
K	51	51	12.72	na	na	na	presence/absence

na – not applicable

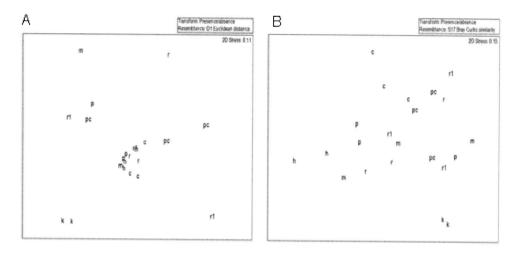

Figure 3. Distance measurements influence interpretation of TRFLP community profiles. Multidimensional scaling (MDS) of a (A) Euclidean distance and (B) Bray-Curtis derived similarity matrix of 16S rRNA gene T-RFLP data from estuarine sediments. The ranked order of similarities between sample DGGE profiles is graphically represented here. The closer two samples are on the ordination the more similar their T-RFLP profiles are.

We then compared a range of different treatments for visualizing the data. Similarity matrices based on presence/absence data were constructed using Euclidean distance or Bray Curtis measures of similarity and a MDS ordination was performed (Figure 3). The two ordinations of the same data show how microbial community structure and diversity is represented in different ways depending on the statistical treatment chosen. When exploring diversity data one must be aware of the assumptions underlying the statistics used and thus modify the interpretation accordingly. Similarly, Enwall and Hallin (2009), demonstrate the differential discriminatory abilities of statistical treatments on T-RFLP and DGGE functional gene profiles derived from agricultural soil microbial communities. Most microbial ecologists may not be fully aware of the extent to which these algorithms affect their interpretation of the data and use maybe determined by convention rather than their specific features.

5. CONCLUSION

The application of molecular biology techniques to microbial ecology in the 20th Century, whilst shedding new light on a previously intransigent subject, has revealed the extent of the problem in attempting to describe community diversity. Now, a new generation of microbial diversity profiling techniques is coming to fruition with the advent of high-throughput sequencing technologies, e.g. pyrosequencing. Each category serves specific purposes when it comes to investigating diversity and community structure. The approach taken must be driven by scientific objectives: What is the biological question being asked? What is the size of the data set and is the quest for 'true diversity' a must or is a comparative assessment of microbial diversity over time and/or space or in the presence of an environmental variable more appropriate.

Plant and animal ecologists have had over 200 years to bend and shape their theory but still have not found the answers to Nature's idiosyncrasy. Likewise, microbial ecology will need time to understand how theory and methodologies fit in with the observations now feasible with available tools. As more observations are made, theories and methodologies will be adopted, adapted, created and understood. Future projects such as the human microbiome project (Turnbaugh et al., 2007), TerraGenome (Vogel et al., 2009) and the National soil survey (Singh et al., 2009) promise the magnitude of data required for laying the foundations for the accurate testing of these available and newly developed statistical models.

It is with a number of caveats and unknowns that we microbial ecologists set out to understand 'how many?', 'what kind?' and 'what for?'. But, perhaps perversely, this is what makes it such a fascinating journey.

REFERENCES

Avaniss-Aghajani, E; et al. A molecular technique for identification of bacteria using small subunit ribosomal RNA sequences. *BioTechniques*, 1994, 17, 144-149.

Bass Becking, LGM. Geobiologie of Inleiding Tot de Milieukunde. Van Stockum & Zoon; *The Hague*, The Netherlands. 1934.

Béjà, O; et al. Construction and analysis of bacterial artificial chromosome libraries from a marine microbial assemblage. *Environ Microbiol*, 2000a, 2, 516-29.

Béjà, O; et al. Bacterial rhodopsin: evidence for a new type of phototrophy in the sea. *Science*, 2000b, 289, 1902-6.

Ben Said, O; Goñi-Urriza, M; El Bour, M; Aissa, P; Duran, R. Bacterial Community Structure of Sediments of the Bizerte Lagoon (Tunisia), a Southern Mediterranean Coastal Anthropized Lagoon. *Microb Ecol*, 2009, [Epub ahead of print]

Bent, SJ; et al. Measuring species richness based on microbial community fingerprints: the emperor has no clothes. *Appl Environ Microbiol*, 2007, 73, 2399-401. author reply 2399-401.

Bent, SJ; Forney, LJ. The tragedy of the uncommon: understanding limitations in the analysis of microbial diversity. *ISME J*, 2008, 2, 689-95.

Bell, T; et al. Larger islands house more bacterial taxa. *Science*, 2005, 308, 1884.

Bell, G. The distribution of abundance in neutral communities. *Am Nat*, 2000, 155, 606-617.

Berger, WH; Parker, FL. Diversity of planktonic foraminifera in deep sea sediments. *Science*, 1970, 168, 1345-1347.

Blackwood, CB; et al. Terminal restriction fragment length polymorphism data analysis for quantitative comparison of microbial communities. *Appl Environ Microbiol*, 2003, 69, 926-32.

Blackwood, CBl; et al. Interpreting ecological diversity indices applied to terminal restriction fragment length polymorphism data: insights from simulated microbial communities. *Appl Environ Microbiol*. 2007, 73, 5276-83.

Bohannan, BJ; Hughes, J. New approaches to analyzing microbial biodiversity data. *Curr Opin Microbiol*, 2003, 6, 282-7.

Bray, SR; Curtis, JT. An ordination of the upland forest communities of Southern Wisconsin. *Ecol. Monogra*, 1957, 27, 325-349.

Britten, RJ; Kohne, DE. Repeated sequences in DNA. Hundreds of thousands of copies of DNA sequences have been incorporated into the genomes of higher organisms. *Science*. 1968, 161, 529-40.

Brodie, EL; et al. Application of a high-density oligonucleotide microarray approach to study bacterial population dynamics during uranium reduction and reoxidation. *Appl Environ Microbiol*, 2006, 72, 6288-98.

Bruce, KD. Analysis of mer gene subclasses within bacterial communities in soils and sediments resolved by fluorescent-PCR-restriction fragment length polymorphism profiling. *Appl Environ Microbiol*, 1997, 63, 4914-4919.

Chao, A. Non-parametric estimation of the number of classes in a population. *Scand J Stat*, 1984, 11, 265-270.

Chao, A; Lee, SM. Estimating the number of classes via sample coverage. *J Am Stat Assoc*, 1992, 87, 210-217. doi:10.2307/2290471

Chatfield, C; Collins, AJ. Introduction to Multivariate Analysis, Chapman and Hall. 1980.

Clarke, KR. Non-parametric multivariate analysis of changes in community structure. *Austr J Ecol*, 1993, 18, 117-143.

Clark, KR; Warwick, RM. Changes in marine communities: an approach to statistical analysis and interpretation. *Natural Environment Research Council, UK*, 1994, 144.

Cohan, FM. What are bacterial species? *Annu. Rev. Microbiol*, 2002, 56, 457-487.

Costello, EK; Lauber, CL; Hamady, M; Fierer, N; Gordon, JI; Knight, R. Bacterial Community Variation in Human Body Habitats Across Space and Time. *Science*. 2009, [Epub ahead of print]

Cox-Foster, DL; et al. A metagenomic survey of microbes in honey bee colony collapse disorder. *Science*, 2007, 318, 283-7.

Curtis, TP; et al. What is the extent of prokaryotic diversity? *Philos Trans R Soc Lond B Biol Sci.*, 2006, 361, 2023-37.

Danovaro, R; et al. Comparison of two fingerprinting techniques, Terminal Restriction Fragment Length Polymorphism and Automated Ribosomal Intergenic Spacer Analysis, for determination of bacterial diversity in aquatic environments. *Appl Environ Microbiol*, 2006, 72, 5982-5989.

DeSantis, TZ; et al. High-density universal 16S rRNA microarray analysis reveals broader diversity than typical clone library when sampling the environment. *Microb Ecol*, 2007, 53, 371-383.

Devereux, R; et al. Diversity and origin of *Desulfovibrio* species : phylogenetic defnition of a family. *J Bacteriol*, 1990, 172, 3609-3619.

Dicksved, J; et al. Molecular analysis of the gut microbiota of identical twins with Crohn's disease. *ISME J*, 2008, 2, 716-27.

Dorer, MS; et al. *Helicobacter pylori*'s Unconventional role in health and disease. PLoS Pathog 2009, 5, e1000544.

Doroghazi, JR; Buckley, DH. Evidence from GC-TRFLP that bacterial communities in soil are lognormally distributed. PLoS One, 2008, 3, e2910.

Dunbar, J; et al. Assessment of microbial diversity in four southwestern United States soils by 16S rRNA gene terminal restriction fragment analysis. *Appl Environ Microbiol*, 2000, 66, 2943-50.

Dunbar, J et al. Empirical and theoretical bacterial diversity in four Arizona soils. *Appl. Environ Microbiol*, 2002, 68, 3035-3045.

Dykhuizen, DE. Santa Rosalia revisited: why are there so many species of bacteria? *Antonie van Leeuwenhoek*, 1998, 73, 25-33.

Eckburg, PB; et al. Diversity of the human intestinal microbial flora. *Science*, 2005, 308, 1635-1638.

Elshahed, MS; et al. Novelty and uniqueness patterns of rare members of the soil biosphere. *Appl Environ Microb*, 2008, 74, 5422-5428.

Engebretson, JJ; Moyer, CL. Fidelity of select restriction endonucleases in determining microbial diversity by terminal-restriction fragment length polymorphism. *Appl Environ Microb*, 2003, 69, 4823-4829.

Enwall, K; et al. Activity and composition of the denitrifying bacterial community respond differently to long-term fertilization. *Appl Environ Microbiol*, 2005, 71, 8335-43.

Enwall, K; Hallin, S. Comparison of T-RFLP and DGGE techniques to assess denitrifier community composition in soil. *Lett Appl Microbiol*, 2009, 48, 145-8.

Faith, DP. Conservation evaluation and phylogenetic diversity. *Biol Conserv*, 1992, 61, 1-10.

Feingersch, R; Béjà, O. Bias in assessments of marine SAR11 biodiversity in environmental fosmid and BAC libraries? *ISME J*, 2009, 3, 1117-9.

Fierer, N. Tilting at windmills: a response to a recent critique of terminal restriction fragment length polymorphism data. *Appl Environ Microbiol*, 2007, 73, 8041; author reply 8041-2.

Fierer, N; et al. Metagenomic and small-subunit rRNA analyses reveal the genetic diversity of bacteria, archaea, fungi, and viruses in soil. *Appl Environ Microbiol*, 2007, 73, 7059-66.

Fisher, RA; et al. The relation between the number of species and the number of individuals in a random sample of an animal population. *J Animal Ecol*, 1943, 12, 42-58.

Fisher, MM; Triplett, E. Automated approach for ribosomal intergenic spacer analysis of microbial diversity and its application to freshwater bacterial communities. *Appl Environ Microbiol*, 1999, 65, 4630-4636.

Forney, LJ; et al. Molecular microbial ecology: land of the one-eyed king. *Curr Opin Microbiol*, 2004, 7, 210-220.

Gamble, MD; et al. Seasonal variability of diazotroph assemblages associated with the rhizosphere of the salt marsh cordgrass, *Spartina alterniflora*. *Microb Ecol*, 2009. [Epub ahead of print]

Gans, J; et al. Computational improvements reveal great bacterial diversity and high toxicity in soil. *Science*, 2005, 309, 1387-1390.

Gilbert, JA; et al. The seasonal structure of microbial communities in the Western English Channel. *Environ Microbiol*, 2009, [Epub ahead of print]

Gill, SR; et al. Metagenomic analysis of the human distal gut microbiome. *Science*, 2006, 312, 1355-9.

Grant, A; Ogilvie, LA. Terminal restriction fragment length polymorphism data analysis. *Appl Environ Microbiol*, 2003, 69, 6342.

Griffiths, BS; et al. The relationship between microbial community structure and functional stability, tested experimentally in an upland pasture soil. *Microb Ecol.*, 2004, 47, 104-113.

Groisman, EA; Ochman, H. Pathogenicity islands: quantum leaps in bacterial evolution. *Cell*, 1996, 87, 791-794.

Hagström, A; et al. Use of 16S ribosomal DNA for delineation of marine bacterioplankton species. *Appl Environ Microbiol*, 2002, 68, 3628-3633.

Hill, TCJ; et al. Using ecological diversity measures with bacterial communities. *FEMS Microbiol Ecol.*, 2003, 43, 1-11.

Hong, SH; et al. Predicting microbial species richness. *Proc. Natl. Acad. Sci. USA*, 2006, 103, 117-122.

Hubbell, SP. The unified neutral theory of biodiversity and biogeography. Princeton Univ. Press, *Princeton*, 2001, New Jersey.

Hughes, JB; et al. Counting the uncountable: statistical approaches to estimating microbial diversity. *Appl Environ Microbiol*, 2001, 67, 4399-406.

Huse, SM; et al. Accuracy and quality of massively parallel DNA pyrosequencing. *Genome Biol.* 2007, 8, R143.

Jaccard, P. Distribution de la flore alpine dans le bassin des Dranses et dans quelques régions voisines. *Bulletin del la Société Vaudoise des Sciences Naturelles*, 1901, 37, 241-272.

Johnston, AWB; et al. Metagenomic marine nitrogen fixation--feast or famine? *Trends Microbiol*, 2005, 13, 416-20.

Kempton, RA; Taylor, LR. The Q-statistic and the diversity of floras. *Nature*, 275, 252-253.

Keswani, J; et al. Relationship of 16S rRNA sequence similarity to DNA hybridization in prokaryotes. *Int J Syst Evol Microbiol.* 2001, 51, 667-78.

Konstantinidis, KT; et al. Microbial diversity and resistance to copper in metal-contaminated lake sediment. *Microb Ecol.* 2003, 45, 191-202.

Krebs, CJ. *Ecological methodology*. Harper & Row, NY, USA. 1989.

Kruskal, JB. Multidimensional scaling by optimizing goodness of fit to a nonmetric hypothesis. Adkins, Dorothy C. and Horst, Paul, editors. Psychometrika. 1964.

Kunin, V; et al. Wrinkles in the rare biosphere: pyrosequencing errors lead to artificial inflation of diversity estimates. *Environ Microbiol.* 2009, [Epub ahead of print]

Lambshead, PJD; et al. The detection of differences among assemblages of marine benthic species based on an assessment of dominance and diversity. *J Natur Hist*, 1983, 17, 859-874.

Liu, WT; et al. Characterization of microbial diversity by determining terminal restriction fragment length polymorphisms of genes encoding 16S rRNA. *Appl Environ Microbiol.* 1997, 63, 4516-22.

Loreau, M; et al. Biodiversity and ecosystem functioning: current knowledge and future challenges. *Science*, 2001, 294, 804-808.

Legendre, P; Legendre, L. Numerical ecology. Developments in Environmental Modelling 20. *Elsevier Science*, 1998, Amsterdam, 853.

Loisel, P; et al. Denaturing gradient electrophoresis (DGE) and single-strand conformation polymorphism (SSCP) molecular fingerprintings revisited by simulation and used as a tool to measure microbial diversity. *Environ Microbiol*, 2006, 8, 720-31.

Lozupone, CA; Knight, R. UniFrac: a new phylogenetic method for comparing microbial communities. *Appl Environ Microbiol*, 2005, 71, 8228-35.

Lozupone, CA; Knight, R. Species divergence and the measurement of microbial diversity. *FEMS Microbiol Rev.* 2008, 32, 557-78.

Lozupone, CA; et al. Quantitative and qualitative beta diversity measures lead to different insights into factors that structure microbial communities. *Appl Environ Microbiol*, 2007, 73, 1576-85.

MacArthur, RH. Fluctuations of animal populations and a measure of community stability. *Ecology.*, 1955, 36, 533-536.

MacArthur, RH. 1957. On the relative abundance of bird species, *PNAS*, 43, 293-295.

MacArthur, RH; Wilson, EO. The theory of Island Biogeography. Princeton Univ. Press, *Princeton*, 1967, New Jersey.

MacLean, D; et al. Application of 'next-generation' sequencing technologies to microbial genetics. *Nat Rev Microbiol*, 2009, 7, 287-96.

Magurran, AE. Measuring biological diversity. Oxford: Blackwell Publishing. ISBN 0-632-05633-9, 2004.

Mayr. E. "The Bearing of the New Systematics on Genetical Problems: The Nature of Species," *Adv Genet*, 1948, 2, 205-237.

Margalef, R. 1958. Information theory in ecology, *General Systems*, 3, 36-71.

Martin, AP. Phylogenetic approaches for describing and comparing the diversity of microbial communities. *Appl Environ Microbiol*. 2002, 68, 3673-82.

McCaig, AE; et al. Molecular analysis of bacterial community structure and diversity in unimproved and improved upland grass pastures. *Appl Environ Microbiol*, 1999, 65, 1721-30.

Mbulaiteye, SM; et al. *Helicobacter pylori* associated global gastric cancer burden. *Front. Biosci*, 2009, 14, 1490-1504.

Miletto, M; et al. 2008. Biogeography of sulfate-reducing prokaryotes in river floodplains. *FEMS Microbiol. Ecol.*, 64, 395-406.

Mills, DK; et al. A comparison of DNA profiling techniques for monitoring nutrient impact on microbial community composition during bioremediation of petroleum-contaminated soils. *J Microbiol Methods*, 2003, 54, 57-74.

Moeseneder, MM; et al. Optimization of terminal-restriction fragment length polymorphism analysis for complex marine bacterioplankton communities and comparison with denaturing gradient gel electrophoresis. *Appl Environ Microbiol*. 1999, 65, 3518-25.

Morales, SE; et al. Extensive phylogenetic analysis of a soil bacterial community illustrates extreme taxon evenness and the effects of amplicon length, degree of coverage, and dna fractionation on classification and ecological parameters. Appl. *Environ. Microbiol*, 2009, 75, 668-675

Motomura, I. A statistical treatment of associations, Jpn. *J. Zool*, 1932, 44, 379-383 (in Japanese)

Muyzer, G; et al. Profiling of complex microbial populations by denaturing gradient gel electrophoresis analysis of polymerase chain reaction-amplified genes coding for 16S rRNA. *Appl Environ Microbiol*. 1993, 59, 695-700.

Narang, R; et al. Modeling bacterial species abundance from small community surveys. *Microb Ecol.*, 2004, 47, 396-406.

Naeem, S; et al. Producer-decomposer co-dependency influences biodiversity effects. *Nature* 2000, 403, 762-764.

Nicol, GW; et al. The influence of soil pH on the diversity, abundance and transcriptional activity of ammonia oxidizing archaea and bacteria. *Environ Microbiol*. 2008, 10, 2966-78.

Nyrén, P; et al. Detection of single-base changes using a bioluminometric primer extension assay. *Anal Biochem*. 1997, 244, 367-73.

Ogilvie, LA; et al. Bacterial diversity of the broadbalk 'classical' winter wheat experiment in relation to long-term fertilizer inputs. *Microb Ecol*, 2008, 56, 525-537.

Ogilvie, LA; Grant, A. Linking pollution induced community tolerance (PICT) and microbial community structure in chronically metal polluted estuarine sediments. *Mar Environ Res*, 2008, 65, 187-98.

Orcutt, B; et al. An interlaboratory comparison of 16S rRNA gene-based terminal restriction fragment length polymorphism and sequencing methods for assessing microbial diversity of seafloor basalts. *Environ Microbiol*. 2009, [Epub ahead of print]

Osborn, AM; et al. An evaluation of terminal-restriction fragment length polymorphism (T-RFLP) analysis for the study of microbial community structure and dynamics. *Environ Microbiol*, 2000, 2, 39-50.

Pace, NR; et al. Ribosomal RNA phylogeny and the primary lines of evolutionary descent. *Cell*, 1986, 45, 325-6.

Parnell, JJ; et al. Biodiversity in microbial communities: system scale patterns and mechanisms. *Mol Ecol*, 2009, 18, 1455-62.

Patra, AK; et al. Effects of management regime and plant species on the enzyme activity and genetic structure of N-fixing, denitrifying and nitrifying bacterial communities in grassland soils. *Environ Microbiol*, 2006, 8, 1005-16.

Peterson, DA; et al. Metagenomic approaches for defining the pathogenesis of inflammatory bowel diseases. *Cell Host Microbe*, 2008, 3, 417-27.

Pielou, EC. The measurement of diversity in different types of biological collections. *J. Theor. Biol*, 1966, 13, 131-144.

Preston, FW. The commonness, and rarity, of species. *Ecology*, 1948, 29, 254-283.

Preston, FW. The canonical distribution of commonness and rarity. *Ecology*, 1962, 43, 185-215.

Prosser, JI et al. The role of ecological theory in microbial ecology. *Nat Rev Microbiol*, 2007, 5, 384-92.

Quince, C; et al. Accurate determination of microbial diversity from 454 pyrosequencing data. *Nat Methods*, 2009, 6, 639-41.

Ramette, A. Multivariate analyses in microbial ecology. *FEMS Microbiol Ecol*, 2007, 62, 142-60.

Ranjard, L; et al. Sampling strategy in molecular microbial ecology: influence of soil sample size on DNA fingerprinting analysis of fungal and bacterial communities. *Environ Microbiol*, 2003, 5, 1111-20.

Roesch, LF; et al. Pyrosequencing enumerates and contrasts soil microbial diversity. *ISME J*, 2007, 1, 283-90.

Rondon, MR; et al. Cloning the soil metagenome: a strategy for accessing the genetic and functional diversity of uncultured microorganisms. *Appl Environ Microbiol*, 2000, 66, 2541-7.

Schloss, PD; Handelsman, J. Introducing DOTUR, a computer program for defining operational taxonomic units and estimating species richness. *Appl Environ Microbiol*, 2005, 71, 1501-1506.

Schloss, PD; Handelsman, J. Toward a census of bacteria in soil. *PLoS Comput Biol*, 2006a, 2, e92.

Schloss, PD; Handlesman, J. Introducing SONS, a tool for operational taxonomic unit-based comparisons of microbial community memberships and structures. *Appl Environ Microbiol*, 2006b, 72, 6773-6779.

Schweiger, F; Tebbe, CC. A new approach to utilise PCR-single-strand-conformation polymorphism for 16S rRNA gene-based microbial community analysis. *Appl Environ Microbiol,* 1998, 64, 4870-4876.

Shannon, AE; Weaver, W. *The mathematical theory of communication.* University of Illinois Press, Urbana, IL 1963.

Simon, C; et al. Rapid identification of genes encoding DNA polymerases by function-based screening of metagenomic libraries derived from glacial ice. *Appl Environ Microbiol,* 2009, 75, 2964-8.

Simpson, EH. Measurement of diversity. *Nature,* 1949, 163, 688.

Singh, BK; et al. Soil genomics. *Nat Rev Microbiol,* 2009, 7, 756.

Sloan, WT; et al. Modeling taxa-abundance distributions in microbial communities using environmental sequence data. *Microb Ecol,* 2007, 53, 443-55.

Sorek, R; et al. Genome-wide experimental determination of barriers to horizontal gene transfer. *Science,* 2007, 318, 1449-1452.

Sogin, ML; et al. Microbial diversity in the deep sea and the underexplored 'rare biosphere'. Proc. *Natl Acad. Sci. USA,* 2006, 103, 12115-12120.

Sørensen, T. A method of establishing groups of equal amplitude in plant sociology based on similarity of species and its application to analyses of the vegetation on Danish commons. Biologiske Skrifter / *Kongelige Danske Videnskabernes Selskab,* 1948, 5, 1-34.

Suzuki, MT; et al. Kinetic bias in estimates of coastal picoplankton community structure obtained by measurements of small-subunit rRNA gene PCR amplicon length heterogeneity. *Appl Environ Microbiol,* 1998, 64, 4522-9.

Suzuki, MT; Giovannoni, SJ. Bias caused by template annealing in the amplification of mixtures of 16S rRNA genes by PCR. *Appl Environ Microbiol,* 1996, 62, 625-630.

Sunagawa, S; et al. Bacterial diversity and White Plague Disease-associated community changes in the Caribbean coral *Montastraea faveolata. ISME J,* 2009, 3, 512-21.

Talbot, G; et al. Multivariate statistical analyses of rDNA and rRNA fingerprint data to differentiate microbial communities in swine manure. *FEMS Microbiol Ecol,* 2009, [Epub ahead of print]

Temperton, B; et al. Bias in assessments of marine microbial biodiversity in fosmid libraries as evaluated by pyrosequencing. *ISME J,* 2009, 3, 792-6.

Todd, JD; et al. Molecular dissection of bacterial acrylate catabolism - unexpected links with dimethylsulfoniopropionate catabolism and dimethyl sulfide production. *Environ Microbiol,* 2009, [Epub ahead of print].

Tokeshi, M. Species abundance patterns and community structure. *Adv Ecol Res.,* 1993, 24, 111-186.

Torsvik, V et al. Comparison of phenotypic diversity and DNA heterogeneity in a population of soil bacteria. *Appl Environ Microbiol,* 1990a, 1990 A.

Torsvik, V; et al. High diversity in DNA of soil bacteria. *Appl Environ Microbiol,* 1990b, 56, 782-7. A.

Tringe, SG; et al. Comparative metagenomics of microbial communities. *Science,* 2005, 308, 554-7.

Turnbaugh, PJ; et al. The human microbiome project. *Nature,* 2007, 449, 804-10.

Tümmler, B; Kiewitz, BC. Cystic fibrosis: an inherited susceptibility to bacterial respiratory infections. *Mol. Med. Today,* 1999, 5, 351-357.

Tyson, GW; et al. Community structure and metabolism through reconstruction of microbial genomes from the environment. *Nature*, 2004, 428, 37-43.

Venter, JC et al. Environmental genome shotgun sequencing of the Sargasso Sea. *Science*, 2004, 304, 66-74.

Vogel, TM; et al. TerraGenome: a consortium for the sequencing of a soil metagenome. *Nat Rev Micro*, 2009, 7, 252.

Warnecke, F; et al. Metagenomic and functional analysis of hindgut microbiota of a wood-feeding higher termite. *Nature*, 2007, 450, 560-5.

Warwick, RM. A new method for detecting pollution effects on marine macrobenthic communities. *Marine Biology*, 1986, 92, 557-562.

Welch, RA; et al. Extensive mosaic structure revealed by the complete genome sequence of uropathogenic *Escherichia coli*. *Proc Natl Acad Sci U S A*, 2002, 99, 17020-4.

Whitman, WB; et al. Prokaryotes: the unseen majority. *Proc. Natl Acad. Sci. U S A*, 1998, 95, 6578-6583.

Whittaker, RH. Evolution and measurement of species diversity. *Taxon*, 1972, 21, 213-251.

Willner, D; et al. Metagenomic analysis of respiratory tract DNA viral communities in cystic fibrosis and non-cystic fibrosis individuals. *PLoS One*, 2009, 4, e7370.

Woese, CR. Bacterial evolution. *Microbial Rev*, 1987, 51, 221-71.

Yergeau, E; et al. Microarray and real-time pcr analyses of the responses of high-arctic soil bacteria to hydrocarbon pollution and bioremediation treatments. *Appl Environ Microb*, 2009, 75, 6258-6267.

Young, JPW; et al. The genome of *Rhizobium leguminosarum* has recognizable core and accessory components. *Genome Biol*, 2006, 7, R34.

Zhou, J; Thompson, DK. Challenges in applying microarrays to environmental studies. *Curr Opin Biotechnol*, 2002, 13, 204-207.

Zhang, Y; et al. Phylogenetic diversity of nitrogen-fixing bacteria in mangrove sediments assessed by PCR-denaturing gradient gel electrophoresis. *Arch Microbiol*, 2008, 190, 19-28.

Zhang, Y; et al. Microarray-based analysis of changes in diversity of microbial genes involved in organic carbon decomposition following land use/cover changes. *FEMS Microbiol Letts*, 2007, 266, 144-51.

Huston, Big questions, small worlds: microbial model systems in ecology. *Trends Ecol. Evol.*, 1996, 19, 189-197, 2004.

Chapter 7

SPATIAL ASSEMBLAGES OF TROPICAL INTERTIDAL ROCKY SHORE COMMUNITIES IN GHANA, WEST AFRICA

Emmanuel Lamptey[*1], *Ayaa Kojo Armah*[1] *and Lloyd Cyril Allotey*[1]

[1]Department of Oceanography and Fisheries,
P. O. Box LG 99, University of Ghana, Legon

ABSTRACT

The Ghana's rocky shore is unique (i.e., being flanked by several kilometres of sandy beaches and backed by several river bodies) along the Gulf of Guinea from Cote d'Ivoire to western Cameroon (West Africa). This suggests the existence of specific biotopes of species assemblages on spatial scales. The variability of environmental factors has been highlighted as the main causes of variations in the rocky intertidal communities. This study tested species assemblage patterns between three geomorphic rocky zones (i.e., western, central and eastern shores of Ghana), and quantified the influence of abiotic factors on the assemblage patterns. The study was carried out from December 2003 to January 2004 (best period of daytime good low tides). In describing the species spatial assemblages, four random sites were located within each geomorphic zone. Further, four belt transects were randomly laid from the lower to upper shores of each site and along which a continuous quadrat (1 square meter) was placed to estimate species percentage cover (macroalgae) and abundance of epibenthic fauna. The species data was standardized before submitted to statistical analyses. Also two replicate water samples were taken at each site for the analyses of nutrients (i.e., nitrate and phosphate), while dissolved oxygen concentration, salinity, water and ambient temperatures were measured at two spots of each site. All together, 86 taxa were found comprising 57 macroalgae and 29 epibenthic fauna. Species assemblage patterns indicated significant ($p=0.001$) differences between western and central, as well as western and eastern shores. In general, the species assemblage was dominated by macroalgae, which showed a spatial

---

[*] Corresponding author: E-mail: elamptey@ug.edu.gh, Tel: +233 24 483 1455, Fax: +233 21 520298.

declivity from west to east shores as opposed to epibenthic fauna. Taxon cumulative dominance decreased from east to west shores indicating high species diversity in the latter. However, spatial differences in the abundance of 24 widespread species influenced the assemblage patterns. Suites of abiotic variables notably nitrate, dissolved oxygen, salinity, and water temperature explained significant variations in Shannon-Wiener species diversity (58.53%), Margalef's species richness (61.49%), and number of species (71.54%). Canonical correspondence analysis showed significant response of individual taxa to nitrate and dissolved oxygen. The results suggest the influence of river flow on the species assemblages. These findings have important consequences for biodiversity and ecosystem functioning as well as conservation.

Keywords: Spatial Assemblage, Species Diversity, Rocky Shore, Abiotic Factors, West Africa, Ghana.

1.0. INTRODUCTION

1.1. Tropical Intertidal Rocky-Bottom Communities

Along the West Africa's coast, Ghana's rocky shore is uniquely flanked by large kilometer stretches of sandy beaches and is backed by several river bodies, which empty their contents (e.g. rich nutrients) onto the intertidal environment. As a sequel, these intertidal rocky habitats constitute heterogeneous environments providing multiple ranges of habitats that support a great variety of organisms. The general pattern of distribution of the intertidal rocky life forms in a particular way, (e.g., on the vertical scales) from the lower to upper shores [Underwood, 1981; Thompson et al. 2002], gave rise to zonation models [Stephenson and Stephenson, 1949; Lewis, 1964] based on emersion/dessication. Zonation here refers to the occurrence of organisms in distinct bands, which are separated by color and morphology of the dominant species.

Organisms living in the intertidal zone experience a suite of physical stresses, including fluctuations in temperature, aerial exposure, salinity, and hydrodynamic forces [Vernberg and Vernberg, 1972; Newell, 1979; Denny, 1988]. The variability of environmental gradient is pronounced in tropical sedimentary environmental [Alongi, 1990]. Consequently increased numbers of hard bottom epifauna occur towards the tropics compared to temperate environments [Warwick and Ruswahyuni, 1987] and also the greatest levels of marine biodiversity are found in tropical countries [Gray, 1997]. Tropical intertidal hard bottom habitats therefore represent an ideal "natural setting" to assess and model taxon-environmental interactions. Although intertidal hard bottom habitats comprise a relatively thin band and form a small fraction of the nearshore benthic habitats, they are characterized by insightful ecological processes.

There is huge body of knowledge on rocky intertidal taxa assemblages and causal processes [Connell, 1972; Underwood and Denley, 1984; Menge, 1995; Raffaelli and Hawkins, 1996; Menge et al., 1997]. The local intertidal taxon varies spatially depending on various abiotic factors, which is usually organized in a complex ways with mechanisms of organization that depend mainly on various environmental factors and biogeographic regions. The spatial and temporal differences in physical and biological processes are the main cause of variation in intertidal communities [Paine and Levin, 1981; Sousa, 1984]. However, the mechanisms of community changes at both large and small spatial scales have still not been appreciatively understood especially in the tropical intertidal environments. Notable quantitative and experimental intertidal studies in the tropics were conducted in Central America, Hong Kong and Australia [Hutchinson and Williams, 2001; Huang et al., 2006]. Relatively very few studies on rocky shores have been documented in other tropical areas including the Galapagos Island, Malaysia and Singapore, representing lower latitude areas [Witman and Smith, 2003; Huang et al., 2006; Vinueza et al., 2006]. In West Africa coast, known intertidal studies included Lawson [1956]; Buchanan [1957]; Gauld and Buchanan [1959]; Bassindale [1961]; Edmunds [1978]; John [1986]; John and Lawson [1990]; Evans *et al.* [1993]; Hardy and Seku [1993]; Yankson and Akpabey [2001]; John et al. [2003]; and Branoff et al. [2009]. Nevertheless, rocky intertidal environment is still riddled with specious controversies of structure mechanisms (i.e., control and regulation factors) of shore biota [Reise, 2002]. Progress to surmount this requires the explanation of pattern in terms of the underlying processes with a consequent reduction in unexplained variation [Benedetti-Cecchi, 2009]. For example, in Ghana spatial differences of rocky intertidal shore (just like any other ecosystem) assemblage has been hinted [Evans et al., 1993] but the underlying mechanism remained unexplained.

1.2. Spatial Patterns of Intertidal Taxa Assemblages

Species richness, species relative abundance and the heterogeneity of their spatial and temporal distribution in a given area have taken center stage in marine ecology [He and Legendre, 2002]. This is because there are substantial knowledge gaps concerning spatial heterogeneity in rocky assemblages, to potential risks in interpreting patterns only in relations to environmental gradient [Terlizzi and Schiel, 2009]. However, analysis of spatial patterns along environmental gradient at different scales is seen as a logical requirement to deal with spatial and temporal confounding [Hurlbert, 1984], and provides tests for generality of models of species assemblages. There are limited studies that have tested the consistency of patterns along sharp environmental gradients at hierarchies of spatial scales [e.g., Benedetti-Cecchi, 2001]. Many environmental factors affecting species performance and interactions vary with spatial scale [Noda, 2009].

Spatial variation in abundance and distribution of rocky shore organisms exist on many scales, both small and large geographical/latitudinal scales [Foster, 1990; Evans et al., 1993; Boaventura et al., 2002]. The mechanisms of local community structure organization of rocky intertidal sessile assemblages depend highly on various environmental factors [Raffaelli and Hawkins, 1996; Menge et al., 1997]. Differences in species response to environmental gradients play a significant role in determining the species composition of local assemblages [Menge et al., 1994]. Spatial heterogeneity in species distribution may enhance productivity, reduce the spread of diseases and increase resistance to disturbance [Hutchinson et al., 2003]. Of significant consequence, patterns of species assemblages provide essential data for measuring potential effects of environmental disturbances and variability. Aside prioritizing conservation areas, understanding spatial patterns in species distribution is critical to estimating role of environmental gradient on possible species extinction.

1.3. Role of Environmental Variability on Taxa Assemblages

The environment (the sum causal factors that show actual interactions) comprises the input and output components, with resources and conditions constituting the input environment. Environmental conditions are all things outside an organism that affect it but, in contrast to resources, are not consumed by it [Begon et al., 1990]. The environment of an organism consists of all those phenomena outside an organism that influence it, whether those factors are physical (abiotic) or are other organisms (biotic) [Olff et al., 2009]. Increasing moderation in environmental conditions leads to increased abundances, more complex trophic structure, and increased influence of species interactions on structure [Menge, 2000; Menge and Branch, 2001].

An important goal in community ecology is to understand the factors contributing to the species assemblage patterns at a variety of spatial scales. Changes in environmental conditions promote changes in species assemblages at a variety of spatial scales. Many environmental factors affecting species performance and interactions vary with spatial scale [Noda, 2009]. There is, however, no particular scales of change emerged that are consistent among taxa [Burrows et al., 2002], congruently demanding ambitious ecological models to decipher spatial patterns.

Models which have been suggested for understanding community dynamics include "environmental stress models" and either "nutrient/productivity models" or the "food chain dynamics hypothesis" [Connell, 1975; Oksanen et al., 1981; Fretwell, 1987; Menge and Olson, 1990; Menge, 2000]. Environmental stress models assume that community structure results from species interactions and disturbances, and how these are modified by underlying gradients of environmental stress (where stress is a consequence of environmental conditions such as temperature, moisture, salinity, etc.) [Menge et al., 2002]. The two models postulate that communities can be ordered along environmental gradient. McGill et al [2006] argued that general principle in community ecology may not be achieved if research continues to focus on pair-wise species interactions independent of the environment. They suggested four research themes: functional traits, environmental gradients, interactions milieu and performance currencies, in order to bring general patterns to community ecology. Relatively few studies have explicitly incorporated structuring abiotic (environmental gradient) and biotic (movement, dispersal) features that are key to species co-existence and vital for the maintenance of species diversity [Loreau et al., 2003].

Global species distributions are generally believed to be determined by abiotic influences related to oceanographic and physiographic properties [Sanders, 1968; Richlefs and Schluter, 1993]. For instance, water motion affects biology by acting as a transport mechanism for organisms and their propagules, as a dynamic boundary between regimes, and as a force to which organisms must adapt or respond, for example, in their feeding and locomotor activities [Nowell and Jumars, 1984; Denny, 1993]. As such, intertidal hard bottom communities are inextricably linked to environment conditions (e.g., oceanic,) for the delivery of food, nutrients, propagules and larvae [Blanchette et al., 2009]. Relatively few studies have documented the importance of nutrition as a local limiting factor [Wooton et al., 1996].

Mechanisms of species assemblages depend strongly on various environmental conditions in rocky intertidal assemblages [Menge et al., 1997, 2003]. Thus, environmental conditions and processes that occur at a variety of spatial scales are critical elements to understand patterns of species assemblages. The influence of environmental conditions and space on species diversity has not been given much prominence in scientific literature. Available scientific evidence shows that oceanographic variability is a strong correlate to the abundance of dominant intertidal hard bottom faunal functional groups [Broitman et al., 2001, Nielsen and Navarrete, 2004]. Variability of physical environmental is important in explaining patterns of intertidal diversity [Zacharias and Roff, 2001]. A better understanding of the role played by abiotic factors is a key prerequisite for forecasting the effects of shifts in environmental conditions on species diversity, as a result of human pressure, and for setting up adequate policies of marine conservation and management [Terlizzi and Schiel, 2009].

1.4. Objectives

The primary objectives of the study were to quantify species assemblage patterns on spatial scales, test statistical differences in taxa assemblages; and model taxon-environment interactions.

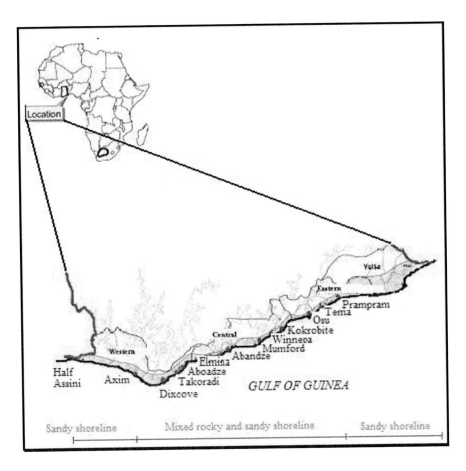

Figure 1. A coastline map of Ghana showing sampling sites.

Materials and Methods

Study Locations

The study was conducted in Ghana, West Africa, located in the western Gulf of Guinea sub-region, about 750 km north off the equator between latitudes 4° and 12° N and longitudes 3° W and 1° E. Ghana has a coastline of 550 km and relatively narrow continental shelf to a depth of around 75–120 m. The coastline of Ghana has been geomorphologically zoned into three: the west, central and east coastlines [Ly, 1980]. The west coast covers 95 km of stable shoreline and extends from Ghana's border with Côte d'Ivoire to the estuary of the Ankobra river (Figure 1). It is basically fine sand with gentle beaches backed by coastal lagoons.

The central zone is 321 km shoreline long (actual study area) and extends from the estuary of Ankobra river near Axim to Prampram (east of Accra). It represents an embayed coast of rocky headlands, rocky shores and littoral sand barriers enclosing a number of coastal lagoons.

The east zone, which is made up of 139 km of shoreline, extends from Prampram eastwards to Aflao, at the border with Republic of Togo. It is characterized by sandy beaches with the deltaic estuary of the Volta river situated halfway in-between. The Volta river is the largest in Ghana with a regulated flow of about 900 cumecs. The east coast has a large proportion of its land covered by the Keta Lagoon, the Volta Estuary and the Songor Lagoon.

The study was undertaken within the central zone, which spanned from Axim to Prampram consisting of rocky coast in-between two long stretches of sandy shores. The rock formations are flat and continuous with some tidal pools.

The climate of the area is tropical. The eastern coastal belt is warm and comparatively dry while the southwest part is hot and humid. It becomes progressively drier from the southwest to the northeast of Accra. Two thirds of the coastal zone falls within the dry coastal savannah strip where annual rainfall ranges from 625 mm to 1000 mm and average of 900 mm (Armah and Amlalo, 1998). There is a double maxima annual distribution of rainfall in the coastal zone with the high maximum occurring in May–June and the low maximum in September–October.

Minimum temperature occurs in July–August and the maximum in February–March. The relatively dry coastal climate of the southeast is believed to be caused by the prevailing winds (south-south-westerlies) blowing almost parallel to the coast and to a cool current of water immediately offshore as a result of a local annual upwelling [Armah and Amlalo, 1998].

Field and Laboratory Methods

Species assemblages in rocky coastline of Ghana differ on a variety of spatial scales mainly due to local environment conditions and resources. In describing and testing the species spatial assemblages, four random sites (based on similar topography) were located within three rocky shores (falls within the central geomorphic zone of 321 km stretch) (Figure 1). Further, four belt transects were randomly laid from the lower (i.e., Sargassum zone) to the upper shores of each site and along which a continuous quadrat (1 square meter) was placed to estimate species percentage cover (macroalgae) and abundance of epibenthic fauna from December 2003 to January 2004. The percentage cover of macroalgae and number of individuals of solitary animals were identified using taxonomic guides and manuals [e.g., John et al. 2003, Edmunds 1978]. The species data was standardized before submitted to statistical analyses. Also two replicate water samples were taken at each site for the analyses of nutrients (i.e., nitrate and phosphate), while dissolved oxygen concentration, salinity, water and ambient temperatures were measured at two spots of each site.

Water samples collected were analyzed for nitrate and phosphate using the HACH DR/2010 spectrophotometer following the methods in A.P.H.A. et al. [1998].

Data Analysis

The abundance data from the four belt transects at each site was pooled, standardized (i.e., abundance/number of quadrats) for all sites before submitting to statistical analyses. Individual macroalgae was computed from the percentage cover data using the formula $MA_{sp(i)}=PC_{ma(i)} \times (MI_{mean}/PC_{mean})$, where: $MA_{sp(i)}$=individual macroalgae, $PC_{ma(i)}$=percentage cover of individual macroalgae, MI_{mean}=mean individual number of species, and PC_{mean}=mean percentage cover of macroalgae. Frequency of occurrence was calculated for all species using the F index [Guille 1970]: $F=p_a/P \times 100$, where: p_a is the number of sites where the species occurred and P is the total number of sites. Using this formula, the species were classified as: constant (F>50%), common (10≤F≤49%) and rare (F<10%) species. In this study, only constant species (total contribution of 94% to the abundance) were used for all statistical analyses (except that F>70% were used for Canonical correspondence analyses) to ensure adequate spatial representation. Differences in community structure within and between sites were quantified using suites of multivariate techniques. Similarity between sites based on dominance distance of cluster analysis using the Bray-Curtis similarity index after fourth-root transformation [Clarke and Green, 1988], and a group-average dendrogram produced [Clark, 1993]. At the same time, a similarity profile test [SIMPROF; Clarke and Gorley, 2006] was performed to test the null hypothesis that a specific subset of samples did not differ from each other in the multivariate structure.To test for differences in community between the study shores; one-way Analysis of Similarity (ANOSIM) was utilized [Clark and Warwick, 2001]. This program computes r-statistics as a measure of discrimination. First, a global R-value was computed to indicate the overall effect of similarity between the study shores. Values of R=1 are obtained when all replicates (sites) within the groups (zones) are more similar. The p-value for the statistics was obtained by simulating all possible permutations of assigning replicates (sites) to study zones. In this study, a random sample of 999 permutations was used in each calculation.

The species that contributed the most and discriminated one study zone from another were investigated using non-metric similarity percentage procedure (SIMPER) [Clark and Warwick, 1994, 2001]. These results assisted in interpretation of the community changes responsible for the observed pattern in the ordination [Clarke, 1993]. The group of species with cumulative contribution above the 50% (dis)similarity threshold were considered important in controlling the taxa assemblages in the studied area.

Univariate analyses of diversity indices were calculated for the Margalef's species richness index (d'), which takes into account species present and number of individuals, and the Shannon–Weiner diversity index (H' using log to base 'e') [Shannon and Weaver, 1963]. Evenness was also estimated as defined by Pielou [1966], which shows how evenly the individuals are distributed among species. Predictive models were developed using stepwise-multiple regressions to explore the relationships between diversity indices and abiotic variables. The model results were obtained having eliminated multicollinearity and autocorrelation (Durbin-Watson statistics). Durbin-Watson statistics is a first order autocorrelation in the residuals of regression analyses, which is approximately, equals 2 (1-R), where R is the correlation coefficient).

A canonical correspondence analysis (CCA) was performed [Ter Braak, 1986] using 27 taxa (total contribution of 85.2%) selected based on (F>70%) [Guille, 1970], using the package CANOCO 4.5 [Ter Braak and Smilauer, 2002], which combines both ordination and regression to ascertain relationships between taxon and abiotic variables [Ter Braak, 1986]. None of the abiotic variables presented an inflation factor >20 [Ter Braak and Smilauer, 2002] and also covaried. All the abiotic variables were transformed (Log (x+1)) to stabilize and normalize the variance.

In the CCA biplot, the first and second axes represent the most important environmental gradient along which the rocky shore taxa are distributed. The direction of each environmental vector represents the maximum rate of change for that particular abiotic variable and its length indicates the relative importance to the ordination. The significance of all primary CCA axes was determined by a Monte-Carlo permutation test (199 permutations) of the eigenvalues [Ter Braak and Smilauer, 2002]. A forward selection procedure ordered the abiotic variables according to the amount of variance they captured in the species data [Ter Braak and Verdonschot, 1995]. In the first step of this method, all abiotic variables were ranked on the basis of the fit for each separate variable. Each variable was treated as the sole predictor variable, and all other variables were ignored; hence the variance explained represents marginal effects. At the end of the first step of the forward selection, the best variable was selected. Hereafter, all remaining abiotic variables were ranked on the basis of the fit (amount of variance explained) that each separate variable gave in conjunction with the variable(s) (covariables) already selected (conditional or unique effects). At each step, the statistical significance of the variable added was tested using a Monte-Carlo permutation test (199 unrestricted permutations) [Ter Braak and Smilauer, 1998].

RESULT

Community Structure Analysis

Several species of macroalgae (57 species) and solitary epibenthic macro-invertebrates (29 species) accounting for 86 taxa were encountered. The records depict higher floristic composition than epibenthic macro-invertebrates. Among others, this may be attributable to the imposition of limitation on the distribution of juvenile gastropods (dominant intertidal epifauna) by macroalgae [Underwood, 1979]. The highest numbers of macroalgal species were observed at western shore, followed by central and eastern, while epibenthic macro-invertebrate showed slight decrease towards the western shore. However, the total and mean densities of the composite biota decreased from the eastern to the western shores (Figure 2).

Apparently, the highest density and number of species was reported at Prampram and Kokrobite respectively in the eastern zone (Figure 3). On average there were less variability in number of species (Figure 3) and mean densities (Figure 2) between sites of western and central zones than the eastern.

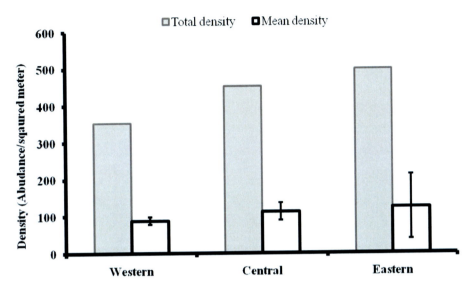

Figure 2. Total and mean densities of the western, central and eastern rocky zones. Vertical bars indicate 95 confidence interval of taxon density.

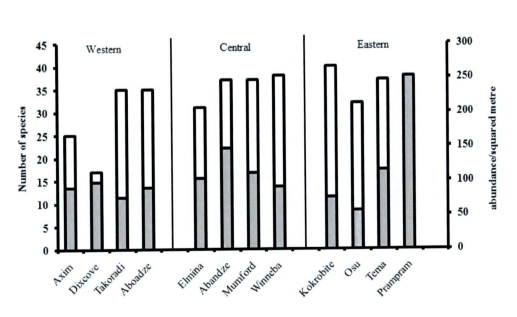

Figure 3. Number of species and total density of study sites.

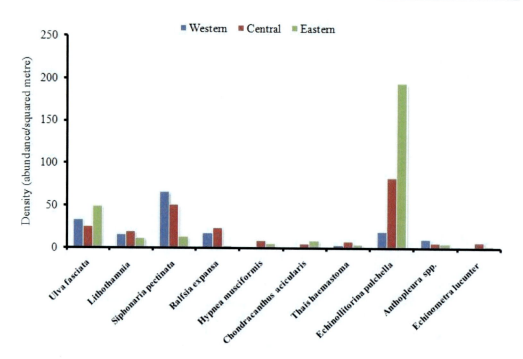

Figure 4. Distribution and density of ten most occurred (F=100%) rocky shore taxa in western, central and eastern zones of Ghana.

The highest ranked numerical dominance of ten most common taxa was *Echinollitorina pulchella,* with increased density from western to eastern shores (Figure 4). Conversely, other epibenthic macro-invertebrates such as *Siphonaria pectinata* and *Anthopleura* spp. decreased from western to eastern shores. The density of *Chondracunthus acicularis* increased eastward in contrast to other macroalgae, *Raftia expansa* and *Lithothmania* (Figure 4). Density of *Lithothamia* (red corraline algae) ranked highest in the central zone likewise *Thais heamastoma*.

The possible differences of taxa assemblages between the zones were tested using one-way Analysis of Similarity (ANOSIM), indicating significant differences. The pairwise test revealed the differences between western and central, and western and eastern zones (Table 2). However, no significant difference occurred between central and eastern zones. R is a measure of variation between samples (zones) compared to variation within samples (sites) in dissimilarities of a suite of species using ranked differences among replicates [Clarke, 1993].

Species cumulative abundance decrease from the eastern to the western shores (Figure 5). The curves revealed 80% saturation in the eastern and central coastline of species dominance with the addition of the tenth (10th) species (Figure 5). The dominance curves begin to near the asymptote with relatively few new species added after the twenty-fifth's (25th) species at which 90% cumulative dominance was attained. Generally, species added beyond the limit of the asymptote are considered rare species and seldom offer additional critical information. The highest ranked dominance in the eastern zone may suggest intensity of physical disturbance (trampling and removal) and physiological stress (high human pollution due to burgeoning population) reducing richness and diversity compared to the western.

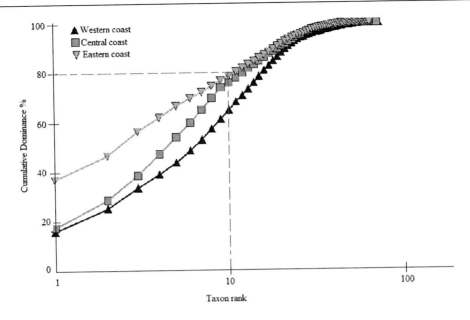

Figure 5. K-dominance curves showing cumulative ranked abundances (density) plotted against log species rank [following Lambshead et al. 1983]. Higher dominance in eastern coast for the first ten species. Lower cumulative dominance in the western zone.

Table 1. Pairwise test of Analysis of Similarity (ANOSIM) between three coasts

Coast	R Statistics	Significant level	Significant difference	Actual Permutation	Number>= Observed
Western & Central	0.219	0.029	Yes	35	1
Western & Eastern	0.25	0.029	Yes	35	1
Central & Eastern	0.302	0.057	No	35	2

Global R=0.287, p=0.001

The group-average hierarchical dendrogram of species dominance distance indicated three significant cluster groups (Figure 6). Cluster 'GP. I' contained sites from all the three zones (Western, Central and Eastern). This is an indication of widespread distribution of dominant species assemblages across sites and may influence the taxa assemblages. Cluster 'GP. II' comprised assemblages between Central and Eastern zones with similar dominance, whereas Cluster GP. III accounted for sites (i.e., Axim and Dixcove) located in the Western zone. Taxa composition and dominance at these sites were significantly different from all the other sampling sites.

SIMPER analysis (Table 2) revealed that between groups dissimilarity on the Western and Central coasts was most attributable to the higher average proportions of *Echinolittorina pulchella*, *Hydropuntia dentata*, *Palythoa* spp., *Nerita atrata*, *Sargassum vulgare*, *Chaetomorpha linum* and *Fissurella nubecula* together contributed 30.10% of the 36.49% average dissimilarity with individual contribution ≥ 3%. The same species together accounted for 28.98% of the 40.73% average dissimilarity between the western and eastern shores (Table 2).

Table 2. SIMPER analysis results: species contributing to the average Bray–Curtis dissimilarity between the western, central and eastern shores based on simultaneous analysis of taxa density data. δ_i: contribution of the i-th faunistic group to the average Bray-Curtis dissimilarity (δ) between the rocky coasts, also expressed as a cumulative percentage ($\sum\delta_i\%$). Diss/SD is the ratio of dissimilarity to standard deviation and F is the frequency of occurrence of the 12 sites. For brevity, only species that contributed to $\geq 2.0\%$ and cumulative percentage of $\geq 70\%$ are listed. The codes in the parenthesis after the species name indicate: 'G' gastropod, 'R' Rhodophyta, 'P' Phaeophyta, 'C' Chlorophyta, 'Cy' Cyanophyta, 'A' anemone, 'Cn' cnidarians, 'E' Echinoderm

Species	Ave. Diss	Diss./SD	(δ_i)	$\sum\delta_i\%$	(F%)
Average dissimilarity between Western and Central =36.49					
Echinolittorina pulchella (G)	1.87	1.53	5.11	5.11	100
Hydropuntia dentata (R)	1.82	1.35	4.99	10.10	83
Palythoa spp. (Cn)	1.78	1.85	4.89	14.98	83
Nerita atrata (G)	1.78	1.58	4.88	19.86	75
Sargassum vulgare (P)	1.53	1.48	4.20	24.07	83
Chaetomorpha linum (C)	1.11	1.40	3.05	27.12	67
Fissurella nubecula (G)	1.10	2.30	3.02	30.13	67
Enteromorpha flexuosa (C)	1.08	1.52	2.95	33.08	75
Bachelotia antillarum (P)	1.07	1.45	2.92	36.00	67
Centroceras clavulatum (R)	1.06	1.16	2.91	38.91	83
Blue green algae (Cy)	1.02	1.47	2.80	41.71	92
Thais nodosa (G)	0.98	0.78	2.67	44.38	67
Padina durvillaei (P)	0.93	1.43	2.56	46.94	92
Bryocladia thyrsigera (R)	0.90	0.99	2.46	49.40	50
Ulva fasciata (C)	0.87	1.34	2.39	51.79	100
Echinometra lucunter (E)	0.86	1.51	2.37	54.16	100
Algal mat (R)	0.86	1.19	2.36	56.52	92
Dictyopteris delicatula (P)	0.86	1.38	2.35	58.86	58
Jania rubens (R)	0.86	1.20	2.34	61.21	75
Cryptonemia crenulata (R)	0.85	1.29	2.33	63.53	50
Dictyota ciliolate (P)	0.82	1.18	2.26	65.79	92
Patella safiana (G)	0.81	1.42	2.23	68.02	92
Hypnea musciformis (R)	0.81	1.78	2.23	70.25	100
Padina tetrastromatica (R)	0.78	1.34	2.15	72.40	75
Caulerpa taxifolia (C)	0.74	1.39	2.03	74.43	75
Average dissimilarity between Western and Eastern =40.73					
Echinolittorina pulchella (G)	2.25	1.45	5.51	5.51	100
Hydropuntia dentata (R)	1.96	1.42	4.81	10.32	83
Nerita atrata (G)	1.60	1.26	3.92	14.24	75
Chaetomorpha linum (C)	1.59	1.28	3.91	18.15	67
Sargassum vulgare (P)	1.53	1.55	3.75	21.91	83
Palythoa spp. (Cn)	1.53	2.32	3.75	25.66	83
Fissurella nubecula (G)	1.35	1.64	3.32	28.98	67

Table 2. (Continued)

Enteromorpha flexuosa (C)	1.29	1.60	3.16	32.14	75
Siphonaria pectinata (G)	1.27	1.81	3.13	22.74	100
Centroceras clavulatum (R)	1.27	1.09	3.11	38.80	83
Caulerpa taxifolia (C)	1.17	1.12	2.87	42.25	75
Thais nodosa (G)	1.17	0.97	2.86	44.11	67
Bachelotia antillarum(P)	1.10	1.52	2.71	46.68	67
Semifusis monrio(G)	1.09	2.23	2.68	49.50	50
Padina durvillaei (P)	1.00	1.18	2.45	51.95	92
Boodlea composita (C)	0.99	1.13	2.44	54.39	50
Anthopleura spp. (Cn)	0.94	1.43	2.31	56.70	100
Ulva fasciata (C)	0.94	1.47	2.30	59.00	100
Hypnea musciformis (R)	0.92	1.61	2.27	61.27	100
Gelidium corneum (R)	0.90	1.70	2.20	63.47	92
Chondracanthus acicularis (R)	0.87	3.49	2.15	65.61	100
Gelidiopsis variabilis (R)	0.84	2.61	2.07	67.68	58
Average dissimilarity between Central and Eastern =28.96					
Nerita atrata (G)	1.40	1.55	4.85	04.85	75
Chaetomorphalinum (C)	1.21	1.65	4.19	09.04	67
Echinolittorina pulchella (G)	1.01	1.22	3.48	12.52	100
Semifusis monrio(G)	0.98	2.93	3.37	15.88	50
Ralfsia expansa (P)	0.95	2.19	3.27	19.15	100
Boodlea composita (C)	0.93	1.26	3.22	22.37	50
Palythoa spp. (Cn)	0.93	1.61	3.20	25.57	100
Siphonaria pectinata (G)	0.91	1.83	3.14	28.71	100
Centroceras clavulatum (R)	0.87	0.95	3.00	31.71	83
Caulerpa taxifolia (C)	0.84	1.53	2.91	34.62	75
Bryocladia thyrsigera (R)	0.81	1.82	2.78	37.40	50
Hypnea cervicornis (R)	0.78	1.64	2.70	40.10	50
Padina tetrastromatica (P)	0.78	1.54	2.70	42.80	75
Dictyopteris delicatula (P)	0.76	1.31	2.63	45.43	58
Enteromorpha flexuosa (C)	0.76	1.41	2.63	48.05	75
Anthopleura spp. (Cn)	0.72	1.77	2.48	50.54	100
Bachelotia antillarum (P)	0.70	1.35	2.41	52.95	67
Cryptonemia crenulata	0.69	1.36	2.39	55.34	50
Patella safiana (G)	0.67	2.19	2.31	57.65	92
Jania rubens (R)	0.67	1.38	2.31	59.96	75
Fissurella nubecula (G)	0.65	2.04	2.25	62.21	67
Echinometra lucunter(E)	0.64	1.51	2.21	64.42	100
Ulva fasciata (C)	0.63	1.47	2.18	66.61	100
Sargassum vulgare (P)	0.59	1.26	2.02	68.63	83

Between the eastern and central, three of these taxa namely: *Nerita atrata, Chaetomorpha linum* and *Echinollitorina pulchella* contributed 12.52% to the average dissimilarity of 28.96%. Of the suite of taxa which contributed greater than 50% to the average dissimilarity between the Western, Central and Eastern shores, *Echinolittorina pulchella, Palythoa* spp., *Nerita atrata, Chaetomorpha linum, Enteromorpha flexuosa, Centroceras clavulatum* were common.

However, ten widespread species occurred at all sites. These were *Ulva fasciata, Lithothamnia, Siphonaria pectinata, Ralfia expansa, Hypnea musciformis, Chondracanthus acicularis, Thais haemastoma, Echinolittorina pulchella, Anthopleura* spp., and *Echinometra lucunter* accounting for 58.18% of the total density (Figure 4). The number of these common taxa discriminating between the shores decreased from the western to eastern (Table 2) with 40% between western and central, 60% between western and eastern, and 70% central and eastern. This possibly suggests the influence of i) oceanographic conditions on larval dispersal from west to east, and ii) the larvae of these species are planktonic and showed less spatial synchrony.

Taxa–Environment Interactions

To test the interactions between diversity indices (species assemblages) and environmental abiotic variable, a multiple predictive regression model was developed (Table 3). The result showed that species assemblages were significantly influenced by suite of abiotic variables ($p<0.05$). Shannon-Wiener species diversity index was influenced by nitrate, salinity and water temperature, and these variables explained 58.53% of the variance (Table 3). The relationships were indirect (negative) for nitrate and water temperature and direct (positive) for salinity. The variations in Margelef's species richness were explained by nitrate, salinity and dissolved oxygen accounting for 61.49%. There were significant negative relations between species richness and abiotic variables (i.e., nitrate and dissolved oxygen) whereas salinity depicted direct relationships (Table 3). Number of species revealed similar relationship with abiotic variables just like observed with Margale'f species richness except for the higher explained variance ($R^2=71.54$). Since none of the explained variances (R squared) were 100%, the unexplained variances constitute surrogate of unmeasured environmental variables. This suggests that other factors such as biological (competition and predation), physical (waves and tides), chemical (pollutants), human disturbances, nature of the topography etc. are critical for a better appreciation of rocky shore species assemblage patterns.

Table 3. Step-wise multiple regression model for taxon assemblages and abiotic variables. Nit.=nitrate, Sal=salinity, DO=dissolved oxygen, Temp. (W)= water temperature

Diversity index	Equation (independent parameters)	R^2
[1]Diversity	-1.43*Nit + 0.01Sal -Temp. (W) + 4.53 ± 0.30	0.5853
[2]Richness	-3.40*Nit +0.47*Sal - 0.40*DO +3.45 ± 1.42	0.6149
Number of species	-3.45*Nit - 2.5*DO + 2.87*Sal. - 37.86 ±5.31	0.7154

[1]Shannon-Wiener species diversity, [2]Margelef's species richness, $p<0.05$

The result of the CCA ordination for 27 taxa and 7 abiotic variables showed that only 57.7% of the variance in taxa abundance was accounted for by the first four ordination axes, with the first two models accounting for 42.6% but the overall model is significant ($p=0.005$ for axis I, and $p=0.04$ for all axes).

The first ordination axis reflected a gradient mostly related to dissolved oxygen, phosphate, pH and ambient temperature (Figure 7). The second axis indicated nitrate and salinity had the largest effect on the species occurrence. Thus, low salinity correlated with high nitrate concentrations. Most epibenthic macro-invertebrates and chlorophytes (green algae) were associated with the first ordination axis, whereas the rhodophytes (red algae) and phaeophytes (brown algae) were associated with second ordination axis (Figure 7B).

Apparently, nitrate and dissolved oxygen were the significant predictors of species distributions (Table 4) located on the second ordination axis (Figure 7). Consequently, taxa that occurred at Tema (= Sakumono), Aboadze and Takoradi showed preference for low nitrate and higher salinity, whereas taxa encountered at Axim, Munford and Elmina depicted strong preference for high nitrate and low salinity (Figure 7A). Most of these sites are either connected to a large river body (Axim), as such receive less saline water rich in nutrients (mostly nitrate) that influence intertidal taxa assemblages.

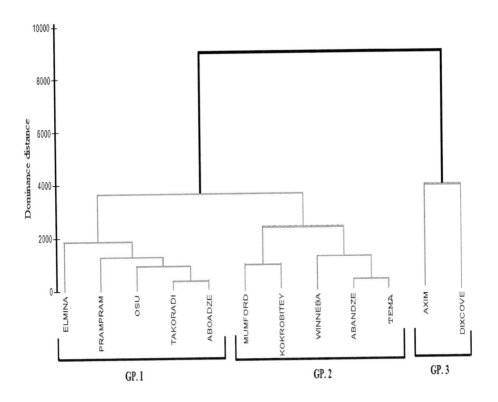

Figure 6. Group average hierarchical dendrogram of species dominance distance at study sites. Thin lines indicate significant evidence of structure (SIMPROF test, p<0.01) and thick lines indicate no evidence of structure. Three significant groups are discernible.

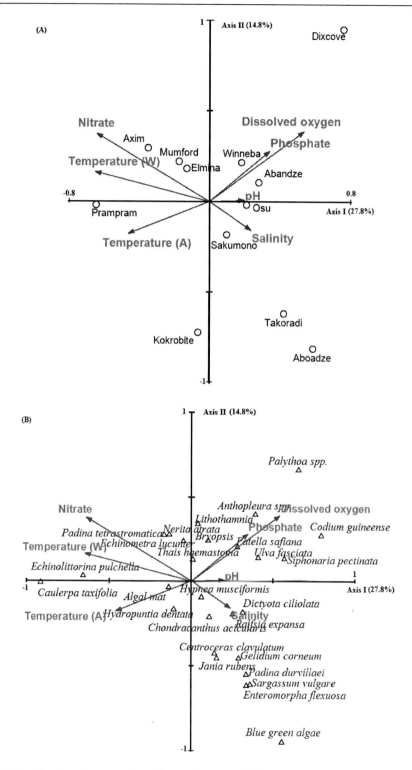

Figure 7. CCA ordination diagrams for **a** site–environment biplot and **b** taxon–environment biplot. Taxon scores along the first and second axes in relation to abiotic variables. Only species with F>70% were used for the biplot.

Table 4. Results of canonical correspondence analysis. Marginal effects denote percentage variance explained (percentage of the total variance in the species data explained) by using each abiotic variable as the sole predictor variable. Conditional (unique) effects denote variance explained by each abiotic variable with the variable (s) already selected and treated as covariable(s) based on forward selection. Environmental variables are listed by the order of their inclusion into the model. Significant levels are based on a Monte Carlo permutation test with 199 restricted permutations

Abiotic variable	Marginal Effect Lambda1	Conditional Effect Lambda1	p-value
Nitrate	0.20	0.20	0.005*
Temperature (W)	0.18	0.07	0.400
Dissolved oxygen	0.16	0.21	0.005*
Temperature (A)	0.11	0.09	0.260
Salinity	0.10	0.11	0.095
Phosphate	0.09	0.05	0.575
pH	0.06	0.09	0.180

CONCLUSION

This contribution provided quantitative evidence of spatial patterns of rocky intertidal assemblages and influence of abiotic factors in the observed patterns. The significant differences (one-way ANOSIM, r=0.287, p=0.001) in species abundance between shores (i.e., western, central and eastern), specifically between i) western and central, and ii) western and eastern (Table 1), were due mainly to the difference in taxa abundance and composition (Figures. 2 & 3), leading to differences in the contributions of discriminating species (Table 2). Ostensibly, these observed differences were the result of gradient in abiotic factors (Table 3 & Figure 6). Ollf et al. [2009] indicated that, short events of extreme conditions (very cold, hot, saline or anoxic conditions) can be fatal for organisms that lack the appropriate adaptations to cope with, or escape from those, and are therefore important for understanding community structure.

The entire rocky shore assemblage was dominated numerically by macroalgae with decreasing densities from west to east coasts corresponding with increased variance (Figure 2). Grazing by epibenthic macro-invertebrates (e.g. sea urchins) and competitive interactions between macroalgae had been highlighted as essential processes regulating the dynamics of macroalgal assemblages in many coastal intertidal areas [Lubchenco and Gaines, 1981], and may influence geographic distribution and abundance of rocky shore biota. Many macroalgal species have very delicate structures and are easily torn or rasped by herbivores. Species such as *Ulva fasciata*, *Enteromorpha flexuosa* and *Chaetomorpha linum* are heavily grazed on and such pressures may not allow for quicker re-establishment thus affecting their abundance and distribution. The observed macroalgal decreasing pattern on west-east axis may suggest interplay of several factors including anthropogenic disturbance (e.g., trampling, removal and pollution), oceanographic regimes (intense upwelling between Cote d'Ivoire and western region of Ghana) and gradient in abiotic factors with declension on west-east axis. These ecological factors may constitute important variables whose gradient simile structure rocky shore assemblages in Ghana. The western shore has been observed to experience moderate natural disturbances whereas anthropogenic impacts (industrial and human pollutions) are characteristic of the eastern shore. The magnitude of such effects may correlate with the organisms' response resulting in a decrease or increase in abundance. A study carried out in Sri Lanka showed that changes in intertidal rocky macro-benthic assemblages were due to human disturbances as opposed to abiotic factors [Deepananda, 2008].

Similarity analysis (SIMPER) indicated differences in the average dissimilarity between the shores, and confirmed the species with the highest contributions (Table 2). Of the taxa that contributed greater than 50% to the average dissimilarity between the shores, seven were common namely: *Echinollittorina pulchella* (Gastropod), *Hydropuntia dentata* (Brown algae), *Palythoa* spp. (Actinia), *Nerita atrata* (Gastropod), *Chaetomorpha linum* (Green algae), *Enteromorpha flexuosa* (Green algae) and *Centroceras clavulatum* (Red algae) (Table 3), indicating widespread distribution and spatial connectivity between the shores. Spatial connectivity assures the maintenance of large scale patterns in metacommunities [Kotta and Witman, 2009]. There were ten species (contributed approximately 52% to the total density) that showed complete/total occurrence (F=100%), which may constitute important subset of species pool with the capacity to tolerate large variations in environmental factors possibly due to their broad range of genotypes, which according to Crisp [1976a, b] allow for continued colonization of heterogeneous environments. Species with small geographical ranges are potentially at a greater risk of extinction than those with larger ranges [Bibby et al., 1992; Mittermeier et al., 1998; Myers et al., 2000].

Despite having relatively similar species, the community structure was remarkably different from the three shores, but pronounced in the western shore. The cumulative taxon dominance curves suggest that, although species assemblages seem similar, the structural differences were the results of dominance of few species (Figure 3). The relative cumulative taxon dominance curves between eastern and central shores begin to disappear after the tenth (10th) taxon is added whereas the addition of the twenty-fifth (25th) taxon caused all the curves to flatten with dissipating taxa dominance (Figure 3). There was progressively higher taxa dominance from western to eastern shores (Figure 3). Increased species dominance acts to decrease total species diversity of smaller subregions (sites), but increases subregional (site) endemic species diversity [Green and Ostling, 2003]. The highest ranked dominance in the eastern shore was associated with low diversity, while the lowest ranked dominance at the western shore was associated with greater species diversity. Greater species diversity (at western shore) may be related to some combination of benign environmental conditions (stability) and periodic (intermediate) disturbance. Interplay of these factors may result in continual succession [Richlefs and Schluter, 1993], but abiotic factors which contribute to scale-dependent taxa dominance may increase the risk of extinctions under habitat loss.

The group-average dominance distance dendrogram indicated three significant site clusters ($p<0.05$). Axim and Dixcove (in western zone) did not show significant relation with other sites (Figure 6) unlike Takoradi and Abodze that showed association with sites of central and eastern shores. The result revealed low species diversity (high dominance) at Axim and Dixcove possibly due to uniqueness of taxa accommodation to environmental conditions. Dixcove (a sheltered bay) was dominated by *Palythoa* spp. which correlated with high dissolved oxygen and phosphate content in the water column (Figure 7). Axim on the other hand, showed high correlation with elevated nitrate and low salinity (Figure 7) due to the riverine incursions (Ankroba river) into the intertidal zone. The riverine discharges carry along with it suspended sediments that affect the composition, structure and dynamics of rocky intertidal assemblages [Airoldi, 2003]. Therefore only species tolerable to these conditions may survive in this environment. Relatively few marine species are able to tolerate low salinity and sand scouring affecting their survival and distribution, which account for the low species diversity in these sites.

Taxon-Environment Interactions

It was possible to describe clear relationships of taxon-environmental factors. Intertidal rocky shore taxa assemblages were mainly influenced by suites of abiotic variables including nitrate, dissolved oxygen, salinity and to a limited extent water temperature (Table 2). Similar effects of elevated temperature and salinity were found in a study carried out at Iture in Ghana (Branoff et al., 2009). Abiotic conditions such as water salinity, water pH and redox highly affect the availability of resources to organisms [Schlesinger, 1991]. Also local variation in nutrient availability influence macrophyte biomass and diversity over small spatial scales [Nielsen, 2001, 2003]. Nitrogen is a limiting nutrient in most marine ecosystems [Vitousek et al. 1997]. The result showed that suite of variables namely nitrate and temperature correlated indirectly with species diversity (e.g. macroalgal) in addition to salinity (direct). The negative nitrate value in the regression model suggests that macroalgae rapidly uptake and utilize nitrate in the oceanic water for growth, consequently depleting the concentration within the immediate boundary layer. According to Valiela [1995] further nitrate uptake is dependent on replenishment from external (allochthonous) sources (rivers in this regard). This observation is corroborated by the CCA model (Figure 7B), which indicated indirect correlation between nitrate and salinity, that influence taxa assemblages (notable macrolagae). This further lends credence to nitrate uptake from river bodies by macroalgae. It is also worth noting that the significant abiotic variable in the CCA model were nitrate and dissolved oxygen (Table 4), further confirming nitrate utilization via photosynthesis (by macroalgae) with the evolution of oxygen.

Apparently, the relationship between nutrient loading and macroalgal diversity relies on species' difference in growth and uptake [Huston, 1994]. Macroalgal diversity peaks at intermediate nutrient availability because nutrient levels are sufficient to permit existence of fast-growing species with high nutrient requirement, but low enough to prohibit their domination [Bracken and Nielsen, 2004]. Growth experiments with unicellular algae show that nitrates are limiting factors to its growth while phosphates do not [Borowitzka, 1971; Ryther and Dunstan, 1971]. Nasr and Aleem [1948] and Waite [1969] explained that added nitrates and phosphates probably favour Chlorophyta growth. *Ulva* spp. generally are known to have a great reproductive capacity and rapid growth rate, thus able to colonize the more polluted areas even after most of the population is killed by extreme conditions [Borowitzka, 1971]. This is consistent with our observations that *Ulva fasciata* occurred in all the sampling sites with appreciable densities especially at the eastern shore (Figure 4) where the gradient of pollution was purportedly high. Coastal eutrophication has been ranked among the most serious threats to species diversity and stability of marine ecosystem worldwide especially in coastal areas that receive large amount of organic and mineral nutrients from municipal waste, agriculture and industrial effluents [Kotta et al., 2007].

Menge et al. [1994] also indicated that fluvial nutrients (e.g., nitrate from rivers) result in increased utilization by macroalgae for growth and biomass bringing about competition in a previously food limited dominance leading to reduced species richness. This explains the negative nitrate load in the regression model for high species richness (Table 3). Also since species richness correlates well with biomass in intertidal environment [Zacharias and Roff, 2001]; higher species richness would mean increased number of predators due to available trophic energy, congruently a reduction in species richness by predators. The regression model indicated that variation in species richness (r^2=61.49) was influenced by nitrate, salinity and dissolved oxygen. In general, marine organisms require submersion in water of fairly high salinity for some period of time [Lewis, 1964]. Species that need high quantities of salt always find themselves at the lower shore while those that are more tolerant of low water salinity due to other influences, especially where freshwater runs into seawater, are found at the upper shore closer to land [Lewis, 1964]. Salinity fluctuation cause changes in osmotic pressure of many intertidal rocky shore organisms especially if they are exposed. For example, growth response of *Enteromorpha intestinalis* and *Ulva expansa* to variable salinities has shown that salinities of 15‰ favored the former species whilst the latter species was favored by salinity of 35‰ when nitrogen was insufficient in supply [Peggy et al., 1996]. This present study has elucidated that rocky shore taxa assemblages (species diversity, species richness and number of species) is favoured with high salinity, low nitrate concentration (due to rapid depletion by macroalgae), low dissolved oxygen (utilization by epifauna), and to a limited extent water temperature (metabolic, growth, survival and reproductive drivers) (Table 3).

The primary influence of temperature on intertidal biota is through the constraint on metabolic rates, as almost all intertidal plants and invertebrates are poikilotherms [Zacharias and Roff, 2001]. Effects of temperature on rocky shore species has important implications of impacts of global warming on intertidal biota. Exposure affects water loss, temperature changes of the thallus, and salinity of cells of algae [Lewis, 1964] influencing their survival and distribution.

In sum, our results indicated that each site contained a distinct assemblage, mainly due to the relative abundances of few species. Most species in the studied area were widely distributed between central and eastern shores but depicted considerable spatial differences in abundances thus discriminating between study shores. Species assemblage pattern in the western shore were significantly different from central and eastern. The community assemblages were controlled by the dominance and widely distributed 24 taxa. The assemblage patterns were significant influenced by suites of abiotic factors notably nitrate, dissolved oxygen, salinity and water temperature. Our results suggest that, higher species richness and diversity of intertidal rocky communities were the result of nitrate concentrations from river bodies coupled with salinity and also dissolved oxygen concentration.

REFERENCES

Airoldi, L. (2003). The effects of sedimentation on rocky coasts assemblage. *Oceanography and Marine Biology: an Annual Review, 41*, 161-236.

Alongi, D. M. (1990). The ecology of tropical soft-bottom benthic ecosystems. *Oceanography and Marine Biology Annual Review*, *28*, 381-496.

A. P. H. A., A. W. W. A., & W. E. F. (1998). Standard methods for the examination of water and wastewater. 20Washington: *American Public Health Association*, (A.P.H.A.).

Armah, A. K. & Amlalo, D. S. (1998). Coastal Zone Profile of Ghana. Gulf of Guinea Large Marine Ecosystem Project. Ministry of Environment, *Science and Technology*, Accra, Ghana. 111.

Bassindale, R. (1961). On the marine fauna of Ghana. *Proc. Zool. Soc. Lond.*, *137*, 481-510.

Begon, M., Harper, J. L. & Townsend, C. R. (1990). *Ecology:* individuals, populations, and communities. Blackwell Scientific Publications, London.

Benedetti-Cecchi, L (2001). Variability in abundance of algae and invertebrates at different spatial scales on rocky sea shores. *Marine Ecological Progress Series*, *215*, 79-92.

Benedetti-Cecchi, L. (2009). Environmental variability: Analysis and ecological implications. In: L. Wahl, (Ed), Marine hard bottom communities: patterns, dynamics, *diversity and change*, 127-139, Berlin Heidelberg Springer-Verlag.

Bibby, C. J., Collar, N. J. Crosby, M. J., Heath, M. F., Imboden, C., Johnson, T. H., Long, A. J., Stattersfield, A. J. & Thirgood, S. J. (1992). Putting biodiversity on the map: priority areas for global conservation. *International Council for Bird Preservation*, Cambridge, UK.

Blanchette, C. A. Raimondi, P. T. & Broitman, B. R. (2009). Spatial patterns of intertidal community structure across the California channel islands and links to ocean temperature. *Proceedings of the California Islands Symposium.*

Boaventura, D., Ré, P., da Fonseca, L. C. & Hawkins, S. J. (2002). Intertidal rocky shore communities of the continental Portuguese coast: analysis and distribution patterns. *Marine ecology*, *23(1)*, 69-90.

Borowitzka, M. A. (1971). Intertidal algal species diversity and the effect of pollution. *Australian Journal of Marine and Freshwater Research*, *23(2)*, 73-84.

Bracken, M. E. S., & Nielsen, K. J. (2004). Diversity of intertidal macroalgae increases with nitrogen loading by invertebrates. *Ecology*, *85(10)*, 2828-2836.

Branoff, B., Yankson, K. & Wubah, D. (2009). Seaweed and associated invertebrates at Iture rocky beach, Cape coast, Ghana. *Journal of Young Investigators*, 1-7.

Broitman, B. R., Navarrete, S. A., Smith, F. & Gaines, S. D. (2001). Geographic variation of southeastern Pacific intertidal communities. *Marine Ecology-Progress Series*, *224*, 21-34.

Buchanan, J. B. (1957). Marine mollusc of the Gold coast. *Journal of West Africa Science Association*, *1*, 30-45.

Burrows, M. T., Moore, J. J. & James, B. (2002). Spatial synchrony of population changes in rocky shore communities in Shetland. *Marine Ecology Progress Series*, *240*, 39-48.

Clarke, K. R. (1993). Non-parametric multivariate analyses of changes in community structure. *Australian Journal of Ecology*, *18*, 117-143.

Clarke, K. R. & Green, R. H. (1988). Statistical design and analysis for 'biological effects' study. *Marine Ecology Progress Series*, *46*, 213-226.

Clarke, K. R. & Warwick, R. M. (1994). Changes in Marine Communities. An approach to statistical analysis and interpretation. *Natural Environment Research Council*, UK. 144.

Clarke, K. R. & Gorley, R. N. (2006). PRIMER v6: User manual/tutorial. Primer-E, *Plymouth*, UK.

Clarke, K. R. & Warwick, R. M. (2001). A further biodiversity index applicable to species lists: variation in taxonomic distinctness. *Marine Ecology Progress Series*, *216*, 265-278.

Connell, J. H. (1972). Community interactions on marine rocky intertidal shores. *Annual Review of Ecology & Systematics*, *3*, 169-192.

Connell, J. H. (1978). Diversity in Tropical Rain Forests and Coral Reefs. *Science*, *199*, 1302-1310.

Connell, J. H. (1975). Some mechanisms producing structure in natural communities: A model and evidence from field experiments. In: M. L. Cody, & J. M. Diamond, (Eds.). *Ecology and Evolution of Communities*, 460-490. Cambridge, Massachusetts Belknap Press.

Crisp, D. J. (1976a). The role of pelagic larva. In: P. S. Davies, (Ed) *Perspectives in Experimental Biology*. vol. 1 145-155, Zoology, Oxford Pergamon Press.

Crisp, D. J. (1976b). Settlement responses in marine organisms. In: R. C. Newell, (Ed), Adaptations to environment: *essays on the physiology of marine animals*. 83-124. London, Butterworths.

Denny, M. W. (1988). Biology and the mechanics of the wave-swept environment. *Princeton, New Jersey: Princeton University Press*.

Denny, M. W. (1993). Air and water: the biology and physics of life's media. Princeton University Press, *Princeton*, N.J.

Deepananda, A. K. H. M. (2008). Community-level analysis of anthropogenic impacts on rocky shore communities in Sri Lanka. Nature Proceedings: doi:10.1038/npre. 2008.2317.1

Edmunds, J. (1978). *Sea shells and other molluscs found on West African Coast and Estuaries*. Ghana Universities Press.

Evans, S. M., Gill, M. E., Hardy, F. G. & Seku, F. O. K. (1993). Evidence of change in some rocky shore communities on the coast of Ghana. *Journal of Experimental Marine Biological Ecology*, *172*, 129-141.

Foster, M. S. (1990). Organization of macroalgae assemblages in the northeast Pacific: the assumption of homogeneity and the illusion of generality. *Hydrobiologia*, *192*, 21-33.

Fretwell, S. D. (1987). Food chain dynamics: The central theory of ecology. *Oikos*, *50*, 291-301.

Gauld, D. T. & Buchanan, J. B. (1959). The principal features of the rocky shore fauna of Ghana. *Oikos*, *10*, 121-132.

Gray, J. S. (1997). Marine biodiversity: patterns, threats and conservation needs. *Biodiversity and Conservation*, *6*, 153-175.

Guille, A. (1970). Benthic bionomy of continental shelf of the French Catalane Coast.II. Benthic communities of the macrofauna. *Vie et Milieu*, *218*, 149-280.

Hardy, F. G. & Seku, F. O. K. (1993). Some notes on collecting sites and field records for marine algae in Ghana. *The Phycologist*, *36*, 2-7.

Hardy, F. G. & Seku, F. O. K. (1993). Some notes on collecting sites and field records for marine algae in Ghana. *The Phycologist*, *36*, 2-7.

He, F. & Legendre, P. (2002). Species diversity patterns derived from species–area models. *Ecology*, *83*, 1185-1198.

Huang, D., Todd, P. A., Chou, L. M., Ang, K. H., Boon, P. Y., Cheng, L. & Ling, H. (2006). Effects of shore height and visitor pressure on the diversity and distribution of four intertidal taxa at Labrador beach, Singapore. *The Raffles Bulletin of Zoology, 54(2)*, 477-484.

Hurlbert, S. H. (1984). Pseudoreplication and the design of ecological field experiments. *Ecological Monograph, 54*, 187-211.

Huston, M. A. (1994). *Biological diversity: the coexistence of species on changing landscapes*. Cambridge, Cambridge University Press, UK. .

Hutchinson, M. J., John, E. A. & Wijesinghe, D. K. (2003). Towards understanding the consequences of soil heterogeneity for plant populations and communities. *Ecology, 84*, 2322-2334.

Hutchinson, N. & Williams. G. A. (2001). Spatio-temporal variation in recruitment on a seasonal, tropical rocky shore: The importance of local versus non-local processes. *Marine Ecology Progress Series, 215*, 57-68.

Kotta, J., Lauringson, V. & Kotta, I. (2007). Role of functional diversity and physical environment on the response of zoobenthos communities to changing eutrophication. *Hydrobiologia, 580*, 97-108.

Kotta, J. & Witman, J. D. (2009). Regional-scale patterns. In: L. Wahl, (Ed.) Marine hard bottom communities: patterns, dynamics, *diversity and change*, 89-99. Springer.

Lawson, G. W. (1956). Rocky shore zonation on the Gold coast. *Journal of Ecology, 44*, 153-170.

Green, J. L. & Ostling, A. (2003). Endemic-Area relationship: The influence of species dominance and spatial aggregation. *Ecology, 84(11)*, 3090-3097.

John, D. M. (1986). Littoral and sub-littoral marine vegetation: In : G. W. Lawson, (Ed.). *Plant Ecology of West Africa*. John Wiley and Sons Ltd.

John, D. M & Lawson, G. W. (1990). Review of mangrove and coastal ecosystem in West Africa and their possible relationship. *Estuarine, Coastal and Shelf Science, 31*, 505-515.

John, D. M. Lawson, G. W. & Ameka, G. K. (2003). The marine macroalgae of the tropical West Africa sub-region. *Beih. Nova Hedw, 125*, 1-219,

Lambshead, P. J. D., Platt, H. M. and. Shaw. K. M. (1983). The detection of differences among assemblages of marine benthic species based on an assessment of dominance and diversity. *Journal of Natural History, 17*, 859-874.

Lewis, J. R. (1964). *The ecology of rocky shores*. English Universities Press Ltd., London, 300.

Loreau, M., Mouquet, N. & Gonzalez, A. (2003). Biodiversity as spatial insurance in heterogeneous landscapes. *Proceedings of the National Academy of Sciences of the USA, 100*, 12765-12770.

Lubcheno, J. & Gaines, S. D. (1981). A unified approach to marine-plant herbivore interactions. I. Populations and communities. *Annual Reviews of Ecology & Systematics, 12*, 405-437.

Ly, C. K. (1980). The role of the Akosombo dam on the Volta River in causing coastal erosion in central and eastern Ghana (West Africa). *Marine Geology, 37*, 323-332.

McGill, B. J., Enquist B. J., Weiher E. & Westoby, M. (2006). Rebuilding community ecology from functional traits. *Trends in Ecological Evolution, 21*, 178-185.

Menge, B. A. (1995). Indirect effects in marine rocky intertidal interaction webs: patterns and importance. *Ecological Monograph, 65*, 21-74.

Menge, B. A., Lubchenco, J., Ashkenas, L. R. & Ramsey, F. (1986). Experimental separation of effects of consumers on sessile prey in the low zone of a rocky shore in the Bay of

Panama: direct and indirect consequences of food web complexity. *J. Mar. Biol. Ecol., 100*, 225-270.

Menge, B. A. (2000). Top-down and bottom-up community regulation in marine rocky intertidal habitats. *J. Exp. Mar. Biol. Ecol., 250*, 257-289.

Menge, B. A. & Branch, G. M. (2001). Rocky intertidal communities. In: M. D., Bertness, S. D. Gaines, & M. E. Hay, (Eds.) *Marine community ecology*, Sunderland, 221-252. Sinauer Associates.

Menge, B. A. & Olson, A. M. (1990). Role of scale and environmental factors in regulation of community structure. *Trends Ecol Evol., 5*, 52-57.

Menge, B. A., Olson, A. M. & Dahlhoff, E. P. (2002). Environmental stress, bottom-up effects, and community dynamics: integrating molecular-physiological and ecological approaches. *Integ. and Comp. Biol., 42*, 892-908.

Menge, B. A., Berlow, E. L., Blanchette, C. A., Navarrete, S. A. & Yamada, S. B. (1994). The keystone species concept: variation in interaction strength in a rocky intertidal habitat. *Ecological Monograph, 64*, 249-286.

Menge, B. A., Daley B. A., Wheeler, P. A., Dahlhoff, E., Sanford, E. & Strub, P. T. (1997a). Benthic-pelagic links and rocky intertidal communities: bottom-up effects or top down control? *Proceedings of the National Academy of Sciences of the USA, 94*, 14530-14535.

Menge, B. A., Daley, B. A., Wheeler, P. A. & Strub, P. T. (1997b). Rocky intertidal oceanography: an association between community structure and nearshore phytoplankton concentration. *Limnology and Oceanography, 42*, 57-66.

Menge, B. A., Lubchenco, J., Bracken M. E. S., Chan, F., Foley, M. M., Freidenburg, T. L., Gaines, S. D., Hudson, G., Krenz, C., Leslie, H., Menge, D. N. L., Russell, R. & Webster, M. S. (2003). Coastal oceanography sets the pace of rocky intertidal community dynamics. *Proceedings of the National Academy of Sciences of the United States of America, 100*, 12229-12234.

Mittermeier, R. A., Myers, N., Thomsen, J. B., da Fonseca, G. A. B. & Olivieri, S. (1998). Biodiversity hotspots and major tropical wilderness areas: approaches to setting conservation priorities. *Conservation Biology, 12*, 516-520.

Myers, N., Mittermeier, R. A., Mittermeier, C. G., da Fonseca, G. A. B. & Kent, J. (2000). Biodiversity hotspots for conservation priorities. *Nature, 403*, 853-858.

Nasr, A. H. & Aleem, A. A. (1948). Ecological studies of some marine algae in Alexandria. *Hydrobiologia, 1*, 251-281.

Newell, R. C. (1979). Biology of intertidal animals. *Marine Ecological Surveys*, Ltd., Kent.

Nielsen, K. J. (2001). Bottom-up and top-down forces in tide pools: test of a food chain model in an intertidal community. *Ecological Monographs, 71*, 187-217.

Nielsen, K. J. (2003). Nutrient loading and consumers: agents of change in open-coast macrophyte assemblages. *Proceedings of the National Academy of Sciences*, (USA) *100*, 7660-7665.

Nielsen, K. J. & Navarrete, S. A. (2004). Mesoscale regulation comes from the bottom-up: intertidal interactions between consumers and upwelling. *Ecology Letters, 7*, 31-41.

Noda, T. (2009). Metacommunity-level coexistence mechanisms in rocky intertidal sessile assemblages based on a new empirical synthesis. *Population Ecology, 51*, 41-55.

Nowell, A. R. M. & Jumars, P. A. (1984). Flow environments of aquatic benthos. *Annual Reviews of Ecology & Systematics, 15*, 303-328.

Oksanen, L. Fretwell, S. D., Arruda, J. & Niemela, P. (1981). Exploitation ecosystems in gradients of primary productivity. *American Naturalist, 118*, 240-261.

Olff, H., Alonso, D., Berg, M. P., Eriksson, B. K. Loreau, M., Piersma, T. & Rooney, N. (2009). Parallel ecological networks in ecosystems. *Philosophical Transactions of Royal Society B, 364*, 1755-1779.

Paine, R. T. & Levin, S. A. (1981). Intertidal landscapes: disturbance and the dynamics of pattern.*Ecological Monograph, 51*, 145-178.

Peggy, F., Boyer, K. E., Desmond, J. S. & Zedler. J. B. (1996). Salinity stress, nitrogen competition and facilitation: what controls seasonal succession of two opportunistic green macroalgae. *Journal of Experimental Marine Biology and Ecology, 206*, 203-221.

Pielou, E. C. (1966). The measurement of diversity in different types of biological collections. *Journal of Theoretical Biology, 13*, 131-144.

Raffaelli, D. & Hawkins, S. (1996). *Intertidal Ecology*. Kluwer Academic Publishers, London. 356.

Reise, K. (2002). Sediment mediated species interactions in coastal waters. *Journal of Sea Research, 48(2)*, 127-141.

Richardson, P. L. & Reverdin, G. (1987). Seasonal cycle of velocity in the Atlantic North Equatorial Countercurrent as measured by surface drifters, current meters, and ship drifts. *Journal of Geophysical Research, 92(4)*, 3691-3708.

Rickleffs, R. E. & Schluter, D. (1993). Species diversity in ecological communities: *historical and geographic perspective*. University of Chicago Press, Chicago.

Ryther, J. H. & Dunstan, W. M. (1971). Nitrogen, phosphorous and eutrophication in the coastal marine environment. *Science, 171*, 1008-1013.

Sanders, H. L. (1968). Marine benthic diversity: a comparative study. *American Naturalist, 102*, 243-282.

Schlesinger, W. H. (1991). Biogeochemistry. An analysis of global change. *San Diego*, CA: Academic Press.

Shannon, C. E. & Weaver, W. (1963). The mathematical theory of communication. 111. Urbana, IL: University of Illinois Press.

Sousa, W. P. (1984). Intertidal mosaics: patch size, propagule availability and spatial variable patterns of succession. *Ecology, 65*, 1918-1935.

Stephenson, T. A. & Stephenson, A. (1949). The universal features of zonation between tidemarks on rocky coasts. *Journal of ecology, 38*, 289-305.

Ter Braak, C. J. F. (1986). Canonical correspondence analysis: A new eigenvector technique for multivariate direct gradient analysis. *Ecology, 67*, 1167-1179.

Ter Braak, C. J. F. & Verdonschot, P. F. M. (1995). Canonical correspondence analysis and related multivariate methods in aquatic ecology. *Aquatic Science, 57*, 255-289.

Ter Braak, C. J. F. & Smilauer, P. (1998). CANOCO Reference Manual and user's guide to Canoco for windows: Software for Canonical Community Ordination (version 4). Ithaca, NY: Microcomputer Power.

Ter Braak, C. J. F. & Smilauer, P. (2002). CANOCO Reference Manual and user's guide to Canoco for windows: *Software for Canonical Community Ordination* (version 4.5). Ithaca, NY: Microcomputer Power.

Terlizzi, A. & Schiel, D. R. (2009). Patterns along environmental gradient. In: L. Wahl (Ed) Marine hard bottom communities: patterns, dynamics, *diversity and change*. 101-109. Springer,

Thompson, R. C., Crowe, T. P. & Hawkins, S. J. (2002). Rocky intertidal communities: past environmental changes, present status and predictions for the next 25 years. *Environmental Conservation, 29*, 168-191.

Underwood, A. J. (1979). The ecology of intertidal gastropods. *Advanced Marine Biology, 16*, 111-210.

Underwood, A. J. (1981). Structure of a rocky intertidal community in New South Wales: patterns of vertical distribution and seasonal changes. *Journal of Experimental Marine Biology, 51*, 57-85.

Underwood, A. J. & Denley, E. J. (1984). Paradigms, explanations and generalizations in models for the structure of intertidal communities on rocky shore. In: D. R. Strong, Jr., D., Simberloff, L. G. Abele, & A. B. Thistle, (Eds). Ecological communities: conceptual issues and the evidence. Princeton University Press, *Princeton*, 151-180.

Valiela, I. (1995). *Marine ecological processes*. Second edition. Springer-Verlag, New York, USA. Welch . 686.

Vernberg, W. B. & Vernberg. F. J. (1972). *Environmental physiology of marine organisms*. Springer-Verlag, New York.

Vinueza, L. R., Branch, G. M., Branch, M. L. & Bustamante, R. H. (2006). Top-down herbivory and Bottom-up El NiNo effects on Galapagos rocky-shore communities. *Ecological Monographs, 76(1)*, 111-131.

Vitousek, P. M., Aber, J. D., Howarth, R. W., Likens, G. E., Matson, P. A., Schindler, D. W., Schlesinger, W. H. & Tilman, D. G. (1997). Human alteration of the global nitrogen cycle: sources and consequences, *Ecological Applications, 7(3)*, 737-750.

Waite, A. (1969). Notes on the growth of *Ulva* as a function of ammonia nitrogen. *Phytologia, 18*, 65-9.

Warwick, R. M. & Clarke, K. R. (1994). Relearning the ABC: taxonomic changes and abundance/biomass relationships in disturbed benthic communities. *Marine Biology, 118*, 739-744.

Warwick, R. M. & Ruswahyuni (1987). Comparative study of the structure of some tropical and temperate marine soft-bottom macrobenthic communities. *Marine Biology, 95*, 641-649.

Witman, J. D. & Smith, F. (2003). Rapid community changes at a tropical upwelling site in the Galápagos Marine Reserve. *Biodiversity and Conservation, 12*, 25-45.

Wootton, J. T., Power, M. E., Paine, R. T. & Pfister, C. A. (1996). Effects of productivity, consumer, competitors, and El Nino on food chain patterns in a rocky intertidal community. *Proceedings of the National Academy of Sciences*, (USA), 93, 13855-13858.

Yankson, K. & Akpabey, F. J. (2001). A Preliminary Survey of the Macro-Invertebrate Fauna at Iture Rocky Beach, Cape Coast, Ghana. *Journal of Natural Sciences, 1(2)*, 11-22.

Zacharias, M. A. & Roff, J. C. (2001). Explanation of patterns of intertidal diversity at regional scales. *Biogeography, 28*, 471-483.

Zacharias, M. A. & Roff, J. C. (2001). The use of focal species in marine conservation and management: a review and critique. *Aquatic Conservation: Marine and Freshwater Systems, 11*, 1-18.

Chapter 8

POLLUTION AND DIVERSITY OF FISH PARASITES: IMPACT OF POLLUTION ON THE DIVERSITY OF FISH PARASITES IN THE TISA RIVER IN SLOVAKIA

Vladimíra Hanzelová[1], Mikuláš Oros[1] and Tomáš Scholz[2]

[1]Parasitological Institute, Slovak Academy of Sciences,
Hlinkova 3, 04001 Košice, Slovakia
[2]Institute of Parasitology, Biology Centre of the Academy of Sciences of the Czech Republic, České Budějovice, Czech Republic

ABSTRACT

An extensive survey of helminth parasites of 1,316 freshwater fish of 31 species from two aquatic ecosystems with different level of environmental pollution in southeastern Slovakia was carried out and a total of 31 gastrointestinal helminths (Trematoda – 11 species, Cestoda – 14, Acanthocephala – 3 and Nematoda – 3) have been found. The Tisa River has been heavily polluted with cyanides and heavy metals after a series of ecological disasters in 2000, whereas its tributary, Latorica River, is less anthropogenically impacted. Even though the fish communities were qualitatively similar (Czekanowski-Sørensen similarity Index - ICS = 81%) and the number of fish examined was approximately the same (676 and 640) in both localities, species richness of helminths and diversity of host-parasite associations were two times lower in the more polluted Tisa River. Helminth communities were also much less abundant in the Tisa River. Based on ICS = 48.8% and the Percentage similarity Index (PI) = 19.5%, the helminth communities were qualitatively and quantitatively different in the two rivers. Four species, the aspidogastrean *Aspidogaster limacoides* Diesing, 1835, the acanthocephalan *Pomphorhynchus tereticollis* (Rudolphi, 1809), and tapeworms *Atractolytocestus huronensis* Anthony, 1958 and *Khawia sinensis* Hsü, 1935 were reported for the first time in Slovakia. Both tapeworms found in wild common carp *Cyprinus carpio carpio* L. have been introduced recently into the Tisa River and their further dissemination to other regions throughout the Danube River basin is probable. Their morphology is briefly described and compared with representatives of other European populations of common carp.

INTRODUCTION

The Tisa River basin plays a unique role in the European ecological network. A series of habitats of the river and its tributaries provides excellent environmental conditions to a wide spectrum of animal species, with several native Carpathian (endemic) species of the first-line importance. Among vertebrate taxa, fishes are the most abundant and important component of this aquatic ecosystem. The Slovak part of the Tisa River course (southeastern Slovakia; 48°23'N, 22°07'E) includes only a 6-km section of the Upper Tisa, forming a border between Slovakia, Ukraine and Hungary (Figure 1). In 2004, this part of the Tisa River was added to the Ramsar list of wetlands of international importance.

Figure 1. Geographic location of rivers Tisa and Latorica, southeastern Slovakia with indication of sampling sites.

Balon (1964) reported the occurrence of 57 fish species in the Tisa River basin which represents about 70% species of the current ichthyofauna of Slovakia. According to the Ordinance No. 93/1999 of the Ministry of Environment of the Slovak Republic, one quarter (14 species) of fishes, which inhabit this area, has a status of endangered or critically endangered species. Danubian longbarbel gudgeon *Romanogobio uranoscopus* (Agassiz, 1828), Danubian roach *Rutilus pigus* (Lacépède, 1803), huchen *Hucho hucho* (L.), mudminnow *Umbra krameri* Walbaum, 1792 and schraetzer *Gymnocephalus schraetser* (L.) are of particular importance as Carpathian endemic species.

Previous parasitological surveys reported a rich parasite fauna in fishes from the Tisa River. Žitňan (1966a, b, 1967) and Ergens et al. (1975) identified as many as 103 gastrointestinal helminths, i.e. 45 trematodes, 24 tapeworms, 7 acanthocephalans and 27 nematode species.

In January 2000, the Tisa River was subjected to a strong anthropogenic pressure, when a wave of cyanide and heavy metals was released on two occasions from a gold-processing factory in Baia Mare, Romania and moved quickly downstream from the Somes River to the Tisa. The resultant water pollution killed tens of thousands of fish and other forms of wildlife.

Even though this ecological disaster was described as the worst since Chernobyl (Cunningham, 2005), the media took almost no interest in this event and the ongoing pollution of the river and its effect on human, animal, and plant life. Based on the initiative of the countries which were directly affected by this disaster, several projects aimed on the environmental safety and strategies in case of disasters have been running. However, none of these projects dealt with the detailed environmental monitoring of the biodiversity of living animals, structure and processes of the restoration of the ecological balance of animal communities in this river.

Nowadays, the Slovak part of Tisa is polluted with heavy metals from industrial and consumer waste sources coming from Romania and Ukraine (Table 1). Heavy metals in aquatic ecosystems are dangerous not only because they tend to accumulate, but they also pose a serious threat to aquatic biodiversity. Pollution can increase parasitism if, for example, host defense mechanisms are negatively affected. However, pollution can also decrease parasitism if parasites are susceptible to pollutants or if pollution drives the necessary intermediate and final hosts to become extinct (Sures, 2004). Moreover, instability of the environment creates proper conditions for so-called biological pollution, the accidental or deliberate introduction of foreign (allochthonous) organisms to an environment (Elliott, 2003). If such introductions lead to a successful colonization, then introduced species could also become a wide-scale environmental problem. During last decades, mostly due to man-made introductions of common carp, new alien parasites, mostly those originally occurring in East Asia, have been introduced to Europe (Bauer et al., 1973; Kennedy, 1994). Among these parasites, the cestode *Bothriocephalus acheilognathi* Yamaguti, 1934 (Bothriocephalidea) is known as the most important pathogen of common carp in aquacultures worldwide, including Slovakia (Žitňan, 1982; Jeney & Jeney, 1995).

Table 1 Heavy metal concentration (µg/l) in the water from the Tisa and Latorica rivers assessed within the investigated period

Heavy metals	October 2003		August 2005		February 2006	
	Tisa	Latorica	Tisa	Latorica	Tisa	Latorica
Cd	1.80	0.40	1.15	0.17	0.95	0.16
Cr	4.20	3.10	2.59	0.85	0.85	0.85
Cu	39.20	16.20	19.60	3.47	6.69	7.84
Hg	0.05	0.05	0.05	0.05	0.05	0.05
Ni	9.50	8.70	-	-	-	-
Pb	34.60	16.50	25.40	7.30	7.60	6.50
Zn	384.10	89.70	149.00	34.20	194.00	105.00

Other two non-indigenous cestodes of the order Caryophyllidea (Lytocestidae), *Atractolytocestus huronensis* and *Khawia sinensis* both parasitizing common carp, have recently been reported from several water reservoirs and rivers in southeastern Slovakia, including the Tisa River (Oros et al., 2004, 2009; Oros & Hanzelová, 2009). Previous comprehensive surveys of fish parasites in the Tisa River basin carried out in the 1970s did not reveal the presence of these two cestodes (Žitňan 1966b; Ergens et al., 1975; Moravec, 2001).

A. huronensis was originally described from common carp from the Huron River, Michigan, USA (Anthony, 1958), and reported only from North America until the late 1990s (Hoffman, 1999; Jones & Mackiewicz, 1969; Amin, 1986; Amin & Minckley, 1996). In Europe, *A. huronensis* was first found in the common carp in England (Chubb et al., 1996; Kirk et al., 2003) and soon also in Hungary (Majoros et al., 2003), Czech Republic and Slovakia (Oros et al., 2004).

K. sinensis Hsü, 1935 was described from common carp from the vicinity of Peking, China (Hsü, 1935) and its original distribution area probably included China and the Russian Far East (see Dubinina, 1971). The tapeworm was then disseminated to several states of the former Soviet Union with the introduction of carp, its hybrids and herbivorous fish, especially grass carp, *Ctenopharyngodon idella* (Valenciennes, 1844). In 1954–1955, the tapeworm appeared in fishponds in Lithuania, Latvia and Byelorussia (Kulakovskaya & Krotas, 1961). A very fast expansion of *K. sinensis* throughout Europe took place mainly in the 1960s and early 1970s when this tapeworm appeared step by step in as many as eight European countries, as well as in Japan and North America (Nakajima & Egusa, 1978; Williams & Sutherland, 1981). Due to its fast colonization (Table 2) and pathogenicity for carp fry, *K. sinensis* caused worldwide concern as a potential danger (Körting, 1975; Oros et al., 2009).

Table 2 First records of *Khawia sinensis* in individual countries

Year	Country/locality	Reference
1935	China	Hsü (1935)
1954-55	Byelorussia	Kulakovskaya and Krotas (1961)
1954-55	Latvia	Kulakovskaya and Krotas (1961)
1954-55	Lithuania	Kulakovskaya and Krotas (1961)
1961	Ukraine	Kulakovskaya and Krotas (1961)
1963	Russia	Musselius et al. (1963)
1965	Czech Republic	Přibyslavský et al. (1965)
1966	Romania	Rădulescu and Georgescu (1966)
1966	Kazakhstan	Akhmetova (1966)
1973	Germany	Kulow (1973)
1974	Poland	Pańczyk and Żelazny (1974)
1975	Hungary	Molnár and Buza (1975)
1975	Japan	Nakajima and Egusa (1978)
1975	USA	Williams and Sutherland (1981)
1977	Bosnia and Herzegovina	Kiškároly (1977)
1980	Bulgaria	Kakacheva-Avramova et al. (1980)
1986	Great Britain	Yeomans et al. (1997)
2004	Slovakia	Oros and Hanzelová (2009)

In the Tisa River basin, another helminth parasite non-indigenous in Europe has recently appeared. Košuthová et al. (2004) reported the presence of the cestode *Nippotaenia mogurndae* Yamaguti & Miyata, 1940 (Nippotaeniidea), a specific parasite of the Chinese sleeper *Perccottus glenii* (Perciformes: Odontobutidae). The distribution area of *N. mogurndae* was originally restricted to Japan and the Far East, but it has been introduced to Europe together with its host, which was first reported from Slovakia in 1998 (Kautman, 1999).

This chapter presents the data on a re-establishment of fish parasite fauna in the Tisa River system after a catastrophic pollution event compared with helminth species diversity in the less polluted Latorica River. Non-indigenous fish parasites, introduced to the Tisa River basin recently, are briefly characterized and data on their morphology, distribution and rates of infection in their fish hosts are provided.

MATERIALS AND METHODS

Sampling sites on the Tisa River were located near the villages of Veľké and Malé Trakany, southeastern Slovakia (Figure 1). As comparative water system with low level of pollution (Figure 1; Table1) the Latorica River (48°28'N, 22°00'E), not affected by cyanide and heavy metal pollution during the accident in the Tisa River basin in 2000, was used. The Latorica River is an indirect tributary of the Tisa River; it flows into the Bodrog River which in turns flows into the Tisa as its main affluent, near the village of Tokay (Hungary), far from the border of Slovakia. Even though the Tisa and Latorica rivers merge through the Bodrog River, their water bodies are not really connected even during the yearly flooding and are, therefore, considered as two separated ecosystems. Consequently, fish and helminth communities of these two rivers are regarded as geographically isolated (allopatric). Sampling sites were located near villages of Beša, Boľ, Soľnička and Zatín, all in southeastern Slovakia.

Fish were caught using gill nets, by angling and electrofishing under a permit issued by the Ministry of Agriculture of the Slovak Republic. Examination of a limited number (5) of year-round protected fish was also included in this permit. Within the years 2002–2006, a total of 1,316 specimens of fish belonging to 31 species (676 fish of 24 species from the Tisa and 640 fish of 28 species from the Latorica River) of nine families (Table 3) were collected and examined for intestinal helminths. Fish were examined by incomplete parasitological necropsy focused on the organs of the digestive system immediately after being caught. The scientific and common names of the fish hosts were used according to the FishBase database (Froese & Pauly, 2009).

Platyhelminthes (trematodes and cestodes) were treated as follows: rinsed in saline solution, immediately fixed in 4% hot formaldehyde, stained with iron acetic carmine according to Georgiev et al. (1986), differentiated in 70% acid ethanol, dehydrated through a graded ethanol series, clarified in clove oil, mounted in Canada balsam as permanent preparation (Scholz & Hanzelová, 1998) and identified using keys of Schmidt (1986), Khalil et al. (1994), Gibson et al. (2002) and Jones et al. (2005). Nematodes were fixed in hot mixture of formaldehyde and saline solution (1: 9); after fixation they were kept in 70% ethanol and examined as temporary microscopic preparations in glycerin (Moravec, 1994).

Acanthocephalans were washed in tap water, fixed in 70% ethanol and examined as temporary microscopic preparations in ethanol-glycerin (Ergens & Lom, 1970). Olympus BX 50 microscope, equipped with differential interference contrast (DIC according to Nomarski) and digital image analysis (Olympus analySIS docu program), was used for identification of parasites. Measurements are in µm unless otherwise indicated. All voucher specimens are deposited at Parasitological Institute SAS in Košice, Slovakia.

In addition, *A. huronensis* tapeworms from carp collected in several fishponds near Třeboň and Veselí nad Lužnicí, South Bohemia (Czech Republic) and near Budapest, (Hungary), processed by the same method, were also studied. Voucher specimens of *A. huronensis* from North America and *Caryophyllaeus fimbriceps* Annenkova-Chlopina, 1923 from Slovakia, deposited in museum collections, were also used for comparison.

Analysis of the parasite community structure was carried out at the component community level (a parasite community of a single host species or intermediate host species) and was characterized by quantitative ecological parameters including prevalence and intensity of infection and was calculated for each parasite and its host. Ecological terms used follow those proposed by Bush et al. (1997). The core, secondary, satellite, and rare species were classified by prevalence (Holmes & Price, 1986). To measure and compare qualitative and quantitative differences of fish and parasite diversity between the Tisa and Latorica rivers, Czekanowski-Sørensen similarity Index (ICS) and Percentage similarity Index (PI); (Brower et al., 1990; Magurran, 1991) were used.

RESULTS

Communities of Fish Helminths of the Tisa River

A total of 676 fish of 24 species of the families Acipenseridae, Cyprinidae, Cobitidae, Ictaluridae, Esocidae, Centrarchidae, Percidae and Odontobutidae were examined for intestinal metazoan parasites from the Tisa River (Table 3). Only 54.2% of fish were infected with gastrointestinal helminths of 13 species (Trematoda – 4 species, Cestoda – 7 and Nematoda – 2). No acanthocephalans were found (Table 4). No endohelminths were found in *Acipenser ruthenus*, *Alburnus alburnus*, *Ameiurus melas*, *A. nebulosus*, *Aspius aspius*, *Blicca bjoerkna*, *Gymnocephalus cernuus*, *Lepomis gibbosus*, *Pelecus cultratus*, *Perccottus glenii* and *Scardinius erythrophthalmus*. Three endangered fish species, namely weatherfish (*Misgurnus fossilis*), ziege (*Pelecus cultratus*) and zingel (*Zingel zingel*), were examined but only weatherfish was infected with metacercariae of the trematode *Clinostomum complanatum*.

Autogenic parasites, i.e. the species in which the entire life cycle is completed within the same aquatic ecosystem, were dominant because they represented 11 species. Encysted larval stages (metacercariae) of two digeneans, *C. complanatum* and *Posthodiplostomum cuticola*, which use fish as intermediate hosts and mature in fish-eating birds, represented the allogenic species. Generalists were most abundant. On the other hand, the tapeworm *Khawia rossittensis* and nematode *Philometroides sanguinea* are strict specialists parasitizing exclusively goldfish (*Carassius auratus*), and the tapeworm *Khawia sinensis* was found only in common carp. Dominant component communities comprised one or two helminth species with the maximum of three helminth species found in common carp, white-eye bream (*Abramis sapa*), common bream (*Abramis brama*), roach (*Rutilus rutilus*), and pikeperch (*Sander lucioperca*). As many as 26 host-parasite combinations were recorded in the Tisa River; six of them (23%) representing new host records in Slovakia (Table 5).

Table 3 Number of fish species examined in the Tisa and Latorica rivers in the period 2002-2006

Fish species	Tisa	Latorica
Acipenseridae		
Acipenser ruthenus L.	3	-
Cyprinidae		
Abramis brama (L.)	88	51
Abramis sapa (Pallas, 1814)	8	30
Alburnus alburnus (L.)	32	10
Aspius aspius (L.)	8	4
Blicca bjoerkna (L.)	4	12
Carassius auratus auratus (L.)	199	51
Chondrostoma nasus (L.)	-	3
Ctenopharyngodon idella (Valenciennes, 1844)	-	2
Cyprinus carpio carpio L.	17	12
Hypophthalmichthys molitrix (Valenciennes, 1844)	-	2
Leuciscus idus (L.)	-	7
Pelecus cultratus (L.)	2	-
Rhodeus sericeus (Pallas, 1776)	19	12
Rutilus rutilus (L.)	67	23
Scardinius erythrophthalmus (L.)	3	1
Squalius cephalus (L.)	9	23
Tinca tinca (L.)	-	4
Cobitidae		
Cobitis taenia L.	27	6
Misgurnus fossilis (L.)	1	-
Ictaluridae		
Ameiurus melas (Rafinesque, 1820)	27	11
Ameiurus nebulosus (LeSueur, 1819)	31	2
Siluridae		
Silurus glanis L.	-	1
Esocidae		
Esox lucius L.	20	7
Centrarchidae		
Lepomis gibbosus (L.)	16	3
Percidae		
Gymnocephalus cernuus (L.)	1	1
Perca fluviatilis L.	25	3
Sander lucioperca (L.)	49	9
Sander volgensis (Gmelin, 1789)	-	1
Zingel zingel (Linnaeus, 1766)	1	1
Odontobutidae		
Perccottus glenii Dybowski, 1877	19	348
Total	676	640

Table 4 List of fish helminth species from the Tisa and Latorica rivers with indication of the similarity between two localities in relation to respective intermediate hosts

Helminth species	Tisa	Latorica	Intermediate host	ICS
Trematoda				
Aspidogaster limacoides Diesing, 1835		+		
Allocreadium markewitschi Koval, 1949		+		
Asymphylodora imitans (Mühling, 1898)	+	+		
Asymphylodora markewitschi Kulakowskaya, 1947		+		
Nicolla testiobliqua (Wisniewski, 1932)	+	+	aquatic snails	36.4%
Nicolla skrjabini (Ivanitzky, 1928)		+		
Orientocreadium siluri Bychowski et Dubinina, 1954		+		
Palaeorchis incognitus Szidat, 1943		+		
Sphaerostomum bramae (Müller, 1776)		+		
Clinostomum complanatum (Rudolphi, 1810)	+			
Posthodiplostomum cuticola (Nordmann, 1832)	+	+	fishes	66.7%
Cestoda				
Atractolytocestus huronensis Anthony, 1958	+	+		
Caryophyllaeus brachycollis Janiszewska, 1951	+			
Caryophyllaeus fimbriceps Annenkova-Chlopina, 1919		+		
Caryophyllaeus laticeps (Pallas, 1781)	+	+	aquatic worms (oligochaetes)	71.4%
Caryophyllaeides fennica (Schneider, 1902)	+	+		
Khawia rossittensis (Szidat 1937)	+	+		
Khawia sinensis Hsü, 1935	+	+		
Paracaryophyllaeus gotoi (Motomura 1927)		+		
Paraglaridacris gobii (Szidat, 1938)		+		
Amurotaenia perccotti Achmerov, 1941		+		
Glanitaenia osculata (Goeze, 1782)		+		
Proteocephalus percae (Müller, 1780)	+	+	zooplankton (copepods)	33.3%
Proteocephalus torulosus (Batsch, 1786)		+		
Triaenophorus nodulosus (Pallas, 1781)		+		
Acanthocephala				
Acanthocephalus anguillae (Müller, 1780)		+		
Acanthocephalus lucii (Müller, 1776)		+	freshwater amphipods	0%
Pomphorhynchus tereticollis (Rudolphi, 1809)		+		
Nematoda				
Camallanus truncatus (Rudolphi, 1814)	+	+	zooplankton (copepods)	50%
Molnaria intestinalis (Dogiel et Bychowsky, 1934)		+		
Philometroides sanguinea (Rudolphi, 1819)	+			
Total	13	28		48.8%

Previous studies reported a rich parasite fauna (283 parasite species, including protists, myxosporeans, monogeneans and parasitic arthropods) in fishes from the Tisa River and its tributaries. Žitňan (1966a, b, 1967) published a synopsis of the fish helminth taxa from the Slovak part of the Tisa River and listed 53 species of intestinal helminths found in a total amount of 497 fish hosts belonging to 10 families and 44 species. A more extensive parasitological survey was carried out by Ergens et al. (1975) on the Tisa River and its tributaries in Ukraine and Hungary. This study, which also included the results of Žitňan (1966a, b, 1967) from Slovakia, listed 103 gastrointestinal helminths, namely 45 trematodes, 24 tapeworms, 7 acanthocephalans and 27 nematode species.

Immediately after the ecological disaster, Sályi et al. (2000) examined 23 fish species in the Hungarian part of the Tisa River. They found 8 parasite species, but only four endoparasitic helminths: the digenean *Azygia lucii* (Müller, 1776), tapeworms *Amphilina foliacea* (Rudolphi, 1819), *Triaenophorus nodulosus*, and the nematode *Schulmanella petruschewskii* (Shulman, 1948). At the beginning of our surveys in 2002, we found seven intestinal helminths belonging to only two systematic groups: *A. huronensis*, *Caryophyllaeus fimbriceps*, *C. laticeps*, *K. rossittensis*, *Proteocephalus* larvae (Cestoda), and *Camallanus truncatus* and larvae of *Raphidascaris acus* (Nematoda) (Oros, 2002). No trematodes and acanthocephalans were found at that time.

Recent investigations in the Slovak part of the Tisa River carried out more than 6 years after the ecological accident did not show a substantial increase of the helminth species richness in the river in comparison to the former studies. Thirteen helminths found by the present authors represent only about one-fourth of the number of helminths listed by Žitňan (1966a, b, 1967) at the same sampling sites. The most remarkable change in the current helminth species diversity of the Tisa River is significant decline in the number of trematodes and total absence of acanthocephalans.

Water invertebrates, such as aquatic molluscs (gastropods and clams) and amphipods, are involved in complex life cycles of digeneans and acanthocephalans, respectively, as their intermediate hosts and thus play an important role in the formation of fish helminth communities. Many of them are highly sensitive to water pollution (Morley, 2007) and are widely used as biological indicators for the assessment of toxicity of water sediments (van Damme et al., 1984; Sheehan et al., 1984; Long et al., 2001). It is, therefore, very probable that the absence of acanthocephalans, which were frequent during previous surveys of fish parasites in the Tisa River (Žitňan, 1967; Ergens et al., 1975), is related to decline of population density or even disappearance of their intermediate hosts (gammarids) from this environment.

The digenean fauna of fishes from the Tisa River is also excessively species-poor. Only two adult digeneans, *Asymphylodora imitans* (Monorchiidae) and *Nicolla testiobliqua* (Opecoelidae), and metacercariae of two species, *C. complanatum* (Clinostomidae) and *P. cuticola* (Diplostomidae), were found. According to the literature (Našincová & Scholz, 1994; Dias et al., 2003; Ondračková et al., 2004), all these digeneans develop through pulmonate snails (Pulmonata) such as *Planorbis*, *Lymnaea* and *Planorbarius*, which are known to be rather tolerant to water pollution (Elder & Collins, 1991; Lefcort et al., 2004). Other groups of digeneans that use prosobranch gastropods (*Bithynia*) or clams (Bivalvia) as intermediate hosts, were not found in Tisa during the current survey. These molluscs are considered to be more sensitive to water quality and pollution (Mouthon & Charvet, 1999).

Table 5 Prevalence (above) and intensity of infection (below - mean with range in parentheses) of gastrointestinal helminths from the Tisa River. New host records for the territory of Slovakia are indicated with asterisks (*)

Helminth species	1	2	3	4	5	6	7	8	9	10	11	12	13
A. imitans													2.1%* / 1
N. testiobliqua											1/1 / 5		
C. complanatum				57.9%* / 5.9 (1-27)				1/1* / 22	29.6% / 3.7 (1-7)				
P. cuticola	22.2% / 1.5 (1-2)						6% / 4.5 (1-15)						
A. huronensis		52.9% / 16.6 (1-60)					1.5% / 1						
C. fennica	22.2% / 2				25% / 20.5 (18-23)	1.2% / 1	23.9% / 3 (1-9)						
K. rossittensis			4.5% / 2.2 (1-10)										
K. sinensis		5.8% / 1											
C. brachycollis					12.5%* / 2								
C. laticeps		11.8% / 1			12.5% / 8	29.5% / 5.8 (1-43)							
P. percae												60% / 14.7 (1-120)	2.1%* / 1
C. truncatus										5% / 1		12% / 4.3 (2-8)	36.7% / 5.4 (1-46)
P. sanguinea			1.5%* / 3.3 (1-5)										

Fish hosts: 1 – *S. cephalus*; 2 – *C. carpio carpio*; 3 – *C. auratus auratus*; 4 – *R. sericeus*; 5 – *A. sapa*; 6 – *A. brama*; 7 – *R. rutilus*; 8 – *M. fossilis*; 9 – *C. taenia*; 10 – *E. lucius*; 11 – *Z. zingel*; 12 – *P. fluviatilis*; 13 – *S. lucioperca*.

On the other hand, fish from the Tisa River harboured a relatively high number of caryophyllidean tapeworms (6 species), which use freshwater oligochaetes (Tubificidae) as intermediate hosts (Mackiewicz, 1972). Oligochaetes are generally considered as pollution-tolerant aquatic organisms (Howmiller & Scott, 1977; Winner et al., 1980) and evidently successfully survived the impact of cyanides and heavy metal pollution in the river.

Comparison of Fish Helminth Communities in Rivers with Different Level of Water Pollution

A total of 24 and 28 fish species were collected and examined from the Tisa and Latorica rivers, respectively. Sterlet (*A. ruthenus*), ziege (*P. cultratus*) and weatherfish (*M. fossilis*) were found only in the Tisa River, whereas ide (*Leuciscus idus*), tench (*Tinca tinca*), sneep (*Chondrostoma nasus*), grass carp (*Ctenopharyngodon idella*), silver carp (*Hypophthalmichthys molitrix*), wels catfish (*Silurus glanis*), and Volga pike-perch (*Sander volgensis*) were caught only in the Latorica River. A total of 21 fish species occurred in both localities (Table 3). Comparison of the qualitative composition of fish communities from the two localities showed a high degree of similarity (ICS = 81%).

Altogether 28 species of gastrointestinal helminths belonging to Trematoda (10), Cestoda (13), Acanthocephala (3), and Nematoda (2) were found in the Latorica River (Table 4). Autogenic parasites dominated, with 26 species found, whereas *C. complanatum* and *P. cuticola* larvae represented only two allogenic species. Similarly as in the Tisa River, generalists were most abundant. One trematode, *Allocreadium markewitschi*, and two acanthocephalans, *Acanthocephalus anguillae* and *Acanthocephalus lucii*, were found in five different hosts. Specialists with just one fish host were caryophyllidean tapeworms *Khawia rossittensis*, *K. sinensis* and *Paracaryophyllaeus gotoi* occurring in goldfish, common carp and spined loach, respectively, as well as another tapeworm, *Glanitaenia osculata*, found only in wels catfish (Table 6). Dominant component communities comprised one to three helminth taxa, with the maximum of nine helminth species found in common bream (*A. brama*). As many as 58 host-parasite combinations were recorded in the Latorica River (Table 6); fourteen of them (26%) represent new host records from Slovakia. Six fish species (*Ameiurus nebulosus*, *Lepomis gibbosus*, *Pelecus cultratus*, *Rhodeus sericeus*, *Sander volgensis*, and *Tinca tinca*) were free of intestinal helminths.

The fish helminth diversity was considerably different in the two rivers. The component communities in the formerly heavily polluted Tisa River were species-poor and consisted of only 13 helminths occurring in 26 host-parasite combinations with the dominance of one to two helminth species (69%) in one fish host. No acanthocepahalans and only four trematodes were found at this river. Core species, with the prevalence exceeding 60%, were not observed and only three host-parasite combinations (12%) involved secondary species with the prevalence ranging between 53–60%, namely tapeworms *Proteocephalus percae* in perch (*Perca fluviatilis*) and *A. huronensis* in common carp, and metacercariae of *C. complanatum* in Amur bitterling (*Rhodeus sericeus*) (Table 5). All other helminths were satellite or rare parasites occurring only occasionally. The caryophyllidean tapeworm *Caryophyllaeides fennica* was found in four fish hosts, whereas other endohelminths occurred in fewer fish hosts; multi-species infections appeared sporadically (Table 5).

Table 6 Prevalence (above) and intensity of infection (below - mean with range in parentheses) of gastrointestinal helminthes from the Latorica River. New host records for the territory of Slovakia are indicated with asterisk (*)

Helminth species	1	2	3	4	5	6	7	8	9	10	11	12	13	14	15	16	17	18	19	20	21	22
A. limacoides									6.6% 26 (2-50)													
A. imitans								8% 55	10% 11 (1-28)	29.4% 14.3 (1-52)												
A. markewitschi			1/1 36		28.6%* 33 (1-65)																	
O. siluri																						0.3 1
P. incognitus				8.7% 9 (4-14)					3.3% 3		4.3% 31											
All. markewitschi	1%* 1			4.4%* 50	28.6%* 9 (2-16)		2%* 11					33.3% 8										
N. skrjabini									3.3% 5													
N. testiobliqua									3.3%* 6									1/1 44				
S. bramae								8.3% 5														
P. cuticola				4.4% 5						17.7% 3.8 (1-15)	8.7% >20		50%* 1	100%* 24.5 (1-48)								
A. perccotti																						43. 2.5
A. huronensis						33.3% 7.5 (1-13)				3.9% 1												
C. fennica				13% 1					6.7% 2.5 (1-4)	2% 1	21.7% 1.8 (1-3)											
K. rossittensis							2%* 12															
K. sinensis						25% 10 (2-25)																

Fish hosts: 1 – *A. alburnus*; 2 – *A. aspius*; 3 – *S. erythrophthalmus*; 4 – *S. cephalus*; 5 – *L. idus*; 6 – *C. carpio carpio*; 7 – *C. auratus auratus*; 8 – *B. bjoerkna*; 9 – *A. sapa*; 10 – *A. brama*; 11 – *R. rutilus*; 12 – *C. nasus*; 13 – *C. idella*; 14 – *H. molitrix*; 15 – *C. taenia*; 16 – *A. melas*; 17 – *S. glanis*; 18 – *E. lucius*; 19 – *Z. zingel*; 20 – *P. fluviatilis*; 21 – *S. lucioperca*; 22 – *P. glenii*.

Table 6 Prevalence (above) and intensity of infection (below - mean with range in parentheses) of gastrointestinal helminthes from the Latorica River. New host records for the territory of Slovakia are indicated with asterisks (*) (continued)

Helminth species	1	2	3	4	5	6	7	8	9	10	11	12	13	14	15	16	17	18	19	20	21	22
C. fimbriceps										9.8% / 1.6 (1-2)												
C. laticeps										17.7% / 4.3 (1-18)												
P. gotoi									3.3% / 3													
P. gobii															16.7% / 1							0.3% / 1
G. osculata																	1/1 / 2					
P. percae																				66.7% / 2.5 (2-3)		
P. torulosus			25% / 1																			
T. nodulosus																		71.4% / 3.8 (2-8)				
A. anguillae					14.3% / 2	8.3% / 1		8.3% / 1		5.9% / 1.7 (1-3)						9.1%* / 2						
A. lucii					14.3% / 1		2.0%* / 1									9.1%* / 1		14.3% / 1		33.3% / 5		
P. tereticollis				4.4% / 5		8.3% / 3				2% / 2												
C. truncatus																			1/1 / 1		33.3% / 22 (2-44)	
M. intestinalis										2% / 3												

Fish hosts: 1 – *A. alburnus*; 2 – *A. aspius*; 3 – *S. erythrophthalmus*; 4 – *S. cephalus*; 5 – *L. idus*; 6 – *C. carpio carpio*; 7 – *C. auratus auratus*; 8 – *B. bjoerkna*; 9 – *A. sapa*; 10 – *A. brama*; 11 – *R. rutilus*; 12 – *C. nasus*; 13 – *C. idella*, 14 – *H. molitrix*; 15 – *C. taenia*; 16 – *A. melas*; 17 – *S. glanis*; 18 – *E. lucius*, 19 – *Z. zingel*; 20 – *P. fluviatilis*; 21 – *S. lucioperca*; 22 – *P. glenii*.

In the less polluted Latorica River, which was not contaminated with high cyanide and heavy metal concentrations during the ecological accident on the Tisa River basin in 2000, the fish helminth community consisted of 28 species found in as many as 58 host-parasite associations. The tapeworms *Triaenophorus nodulosus* in northern pike (*Esox lucius*; prevalence 71%) and *P. percae* in perch (67%) represented core species, i.e. with prevalence above 60%. The secondary species were represented by the digenean *Sphaerostomum bramae* in grass carp (50%) and the tapeworm *Nippotaenia mogurndae* in the Chinese sleeper (*P. glenii*; 43%). All other recorded host-parasite associations involved satellite or rare parasite species (Table 6). The digenean *Allocreadium markewitschi* and acanthocephalans *A. anguillae* and *A. lucii* parasitized five fish hosts. Common bream harboured the maximum of nine species of endohelminths (Table 6).

Based on the Czekanowski-Sørensen similarity index (ICS = 48.8%) and the percentage similarity index (PI = 19.5%), the helminth communities were qualitatively and quantitatively considerably different in the two rivers. Even though the fish community structure were relatively similar (ICS = 81%) and the number of fish examined was approximately the same (676 and 640), the helminth species richness and the diversity of host-parasite combinations were two times higher in the Latorica River. The helminth communities were also much more abundant and, in contrast to the Tisa River, species of all principal groups of helminths were present.

New Geographic and Host Records of Parasites

Among 31 gastrointestinal helminths found from 2002 to 2006 in the Tisa and Latorica Rivers, the four following species were reported from Slovakia for the first time (Oros, 2002; Oros & Hanzelová, 2009): the trematode *Aspidogaster limacoides* found in the white-eye bream (*A. sapa*), acanthocephalan *Pomphorhynchus tereticollis* in European chub (*Squalius cephalus*) and common carp (*C. carpio carpio*), and tapeworms *Atractolytocestus huronensis* and *Khawia sinensis* in wild populations of common carp (Figure 2).

The aspidogastrean *A. limacoides* was described by Diesing (1835) from European chub in Austria. It has been reported from a broad host spectrum of fish hosts (21 fish – Evlanov, 1990; Reimer, 2002; Schludermann et al., 2005). This relatively infrequent parasite can mature also in bivalves (*Adacna*, *Cardium*, *Dreissena*, *Pisidium*, and *Sphaerium*) (Nagibina & Timoveeva, 1971).

The acanthocephalan *P. tereticollis* is a parasite of a wide range of fish hosts occurring in fresh and brackish waters of Europe. It often occurs sympatrically with a morphologically similar and more common congeneric species, *Pomphorhynchus laevis* (M. Špakulová – personal communication) from which it was differentiated by molecular methods (Bombarová et al., 2007). During our study, this species was found in three fish hosts in the Latorica River, namely in common carp, common bream and European chub.

As many as 20 new host-parasite records for the territory of Slovakia were registered, more than two thirds of them from the Latorica River (Tables 5 and 6; see above) (Oros & Hanzelová, 2009).

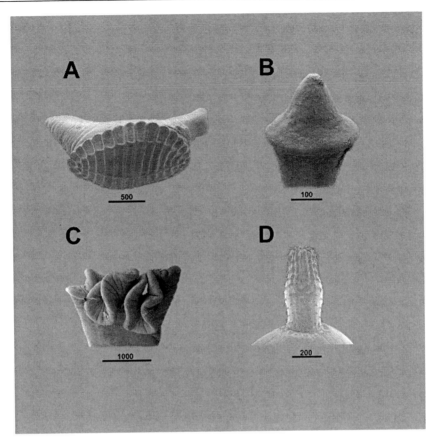

Figure 2. New species of the Slovak fauna. SEM microphotogaphs. A – *Aspidogaster limacoides* (ventral view); B – *Atractolytocestus huronensis* (scolex); C – *Khawia sinensis* (scolex); D – *Pomphorhynchus tereticollis* (detail of proboscis).

Newly Introduced Parasite Species

Many parasites have a good ability to colonize new regions and there is a high number of helminth species that have been introduced with their hosts to new environments (Kennedy, 1994). The most successful colonizers were found among parasites of economically important fish hosts, such as salmonids and carp, because they are most frequently transported to new areas, often without adequate veterinary inspection of imported stocks. Successful colonization depends on numerous factors, many of them being related to the life-cycle and ecology of a parasite (Kennedy, 1994). A low specificity to final host and availability of potential intermediate hosts can also facilitate successful colonization of parasites.

Common carp is probably the most widely spread fish species cultured on almost all continents under a wide range of geographic, climatic and technological conditions. This fish host, an extraordinarily rich parasite fauna with as many as 226 parasite species was reported (Landsberg, 1989). Many of these parasites, including serious pathogens, have been disseminated to new regions (Jeney & Jeney, 1995). Among tapeworms, *Bothriocephalus acheilognathi*, *Khawia sinensis*, and *Ligula intestinalis* (Linnaeus, 1758) (Diphyllobothriidea;

apparently misidentification of *Digramma interrupta*, see Dubinina 1980) were listed as the most important (Jeney & Jeney, 1995).

Two tapeworms parasitic in carp, *A. huronensis* and *K. sinensis*, which are not native to Slovakia have been found just recently in the Tisa River basin (Oros et al., 2004, 2009; Oros & Hanzelová, 2009). Since parasites may become pathogenic for their fish hosts when introduced to new areas, attention should be paid to these colonizers to prevent possible outbreaks of parasitic diseases, as it happened in Norway when the monogenean *Gyrodactylus salaris* Malmberg, 1957 was introduced from Sweden and caused massive mortalities of Atlantic salmon (*Salmo salar*) (Sterud et al., 2002) or in Central Europe after introduction of parasites of eel (*Anguilla anguilla*), such as the nematode *Anguillicoloides crassus* and monogeneans *Pseudodactylogyrus anguillae* Gussev, 1965 and *P. bini* Gussev, 1965 (Kennedy, 2007).

Atractolytocestus huronensis

A. huronensis (Figures 2 and 3) was predominantly found in common carp, but it was recorded also in other cyprinids (common bream and roach), which may represent accidental hosts (Tables 5 and 6). Prevalence in principal host ranged from 33% to 57% and intensity of infection varied between 1 and 27 specimens per fish.

The tapeworm possesses the typical afossate bulboacuminate scolex (Figures 2 and 3; see also Figure 1A in Oros et al., 2010), vitelline follicles cortical, uninterrupted along the ovarian lobes, and separate gonopores opening into a shallow common genital atrium. The tapeworms had 14 to 20 testes that begin far posterior to the scolex and always posterior to the first vitelline follicles (Figure 3).

Comparative measurements of tapeworms from feral and cultured carp from Slovakia, Czech Republic, Hungary and USA, together with published data (Anthony, 1958; Majoros et al., 2003), are summarized in Table 7. Tapeworms from Slovakia most correspond to American specimens. Slovak specimens have from 9 to 20 testes, which well corresponds to the number of testes in *A. huronensis* from USA (6–18 testes) reported by Anthony (1958) (Table 7). On the other hand, worms from Hungary were rather different, possessing only 3 to 5 testes or, allegedly, even none in some specimens (Majoros et al., 2003). However, a study of a few *A. huronensis* individuals from fishpond carp from another locality in Hungary has revealed that the actual number of testes in Hungarian material may be higher than that reported by Majoros et al. (2003). Although the testes were difficult to count, 6 to 10 testes were found in five specimens (Table 7).

Extremely low number of testes may be related to the fact that *A. huronensis* was described to have triploid and parthenogenetic populations (Jones & Mackiewicz, 1969; Kráľová-Hromadová et al., 2010) and it is also typified by the presence of ITS paralogues in its genome (Kráľová-Hromadová et al., 2010). However, it is not yet known whether these genetic and karyological peculiarities actually represent some advantage for *A. huronensis* in colonization of new regions and competing with other intestinal parasites of carp, including another caryophyllidean cestode, *K. sinensis* (see below).

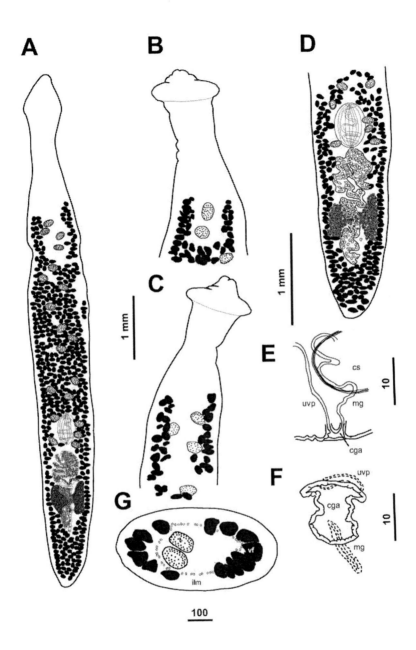

Figure 3. *Atractolytocestus huronensis* from river carp (*Cyprinus carpio carpio*) in Slovakia. A – total view; B, C – anterior part of the body; D – posterior part of the body; E – lateral view and F – frontal view of terminal genitalia; G – arrangement of testes and vitellaria in cross section. Abbreviations: cs = cirrus sac; cga = common genital atrium; ilm = inner longitudinal musculature; mg = male gonopore; t = testes; uvp = utero-vaginal pore; vf = vitelline follicles.

The presence of *A. huronensis* in common carp from the Tisa River at the border region between Slovakia, Ukraine and Hungary is important from an epizootiological point of view. The Tisa River drains large parts of Romania, Ukraine, Slovakia and Hungary, and enters the Danube River as a main tributary within the territory of Serbia. The occurrence of the tapeworm in wild fish populations in this river thus may facilitate its current dispersion. The most recent finding of *A. huronensis* was reported from Croatia (Gjurcevic et al., 2009).

Khawia sinensis

Despite a long history of colonization of Asia, Europe and North America with *K. sinensis* summarized above (see Table 2), including the findings of this parasite in the western part of the former Czechoslovakia in 1965 (now the Czech Republic, Přibyslavský et al., 1965), *K. sinensis* has not been reported from Slovakia up to the present. However, during the most recent monitoring of fish parasites in Slovakia, *K. sinensis* was found many times in common carp from both the Tisa and Latorica rivers, with prevalence ranging from 6% to 53% and intensity of infection from 1 to 39 tapeworms per fish (Oros & Hanzelová, 2009).

The absence of previous records of *K. sinensis* in Slovakia is difficult to explain because this parasite was very common in almost all neighbouring countries (Table 2). It is possible that it was not detected during routine examinations of cultured carp, even if the tapeworm is difficult to overlook even with the naked eye due to its considerable size. Another explanation could be its misidentification with the caryophyllidean *Caryophyllaeus fimbriceps*, which occurred frequently in common carp in Europe in the past and, like *K. sinensis*, was also considered to be a serious pathogen of this fish (Bauer et al., 1973). However, no *K. sinensis* specimens were found among 62 preparations of *C. fimbriceps* tapeworms from Slovakia (Oros et al., 2009).

The specimens of *K. sinensis* from Slovakia (Figure 4) were phenotypically rather uniform. They possessed an afossate, festoon-like scolex with deep folds on its anterior margin (Figure 4; see also Figure 1Q in Oros et al., 2010); the morphology of the genital system was identical to that reported by previous authors, e.g., Scholz (1989). Measurements of tapeworms from Slovakia are summarized in Table 8 and they are compared with published data from Europe and USA.

Reports on *K. sinensis* published at the beginning of the 1970s pointed out that this parasite might represent serious veterinary problem in fisheries, with negative impact on the health and productivity of common carp. Khawiosis was listed among helminth infections of epizootiological importance with a potentially heavy impact on commercial pond fisheries (Bauer et al., 1973; Körting, 1975). Some histopathological studies demonstrated that *K. sinensis* can cause serious inflammations of the host intestine, mechanical destructions of its intestinal epithelium and negative changes of biochemical indices (decrease of erythrocyte number and leukocyte activity) (Jara & Szerow, 1981; Morley & Hoole, 1995, 1997). Between 1960 and 1970, cases of mortality in carp fry were reported (Kulakovskaya, 1963; Bauer et al., 1973; Odening, 1989).

Table 7 *Atractolytocestus huronensis*. Characteristics of the cestode from different geographical regions. Measurements in μm unless otherwise stated

Host	Slovakia		Czech Republic		Hungary		USA
	Present data Common carp (n=52)	Present data Farmed carp (n=15)	Present data Farmed carp (n=16)	Majoros et al. (2003) Farmed carp (n=5)	Present data Farmed carp (n=5)	Anthony (1958)[1] Common carp (n=10)	Present data Common carp (n=11)
Body length (mm)	5.5-12.0	5.0-8.0	4.0-9.0	3.0-9.0	3.0-7.5	5.0-18.0	4.0-8.5
Body width (mm)	0.6-1.2	0.5-0.8	0.5-0.9	0.5-1.2	0.4-0.6	0.9-2.0	0.6-1.5
Scolex width	484-974	516-741	415-938	-	342-497	-	414-900
Distance between first vitelline follicles to anterior extremity (mm)	1.3-2.3	0.8-1.1	0.9-1.9	-	0.8-1.8	-	0.8-1.6
Distance between first testes to anterior extremity (mm)	1.4-2.6	1.1-1.5	1.0-2.6	-	1.0-2.2	-	1.1-2.6
Distance between first vitelline follicles and first testes	135-579	250-384	213-703	-	156-407	-	120-999
Testis number	14-20	16-20	9-14	0-5	6-10	6-18	12-17
Cirrus-sac length	405-780	506-617	350-555	350-450	232-417	256-457	320-416
Cirrus-sac width	260-456	299-374	214-347		210-276	-	205-264
Ovary width	470-720	421-574	297-482	600-900	237-440	-	266-885
Ovarian arm length	360-1024	446-703	365-691	800-1100	203-466	670-920	167-847
Ovarian arm width	128-240	134-210	117-172	-	83-147	-	69-361
Extent of uterus in relation to length of testicular area	1/3-1/2	1/3-1/2	1/3-1/2	-	1/4-1/3	1/2	1/3-1/2
Egg length	47-65	48-52	46-67	-	47-50	46-57	40-52
Egg width	28-39	35-36	30-39	-	28-31	30-37	26-35
Prevalence (%)	56	67	38				
Mean intensity of infection (range)	16.0 (1-85)	4.6 (1-12)	10.4 (1-28)				

[1] Original description

Table 8 *Khawia sinensis*. Characteristics of the cestode from cultured and feral carp. Measurements in µm unless otherwise stated

Host	*Cyprinus carpio*	*Cyprinus carpio*	*Cyprinus carpio*	*Cyprinus carpio*	*Cyprinus carpio carpio*
Country	Ukraine	Czech Republic	UK	USA	Slovakia
Author	Kulakovskaya and Krotas	Scholz	Chubb and Yeomans	Williams and Sutherland	Present data
Year	1961	1989	1995	1981	2009
Water body	Fish ponds	Reservoirs	Fish ponds	Reservoirs	Reservoirs and rivers
Body length (mm)	45–83	35–115	27.3–55	62–69	41–112
Body width (mm)	2.5–3	2.1–3.7	2.1–3.7	1.1–1.2	1.3–2.3
Scolex width (mm)		2.7–5.2	1.1–1.7		2–2.4
First testes[a] (mm)		0.1–4	0.4–2.7		0.1–2.4
Testis size	160–190 × 140–180	150–370 × 140–330	110–290 × 90–180		128–163 × 159–179
Cirrus-sac size (mm)	0.7–0.8	1.0–1.9 × 0.7–1.3	0.6–0.8 × 0.5–0.9	0.5 × 0.7	0.5–0.7 × 0.7–1
Ovary shape	H-shaped	H-shaped	H-shaped, but lateral arms arched inwards posteriorly		H-shaped
Ovary length (mm)	3.2–3.4	2.2–5.2	1.4–2.9		2.2–4.1
Ovary width (mm)		1.6–2.9	0.9–2		1.1–2.1
Vitelline follicle size	128–135	110–250 × 90–210	40–70 × 110–180		93–108 × 116–133
First vitellaria[b] (mm)		3–11	1.6–3.7		1.9–6.3
Size of intrauterine eggs	42–48 × 25–27	30–54 × 26–46	38–49 × 22–30		26–23 × 45–41

[a] Distance from first (anteriormost) vitelline follicles; [b] Distance from anterior extremity of scolex.

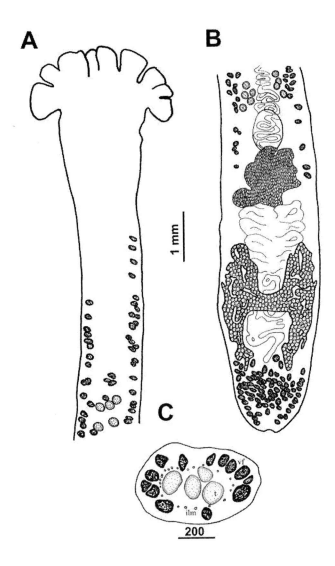

Figure 4. *Khawia sinensis* from river carp (*Cyprinus carpio carpio*) in Slovakia. A – scolex; B – posterior part of the body; C – cross-section. Scale bars: A and B = 1 mm; C = 200 μm. Abbreviations: ilm = inner longitudinal musculature; t = testes; vf = vitelline follicles.

On the contrary, other studies indicated that the veterinary and economic significance of *K. sinensis* may be lower (Weirowski, 1979) and that its pathogenic effect on its fish host may be negligible even in heavy infections (Chubb & Yeomans, 1995). The findings of this tapeworm in eastern Slovakia correspond to the latter observations because neither pathological changes such as enteritis were found in the site of the attachment of tapeworms, nor occlusion of the intestinal lumen or poor physiological state of infected fish were observed during the 6-year study (Oros et al., 2009).

By 1986, *K. sinensis* was registered in the following countries: Bosnia and Herzegovina, Bulgaria, Byelorussia, China, Czech Republic, Germany, Great Britain, Hungary, Japan, Kazakhstan, Latvia, Lithuania, Poland, Romania, Russia, Slovakia, Ukraine and USA (Figure 5; Table 2). The parasite was particularly well established and common in Russia and Central Europe (Přibyslavský et al., 1965; Kulow, 1973; Pańczyk & Żelazny, 1974; Molnár & Buza, 1975; Protasova et al., 1990). The current decline of literature records on *K. sinensis* (Figure 6) may be related to the fact that veterinary importance of this cestode diminished rather soon after it had colonized new areas of Europe, partly also due to successful control and effective treatment with piperazine salt of niclosamide (Muzykovskii et al., 1971; Schrenkenbach, 1975; Weirowski, 1979; Żelazny & Pańczyk, 1984). The subsidence of khawiosis in Europe and gradual disappearance of its causative agents may also be a result of the recent introduction *A. huronensis* to Central Europe and Great Britain (Majoros et al., 2003; Oros et al., 2004; Kappe et al., 2006). Both cestodes have a similar life-cycle that includes aquatic annelids (genera *Ilyodrilus*, *Limnodrilus*, *Psammoryctes* and *Tubifex*) as intermediate hosts (Mackiewicz, 1982; Protasova et al., 1990) and they differ only slightly from each other in the spectrum of definitive hosts. Besides their principal host, common carp, *K. sinensis* has also been recorded from grass carp (*C. idella*), black carp (*Mylopharyngodon piceus* (Richardson, 1846)) and, probably accidentally, from *Hemibarbus barbus* (Temminck & Schlegel, 1846) in Japan (Rădulescu & Georgescu, 1966; Dubinina, 1971; Demshin & Dvoryadkin, 1981; Protasova et al., 1990; Scholz et al., 2001), whereas *A. huronensis* occurs only in common carp (Oros et al., 2004). Recent records of the tapeworm from common bream and roach in Slovakia are considered to be accidental (Oros & Hanzelová, 2009).

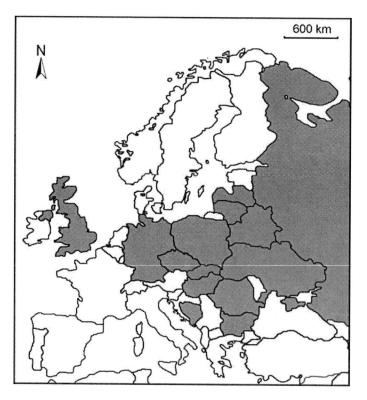

Figure 5. Distribution of *Khawia sinensis* in Europe. States with records of the tapeworm are shaded.

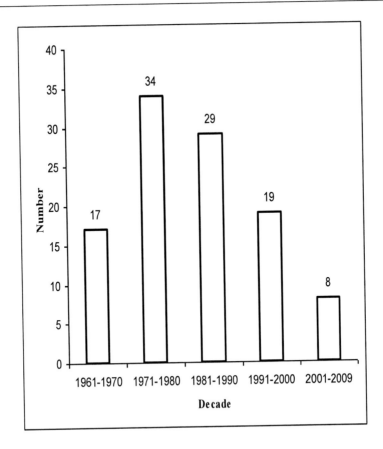

Figure 6. Published records on *Khawia sinensis*.

To date, no precise data on interspecific competition between *K. sinensis* and *A. huronensis* are available, but multiple infections of carp with cestodes are not very common (unpublished data). However, gradual disappearance of *K. sinensis* is indicated by observations of one of the authors of this chapter (T.S.) in South Bohemia, where *K. sinensis* was the only caryophyllidean cestode found in carp from fishponds in the late 1980s and early 1990s (see papers by Scholz et al., 1990; Scholz, 1991a, b). After *A. huronensis* appeared in carp from this region in the early 2000s, *K. sinensis* became much less common than in the previous decades in the Czech Republic (T. Scholz, unpublished data). Similar situation was observed in England and Wales (C. Williams, personal communication).

Similarly to *A. huronensis*, the finding of *K. sinensis* in Slovakia (Oros & Hanzelová, 2009; Oros et al., 2009) is important from an epizootiological point of view because it is the first report of the parasite in wild populations of carp in two rivers of the Tisa River drainage system. This river and its tributaries drain a large area of eastern Europe (Ukraine, Romania, Slovakia, Hungary and Serbia (Figure 1) and thus another dissemination of this parasite of carp to new regions cannot be ruled out. It is obvious that further monitoring of the occurrence of *K. sinensis* is necessary to assess its potential to survive and disseminate in free waters.

Nippotaenia mogurndae

As a consequence of expansion of Chinese sleeper, *Perccottus glenii* (Perciformes: Odontobutidae), another alien tapeworm, *Nippotaenia mogurndae*, has lately emerged in the Latorica River (Košuthová et al., 2004, 2008). The host of this tapeworm, Chinese sleeper, is a freshwater predatory perciform fish of minor commercial importance but adverse ecological impact after its introduction to new territories (Koščo et al., 2003). Recently, Chinese sleeper is rapidly spreading from its original territory in northeastern Korea and the Amur River basin to Central Europe. During the last decade it colonized Poland, Latvia, Hungary, Romania, Serbia and Bulgaria (Antychowicz, 1994; Harka, 1998; Plikašs and Aleksejevs, 1998; Nalbant et al., 2004; Šipoš et al., 2004; Jurajda et al., 2006). Kautman (1999) first recorded this fish species from the Latorica River, Slovakia in 1998 and *N. mogurndae* was found in the same locality a few years later (Košuthová et al., 2004). Currently, the parasite is widely distributed in eastern Slovakia. Košuthová et al. (2008, 2009) found this tapeworm in numerous sites of the Bodrog, Laborec and Latorica rivers (all tributaries of the Tisa River), but the parasite has not yet been reported from the Tisa River itself (Oros & Hanzelová, 2009; Košuthová et al., 2009).

N. mogurndae is a small-sized tapeworm characterized by hyperapolytic proglottides, which detach from the anterior part of the body with the scolex and proliferation zone while immature and eggs are formed in individual detached proglottides (Hine, 1977; Bray, 1994). This parasite, however, is not likely to represent a major risk to potential native freshwater fish hosts because it is strictly specific to Chinese sleeper.

CONCLUSION

It has been documented that fish helminth diversity is reduced and the structure of helminth communities is considerably changed in polluted or otherwise artificially affected environments (Marcogliese and Cone, 1997; MacKenzie, 1999). On the other hand, it has been also showed in a variety of helminth assemblages that improvement of ecosystem health in restored sites is followed by increase of prevalence and species richness of parasites (Marcogliese 2005). However, only subtle positive changes in the parasite community structure have been observed recently in the Tisa River compared with the situation immediately after ecological catastrophe in 2000. This indicates that the river still suffers from cyanide and heavy metal load and that the restoration process at this locality is evidently only at the beginning. This chapter documents the detrimental and long-term effect of chemical pollution on the communities of endoparasitic helminths with complex life cycles, which occur in freshwater fish in the Tisa River. Rapid dissemination of non-indigenous fish parasite species (*A. huronensis, K. sinensis* and *N. mogurndae*) in this area indicates their good adaptability and ability to colonize new regions. It remains to be studied whether or not this rapid colonization of new parasites was facilitated by deterioration of the ecosystems in the Tisa River, the fish of which now harbour depauperate communities of native intestinal helminths. The transboundary location of the Tisa River represents a potential danger because of possibility of the further rapid dissemination of these parasites, potential pathogens of farmed fish, to other regions throughout the Danube River basin.

ACKNOWLEDGMENTS

Thanks are due to Dr. S. L. Gardner, University of Nebraska State Museum, Lincoln and Dr. P. Pillit, US National Parasite Collection, Beltsville, Maryland, USA, and Dr. M. Fulín, East-Slovakian Museum, Košice, Slovakia for the loan of museum specimens; Dr. Cs. Székely and Prof. K. Molnár, Veterinary Medical Research Institute, Budapest, Hungary, for collecting *Atractolytocestus huronensis* tapeworms from Hungary. The authors are grateful to C. Williams, Environment Agency, Brampton, UK, for providing unpublished data on *K. sinensis* and *A. huronensis*; Dr. M. Špakulová, Dr. Ľ. Turčeková and Bc. Ľ. Burik, Parasitological Institute SAS, Košice for their invaluable help in field collections of fish helminths; Ing. Blanka Škoríková and Martina Borovková, Institute of Parasitology, Biology Centre of the AS CR, České Budějovice, Czech Republic for technical assistance and Ing. N. Rozdobudková, Slovak Enterprise of Water Management Košice, for providing data on heavy metal load of the investigated rivers. T.S. and M.O. are grateful for the financial support to the National Science Foundation, USA (PBI award Nos. 0818696 and 0818823). This study was supported by the Grant Agency VEGA (project No. 2/0080/10), Slovak Research and Development Agency (projects Nos. APVV-LPP-0151-07, SK-BG-0031-08, LPP-0171-09), Institute of Parasitology (project Nos. Z60220518 and LC 522) and Grant Agency of the Czech Republic (project No. 524/08/0885).

REFERENCES

Akhmetova, B. (1966). Epizootiology of *Khawia sinensis* of carp on the Alma-Ata fish farms, *Scientific & Production Conference on the Control of Diseases of Fish in Kazakhstan and Republics of Central Asia Alma-Ata*, 15-19. (In Russian)

Amin, O. M. (1986). Caryophyllaeidae (Cestoda) from lake fishes in Wisconsin with a description of *Isoglaridacris multivitellaria* sp. n. from *Erimyzon sucetta* (Catostomidae). *Proceedings of the Helminthological Society of Washington*, 53, 48-58.

Amin, O. M. & Minckley, W. L. (1996). Parasites of some fish introduced into an Arizona reservoir, with notes on introductions. *Journal of the Helminthological Society of Washington*, 63, 193-200.

Anthony, J. D. (1958). *Atractolytocestus huronensis* n. gen., n. sp. (Cestoda: Lytocestidae) with notes on its morphology. *Transaction of the American Microscopical Society*, 77, 383-390.

Antychowicz, P. (1994). *Perccottus glenii* in our water bodies. *Komunikaty Rybackie*, 2, 21-22. (In Polish)

Balon, K. (1964). A preliminary list of the Slovakian lampreys and fishes (Petromyzones et Teleostomi). *Biológia*, 19, 343-358.

Bauer, O. N. (1987). Key to the parasites of freshwater fishes of the USSR. Leningrad: Nauka. (In Russian)

Bauer, O. N., Musselius, V. A. & Strelkov, Y. A. (1973). Diseases of pond fishes. Jerusalem: Israel Programme for Scientific. (In Russian)

Bombarová, M., Marec, F., Nguyen, P. & Špakulová, M. (2007). Divergent location of ribosomal genes in chromosomes of fish thorny-headed worms, *Pomphorhynchus laevis* and *Pomphorhynchus tereticollis* (Acanthocephala). *Genetica*, *131*, 141-149.

Bray, R. A. (1994). Order Nippotaeniidea Yamaguti, 1939. In L. F., Khalil, A. Jones, & R. A. Bray, (Eds.), *Keys to the Cestode Parasites of Vertebrates*, (253-255). Wallingford: CAB International.

Brower, J., Zar, J. & von Ende, C. (1990). *Field and laboratory methods for general ecology.* Brown, Iowa: Dubuque.

Bush, A. O., Lafferty, K. D., Lotz, J. M. & Shostak, A. W. (1997). Parasitology meets ecology on its own terms: Margolis et al. revised. *Journal of Parasitology*, *83*, 575-83.

Chubb, J. C. & Yeomans, W. E. (1995). *Khawia sinensis* Hsü, 1935 (Cestoda: Caryophyllidea), a tapeworm new to the British Isles: a treat to carp fisheries? *Fisheries Management and Ecology*, *2*, 263-277.

Chubb, J. C., Kirk, R. & Wellby, I. (1996). Caryophyllaeid tapeworm *Atractolytocestus huronensis* Anthony, 1958 (= *Markevitschia sagittata* Kulakovskaya et Akhmerov, 1965) in carp *Cyprinus carpio* L. in British Isles—another translocation? *Abstracts of the Spring Meeting of the British Society of Parasitology*, University of Wales, Bangor, *66*.

Cunningham, S. A. (2005). Incident, accident, catastrophe: cyanide on the Danube. *Disasters*, *29*, 99-128.

van Damme, D., Heip, C. & Willems, K. A. (1984). Influence of pollution on harpacticoid copepods of two North Sea estuaries. *Hydrobiologia*, *112*, 143-160.

Demshin, N. I. & Dvoryadkin, V. A. (1981). The development of *Markevitschia sagittata* (Cestoidea: Caryophyllidae), a parasite of the Amur wild carp, in the external conditions and intermediate host. *Parazitologiya*, *15*, 113-117. (In Russian)

Dias, M. L. G. G., Eiras, J. C., Machado, M. H., Souza, G. T. R. & Pavanelli, G. C. (2003). The life cycle of *Clinostomum complanatum* Rudolphi, 1814 (Digenea, Clinostomidae) on the floodplain of the high Paraná river, Brazil. *Parasitology Research*, *89*, 506-508.

Diesing, C. M. (1835). *Aspidogaster limacoides*, eine neue Art Binnenwurm. *Medizinische Jahrbücher des k.k. österreichischen Staates*, *7*, 420-431.

Dubinina, M. N. (1971). Cestodes from fishes of the Amur River basin. *Parasitologicheskii Sbornik*, *255*, 77-119. (In Russian)

Dubinina, M. N. (1980). Importance of attachment organs for phylogeny of tapeworms. *Parazitologiya*, *29*, 65-83. (In Russian)

Elder, J. F. & Collins, J. J. (1991). Freshwater molluscs as indicators of bioavailability and toxicity of metals in surface-water systems. *Reviews of Environmenatal Contamination & Toxicology*, *122*, 37-79.

Elliott, M. (2003). Biological pollutants and biological pollution—an increasing cause for concern. *Marine Pollution Bulletin*, *46*, 275-280.

Ergens, R. & Lom, J. (1970). *Causative agents of parasitic diseases of fish*. Praha: Academia. (In Czech)

Ergens, R., Gussev, V. A., Izyumova, N. A. & Molnár, K. (1975). *Parasite fauna of fishes of the Tisa River Basin*. Praha: Academia.

Evlanov, I. A. (1990). Distribution of the trematode *Aspidogaster limacoides* in a population of roach (*Rutilus rutilus*) as a function of the age and sex of the host. *Zoologicheskii Zhurnal*, *69*, 132-134. (In Russian)

Froese, R. & Pauly, D. (2009). FishBase. World Wide Web electronic publication, www.fishbase.org, version (11/2009)

Georgiev, B., Biserkov, V. & Genov, T. (1986). The staining method for cestodes with iron acetocarmine. *Helminthologia, 23*, 279-281.

Gibson, D. I., Jones, A. & Bray, R. A. (2002). *Keys to the Trematoda. Vol. 1*. Wallingford: CAB International.

Gjurcevic, E., Bambir, S. & Beck, A. (2009). *Atractolytocestus huronensis* Anthony, 1958 from farmed common carp in Croatia. Abstract Book, 14th EAFP International Conference, *Diseases of Fish and Shellfish, Prague*, Czech Republic, 235.

Harka, Á. (1998). New fish species in the fauna of Hungary. *Perccottus glenii* Dybowski, 1877. *Halászat, 91*, 32-33. (In Hungarian)

Hine, P. M. (1977). New species of *Nippotaenia* and *Amurotaenia* (Cestoda: Nippotaeniidae) from New Zealand freshwater fishes. *Journal of the Royal Society of New Zealand, 7*, 143-155.

Hoffman, G. L. (1999). *Parasites of North American freshwater fishes*. 2nd ed. Ithaca, NY: Comstock Publishing Associates.

Holmes, J. C. & Price, P. W. (1986). Communities of parasites. In D. J. Anderson, & J. Kikkawa, (Eds.), *Community Biology: Patterns and Processes*, (186-213). Oxford: Blackwell Scientific Publications.

Howmiller, R. P. & Scott, M. A. (1977). Environmental index based on relative abundance of oligochaete species. *Journal of the Water Pollution Control Federation, 49*, 809-815.

Hsü, H. F. (1935). Contribution à l'étude des cestodes de Chine. *Revue Suisse de Zoologie, 42*, 485-492.

Jara, Z. & Szerow, D. (1981). Histopathological changes and localisation of tapeworm *Khawia sinensis* Hsü, 1935 in the intestine of carps (*Cyprinus carpio* L.). *Wiadomości Parazytologiczne, 27*, 695-703.

Jeney, Z. & Jeney, G. (1995). Recent achievements in studies on diseases of common carp (*Cyprinus carpio* L.). *Aquaculture, 129*, 397-420.

Jones, A. W. & Mackiewicz, J. S. (1969). Naturally occurring triploidy and parthenogenesis in *Atractolytocestus huronensis* Anthony (Cestoidea: Caryophyllidea) from *Cyprinus carpio* L. in North America. *Journal of Parasitology, 55*, 1105-1118.

Jones, A., Bray, R. A. & Gibson, D. I. (2005). *Keys to the Trematoda. Vol. II*. Wallingford: CAB International.

Jurajda, P., Vassilev, M., Polačik, M. & Trichková, T. (2006). A first record of *Perccottus glenii* (Perciformes: Odontobutidae) in Danube River in Bulgaria. *Acta Zoologica Bulgarica, 58*, 279-282.

Kakacheva-Avramova, D., Menkova, I. & Karanikolov, I. (1980). *Khawia* in carp in Bulgaria, *Ribnoe Stopanstvo, 27*, 17-18 (In Bulgarian)

Kappe, A., Seifert, T., El-Nobi, G. & Brauer, G. (2006). Occurrence of *Atractolytocestus huronensis* (Cestoda, Caryophyllaeidae) in German pond-farmed common carp *Cyprinus carpio*. *Diseases of Aquatic Organisms, 70*, 255-259.

Kautman, J. (1999). *Perccottus glenii* Dybowski, 1877 from East Slovakian water bodies. *Protected areas of Slovakia, 40*, 20-22. (In Slovak)

Kennedy, C. R. (1994). The ecology of introductions. In A. W. Pike, & J. W. Lewis, (Eds.), *Parasitic Diseases of Fish* (189-209). Tresaith, UK: Samara.

Kennedy, C. R. (2007). The pathogenic helminth parasites of eels. *Journal of Fish Diseases, 30*, 319-334.

Khalil, L. F., Jones, A. & Bray R.A. (1994). *Keys to the Cestode Parasites of Vertebrates*. Wallingford: CAB International.

Kirk, R. S., Veltkamp, C. J. & Chubb, J. C. (2003). Identification of *Atractolytocetus huronensis* (Caryophyllidea: Lytocestidae) from carp (*Cyprinus carpio*) using histological and ashing techniques. *Abstracts of the Spring Meeting of the British Society of Parasitology*, Manchester, 45-46.

Kiškároly, M. (1977). Study of the parasite fauna of freshwater fishes from fish ponds of Bosnia and Herzegovina. I - Cyprinid fish ponds. B. Cestodes 1 (systematics and morphology). *Veterinaria Yugoslavia, 26*, 477-483.

Körting, W. (1975). Aspekte zum Bandwurmbefall der Fische. Die Bedeutung der Parasiten für die Produktion von Süsswasserfischen. *Fisch Umwelt, 1*, 81-87.

Koščo, J., Lusk, S., Halačka, K. & Lusková, W. (2003). The expansion and occurrence of Amur sleeper (*Perccottus glenii*) in eastern Slovakia. *Folia Zoologica, 52*, 329-336.

Košuthová, L., Letková, V., Koščo, J. & Košuth, P. (2004). First record of *Nippotaenia mogurndae* Yamaguti and Miyata, 1940 (Cestoda: Nippotaeniidea), a parasite of *Perccottus glenii* Dybowski, 1877, from Europe. *Helminthologia, 41*, 54-56.

Košuthová, L., Koščo, J., Miklisová, D., Letková, V., Košuth, P. & Manko, P. (2008). New data on an exotic *Nippotaenia mogurndae* (Cestoda), newly introduced to Europe. *Helminthologia, 45*, 81-85.

Košuthová, L., Koščo, J., Letková, V., Košuth, P. & Manko, P. (2009). New records of endoparasitic helminths in alien invasive fishes from the Carpathian region. *Biológia, 64*, 776-780.

Kráľová-Hromadová, I., Štefka, J., Špakulová, M., Orosová, M., Bombarová, M., Hanzelová, V., Bazsalovicsová, E. & Scholz T. (2010). Intraindividual ITS1 and ITS2 ribosomal sequence variation linked with multiple rDNA loci: a case of triploid *Atractolytocestus huronensis*, the monozoic cestode of common carp. *International Journal of Parasitology, 40*,175-181.

Kulakovskaya, O. P. (1963). On the biology and life cycle of the tapeworm *Khawia sinensis* Hsü, 1935. *Problemy Parazitologii, 2*, 200-205. (In Russian)

Kulakovskaya, O. P. & Krotas, R. A. (1961). About the parasite *Khawia sinensis* Hsü, 1935 (Caryophyllaeidae, Cestoda) introduced to carp ponds of western regions of USSR from the Far-East. *Doklady Akademii Nauk SSSR, 137*, 1253-1255. (In Russian)

Kulow, H. (1973). Sowjetische Erfahrungen über die Bothriocefalosis, Khawiosis und Philometrosis. *Zeitschrift für die Binnenfischerei der DDR, 20*, 263-268.

Landsberg, J. H. (1989). Parasites and associated diseases in fish in warm water aquaculture, with special emphasis on intensification. In M. Shilo, & S. Sarig, (Eds.), *Fish Culture in Warm Water Systems: Problems and Trends*, (195-252), Cleveland, OH: CRC Press.

Lefcort, H., Abbott, D. P., Cleary, D. A., Howell, E., Keller, N. C. & Smith, M. (2004). Aquatic snails from mining sites have evolved to detect and avoid heavy metals. *Archives of Environmental Contamination and Toxicology, 46*, 478-484.

Long, E. R., Hong, C. B. & Severn, C. G. (2001). Relationships between acute sediment toxicity in laboratory tests and abundance and diversity of benthic fauna in marine sediments: a review. *Environmental Toxickology and Chemistry, 20*, 46-60.

MacKenzie, K. (1999). Parasites as pollution indicators in marine ecosystems: a proposed early warning system. *Marine Pollution Bulletin, 38*, 955-959.

Mackiewicz, J. S. (1972). Caryophyllidea (Cestoidea): a review. *Experimental Parasitology, 34*, 417-512.

Mackiewicz, J. S. (1982). Caryophyllidea (Cestoidea): perspectives. *Parasitology, 84*, 397-417.

Magurran, A. E. (1991). *Ecological diversity and its measurement*. London, UK: Chapman and Hall.

Majoros, G., Csaba, G. & Molnár, K. (2003). Occurrence of *Atractolytocestus huronensis* Anthony, 1958 (Cestoda: Caryophyllaeidae), in Hungarian pond-farmed common carp. *Bulletin of the European Association of Fish Pathologists, 23*, 167-175.

Marcogliese, D. J. (2005). Parasites of the superorganism: are they indicators of ecosystem health? *International Journal of Parasitology, 35*, 705-716.

Marcogliese, D. J. & Cone, D. K. (1997). Parasite communities as indicators of ecosystem stress. *Parassitologia, 39*, 227-232.

Molnár, K. & Buza, L. (1975) Infection of carps with the tapeworm *Khawia sinensis* Hsü, 1935 in Hungary. *Halászat, 3*, 24-124. (In Hungarian)

Moravec, F. (1994). *Parasitic nematodes of freshwater fishes of Europe*. Prague: Academia.

Moravec, F. (2001). *Checklist of the metazoan parasites of fishes of the Czech Republic and Slovak Republic* (1873–2000). Prague: Academia.

Morley, N. J. (2007). Anthropogenic effects of reservoir construction on the parasite fauna of aquatic wildlife. *EcoHealth, 4*, 374-383.

Morley, N. J. & Hoole, D. (1995). Ultrastructural studies on the host-parasite interface between *Khawia sinensis* (Cestoda: Caryophyllidea) and carp *Cyprinus carpio*. *Diseases of Aquatic Organisms, 23*, 93-99.

Morley, N. J. & Hoole, D. (1997). The in vitro effect of *Khawia sinensis* on leukocyte activity in carp (*Cyprinus carpio*). *Journal of Helminthology, 71*, 47-52.

Mouthon, J. & Charvet, S. (1999). Compared sensitivity of species, genera and families of Molluscs to biodegradable pollution. *International Journal of Limnology, 35*, 31-39.

Musselius, V., Ivanova, N., Laptev, V. & Apazidi, L. (1963). *Khawia sinensis* Hsü, 1935 of carp. *Ribovodstvo i Ribolovstvo, 3*, 25-27. (In Russian)

Muzykovskii, A. M., Sapozhnikov, G. I. & Nazarova, N. S. (1971). Treatment of carp for *Khawia* infection. *Veterinaria Moscow, 48*, 75-76. (In Russian)

Nagibina, L. F. & Timoveeva, T. A. (1971). True hosts of *Aspidogaster limacoides* Diesing, 1834 (Trematoda, Aspidogastrea). *Doklady Biologicheskikh Nauk, 200*, 677-678. (In Russian).

Nakajima K. & Egusa, K. (1978). Notes of *Khawia sinensis* Hsü found in cultured carp. *Fish Pathology, 12*, 261-263.

Nalbant, T., Battes, K., Pricope, F. & Ureche, D. (2004). First record of the Amur sleeper *Perccottus glenii* (Pisces: Perciformes: Odontobutidae) in Romania. *Travaux du Museum national d'Histoire Naturelle, "Grigore Antipa", 47*, 279-284.

Našincová, V. & Scholz, T. (1994). The life cycle of *Asymphylodora tincae* (Modeer 1790) (Trematoda: Monorchiidae): a unique development in monorchiid trematodes. *Parasitology Research, 80*, 192-197.

Odening, K. (1989). New trends in parasitic infections of cultured freshwater fish. *Veterinary Parasitology, 32*, 73-100.

Ondračková, M., Šimková, A., Gelnar, M. & Jurajda, P. (2004). *Posthodiplostomum cuticola* (Digenea: Diplostomatidae) in intermediate fish host: factors contributing to the parasite infection and prey selection by the definitive bird host. *Parasitology, 129*, 761-770.

Oros, M. (2002). Parasite fauna of fishes in the Slovak part of the Tisa River. *Parasitological Institute, SAS, Košice* (MSc. Thesis).

Oros, M. & Hanzelová, V. (2009). Re-establishment of the fish parasite fauna in the Tisa River system (Slovakia) after a catastrophic pollution event. *Parasitology Research, 104*, 1497-1506.

Oros, M., Hanzelová, V. & Scholz, T. (2004). The cestode *Atractolytocetus huronensis* (Caryophyllidea) continues to spread in Europe: new data on the helminth parasite of common carp. *Diseases of Aquatic Organisms, 62*, 115-119.

Oros, M., Hanzelová, V. & Scholz, T. (2009). Tapeworm *Khawia sinensis*: review of the introduction and subsequent decline of a pathogen of carp, *Cyprinus carpio*. *Veterinary Parasitology, 164*, 117-122.

Oros, M., Scholz, T., Hanzelová, V. & Mackiewicz, J. S. (2010). Scolex morphology of monozoic cestodes (Caryophyllidea) from the Palaearctic Region: a useful tool for species identification. *Folia Parasitologica, 57* (In press)

Pańczyk, J. & Żelazny, J. (1974). *Khawia sinensis* and *Bothriocephalus gowkongensis* infections in carp – new parasitic diseases found in Poland. *Gospodarka Rybna, 26*, 10-13.

Plikšs, M. & Aleksejevs, E. (1998). *Fish of Latvia*. Zivis Riga: Gandrs.

Protasova, E. P., Kuperman, B. I., Roitman, V. A. & Poddubnaya, L. G. (1990). *Caryophyllid Tapeworms of the Fauna of the USSR*. Moscow: Nauka. (In Russian)

Přibyslavský, J., Jílek, J. & Lucký, Z. (1965). Die Helminthen der Fische des Stauses Kníničská, neu für die Fauna ČSSR. *Věstník Československé Společnosti Zoologické, 29*, 5-8. (In Czech)

Reimer, L. W. (2002). *Aspidogaster limacoides* – ein Neozoe aus einer Plötze der mittleren Weser. *Fischer and Teichwirt, 1*, 10-11.

Rădulescu, I. & Georgescu, R. (1966). New contributions to the knowledge of the parasite fauna of the species *Ctenopharyngodon idella* during its acclimatization in Romania. *Buletinul Institutuli de Cercetari Proiectari Piscicole, 25*, 48-51.

Sályi, G., Csaba, G., Gaálné, D. E., Orosz, E., Láng, M., Majoros, G., Kunsági, Z. & Niklcsz, C. (2000). Effect of the cyanide and heavy metal pollution passed in river Szamos and Tisza on the aquatic flora and fauna with special regard to the fish. *Magyar Állatorvosok Lapja, 122*, 493-500. (In Hungarian)

Schludermann, C., Laimgruber, S., Konečný, R. & Schabuss, M. (2005). *Aspidogaster limacoides* Diesing, 1835 (Trematoda, Aspidogastridae): a new parasite of *Barbus barbus* (L.) (Pisces, Cyprinidae) in Austria. *Annales der Naturhistorisches Museum Wien, 106*, 141-144.

Schmidt, G. D. (1986). *Handbook of tapeworm identification*. Florida: CRC Press, Boca Raton.

Scholz, T. (1989). Amphilinida and Cestoda, parasites of fish in Czechoslovakia. *Acta Scientarium Naturalium Academiae Scientarium Bohemoslovacae Brno, 23*, 1-56.

Scholz, T. (1991a). Early development of *Khawia sinensis* Hsü, 1935 (Cestoda: Caryophyllidea). *Folia Parasitologica, 38*, 133-142.

Scholz, T. (1991b). Development of *Khawia sinensis* Hsü, 1935 (Cestoda: Caryophyllidea) in the definitive host. *Folia Parasitologica, 38*, 225-234.

Scholz, T. & Hanzelová, V. (1998). *Tapeworms of the genus* Proteocephalus *Weinland, 1858 (Cestoda: Proteocephalidae) parasites of fishes in Europe*. Prague: Academia, Studie AV ČR.

Scholz, T., Shimazu, T., Olson, P. D. & Nagasawa, K. (2001). Caryophyllidean tapeworms (Platyhelminthes: Eucestoda) from freshwater fishes in Japan. *Folia Parasitologica, 48*, 275-288.

Scholz, T., Špeta, V. & Zajíček, J. (1990). Life history of the tapeworm Khawia sinensis Hsü, 1935, a carp parasite, in the pond Dražský Skaličany near Blatná, Czechoslovakia. *Acta Veterinaria Brno, 59*, 51-63.

Schrenkenbach, K. (1975). Research problem in the control of fish diseases to ensure and improve fish production. *Monatshefte der Veterinärmedizin, 30*, 699-703.

Sheehan, P. J., Miller, D. R., Butler, G. C. & Bordeau, P. (1984). *Effects of pollutants at the ecosystem level*. Chichester: Wiley.

Sterud, E., Mo, T. A., Collins, C. M. & Cunningham, C. O. (2002). The use of host specificity, pathogenicity, and molecular markers to differentiate between *Gyrodactylus salaris* Malmberg, 1957 and *G. thymalli* Zitnan, 1960 (Monogenea: Gyrodactylidae). *Parasitology, 124*, 203-213.

Sures, B. (2004). Environmental parasitology: relevancy of parasites in monitoring environmental pollution. *Trends in Parasitology, 20*, 170-177.

Šipoš, Š., Miljanović, B. & Pelčić, L. J. (2004). The first record of Amur sleeper (*Perccottus glenii* Dybowski, 1877, fam. Odontobutidae) in Danube River. *International Association for Dunube Research, 35*, 509-510.

Weirowski, F. (1979). Die wirtschaftliche Bedeutung und Verbreitung von *Khawia sinensis* Hsü, 1935 in der Karpfenproduktion der DDR. *Zeitschrift für die Binnenfischerei der DDR, 26*, 373-376.

Williams, D. D. & Sutherland, D. R. (1981). *Khawia sinensis* (Caryophyllidea: Lytocestidae) from *Cyprinus carpio* in North America. *Proceedings of the Helminthological Society of Washington, 48*, 253-255.

Winner, R. W., Boesel, M. W. & Farrell, M. P. (1980). Insect community structure as an index of heavy-metal pollution in lotic ecosystems. *Canadian Journal of Fisheries and Aquatic Sciences, 37*, 647-655.

Yeomans, W. E., Chubb, J. C. & Sweeting, R. A. (1997). *Khawia sinensis* (Cestoda: Caryophyllidea) – an indicator of legislative failure to protect freshwater habitats in British Isles? *Journal of Fish Biology, 51*, 880-885.

Żelazny J. & Pańczyk, J. (1984). Anthelmintic effect of some niclosamide preparations against *Khawia sinensis* Hsü, 1935 in carp. *Medycyna Weterynaryjna, 40*, 365-368. (In Polish)

Žitňan, R. (1966a). Monogenoidea of fishes from the flowing waters of the Tisa River lowlands (Helminthes of fishes of the Tisa River lowlands (Czechoslovakia) I). *Biológia, 21*, 681-692.

Žitňan, R. (1966b). Cestoidea and Trematoidea of fishes from the flowing waters of the Tisa River lowlands (Helminthes of fishes of the Tisa River lowlands (Czechoslovakia) II). *Biológia, 21*, 763-770.

Žitňan, R. (1967). Nematoda, Acanthocephala and Hirudina of fishes from the flowing waters of the Tisa River lowlands (Helminthes of fishes of the Tisa River lowlands (Czechoslovakia) III). *Biológia, 22*, 381-385.

Žitňan, R. (1982). Current helminthoses of farmed carp. *Zprávy Československé Společnosti Parazitologické, 22*, 22-23. (In Slovak).

In: Species Diversity and Extinction
Editor: Geraldine H. Tepper, pp. 297-316
ISBN: 978-1-61668-343-6
© 2010 Nova Science Publishers, Inc

Chapter 9

FRESHWATER ENDEMICS IN PERIL: A CASE STUDY OF SPECIES *ECHINOGAMMARUS CARI* (AMPHIPODA: GAMMARIDAE) THREATENED BY DAMMING

Krešimir Žganec[1]*, Sanja Gottstein[1], Nina Jeran[2], Petra Đurič[3] and Sandra Hudina[1]

[1]Department of Biology, University of Zagreb, Rooseveltov trg 6, 10000 Zagreb, Croatia
[2] Držićeva 6, 10000 Zagreb, Croatia
[3]Zelena Akcija/ Friends of the Earth Croatia, Frankopanska 1, 10000 Zagreb, Croatia

ABSTRACT

Freshwaters, especially running waters, have been impacted globally by suite of pressures among which pollution, overexploitation, physical alternation and damming, water abstraction and introduction of non-native species have caused the most severe degradations. Damming of rivers and creation of impoundments have been the most important causes of habitat loss and hydrological alternation in running waters. Moreover, as many freshwater organisms have restricted geographical distributions, any modification of their habitats can have a great impact on species and cause reduction or complete loss of specific local biodiversity.

Echinogammarus cari (S. Karaman 1931) is Croatian endemic species whose presently known distribution is restricted to only 25 km of watercourse length in the upper canyon part of the Gojačka Dobra River and its two tributaries. After the completion of a 52.5 m high dam at the end of the canyon part of the Gojačka Dobra in 2010, about 60% of presently known distribution area of *E. cari* will be lost and the species will become endangered with great probability of extinction.

This study was conducted to determine the extent of *E. cari* distribution in the Dobra River and its tributaries. Also, some aspects of its ecology like its microdistribution and relationship to physicochemical parameters were examined. Hydrological and physicochemical conditions were examined in detail in order to establish the pre-damming state in this river system. Accordingly, impacts of the damming are predicted and possible

---

* Corresponding author: (e-mail: kreso@biol.pmf.hr)

conservation measures for the species are proposed and discussed. We argue that studies of biodiversity in freshwaters should be more focused on endemic species as they are the most likely to be threatened by damaging activities. This fact can be used as the strong argument for the conservation of extremely endangered freshwater ecosystems.

INTRODUCTION

Freshwaters cover only a tiny fraction (0.01%) of the total surface of the globe. Small surface area of freshwater bodies, their insular nature and dependence on surrounding terrestrial ecosystems from which they drive their specific character, are important characteristics responsible for specific nature of these ecosystems (Strayer, 2006). However, although small in size, freshwaters have enormous importance for global biodiversity. It is estimated that 126 000 animal species inhabit or depend on freshwaters, which makes 9.5% of the total number of animal species recognized globally, although this figure significantly underestimates real animal diversity of freshwaters (Balian et al., 2008). Though freshwaters have fewer species than vastly bigger terrestrial and marine realms they stand out as systems with the highest species richness per unit area on the planet, given the small percentage of territory they cover. Moreover, insular nature and small size of freshwater habitats with limited ability of many freshwater species to disperse often accompanied with adaptation to narrow habitat conditions, have all contributed to high degree of endemism among freshwater taxa (Dudgeon et al., 2006). Unfortunately, these characteristics make freshwaters systems considerably more susceptible to huge and increasing human pressure which has often caused irreversible changes such as loss of species and the consequential loss of their structural and/or functional roles in freshwater communities.

Throughout human history freshwaters and especially running waters have had great importance for humankind. People lives were always closely dependent on freshwaters which have provided water for drinking, agriculture and industry, harvestable plants and animals, routes for navigation, means of waste removal and energy that was used for watermills or electric power generation in recent times. These services made freshwaters, especially running waters, magnets for human settlements and made them the most densely populated parts of landscapes. However, great importance of running waters for human subsistence turned out to their misfortune, as there are now only few river systems in the World left that are not significantly altered from the natural state by human activities (Revenga et al., 2005). Among many pressures to running waters and their rich biodiversity, land-use change, overexploitation, pollution, physical alternation and damming, water abstraction and introduction of non-native species have caused the most severe degradations (reviewed in Allan and Flecker, 1993; Allan, 1996; Malmqvist and Rundle, 2002; Dudgeon et al., 2006; Strayer, 2006). Today human pressure on freshwaters has risen to unprecedented level and it is about to increase even further with increasing human population. Moreover, multiple pressures to which almost every river is exposed make the running water probably the most impacted ecosystems of all.

Habitat destruction and alteration in running waters have to a great extent been caused by damming. Worldwide construction of large dams (either >15 m high or 5-15 m high and with an impoundment volume >3,000,000 m^3) and impoundments occurred largely during the twentieth century. Reservoir created after damming a river can be used for water supply, irrigation, flood control, navigation or generation of electricity, often having multiple purposes. Accordingly, impacts of dams on river systems can differ widely depending on their size, purpose and operation (Allan, 1996). Today, after more than a century of intensive damming, global state of the running waters is affected with >45 000 large dams (World Commission on Dams, 2000). Nilsson et al. (2005) analyzed 292 large river systems that drain more then a half (54%) of world's land area and showed that 65% of these river systems are fragmented by dams. Moreover, several million smaller, but ecologically significant dams have been built on streams and rivers across the globe (Strayer, 2006). The only remaining large free-flowing rivers in the world are found in tundra regions and in smaller coastal basins in Africa and Latin America (Revenga et al., 2005).

Dams transform naturally free-flowing and continuous river courses into river segments interrupted by impoundments. Numerous studies showed multiple negative effects that dams have on freshwater communities and ecosystem function (reviewed in Allan, 1996). Reservoirs created upstream of the dam replace natural free-flowing stream habitats, making conditions that are unsuitable for the majority of running-water species. Thus, most of the running-water species are extirpated after their habitat becomes flooded during the transformation of river segments into reservoirs. Furthermore, dams and impoundments, especially chain of dams, break the longitudinal connectivity of rivers by blocking migrations of diadromous fishes to their spawning habitats causing strong population declines of those species in affected rivers (Larinier, 2000). In the reaches below the dams the changes of physical conditions very often include great modifications of natural flow and temperature regime that few native species can tolerate (Bunn and Arthington, 2002). Due to all these adverse effects, dams and impoundments have been recognized as the most important cause of fragmentation and habitat loss in running waters (Strayer, 2006) and together with pollution and the introduction of non-native species they present the most important cause of threats to freshwater fauna (Richter at al., 1997).

Great impacts of human activities on freshwaters have led to one of probably the most severe biodiversity crisis in the history of life on Earth. According to some conservative estimates, around 20% of freshwater fish species are extinct or in serious decline (Moyle and Leidy, 1992), while around 10 000 species of freshwater invertebrates around the world may already be extinct or imperiled (Strayer, 2006). Although global extent of freshwater biodiversity crisis is very difficult to asses, scale of the problem can be realized from the status of aquatic biodiversity in USA where data for few groups are relatively well documented: 34% of fish species, 65% of crayfishes and 75% of unionid mussels are regarded as rare or extinct (Master, 1990). Extinction rate of North America's freshwater fauna is estimated to be five times higher than that of terrestrial fauna, with 123 freshwater animal species recorded as extinct since 1990 (Ricciardi and Rasmussen, 1999). These authors also predicted future extinction rates for North America's temperate freshwater fauna to be 4% per decade, which is comparable to the species loss in tropical forests. Moreover, according to Millennium Ecosystem Assessment (2005) freshwater biodiversity declined about 55% in the period between 1970 and 2002 while that of terrestrial and marine systems each declined ~32%. Thus, one more specific characteristic of freshwater systems can be added to the list; ongoing rapid depletion of its biodiversity due to unsustainable use by human species.

Interestingly, although more than a half of the world's river systems are affected by dam-induced habitat loss (Nilsson et al., 2005) biological studies on river systems before damming are very scarce (e.g. Cline and Ward, 1984; Stanford and Ward, 1984). This is mainly due to the fact that biological studies of rivers before damming in the past were not obligatory to dam constructors and therefore were considered to be only an unnecessary cost. That has resulted in chronic lack of pre-damming data that seriously hampers urgent work on mitigating negative impacts caused by damming (Malmqvist and Rundle, 2002). Numerous studies conducted after the damming showed that dams have caused severe population declines in or the extinction of many freshwater mussels, snail and fish species worldwide (e.g. Ikonen, 1984; McAllister et al., 2001; Sheldon and Walker, 1997; Vaughn and Taylor, 1999). However, in the case of endemic species which inhabit river system that are about to be dammed and for which no ecological data are available, only studies conducted before damming could provide understanding of their ecology and distribution needed for their conservation. Our study of the Croatian endemic species *Echinogammarus cari* (S. Karaman 1931) in the Dobra River that will be dammed in 2010, thus represents a rare case study of the species ecology and its distribution before a major disturbance event.

Endemic freshwater organisms are very vulnerable to habitat destruction and other anthropogenic modifications to running waters due to their restricted geographical distribution. Many species, especially among better known higher taxa of freshwater organisms, like fishes (Lévêque et al., 2008), molluscs (Bogan, 2008; Strong et al., 2008) and crustaceans (Crandall and Buhay, 2008; De Grave et al., 2008; Yeo et al., 2008; Väinölä et al., 2008) are known to have restricted distribution and many are narrow-ranged endemics. For example, the majority of the native European amphipod species of the genus *Echinogammarus* (35 or 66%) have restricted distribution with 20 species restricted to the Iberian Peninsula and the other 15 species scattered through Italy, Sicily, Sardinia, Greece and on the eastern Adriatic coast (Pinkster, 1993). Also, among 53 species of the genus *Gammarus* listed for Europe, Middle East and North Africa by Karaman and Pinkster (1977a, b, 1987), 41 or 77% are known only from few localities. Distribution and ecology of the majority of these restricted species are poorly studied. Many are known only from few localities or only from type locality and have not been documented in more detail than species description. The Croatian endemic species *E. cari* was originally described on the basis of specimens collected in the main source of the Bistrica stream (right tributary of the Gojačka Dobra), and until our recent study (Žganec and Gottstein, 2009) no information existed about its distribution and ecology (Karaman, 1931; Karaman, 1973; Pinkster, 1993). In our study we established the distribution and some aspects of ecology of this species in the drainage area of the Dobra River before damming and estimated that the species will lose about 60% of presently known area after the creation of the impoundment in 2010. In this chapter some further insights into physico-chemical factors that limit its distribution are given. Furthermore, hydrological and physico-chemical conditions are examined in detail in order to establish the pre-damming state of this river system. Accordingly, predictions of damming impacts and means for their mitigation are proposed. Potential habitats for the species *E. cari*, like springs and upper courses in surrounding watercourses were also examined because possible conservation measures for the species could include introduction of the species into those habitats.

STUDY AREA

The investigated area is located in the central Croatia within the Dinaric Mountain chain with the typical karst characteristics, being composed mainly of limestone and dolomite. This part of Croatia represents a transition zone between the deep and shallow karst belts. The hydrogeology of the area is characterized by sinking streams and rivers, underground flow, reappearing waters and irregularities between subterranean and surface watersheds (Bahun, 1968). The Gojačka Dobra and the streams Bistrica and Ribnjak are located in the area that consists mainly of pervious (limestone and dolomite) rocks where the ground water level is at 200-220 m a.s.l. These relationships create conditions that support permanent superficial flow of watercourses toward the north (Bahun, 1968).

The Dobra River (total length of 104 km) is a tributary of the Kupa River (Danube drainage area), and the second largest river in mid-west Croatia (Figure 1c). Its upper part (Gornja Dobra) begins after joining of several small streams in the mountain area of Gorski kotar and it flows in a SE direction to the town of Ogulin where it disappears in Đula sinkhole. Since 1959 waters of Gornja Dobra and small sinking river Zagorska Mrežnica (where impoundment Sabljaki was created) have been redirected through pipe-lines to the Gojak Power Plant and now only a small amount of water flows into the Đula sinkhole. Đula sinkhole and Medvedica cave are connected into the longest subterranean system in Croatia (total length of 16396 m) with rich and endemic underground biodiversity. However, pollution from the town Ogulin and mentioned water redirection strongly affected this subterranean ecosystem, and due to these reasons Karst Waters Institute (West Virginia, USA) included Ogulin area in the "Top ten list of endangered karst ecosystems" (Tronvig and Belson, 2000/2001). Water from the watershed of Gornja Dobra and surrounding area reappears about 4.5 km NW as a powerful spring at the beginning of an about 15 km-long canyon of the lower part of the Gojačka Dobra which flows in NE direction and after about 50 km joins the Kupa River near the town of Karlovac (Figure 1c). After redirection of water from the Zagorska Mrežnica and Gornja Dobra to Gojak Hydroelectric Power Plant (Gojak HPP) located at the source of the Gojačka Dobra, average annual discharge has been doubled and hydrology in this part of the Gojačka Dobra has been completely changed, causing at the same time drastic reduction of the discharge at the spring of Tounjčica River that had been receiving water from the sinking Zagorska Mrežnica River until 1959.

MATERIALS AND METHODS

Sampling Methods

Samples were collected at least once at 23 study sites in the drainage area of the Gornja Dobra and at 24 sites in the Gojačka Dobra and its tributaries in the period from 2005 to 2009. In the drainage area of the Gornja Dobra qualitative samples from various microhabitats were collected, while at 24 study sites in the Gojačka Dobra River and its tributaries, semiquantitative samples were collected by D-net (mesh size 200 μm). In the latter sampling procedure, approximately 3 min were spent at two dominant microhabitats: 1) the moss growing on falls or larger stones and 2) stony substrate, mainly pebble and cobble.

Quantitative samples on moss (sampling area 0.0064 m^2) and on stony substrate (0.1 m^2) at site D6 (Figure 1d) were collected in July 2007. Samples were preserved in 70% ethanol. Samples of amphipods at all other sites across whole Croatia were collected in the period from 1999-2009 during different studies using different (qualitative and quantitative) methods of sampling. Altogether, amphipod samples from 412 sites ranging from springs, small and medium streams to large rivers were analyzed and included in this study (Figure 1b). Study sites on watercourses in the surrounding area where the species *Echinogammarus cari* is distributed are more thoroughly analyzed in order to identify potential new habitats where the species could be introduced (Figure 1c).

During the sampling in the drainage area of the Gornja and Gojačka Dobra, five environmental factors were measured. Water velocity was measured with a SonTek/YSI FlowTracker. Temperature, dissolved oxygen, conductivity and pH were measured with WTW probes. These parameters were measured at least once at each study site. From June 2008 to May 2009 all mentioned parameters (except current velocity) were measured once a month at sites D2 and D5 (Figure 1d) while at site B2 the same parameters were measured once a month from October 2007 to September 2008. At other sites in Gojačka Dobra, Bistrica and Ribnjak streams parameters were measured 1-9 times during different months. Daily water temperature measurements in the period from 1964-1980 and measurements of water flow from 1948-2005 at the study site Trošmarija on the Gojačka Dobra (study site D4 in Figure 1d) were obtained from the Croatian Meteorological and Hydrological Service. Monthly measurements of all above mentioned parameters at the Lešće study site (D9 in Figure 1d) were obtained from Croatian Waters for the period from 2000 to 2006.

Collected fauna was separated in the laboratory from mineral particles, mosses and plant debris using saturated solution of calcium chloride (Hynes, 1954). Gammmarids below three millimeters of total body length (TBL) were not taken into account in this study. Relative abundance of *E. cari* was calculated as its proportion (number of specimens with TBL>3 mm) in the total number of amphipods in the sample. Total length of juvenile and adult specimens (n=530) of *E. cari* was measured under the stereoscopic microscope equipped with an ocular micrometer, from the tip of the rostrum to the tip of the telson after the straightening of the body with fine forceps. Adults were sexed and the number of eggs/juveniles per ovigerous female was determined.

Data Analyses

Differences between the relative abundance of gammarids at two examined microhabitats were tested using the nonparametric Mann-Whitney U-test, while differences in log-transformed densities were tested by t-test. In order to test potential difference in average conductivity between Bistrica stream study sites (4 sites) and all other sites in the Gojačka Dobra and the stream Ribnjak, the unpaired t-test was used. Changes of average, minimum and maximum yearly flow at the site Trošmarija in the Gojačka Dobra were tested with repeated-measures analysis of variance (RM ANOVA) for the period of 11 years (1948-1958) before and 11 years (1960-1970) after the Gojak HPP was built. Pearson's correlation coefficient was also used to determine the relationship between different parameters (temperature mean, maximum and range, conductivity and calcium ions concentration) and relative abundance of *E. cari*. Statistical software packages Statistica 7.1 (Systat Software Inc., Richmond CA, USA) and SPSS 10.0 (StatSoft Inc., Tulsa, USA) were used. Significant differences for all tests were accepted at the 5 % level.

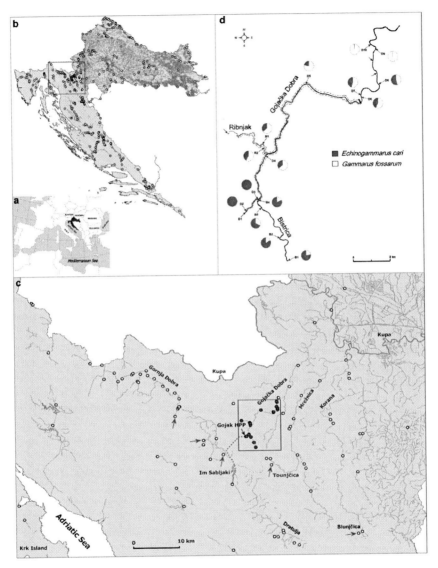

Figure 1. (a) Position of Croatia in Europe; (b) Croatia with all watercourses and all study sites (n=412), red circles represent the sites where *Echinogammarus cari* was recorded and empty circles all other study sites; (c) enlarged mid-west part of Croatia with all watercourses surrounding Gojačka Dobra and its two tributaries where *E. cari* occurs (red circles); dashed line shows the pipe-line between impoundment Sabljaki, Gornja Dobra and Gojak HPP (thinner dashed line represents dray canal of formerly sinking Zagorska Mrežnica River, that was connected to springs of Tounjčica River); red arrows represent potential habitats for the introduction of *E. cari*; (d) sampling study sites in the Gojačka Dobra, Bistrica and Ribnjak streams with relative abundance of the species *E. cari* and *Gammarus fossarum* at all sites except D1 where only *Niphargus* spp. where found. The dashed line represents the boundary of the impoundment that will be created after the dam construction in 2010 (smaller arrows represent sites where the species was found but for which relative abundance was not determined).

RESULTS AND DISCUSSION

Distribution in Relation to Environmental Factors

In the samples of amphipods collected from total of 412 sites distributed across the whole Croatia, the species *Echinogammarus cari* was recorded only at 17 study sites in the Gojačka Dobra River and its two tributaries, Bistrica and Ribnjak streams (Figure 1b, c, d). The species was not found at 23 sites in the Gornja Dobra River and its tributaries, where only *Gammarus fossarum* and *Niphargus* spp. (in springs) were present. Also, it was not recorded at any of sites in surrounding watercourses (Figure 1c). Only the species *Niphargus longiflagellum* S. Karaman and *N. licanus* S. Karaman were present in the Gojak spring (D1 in Figure 1d). At all study sites where it was recorded, *E. cari* was found in coexistence with *Gammarus fossarum* (Koch). The relative abundance of these two species varies longitudinally, from dominance of *E. cari* in the Bistrica and headwaters of the Gojačka Dobra River with relative abundance of 67.5-98.5% to higher relative abundance of *G. fossarum* in the Ribnjak and the downstream study sites D4-D10 in the Gojačka Dobra where relative abundance of *E. cari* was 21.9-45.1% (Figure 1d). The species was not found in the glide at study site D9, while on the fall at site D10, 500 m downstream from site D9, it had very low relative abundance (1.4%). At one site in the Globornica stream (longer right tributary of Gojačka Dobra in Figure 1c) and at four downstream sites in the Gojačka Dobra, only *G. fossarum* was found. The study site D10 represents the most downstream site where *E. cari* was recorded. Thus, *E. cari* is restricted to an about 15 km-long part of the Gojačka Dobra River and its tributaries, the streams Bistrica and Ribnjak, or the total of about 25 km of respective watercourses.

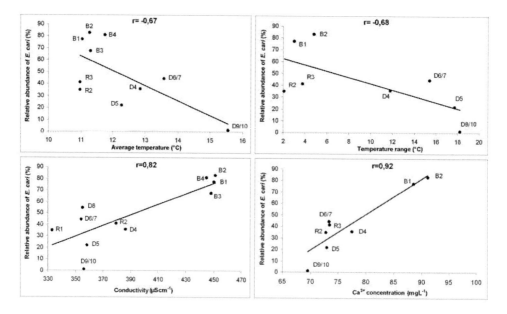

Figure 2. Relationship between (a) mean temperature, (b) temperature range, (c) conductivity, (d) calcium ion concentration and relative abundance of the species *Echinogamarus cari* at different study sites in the Gojačka Dobra and its tributaries with values of Pearson's correlation indices shown. Lines represent linear regression for these relationships.

Downstream decrease of relative abundance of *E. cari* is negatively correlated with increase of mean annual temperature and temperature range (Figure 2a and b), while at the same time it is positively correlated with decrease of conductivity and concentration of calcium ions (Figure 2c and d). Statistically significant negative correlation between relative abundance of *E. cari* and mean annual temperature was established (r= -0.67, p=0.03, n=10), while negative correlation between temperature range (r= -0.68, p=0.06, n=8) and maximum temperatures (r= -0.66, p=0.08, n=8) were not significant although p-values were very close to 0.05.

These findings indicate that restricted distribution of *E. cari* could be the adaptation to specific temperature conditions in the canyon of the Gojačka Dobra and its tributaries. In previous paper (Žganec and Gottstein, 2009) it was shown that this part of the Gojačka Dobra has characteristics of a 'summer-cool' river in which summer temperatures seldom exceed 20 °C. Measurements of water temperatures once a month from June 2008 to June 2009 at the sites D2, B2 and D5 confirm these findings. Thus, the most of canyon part of Gojačka Dobra, whole Bistrica and Ribnjak streams have characteristic of 'summer-cool' streams. The species *Gammarus fossarum* finds optimal thermal conditions in 'summer-cool' streams, while in the 'summer-warm' streams (summer temperatures exceed 20 °C) this species should be scarce or absent (Pöckl and Humpesch, 1990; Pöckl, 1993). This is in agreement with our results, which show that *G. fossarum* had lower abundance in the semiquantitative samples that were collected at the most downstream study sites (D9 and D10), while *E. cari* is a rare species here. During the summer, at site D9 temperature regularly exceeds 20 °C, so this part of the Gojačka Dobra has characteristics of a 'summer-warm' river. Also, it was clearly shown that downstream decreasing relative abundance of *E. cari* is in strong and statistically significant negative correlation with mean and maximum annual temperatures, as well with temperature ranges. Thus, higher summer and probably lower winter temperatures, i.e. longitudinally changing temperature regime, are important restricting factors for further downstream distribution of *E. cari* in the Gojačka Dobra.

It was also shown that downstream decrease of relative abundance of *E. cari* is positively correlated with decrease of conductivity and concentration of calcium ions (Figure 2c and d). Pearson's correlation coefficients for both relationships are positive and highly significant (conductivity: r=0.82, p=0.002, n=11; conc. Ca^{2+}: r=0.92, p=0.001, n=8). In the Bistrica stream the mean conductivity was significantly higher than at sites in the Gojačka Dobra and in the Ribnjak stream, which were grouped together (t-test, p<0.001). The highest conductivity was recorded in the Bistrac spring (B1) and the Gojačka Dobra spring (D1), and downstream decrease of conductivity was observed in both watercourses. Negative correlation between distance from the source and average conductivity in the Gojačka Dobra River was significant (r= -0.78, p=0.023), while in the Bistrica this correlation was -0.87 but it was not significant (p=0.128, n=4). The large travertine waterfall located at the mouth of the Bistrica stream, as well as the river bottom between sites D6 and D9 covered by about 10 centimeters tick layer of travertine, on which dense stands of mosses and macrophytes are growing, indicate that Gojačka Dobra is in this respect similar to nearby Korana and Mrežnica Rivers in which process of travertine forming is much more intense. Thus, downstream decrease of conductivity and calcium concentration along the river course could be explained by the process of calcium carbonate precipitation and formation of travertine at waterfalls and on the river bottom. Similar downstream decrease of conductivity caused by the process of travertine formation on waterfalls and barriers between lakes was established for the watercourses and lakes of the world famous National Park Plitvice Lakes located at headwaters of Korana River (Kempe and Emeis, 1985). Dissolved calcium is one of the most commonly reported chemical parameters limiting distribution of gammarids. Concentration of Ca^{2+} is usually highly correlated with conductivity (e.g., Glazier et al., 1992) and that was also established in our study (r=0.95, p<0.001, n=9). This explains strong positive relationship between conductivity and relative abundance of *E. cari* that we observed. Shrimpff and Foeckler (1985) established that *Gammarus fossarum* and *G. roeseli* (Gervais) can only survive in streams with fairly high concentration of Ca^{2+} and Mg^{2+} ions combined with fairly high pH values. Similarly, Glazier et al. (1992) established that *Gammarus minus* (Say) is absent from springs with low conductivity and pH values, while its density is positively correlated with calcium and magnesium hardness. Direct evidence has been provided that calcium is a limiting factor in the distribution of *Gammarus pseudolimnaeus* (Bousfield) (Zehmer at al., 2002) and *Gammarus lacustris* Sars (Rukke, 2002). Pinkster (1993) also reported that the majority of the freshwater species of the genus *Echinogammarus* are found in Ca^{2+} rich waters. Our results are in accordance with findings of those studies. Thus, downstream decrease of calcium ions concentration in combination with changing temperature regime restricts further downstream distribution of *E. cari* in the Gojačka Dobra, because this species is adapted to relatively colder temperature conditions in headwaters of Gojačka Dobra that also have higher concentration of calcium ions.

Microdistribution

In previous analysis (Žganec and Gottstein, 2009) it was shown that *E. cari* clearly prefers moss microhabitats where it had significantly higher average relative abundance (82.4%) than in stony substrate where *G. fossarum* dominated and *E. cari* had an average relative abundance of 29.4% (Mann-Whitney U-test: p<0.01). Density and relative abundance of both species were estimated with replicate sampling on moss and stony substrate (n=4) at site D6 (first waterfall downstream from the dam) in July 2007 (Figure 3). Both species had much higher density in moss then on stony substrate and these differences were significant (t-test of log transformed density, p<0.05). However, the density of *E. cari* in moss was about six times higher then the density of *G. fossarum*, while the density of *G. fossarum* was about 37 times higher on stony substrate. Accordingly, relative abundance of *E. cari* was statistically significantly higher in moss (t-test, p=0.002, Figure 3c), while vice versa was established for relative abundance of *G. fossarum* on stony substrate (t-test, p<0.001, Figure 3d).

E. cari is smaller species then *G. fossarum*, with the ranges of total body length of adults 4.1-8.4 and 5.5-14.0, respectively. Thus, the largest adults of *E. cari* are as big as medium sized adults of *G. fossarum*. The adults of larger species *G. fossarum* probably have a smaller density and relative abundance in moss due to smaller space available between mosses and probably inadequate food supply present there. On the other hand it is possible that *E. cari* prefers moss microhabitats because it has specialized for feeding on particulate organic matter that becomes trapped in mosses. It was shown that predation between different species of gammarids, i.e. intraguild predation can strongly affect their microdistribion (MacNeil et al., 1999; MacNeil and Prenter, 2000) and that these predation can even be reciprocal with moulting females being the most sensitive to predation by larger males of sintopic gammarid species (Dick, 2008). Thus, in our case it is possible that the larger *G. fossarum* confines the smaller *E. cari* to the moss microhabitat by predation, but also that this predatory impact is reciprocal, causing much higher density and relative abundance of *E. cari* in moss microhabitats.

Hydrological Conditions and Impact of Gojak HPP

Redirection of water from the Sabljaci impoundment at the Zagorska Mrežnica River and from the Gornja Dobra to the Gojak HPP drastically increased inflow of water to the Gojačka Dobra (Figure 4). The average annual flow at the study site Trošmarija was 13.9 m^3s^{-1} (range 8.5-18.9 m^3s^{-1}) in the period from 1948-1958, before the Gojak HPP was built. After that, in the period from 1959-2005, the average annual flow increased to 27.9 m^3s^{-1} (range 17.6–39.5 m^3s^{-1}), and this difference was significant (RM ANOVA: p<0.001). Average maximum annual flow for the same periods also significantly increased from 95.6 m^3s^{-1} to 159.2 m^3s^{-1} (RM ANOVA: p<0.001), while the average minimum annual flow had slightly but not significantly decreased from 1.4 m^3s^{-1} to 1.0 m^3s^{-1} (RM ANOVA: p=0.06).

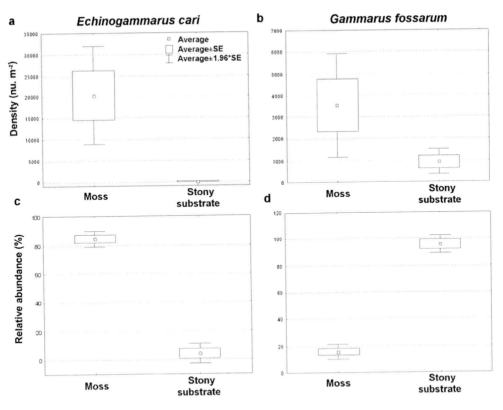

Figure 3. Density (box-SE, whiskers-95% confidence limits) of adults of the species *Echinogamarus cari* and *Gammarus fossarum* in two microhabitats sampled (n=4) at the site D6 in July 2007.

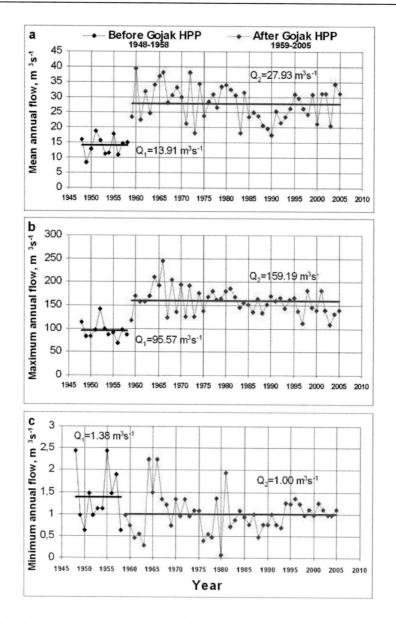

Figure 4. Series of mean (a), maximum (b) and minimum (c) annual discharges measured at the Trošmarija gauging station (site D4 in Figure 1d) before and after Gojak HPP was built in 1959 with averages for both periods shown.

Mean monthly discharges were also greatly increased due to operation of Gojak HPP (Figure 5a), while seasonal changes of flow remained similar. However, when daily discharges changes (measured every hour) at the Trošmarija gauging station were considered for the year 2001, great daily changes were established (Figures 5b-d) that were different in different months and seasons. During the winter and spring months with high discharges, these changes often included two reductions in flow (Figure 5b). The majority of the days without noticeable changes of flow occurred during July and August at time of low summer discharges, but also during high winter flows in January and February. Strong and sudden increases of flow in the middle of the day, or rarely during the night, happened most often during summer months (Figure 5c). In all other months days with changing flows were more frequent than those with stable flow. Irregular daily changes of flow most often occurred from September (Figure 1d) to December.

Interestingly, the species *E. cari* seems to be very well adapted to described great hydrological changes due to operation of the Gojak HPP since it thrives in great abundance in moss microhabitats almost along whole length of canyon part of Gojačka Dobra. This is probably the consequence of the fact that species prefers moss microhabitats where it is very well protected from high current velocity and sudden changes of flow. Massive upstream movements of adult specimens of *E. cari* were observed in the springbrook of Gojačka Dobra (site D2) in November 2006 during very low flow (Žganec and Gottstein, 2009). It is possible that these massive upstream movements are compensation for increased drift during periods of high hydrological disturbances and that this behaviour also enables the species to thrive in conditions with high hydrological disturbance.

Impact of the Damming

According to our present knowledge, the total length of the Gojačka Dobra River and its tributaries where *E. cari* is distributed is approximately 25 km. After the 52.5 m high dam is built and impoundment filled at the begining of 2010, running water habitats upstream from the dam will be flooded. Since the species will not be able to survive in 30-50 m deep impoundment that will be created (Figure 1d), it will be extirpated from about 60% of the watercourse length of its current distribution. Thus, this narrow-ranged endemic species will experience loss of more then a half of its present distribution area.

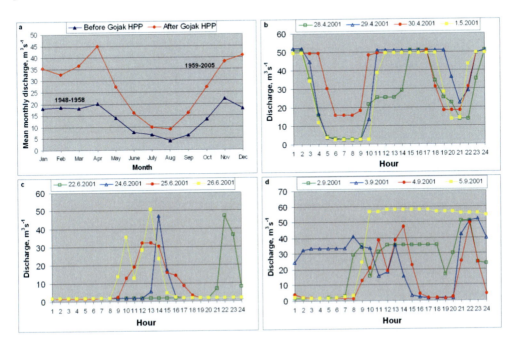

Figure 5. Mean monthly discharge at the Trošmarija gauging station before and after Gojak HPP was built (a); daily changes of discharge for four days in April/May (b), June (c) and September 2001 (d).

In the reaches below the dams the changes of physical conditions include great modifications of natural flow and temperature regime that few native species can tolerate (Bunn and Arthington, 2002). We showed that the natural flow regime of Gojačka Dobra has greatly changed as a result of Gojak HPP impact. Redirection of water from Zagorska Mrežnica and Gornja Dobra caused 100% increase of average annual discharge. Water from reservoirs on those rivers is released to turbines of Gojak HPP continuously during high winter and early spring inflows. However, most often water accumulates for some time and is then released, mainly when the need for electricity is at its maximum. That causes sudden and often severe daily changes of flow in Gojačka Dobra. Operation of the new Lešće HPP with much greater impoundment will cause even more severe daily changes of flow in the reach downstream from the dam, where *E. cari* currently inhabits about 2.5 km long part of the river. The 52.5 m high deep-release dam of Lešće HPP will drastically reduce annual temperature ranges in downstream reach. Such changes under deep-release dams are well known, but changes could also include altered water chemistry (Allan, 1996). In a worst-case scenario, if downstream changes include a decrease in the conductivity and destruction of moss microhabitats due to severe daily changes of flow, *E. cari* could be completely eliminated from the downstream part of the river after damming. In that case, the species distribution would be restricted to a length of about 5 km (20% of present area) of remaining free flowing parts of Bistrica and Ribnjak streams (Figure 1d). In a best-case scenario conductivity would not decrease downstream from the dam, and the changed temperature regime with reduced temperature range would create good conditions for the distribution of the species further downstream. Nevertheless, the impact of the damming will greatly increase species sensitivity to all future changes, like water pollution and alternation of hydro-morphology of the watercourses that it inhabits and it could easily become critically endangered or even extinct.

Conservation Measures

The destiny of the endemic species *E. cari* in the reach downstream from the dam depends on the ability of moss microhabitats to persist in conditions of very high hydrological disturbance after the Lešće HPP starts operation, but also on the resulting changes of water quality. On the other hand, changes of temperature regime will create more suitable conditions for the species in the downstream reach. Further monitoring of plant and animal communities downstream from the dam, with special emphasis on *E. cari* and other rare and threatened species, should be continued for at least one year after Lešće HPP starts operating. These should provide answers whether the species will be extirpated from these reach or, as mentioned above, will increase its distribution further downstream. Whatever happens, conservation measures that would ensure its survival should focus on management activities aimed to restrain further destruction and pollution of the remaining, free flowing parts of Bistrica and Ribnjak streams. In the worst-case scenario introduction of the species into new habitats should be considered. Five potential new sites for species introduction are shown in Figure 1c. These springs and springbrooks seem to be suitable habitats as they have very similar temperature regime and other physicochemical conditions like springs of the Bistrica and Ribnjak streams. They also have dense cover of mosses and only one gammarid species,

Gammarus fossarum, with which *E. cari* is coexisting. Springs of the sinking Dretulja River could also be suitable for the species, but as two species of gammarids, *G. fossarum* and *G. balcanicus*, are coexisting there, it is much less likely that introduction of species would be successful. Therefore, after ongoing study provides new insights into the ecology of the species and more data about watercourses of the surrounding area, conservation strategy for the species could be more closely focused on identification of more potential inhabitable streams, where the species could be introduced. Also, with more data on other rare and threatened species, and by use of range variability approach (Richter, et al., 1996; Richter, et al., 1997) based on natural flow paradigm (Poff et al., 1997) it would be possible do develop environmental flow recommendations (Arthington et al., 2006) for the Gojačka Dobra River.

Endemic Species and Freshwater Biodiversity Conservation

Information on freshwater species and their population change trends are, in general, poor for the most freshwater taxa, including even economically important groups such as fishes. Without this information it is hard to asses the effects of pressures or the risk of extinction of particular species (Revenga et al., 2000). However, because monitoring and assessment of all freshwater species, especially numerous invertebrates, would be a daunting task even at a small geographical scale, it is widely thought that species orientated approach is not feasible conservation strategy for freshwater ecosystems (Strayer, 2006), moreover it is considered even to be counterproductive (Moss, 2000). Due to multiple pressures from human activities, the function of 'receivers' in terrestrial ecosystems, rich biodiversity with often high level of endemism and, thus, non-substitutability, running waters pose ultimate conservation challenge ("conservation nightmare"). There are also problems with longitudinal and lateral connectivity, great natural dynamics and spatio-temporal heterogeneity of these systems (Ward, 1998). Thus, to effectively protect species and habitats in running waters, conservation typically has to protect landscapes or whole watersheds, not just local sites where rare and threatened species occur (Ward, 1998; Moss, 2000; Strayer, 2006). However, in most cases protection of the whole watershed is problematic, as large areas of land need to be managed (Dudgeon et al., 2006). Also, to be effective, conservation measures often require ecosystem management that has to re-establish environmental gradients along longitudinal, lateral and vertical dimension while at the same time reconstituting some semblance of the natural dynamics of the system (Ward, 1998). Saunders et al. (2001) proposed creation of freshwater protected areas as one way of protecting freshwater habitats, but based on three strategies, whole-catchment management, natural flow maintenance and active exclusion of non-native species, and two basic principles, protection of surrounding land and of headwaters. Therefore, it seems most reasonable to invest often limited amount of money into freshwater conservation based on the development of whole-catchment management plans. However, there is also an urgent need for monitoring of individual species selected to represent wider biodiversity or some other valued attributes of the system (Revenga et al., 2005). Also, arguments for the natural flow maintenance in rivers segments fragmented by dams could be obtained by monitoring specifically selected species by using population viability analysis to predict their long-term survival (Saunders et al., 2001).

Endemic species, to our knowledge, have never been identified as a potential group for monitoring and conservation of freshwaters, probably because many of these species are among the least studied. For many endemic species, as was the case with Croatian endemic gammarid *Echinogammarus cari*, data about distribution and ecology are very limited or non existing. Very often that means that biodiversity of river systems inhabited by such species was never studied. When no data exist about biodiversity of few remaining pristine natural, free-flowing rivers, that are about to be heavily altered by human activities, all those interested in protection of such rivers are left to fight with little or no arguments. In such cases the best arguments that could stop damaging project, or at least make it less damaging, are the existence and number of threatened and endemic species. From this perspective endemic species seem to have great importance. Therefore, we propose optional conservation strategy for freshwaters based on identification and monitoring of one or few endemic species with narrow distribution. In most cases such studies would not need to be expensive or time consuming, and represent good alternative when the resources for the conservation are limited as is often the case in developing countries with high biodiversity and levels of endemism in freshwaters. Further argument, as Tisdell (2006) established, is that when threatened (or in our case endemic) species becomes more endangered and this becomes known, public can be expected to increase the level of their donation and support for its conservation.

CONCLUSION

The endemic species *Echinogammarus cari* is confined to the first 15 km of the canyon part of the Gojačka Dobra River and its tributary streams Bistrica and Ribnjak, while it is absent from all other examined sites in surrounding watercourses. The species had a significantly higher density and relative abundance on moss microhabitats. Its relative abundance is in significant positive correlation with calcium ion concentration, i.e. its relative abundance decreases while that of the species *Gammarus fossarum* increases with the downstream decrease in calcium concentration. On the other hand, relative abundance of the species is in significant negative correlation with water temperature range, mean and maximum temperatures. Restricted distribution of this species is, thus, consequence of its adaptation to relatively colder temperature conditions in headwaters of Gojačka Dobra that have characteristics of 'summer-cool' streams and higher concentration of calcium ions.

Redirection of water from Zagorska Mrežnica and Gornja Dobra to Gojak Hydropower Plant after 1959 has caused 100% increase of average annual discharge of Gojačka Dobra River. Operation of Gojak HPP has also caused severe and sudden daily changes of flow. *E. cari* seams to be very well adapted to such conditions, since it has high abundance along the canyon part of the river affected by those changes. Adaptation for life in moss where it is protected from changing water current and the ability to compensate increased drift during periods of high hydrological disturbances with upstream movements, are important traits of this species enabling it to survive in conditions of high hydrological disturbance.

After construction of a 52.5 m high dam of the Lešće HPP in 2010, about 60% of the present area of this endemic species will be flooded. The status of *E. cari* species (endemic and very likely to become endangered) requires development of a conservation strategy and further monitoring of river downstream from the dam as well as Bistrica and Ribnjak streams.

Immediate conservation measures should focus on management activities required to restrain further destruction and/or pollution of the Bistrica and Ribnjak streams, because they should remain as only natural and undisturbed habitats of the species. Preliminary identification was made of five potentially inhabitable sites (springs and headwaters) for streams of the surrounding area where the species could be introduced if worst-case scenario should occur (reduction of species to 20% of present area).

ACKNOWLEDGMENTS

We would like to acknowledge the assistance of Ana Slavikovski during the field work. We also thank people from the Croatian Meteorological and Hydrological Service and Croatian Waters who made temperature and flow data available for our study.

REFERENCES

Allan, J. D. (1996). Stream ecology. *Structure and function of running waters*, (First edition). London, UK: Chapman & Hall.

Allan, J. D. & Flecker, A. S. (1993). Biodiversity conservation in running waters. Identifying the major factors that threaten destruction of riverine species and ecosystems. *Bioscience*, *43*, 32-43.

Arthington, A. H., Bunn, S. E., Poff, L. N. & Naiman, R. J. (2006). The challenge of providing environmental flow rules to sustain river ecosystems. *Ecological Applications*, *16*, 1311-1318.

Bahun, S. (1968) Geologic basis of hydrogeologic relations of the karst area between Slunj and Vrbovsko (Croatia). *Geološki vjesnik*, *21*, 19-82.

Balian, E. V., Segers, H., Lévèque, C. & Martens, K. (2008). The freshwater animal diversity assessment: an overview of the results. *Hydrobiologia*, *595*, 627-637.

Bogan, A. E. (2008). Global diversity of freshwater mussels (Mollusca, Bivalvia) in freshwater. *Hydrobiologia*, *595*, 139-147.

Bunn, S. E. & Arthington, A. H. (2002). Basic principles and ecological consequences of altered flow regimes for aquatic biodiversity. *Environmental Management*, *30*, 492-507.

Cline, L. D. & Ward, J. V. (1984). Biological and physicochemical changes downstream from construction of a subalpine reservoir, Colorado, U.S.A. In Regulated Rivers (A. Lillehammer & S. J. Saltveit, Eds), 231-243. Universitetsforlaget AS, Oslo.

Crandall, K. A. & Buhay, J. E. (2008). Global diversity of crayfish (Astacidae, Cambaridae, and Parastacidae - Decapoda) in freshwater. *Hydrobiologia*, *595*, 295-301.

De Grave, S., Cai, Y. & Anker, A. (2008) Global diversity of shrimps (Crustacea: Decapoda: Caridea) in freshwater. *Hydrobiologia*, *595*, 287-293.

Dick, J. T. A. (2008). Role of behaviour in biological invasions and species distributions; lessons from interactions between the invasive *Gammarus pulex* and the native *G. duebeni* (Crustacea: Amphipoda). *Contributions to Zoology*, *77*, 91-98.

Dudgeon, D., Arthington, A. H., Gessner, M. O., Kawabata, Z. I., Knowler, D. J., Lévêque, C., Naiman, R. J., Prieur-Richard, A. H., Soto, D., Stiassny, M. L. J. & Sullivan, C. A.

(2006). Freshwater biodiversity: importance, threats, status and conservation challenges. *Biological Reviews, 81*, 163-182.

Glazier, D. S., Horne, M. T. & Lehman, M. E. (1992). Abundance, body composition and reproductive output of *Gammarus minus* (Crustacea: Amphipoda) in ten cold springs differing in pH and ionic content. *Freshwater Biology, 28*, 149-163.

Hynes, H. B. N. (1954). The ecology of *Gammarus duebeni* Lilljeborg and its occurrence in freshwater in Western Britain. *J. Anim. Ecol., 23*, 38-84.

Ikonen, E. (1984). Migratory fish stocks and fishery management in regulated finnish rivers flowing into the Baltic sea. In A. Lillehammer, & S.J. Saltveit (Eds), *Regulated Rivers*, (First edition, 437-451). Oslo, Norway: Universitetsforlaget AS.

Karaman, G. S. (1973). 51. Contribution to the Knowledge of the Amphipoda. Two members of *Echinogammarus simoni* group from southern Europe, *E. cari* (S. Kar. 1931) and *E. roco*, n. sp. (fam. Gammaridae). *Poljoprivreda i šumarstvo, 19*, 1-21.

Karaman, G. S. & Pinkster, S. (1977a). Freshwater *Gammarus* species from Europe, North Africa and adjacent regions of Asia (Crustacea - Amphipoda) Part I. *Gammarus pulex* - group and related species. *Bijdragen tot de Dierkunde, 47*, 1-97.

Karaman, G. S. & Pinkster, S. (1977b). Freshwater *Gammarus* species from Europe, North Africa and adjacent regions of Asia (Crustacea - Amphipoda) Part II. *Gammarus roeseli* - group and related species. *Bijdragen tot de Dierkunde, 47*, 165-196.

Karaman, G. S. & Pinkster, S. (1987). Freshwater *Gammarus* species from Europe, North Africa and adjacent regions of Asia (Crustacea-Amphipoda). Part III. *Gammarus balcanicus* - group and related species. *Bijdragen tot de Dierkunde, 57*, 207-260.

Karaman, S. L. (1931). *Gammarus cari* n. sp. aus Westjugoslawien. *Zoologischer Anzeiger, 9*, 265-268.

Kempe, S. & Emeis, K. (1985). Carbonate chemistry and formation of Plitivce Lakes. Mitt. Geol.-Paläont. Inst. Univ. Hamburg, *SCOPE/UNEP Sonderband, 58*, 351-383.

Larinier, M. (2000). Dams and fish migration. *World Commission on Dams, Environmental Issues*, Prepared for Thematic Review II.1: Dams, ecosystem functions and environmental restoration.

Lévêque, C., Oberdorff, T., Paugy, D., Stiassny, M. L. J. & Tedesco, P. A. (2008). Global diversity of fish (Pisces) in freshwater. *Hydrobiologia, 595*, 545-567.

MacNeil, C., Elwood, R. W. & Dick, J. T. A. (1999). Differential microdistributions and interspecific interactions in coexisting *Gammarus* and *Crangonyx* amphipods. *Ecography, 22*, 415-423.

MacNeil, C. & Prenter, J. (2000). Differential microdistributions and interspecific interactions in coexisting native and introduced *Gammarus* spp. (Crustacea: Amphipoda). *J. Zool. Lond., 251*, 377-384.

Malmqvist, B. & Rundle, S. (2002). Threats to the running water ecosystems of the world. *Environmental Conservation, 29*, 134-153.

Master, L. (1990). The imperiled status of North American aquatic animals. *Biodiversity Network News, 3*, 1-2, 7-8.

McAllister, D. E., Craig, J. F., Davidson, N., Delany, S. & Seddon, M. (2001). *Biodiversity impacts of large dams*, Background Paper nr. 1, Prepared for IUCN / UNEP / WCD.

Millennium Ecosystem Assessment, (2005). Washington, DC: Island Press.

Moss, B. (2000). Biodiversity in fresh waters - an issue of species preservation or system functioning? *Environmental Conservation, 27*, 1-4.

Moyle, P. B. & Leidy, R. A. (1992). Loss of biodiversity in aquatic ecosystems: evidence from fish faunas. In P. L. Fielder, & S. K. Jain, (Eds) Conservation biology: the theory and practice of nature conservation, preservation, and management (First edition, 127-169). New York, USA: Chapman and Hall.

Nilsson, C., Reidy, C. A., Dynesius, M. & Revenga, C. (2005). Fragmentation and flow regulation of the world's large river systems. *Science, 308*, 405-408.

Pinkster, S. (1993). A revision of the genus *Echinogammarus* Stebbing, 1899 with some notes on related genera (Crustacea, Amphipoda). *Memorie del Museo di Storia Naturale, 10*, 1-185.

Pöckl, M. (1993). Reproductive potential and lifetime potential fecundity of the freshwater amphipods *Gammarus fossarum* and *G. roeseli* in Austrian streams and rivers. *Freshwater Biology, 30*, 73-91.

Pöckl, M. & Humpesch, U. H. (1990). Intra- and inter-specific variation in egg survival and brood development time for Austrian populations of *Gammarus fossarum* and *G. roeseli* (Crustacea: Amphipoda). *Freshwater Biology, 23*, 441-455.

Poff, N. L., Allan, J. D., Bain, M. B., Karr, J. R., Prestegaard, K. L., Richter, B. D., Sparks, R. E. & Stromberg, J. C. (1997). The natural flow regime: a paradigm for river conservation and restoration. *BioScience, 47*, 769-784.

Revenga, C., Campbell, I., Abell, R., de Villiers, P. & Bryer, M. (2005). Prospects for monitoring freshwater ecosystems towards the 2010 targets. *Philosophical Transactions of the Royal Society B, 360*, 397-413.

Ricciardi, A. & Rasmussen, J. B. (1999). Extinction rates of North American freshwater fauna. *Conservation Biology, 13*, 1220-1222.

Richter, B. D., Baumgartner, J. V., Powell, J. & Braun, D. P. (1996). A method for assessing hydrological alternation within. *Conservation Biology, 10*, 1163-1174.

Richter, B. D., Baumgartner, J. V., Wigington, R. & Braun, D. P. (1997). How much water does a river need? *Freshwater Biology, 37*, 231-249.

Richter, B. D., Braun, D. P., Mendelson, M. A. & Master, L. L. (1997). Treats to imperiled freshwater fauna. *Conservation Biology, 11*, 1081-1093.

Rukke, N. A. (2002). Effects of low calcium concentrations on two common freshwater crustaceans, *Gammarus lacustris* and *Astacus astacus*. *Functional Ecology, 16*, 357-366.

Saunders, D. L., Meeuwig, J. J. & Vincent, A. C. (2001). Freshwater protected areas: strategies for conservation. *Conservation Biology, 16*, 30-41.

Schrimpff, E. & Foeckler, F. (1985). Gammarids in streams of Northeastern Bavaria, F. R. G. I. Prediction of their general occurence by selected hydrochemical variables. *Archiv für Hydrobiologie, 103*, 479-495.

Sheldon, F. & Walker, K. F. (1997). Changes in biofilms induced by flow regulation could explain extinctions of aquatic snails in the lower River Murray, *Australia. Hydrobiologia, 347*, 97-108.

Stanford, J. A. & Ward, J. V. (1984). The effects of regulation on the limnology of the Gunnison River: A North American case history. In A. Lillehammer, & S. J. Saltveit, (Eds), Regulated Rivers, (First edition, 467-480). Oslo, Norway: Universitetsforlaget AS.

Strayer, D. L. (2006). Challenges for freshwater invertebrate conservation. *Journal of North American Benthological Society, 25*, 271-287.

Strong, E. E., Gargominy, O., Ponder, W. F. & Bouchet, P. (2008). Global diversity of gastropods (Gastropoda; Mollusca) in freshwater. *Hydrobiologia, 595*, 149-166.

Tisdell, C. (2006). Knowledge about a species' conservation status and funding for its preservation: Analysis. *Ecological modelling*, *198*, 515-519.

Tronvig, T. A. & Belson, C. S. (2000/2001). Top ten list of endangered karst ecosystems. Karst Waters Institute. 2007 [cited 22 Feb 2007]. Available from: URL: http://www.karstwaters.org/TopTen3/topten3.htm#OGULIN

Väinölä, R., Witt, J .D. S., Grabowski, M., Bradbury, J. H., Jażdżewski, K. & Sket, B. (2008). Global diversity of amphipods (Amphipoda; Crustacea) in freshwater. *Hydrobiologia*, *595*, 241-255.

Vaughn, C. C. & Taylor, C. M. (1999). Impoundments and the decline of freshwater mussels: a case study of an extinction gradient. *Conservation Biology*, *13*, 912-920.

Ward, J. V. (1998). Riverine landscapes: biodiversity patterns, disturbance regimes, and aquatic conservation. *Biological Conservation*, *83*, 269-278.

World Commission on Dams, (2000). Dams and development. A new framework for decision-making, London UK: Earthscan Publications.

Yeo, D. C. J., Ng, P. K. L., Cumberlidge, N., Magalhães, C., Daniels, S. R. & Campos, M. R. (2008). Global diversity of crabs (Crustacea: Decapoda: Brachyura) in freshwater. *Hydrobiologia*, *595*, 275-286.

Zehmer, J. K., Mahon, S. A. & Capelli, G. M. (2002). Calcium as a limiting factor in the distribution of the amphipod *Gammarus pseudolimnaeus*. *American Midland Naturalist*, *148*, 350-362.

Žganec, K. & Gottstein, S. (2009). The river before damming: distribution and ecological notes on the endemic species *Echinogammarus cari* (Amphipoda: Gammaridae) in the Dobra River and its tributaries, Croatia. *Aquatic Ecology*, *43*, 105-115.

In: Species Diversity and Extinction
Editor: Geraldine H. Tepper, pp. 317-339

ISBN: 978-1-61668-343-6
© 2010 Nova Science Publishers, Inc

Chapter 10

THREATENED TEMPERATE PLANT SPECIES: CONTRIBUTIONS TO THEIR BIOGEOGRAPHY AND CONSERVATION IN MEXICO

Isolda Luna-Vega, Othón Alcántara Ayala and Raúl Contreras-Medina

Departamento de Biología Evolutiva, Facultad de Ciencias, Universidad Nacional Autónoma de México (UNAM), Apartado Postal 70-399, 04510 México, D.F., Mexico

ABSTRACT

Many species of vascular plants inhabiting the Mexican temperate forests are threatened by continuous environmental impact, mainly due to human activities. Some are recorded in some risk category in the Mexican official publication named 'Norma Oficial Mexicana 059' (NOM-059-ECOL-2001), in the IUCN red lists (the World Conservation Union) or in CITES (Convention on International Trade in Endangered Species). Distribution maps of 31 threatened species of vascular plants were generated or updated with information obtained from institutional databases, herbarium specimens, field work, and specialized literature; with this information, we analyzed species richness, endemism and distributional patterns. Mexican territory was divided using a grid system based on a chart index (scale 1:50,000 composed system by grids of 15' x 20'). Factors that represent a threat for these species, to their current habitats and natural populations are discussed, and some strategies are proposed to prevent their extinction. Regarding conservation, we evaluated their geographic distribution based on current Mexican National Protected Areas (NPAs) and Mexican Priority Terrestrial Regions for Conservation (PTRs). We also suggest the urgent incorporation of some of these species in the recent IUCN Red Data List of Threatened Species (2009) and in the NOM-059. Most of the grid-cells with high diversity and endemism are located in eastern Mexico, and are especially associated with mountain landscapes. Among the factors that represent a threat, we concluded that the main problem for all species is habitat destruction, due to increase in agricultural areas, forest exploitation, and animal husbandry, followed by expansion of human settlements, illicit extraction, and traffic of plants. Most of the species studied require special policies for their conservation due to problems that affect their natural populations. Conservation strategies include: (1) demographic and ecological studies of these species; (2) collection of seeds and vegetative propagation for ex situ conservation in botanical gardens; (3) targeting of specific areas where these species inhabit, in order to include them in future conservation plans. With few exceptions, these species are underrepresented in the current Mexican System of Natural Protected Areas, especially those with restricted distributions. Only six species are represented in five or more NAPs, allowing their populations to be protected. In the case of PTRs, most of the species listed are included, but these areas are not official nor are they regulated by federal laws.

INTRODUCTION

In the last years, interest in studying species richness and conservation in Neotropical temperate forests has been raised (Churchill et al. 1995; Kappelle and Brown 2001; Luna et al. 2006a; Téllez-Valdés et al. 2006). Reasons for these interests are based on high rates of deforestation and loss of these forests due to agricultural activities, as well as forestry and cattle farming (Luna et al. 2006a; Téllez-Valdés et al. 2006). Fortunately, some of these forests are located in inaccessible or restricted sites in the Mexican mountains and consequently, they are still present and reasonably well conserved. In contrast, those located in sites where humans have access have been drastically transformed to secondary pastures and cultivated lands (Téllez-Valdés et al. 2006). For these reasons, these forests are considered among the most threatened and, in most of the cases, lack conservation policies (Luna et al. 2006a).

Former studies have attempted priority areas identification for Mexican temperate forest conservation using parsimony analysis of endemicity combined with track analysis (Luna et al. 2000), and climate change associated to predictive models (Téllez-Valdés et al. 2006). Other studies have been focused on species that inhabit these temperate forests applying endemism indices and recognition of areas with high species richness and endemism (Luna et al. 2004, 2006a; Contreras-Medina and Luna 2007).

Many species of vascular plants inhabiting Mexican temperate forests are threatened by continuous environmental impact, mainly due to human activities. Some of them are recorded in some risk category in the Mexican official publication named 'Norma Oficial Mexicana NOM-059-ECOL-2001 (SEMARNAT 2002), in the IUCN red lists (the World Conservation Union) or in CITES (Convention on International Trade in Endangered Species).

This information is relevant in order to focus our efforts and resources to undertake accurate long-term conservation actions that can guarantee the survival of these temperate plant communities and some of the threatened species inhabiting them, and also to evaluate if they are present in the current Mexican protected areas. Jackson et al. (2009) recently considered that protected areas must include a sample of biodiversity, emphasizing those species that are rare, threatened or with some conservation concern, and protect that sample from threatening present or future processes.

Our aim is to offer a comprehensive review of the current risk situation of 31 selected vascular plant species and to provide ecological and biogeographic information on these species, which are frequent elements of the Mexican temperate forests (Rzedowski 1996). We thus attempted to undertake the following actions: (1) to document the current recorded distribution of these species in Mexico; (2) to make suggestions for their management and conservation; (3) to evaluate their geographic distribution in relation to the current Mexican System of Natural Protected Areas (NAPs) and with the Mexican Priority Terrestrial Regions for Conservation (PTRs); (4) to evaluate threatening factors affecting these species and suggest the incorporation of some of them in the more recent National and International lists of threatened species.

Table 1. Geographic distribution of the 31 species considered in this chapter (number of Mexican states/number of grid-cells). Vegetation types follow Rzedowski (1981)

Species	Distribution	Vegetation type	Altitudinal range (m asl)
Acer negundo L.	Mx (19/27)	CF, QF, PQF	1500-2400
Aporocactus flagelliformis (L.) Lem.	Mx (5/12)	CF, QF	1600-2450
Bernardia mollis Lundell	Mx (1/2), NCA	CF	1550-2430
Bouvardia xylosteoides Hook et Arn.	Mx (1/4)	QF, TDF	1615-2300
Carpinus caroliniana Walter	Mx (13/81), NCA	CF, QF, CONF	1350-2500
Ceratozamia kuesteriana Regel	Mx (1/2)	CF, QF, PQF, TDF	800-1450
Ceratozamia mexicana Brongn.	Mx (5/15)	CF, PQF	880-1970
Ceratozamia sabatoi Vovides, Vázq. Torres, Schutzman et Iglesias	Mx (2/3)	CF, QF, PQF	1650-1900
Cleyera cernua (Tul.) Kobuski	Mx (1/3)	CF	1500-1550
Cleyera theaeoides (Sw.) Choisy	Mx (6/55), CA	CF, QF, PQF	900-3000
Cleyera velutina B.M. Barthol.	Mx (2/8)	CF, PQF	2100-3000
Cupressus lusitanica Mill.	Mx (19/75), NCA	CF, PQF, CONF	1400-3600
Cyathea mexicana Schltdl. et Cham.	Mx (8/27)	CF, PQF, TEF	800-2500
Deppea hernandezii Lorence	Mx (1/1)	CF	1850
Diospyros riojae Gómez Pompa	Mx (4/11)	CF, QF, PQF	740-1900
Juglans pyriformis Liebm.	Mx (5/10)	CF, PQF	1200-1400
Litsea glauscescens Kunth	Mx (19/46), CA	CF, QF, PQF, CONF, TDF	800-2830
Magnolia dealbata Zucc.	Mx (4/11)	CF, QF, PQF	1500-1820

Table 1. (Continued)

Species	Distribution	Vegetation type	Altitudinal range (m asl)
Magnolia schiedeana Schltdl.	Mx (5/20)	CF, QF, PQF, TEF	1000-2000
Matudaea trinervia Lundell	Mx (7/15)	CF, QF, PQF	500-2000
Nopalxochia phyllanthoides (DC.) Britton et Rose	Mx (3/6)	CF, QF	1200-1850
Ostrya virginiana (Mill.) K. Koch	EUSA, Mx (15/74), NCA	CF, QF, PQF, CONF	1200-2800
Podocarpus reichei J. Buchholz et N.E. Gray	Mx (13/32), NCA	CF, QF, PQF	800-2300
Symplocos coccinea Humb. et Bonpl.	Mx (5/22)	CF, QF, PQF	1750-2700
Taxus globosa Schltdl.	Mx (9/32), CA	CF, QF, PQF	1000-2950
Ternstroemia dentisepala B.M. Barthol.	Mx (3/4)	CF, QF, CONF	1000-2160
Ternstroemia huasteca B.M. Barthol.	Mx (5/13)	CF, QF, PQF	900-2200
Tilia mexicana Schltdl.	Mx (15/51)	CF, QF, PQF, CONF	1100-2500
Zamia vazquezii D.W. Stev., Sabato et Moretti	Mx (4/5)	CF, QF, TDF	270-950
Zamia loddigesii Miq.	Mx (7/22), NCA	CF, QF, TDF	260-830
Zinowiewia concinna Lundell	Mx (3/8)	CF, QF, PQF	1800-2000

Abbreviations: Distribution: Mx = Mexico; NCA = northern Central America; CA = Central America; EUSA = eastern United States. Vegetation types: CF = cloud forest; QF = *Quercus* forest; PQF = *Pinus-Quercus* Forest; CONF = Coniferous forest; TDF = Tropical deciduous forest; TEF = tropical evergreen forest.

DISTRIBUTIONAL DATA OF VASCULAR PLANT SPECIES

A set of 31 species of vascular plants was chosen for this study, including 22 angiosperms, eight gymnosperms, and one tree fern (Table 1), many of them endemic to Mexico and well represented in the Mexican mountains. All of these species are cited as frequent elements in the Mexican temperate forests.

Distributional data used for this study were obtained from different sources: institutional databases, herbarium specimens, field work, and specialized literature. Based on our Mexican plant database of temperate forests, which is the result of several floristic projects, we built a database comprising nearly 2000 records of these 31 species. Herbarium specimens of the species studied were also revised from the following collections: MEXU, ENCB, XAL, IEB, FCME, IBUG, CHAP, INIF, MO, NY, and K (acronyms sensu Holmgren et al. 1990). Additionally, botanical field exploration was carried out in the Mexican states of Hidalgo, Puebla, Querétaro, México, Oaxaca, and Veracruz in order to collect, take photographs, and observe natural populations of some of these species.

Finally, information published in regional or state floras of Mexico and revisionary studies were used to assess the geographic and ecological distribution and other characteristics of these species (Standley and Steyermark 1946-1976; Meyrán 1962; Bravo 1978; Hernández-Cerda 1980; Nee 1981; Pacheco 1981; Rzedowski 1981; Vovides 1981, 1985, 1988, 1995, 1999; Moretti et al. 1982; Sánchez-Mejorada 1982; Zanoni 1982; Narave 1983; Vovides et al. 1983; Cabrera 1985; Breedlove 1986; Pattison 1986; Puig and Bracho 1987; Luna et al. 1989, 1994; Angulo and Soto 1990; Malda 1990; Niembro 1990; Soto and Gómez-Pompa 1990; McVaugh 1992; Zamudio 1992, 2002; Gómez 1993; Zulueta and Soto 1993; Cuevas and Nuñez 1994; Dieringer and Espinosa 1994; Martínez and Chacalco 1994; Zamudio and Carranza 1994; Bravo and Sheinvar 1995; Chávez and Rubluo 1995; Osborne 1995; Pérez-Escandón and Villavicencio 1995; Vázquez 1994; Vázquez et al. 1995; Alcántara and Luna 1997, 2001; Arreguín et al. 1997; Gutiérrez and Vovides 1997; Van der Werff and Lorea 1997; Mayorga et al. 1998; Sosa et al. 1998; Carranza 2000; García and Castillo 2000; Hilton-Taylor 2000; Pavón and Rico-Gray 2000; Calderón de Rzedowski 2001; Contreras-Medina and Luna 2001; Diego et al. 2001; Fonseca et al. 2001; Cartujano et al. 2002; Cervantes 2002; Godínez-Ibarra and López-Mata 2002; Contreras-Medina et al. 2003, 2006; Guzmán et al. 2003; Lozada et al. 2003; Sánchez-González and López-Mata 2003; Borhidi 2006; García-Franco et al. 2008; Juárez 2008; Yberri 2009).

BIOGEOGRAPHIC ANALYSIS APPLIED TO MEXICAN TEMPERATE FOREST PLANT DISTRIBUTION

A more objective way to determine the size of the geographic range of the species may be to use a standard and minimal area unit that is easily recognizable (Luna et al. 2004); in our case, we used a chart index scale of 1:50,000 formulated by the inventory of geographic information produced by the Instituto Nacional de Estadística, Geografía e Informática (INEGI 1992). This index makes reference to published maps on a scale of 1:50,000, a system based on 2313 grids of 15' x 20' (at a spatial resolution of approximately 27.75 km x 36.75 km), which shows real sections of the Mexican territory, so the circumscription of the areas with greater species richness and/or that contain certain species or taxa with biological meaning is easily located for the country and can be easily and objectively quantified (Luna et al. 2004). This chart index, composed by 2313 grid-cells, was digitized by the Comisión Nacional para el Conocimiento y Uso de la Biodiversidad (CONABIO 1999). The names of each topographic chart were used to identify the grid-cells.

We obtained distributional maps of each of the 31 species using the software ArcView GIS (ESRI 1999), which were then superimposed on digital 1:1,000,000 scale maps of Mexico (CONABIO 1998). The percentage of species distribution area related to the Mexican territory was calculated, in order to evaluate their geographic range under the risk evaluation method (MER), which is included in the NOM-059-ECOL-2001 (SEMARNAT 2002). The

MER includes four points that must be evaluated for species included in the NOM-059-ECOL-2001 (SEMARNAT 2002), and these are: (a) species' geographic range, (b) habitat status regarding natural species development, (c) species' intrinsic biological vulnerability, and (d) human activity impact on the natural populations of species. In the first point, geographic range is divided into four categories: (1) highly restricted (distribution less than 5% of the Mexican territory); (2) restricted (between 5 and 15%); (3) moderately restricted to widely distributed (more than 15% but less than 40%); and (4) widely distributed (more than 40% of the Mexican territory) (SEMARNAT 2002).

Using ArcView GIS (ESRI 1999), we intersected the species maps of localities with the grid system scale 1:50,000, first including all the species to obtain the richness of each grid-cell and then, using only the species with restricted distribution (those represented in less than 15 grid-cells), to obtain the areas with high concentration of restricted species. Then we selected those areas with more species richness and/or more concentration of restricted species (more than 10 living sympatrically) and drew them on the map of Natural Protected Areas of Mexico (NAPs) produced in digital format by SEMARNAT (CONANP 2007), as well as on the map of Mexican Priority Terrestrial Regions for Conservation (PTRs) produced in digital format by CONABIO (2000), to find out if the areas occupied by these species are under some type of special protection. PTRs represent areas with high biodiversity, formulated by an expertise group of national researchers coordinated by CONABIO, and comparatively these areas have high values of ecosystem and species richness in relation to other areas of Mexico, as well as a functional ecological integrity where real opportunities for conservation exist (Arriaga et al. 2000), but unfortunately these areas are not official nor are they regulated by federal laws.

In order to evaluate species presence in the NAPs and PTRs, we intersected the geographic distribution map of each species with the above mentioned digital maps of these areas.

DISTRIBUTIONAL PATTERNS OF THREATENED TEMPERATE PLANT SPECIES

The species of vascular plants studied here are mainly distributed in the Región Mesoamericana de Montaña sensu Rzedowski (1981), which is a region composed by the following floristic provinces: Sierra Madre Occidental, Sierra Madre Oriental, Serranías Meridionales, and Serranías Transístmicas, characterized by its mountain temperate humid conditions; these species mainly inhabit cloud forests, as well as oak and pine forests (Figure 1).

The species studied herein mainly inhabit temperate forests, especially in areas with humid subtropical to humid mild climates, in a preferential altitudinal range of 800 to 3000 m. A few species inhabit both tropical and temperate forests with different vegetation types, along broad altitudinal and latitudinal ranges (i.e. *Litsea glaucescens*). Most of these plant species inhabit humid slopes with abundant atmospheric moisture and protected from direct sun exposure; some tree species are important elements in the vegetation structure of the arboreal stratum of the forests in which they grow, such as *Carpinus caroliniana*, *Ostrya virginiana*, *Zinowiewia concinna* and *Symplocos coccinea* (Luna et al. 2006b; García-Franco et al. 2008) in the case of angiosperms, and *Podocarpus reichei* and *Cupressus lusitanica* (Ruiz-Jiménez et al. 2000, Contreras-Medina et al. 2001) for gymnosperms, whereas among epiphytes *Nopalxochia phyllanthoides* represents an important element (García-Franco et al. 2008).

Most of the grid-cells with high diversity and endemism are located in eastern Mexico, and are especially associated with mountainous landscapes. Figure 2 shows that the richest area (16 species) is located in eastern Mexico, in the Zacualtipán area, in the states of Hidalgo-Veracruz. The geographic distribution of the taxa included in this study show different patterns, from wide to narrow ranges assessed by their presence in different Mexican states or grid-cells (Table 1). For example, *Ceratozamia kuesteriana* is restricted to the state of Tamaulipas, *Cleyera cernua* to the state of Oaxaca, and *Deppea hernandezii* to the state of Hidalgo, and some species are restricted to two or three states, as are *Ceratozamia sabatoi*, *Cleyera velutina*, *Nopalxochia phyllanthoides*, and *Ternstroemia dentisepala*, while others are found in more than 12 states, such as *Carpinus caroliniana*, *Cleyera theaeoides*, *Cupressus lusitanica*, *Ostrya virginiana*, and *Tilia mexicana*; these species also were recorded in more than 50 grid-cells, whereas *Bernardia mollis*, *Bouvardia xylostoides*, *Ceratozamia kuesteriana*, *Ceratozamia sabatoi*, *Cleyera cernua*, *Deppea hernandezii*, *Nopalxochia phyllanthoides*, *Ternstroemia dentisepala*, and *Zamia vazquezii* were only recorded in five grid-cells or less (Figure 3 and Table 1).

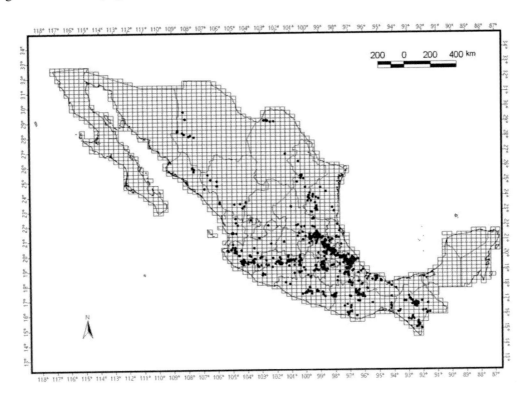

Figure 1. Geographic distribution of the 31 species studied using a grid system of Mexico by INEGI (1992).

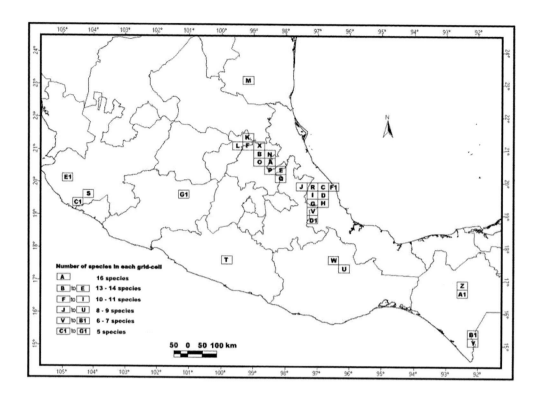

Figure 2. Representation of the Mexican richest grid-cells based on the species studied. A. Zacualtipán, Hgo., Ver.; B. Molango, Hgo.; C. Misantla, Ver.; D. Xalapa, Ver.; E. Pahuatlán, Hgo., Ver., Pue.; F. Jacala, Hgo., Qro.; G. Xico, Pue., Ver.; H. Coatepec, Ver.; I. Perote, Pue., Ver.; J. Teziutlán, Pue.; K. Ahuacatlán, Qro.., S.L.P.; L. Jalpan, Qro., Hgo.; M. Gómez Farías, Tamps.; N. Calnali, Hgo., Ver.; O. Metztitlán, Hgo.; P. Carbonera Jacales, Hgo., Ver.; Q. Huachinango, Hgo., Pue.; R. Altotonga, Pue., Ver.; S. El Chante, Jal.; T. Chichihualco, Gro.; U. San Miguel Talea de Castro, Oax.; V. Coscomatepec, Pue., Ver.; W. San Juan Quiotepec, Oax.; X. Chapulhuacán, Hgo., S.L.P., Qro.; Y. Pavincul, Chis.; Z. Oxchuc, Chis.; A1. San Cristóbal de las Casas, Chis.; B1. Motozintla, Chis.; C1. La Huerta, Col., Jal.; D1. Orizaba, Ver., Pue. E1. Llano Grande, Jal.; F1. Villa Emilio Carranza, Ver.; G1. Morelia, Mich.

All the Mexican plant species studied belong to the category of very restricted geographic distribution (less than 5% of Mexican territory) that is established by the MER, which means that all species inhabit less than 115 grid-cells.

Figure 3 shows that when we superimpose localities of species with restricted distribution on the grid system, grid-cells with a high number of restricted species are located at Jacala grid-cell, within the limits of the states of Querétaro and Hidalgo, with five species.

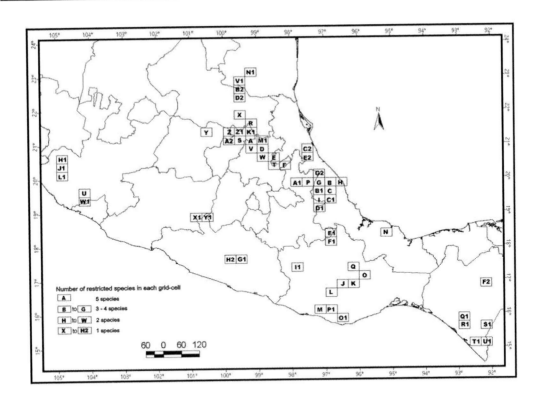

Figure 3. Grid-cells containing more restricted species. A. Jacala, Hgo., Qro.; B. Misantla, Ver.; C. Xalapa, Ver.; D. Molango, Hgo.; E. Zacualtipán, Hgo., Ver.; F. Pahuatlán, Hgo., Ver., Pue.; G. Altotonga, Pue., Ver.; H. Villa Emilio Carranza, Ver.; I. Xico, Pue., Ver.; J. Tlacolula de Matamoros, Oax.; K. San Pedro Quiantón, Oax.; L. Ejutla de Crespo, Oax.; M. Santa Catarina Juquila, Oax.; N. San Andrés Tuxtla, Ver.; O. San Juan Cotzocón, Oax.; P. Teziutlán, Pue.; Q. San Miguel Talea de Castro, Oax.; R. Aquismón, S.L.P., Qro.; S. Jalpan, Qro., Hgo.; T. Carbonera Jacales, Hgo., Ver.; U. El Chante, Jal.; V. San Nicolás, Hgo.; W. Metztitlán, Hgo.; X. Tamasopo, S.L.P.; Y. San Luis de la Paz, Gto.; Z. El Carricillo, Gto., Qro., S.L.P.; A1. Zacatlán, Pue.; B1. Perote, Pue., Ver.; C1. Coatepec, Ver.; D1. Coscomatepec, Pue., Ver.; E1. Coyomeapan, Pue., Oax.; F1. Huautla, Pue., Oax.; G1. Chilpancingo, Gro.; H1. Mascota, Jal; I1. Tlaxiaco, Oax.; J1. Talpa de Allende, Jal.; K1. Ahuacatlán, Qro., S.L.P.; L1. Llano Grande, Jal.; M1. Chapulhuacán, Hgo., S.L.P., Qro.; N1. Gómez Farías, Tamps.; O1. San José Chacalapa, Oax.; P1. San Baltasar Loxicha, Oax.; Q1. Angel Albino Corzo, Chis.; R1. Samule León Brindis, Chis.; S1. Chicomuselo, Chis.; T1. Huixtla, Chis.; U1. Pavincul, Chis.; V1. Ocampo, Tamps.; W1. Minatitlán, Jal., Col.; X1. Tiquicheo, Mich.; Y1. Bejucos, Méx., Mich.; Z1. Conca, Qro., S.L.P.; A2. Peña Miller, Gto., Qro.; B2. Salto del Agua, S.L.P., Tamps.; C2. Tuxpan, Ver.; D2. Cd. del Maíz, S.L.P.; E2. Poza Rica, Ver., Pue.; F2. Ocosingo, Chis.; G2. Martínez de la Torre, Pue., Ver. H2. Chichihualco, Gro.

Based on the above analyses, we detected the following types of areas to be conserved (the Mexican states where are located are showed in parentheses): (a) areas with high taxonomic richness; among the most important areas are Zacualtipán (Hidalgo-Veracruz) that contain 16 of the 31 species considered; Misantla (Veracruz), with 14 species, and Molango (Hidalgo), Xalapa (Veracruz), and Pahuatlán (Hidalgo-Puebla-Veracruz) with 13 species each; Jacala (Querétaro-Hidalgo) and Xico (Veracruz) with 11 species each. The area occupied by these grid-cells conforms the largest and most continuous fragment of cloud forest in eastern Mexico (Figure 2); (b) areas with a high concentration of restricted species; in this case, the richest grid-cells are Jacala (Querétaro-Hidalgo) with five restricted species, and Molango (Hidalgo) Zacualtipán (Hidalgo-Veracruz), and Misantla (Veracruz) with four species each, followed by Altotonga (Puebla), Pahuatlán (Puebla), and Xalapa (Veracruz) with three species each (Figure 3).

Table 2. Presence of species studied in the Mexican Natural Protected Areas and Priority Terrestrial Regions (NAPs and PTRs)

SPECIES	NAP	PTR
Acer negundo	8 and 14	5, 6, 11, 20, 25, 32, 48, 49, and 50
Aporocactus flagelliformis	18 and 24	5, 25, 40, 47, and 49
Bernardia mollis	---	---
Bouvardia xylosteoides	---	---
Carpinus caroliniana	3, 10, 13, 19, 20, 21, 23, and 24	5, 6, 11, 12, 14, 17, 18, 23, 25, 30, 32, 33, 34, 36, 40, 42, 43, 46, 47, 48, and 49
Ceratozamia kuesteriana	---	12
Ceratozamia mexicana	24	5 and 40
Ceratozamia sabatoi	17 and 24	40
Cleyera cernua	---	47
Cleyera theaeoides	1, 9, 16, 18, 19, 24, 25, 26, and 27	1, 5, 6, 7, 12, 13, 14, 17, 20, 25, 27, 30, 32, 40, 42, 45, 47, 48, 49, and 52
Cleyera velutina	---	7, 42, and 48
Cupressus lusitanica	1, 3, 6, 8, 11, 12, 15, 19, 22, 23, and 24	1, 4, 5, 10, 11, 12, 17, 19, 23, 25, 27, 29, 30, 32, 33, 39, 40, 44, and 47
Cyathea mexicana	20 and 21	5, 6, 11, 13, 22, 30, 34, 40, 42, 47, and 49
Deppea hernandezii	---	5
Diospyros riojae	24	11, 34, and 40
Juglans pyriformis	---	4, 5, and 40
Litsea glaucescens	4, 9, 20, 23, and 24	1, 5, 9, 10, 12, 13, 19, 20, 21, 23, 25, 29, 35, 40, 45, 47, 48, 49, and 51
Magnolia dealbata	24	5, 40, and 47
Magnolia schiedeana	24	5, 18, 25, 40, 42, and 47
Matudaea trinervia	19 and 23	9, 11, 17, 23, 43, and 45
Nopalxochia phyllanthoides	---	5 and 11
Ostrya virginiana	2, 3, 7, 21, 23 and 24	3, 5, 7, 8, 11, 13, 15, 16, 20, 23, 25, 27, 30, 31, 34, 38, 40, 41, 42, 47, 48, and 49
Podocarpus reichei	23, 24 and 26	5, 9, 11, 12, 23, 25, 33, 40, 45, and 47
Symplocos coccinea	18	5, 6, 11, 16, 25, 34, and 47
Taxus globosa	7, 8, and 24	5, 12, 15, 25, 26, 28, 35, 40, and 47
Ternstroemia dentisepala	23	23 and 27
Ternstroemia huasteca	18 and 24	5, 11, and 40
Tilia mexicana	2, 5, 7, 18, 23, and 24	2, 3, 5, 12, 15, 23, 25, 38, 40, 41, and 47
Zamia vazquezii	24	40
Zamia loddigesii	24	11, 18, 34, 37, and 40
Zinowiewia concinna	9 and 23	1, 23, 24, 46, 47, and 48

NAP

FLORA AND FAUNA PROTECTED AREAS: 1. Corredor Biológico Chichinautzin (Morelos); 2. Maderas del Carmen (Coahuila).

NATIONAL PARKS: 3. Cañón del Río Blanco (Veracruz); 4. Cañón del Sumidero (Chiapas); 5. Cascada de Bassaseachic (Chihuahua); 6. Cofre de Perote (Veracruz); 7. Cumbres de Monterrey (Nuevo León); 8. El Chico (Hidalgo); 9. El Tepozteco (Morelos-Distrito Federal); 10. Insurg. José María Morelos (Michoacán); 11. Iztaccíhuatl-Popocatépetl (México-Morelos-Puebla); 12. Lagunas de Zempoala (Morelos-México); 13. Volcán Nevado de Colima (Jalisco); 14. Xicoténcatl (Tlaxcala); 15. Zoquiapan and vicinities (México-Puebla); 16. Lagunas de Montebello (Chiapas); 17. Los Mármoles (Hidalgo).

BIOSPHERE RESERVES: 18. Barranca de Metztitlán (Hidalgo); 19. El Triunfo (Chiapas); 20. La Sepultura (Chiapas); 21. Los Tuxtlas (Veracruz); 22. Mariposa Monarca (Michoacán-México); 23. Sierra de Manantlán (Jalisco-Colima); 24. Sierra Gorda (Querétaro-Hidalgo); 25. Volcán Tacaná (Chiapas).

NATURAL RESOURCES PROTECTED AREAS: 26. Cuenca hidrográfica del Río Necaxa (Hidalgo-Puebla); 27. Cuencas de los ríos Valle de Bravo y Malacatepec (Michoacán-México).

TPR

1. Ajusco-Chichinautzin (Distrito Federal-México-Morelos); 2. Alta Tarahumara-Barrancas del Cobre (Chihuahua); 3. Bassaseachic (Chihuahua); 4. Bavispe-El Tigre (Chihuahua-Sonora); 5. Bosques mesófilos de la Sierra Madre Oriental (Hidalgo-Puebla-Veracruz); 6. Bosques mesófilos de los Altos de Chiapas (Chiapas); 7. Cañón del Zopilote (Guerrero); 8. Cerro Viejo-Sierras de Chapala (Jalisco-Michoacán); 9. Chamela-Cabo Corrientes (Jalisco); 10. Cuenca del río Jesús María (Durango-Jalisco-Nayarit-Zacatecas); 11. Cuetzalan (Puebla-Veracruz); 12. El Cielo (Tamaulipas); 13. El Momón-Montebello (Chiapas); 14. El Mozotal (Chiapas); 15. El Potosí-Cumbres de Monterrey (Coahuila-Nuevo León); 16. El Tlacuache (Oaxaca); 17. El Triunfo-La Encrucijada-Palo Blanco (Chiapas); 18. Encinares tropicales de la planicie costera veracruzana (Veracruz); 19. Guacamayita (Durango); 20. Huitepec-Tzontehuitz (Chiapas); 21. La Chacona-Cañón del Sumidero (Chiapas); 22. Lacandona (Chiapas-Tabasco); 23. Manantlán-Volcán de Colima (Colima-Jalisco); 24. Nevado de Toluca (México); 25. Pico de Orizaba-Cofre de Perote (Puebla-Veracruz); 26. Puerto Purificación (Nuevo León-Tamaulipas); 27. Río Presidio (Durango-Sinaloa); 28. San Antonio-Peña Nevada (Nuevo León-Tamaulipas); 29. San Juan de Camarones (Durango-Sinaloa); 30. Selva Zoque-La Sepultura (Chiapas-Oaxaca-Veracruz), 31. Sierra Bustamante (Coahuila-Nuevo León); 32. Sierra de Chincua (México-Guanajuato-Michoacán); 33. Sierra de Coalcomán (Jalisco-Michoacán); 34. Sierra de los Tuxtlas-Laguna del Ostión (Veracruz); 35. Sierra de Álvarez (San Luis Potosí); 36. Sierra de San Carlos (Nuevo León- Tamaulipas); 37. Sierra de Tamaulipas (Tamaulipas); 38. Sierra El Burro-río San Rodrigo (Coahuila); 39. Sierra Fría (Aguascalientes-Zacatecas); 40. Sierra Gorda-Río Moctezuma (Guanajuato-Hidalgo-Querétaro-San Luis Potosí); 41. Sierra Maderas del Carmen (Coahuila); 42. Sierra Madre del Sur de Guerrero (Guerrero); 43. Sierra Nanchititla (México-Guerrero); 44. Sierra Nevada (México-Morelos-Puebla-Tlaxcala); 45. Sierra sur y costa de Oaxaca (Oaxaca); 46. Sierras de Taxco-Huautla (México-Guerrero-Morelos-Puebla); 47. Sierras del norte de Oaxaca-Mixe (Oaxaca-Puebla-Veracruz); 48. Sierras Triqui-Mixteca (Guerrero-Oaxaca); 49. Tacaná-Boquerón (Chiapas); 50. Tokio (Coahuila-Nuevo León-San Luis Potosí-Zacatecas); 51. Valle de Jaumave (Tamaulipas); 52. Tancítaro (Michoacán).

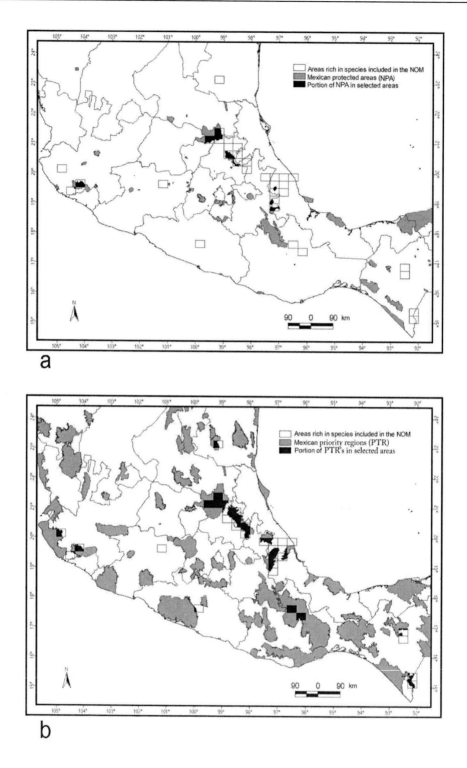

Figure 4. Superimposition of the richest grid-cells on (a) the map of the Mexican protected areas; and (b) Mexican Priority Terrestrial Regions for conservation by CONABIO (2000).

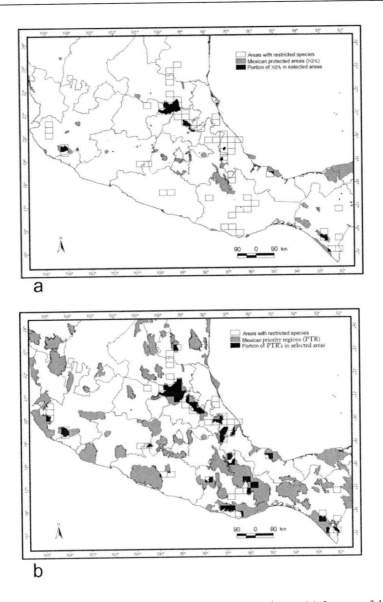

Figure 5. Superimposition of the grid-cells with more restricted species on (a) the map of the Mexican protected areas; and (b) Mexican Priority Terrestrial Regions for conservation by CONABIO (2000).

DISTRIBUTIONAL RANGE OF TEMPERATE PLANT SPECIES IN RELATION TO NAPS AND PTRS

When we compared both results with the map of the Mexican Natural Protected Areas (NAPs) of SEMARNAT (CONANP 2007), we detected that many of the grid-cells considered as important areas for conservation of these plants are not included in any protected area (Figures 4a and 5a). When we followed the same procedure on the map of the Mexican Priority Terrestrial Regions for Conservation (PTRs) of CONABIO (2000), we observed that many of these grid-cells are included in these regions (Figures 4b and 5b). Unfortunately, many of these grid-cells are not included in either system; we found that among the current Natural Protected Areas of Mexico, the known records are located within the limits of 15 National Parks, eight Biosphere Reserves, two protected areas of flora and fauna, and two protected areas of natural resources (Table 2). In Mexico, some Natural Protected Areas may prevent habitat degradation and overexploitation, or at least reduce their rates of occurrence, and thus reduce extinction rates and promote species richness (Jackson et al. 2009).

Of the 31 species analyzed, 23 (74 %) were represented at least once within protected areas across Mexican territory. Despite relatively good overall species represented, natural protected areas across Mexico cover less than one fifth of the total number of occurrence records for threatened plant species analyzed, and the level of multiple representation is rather low, because near 70% of species studied are represented less than 10 times. This situation increases the importance of ensuring the continued persistence of these species within those natural protected areas in which they do occur.

Our data indicate that the representation of these 31 plant species in the Mexican Natural Protected Areas system are located along two extremes, due that some (*Bernardia mollis, Bouvardia xylostoides, Ceratozamia kuesteriana, Deppea hernandezii, Juglans pyriformis, Cleyera cernua, Cleyera velutina,* and *Nopalxochia phyllanthoides*) are not included in any protected area, whereas other species are present in a wide geographic distribution in the country (*Carpinus caroliniana, Cleyera theaeoides, Cupressus lusitanica, Litsea glaucescens, Ostrya virginiana,* and *Tilia mexicana*), represented in five or more protected areas (Table 2). We consider that at least for the last six species, their populations are ensured by these protected areas.

THREATENED CATEGORIES IN IUCN, CITES AND NOM-059-ECOL-2001

A comparison among the risk categories of IUCN, CITES, and NOM-059-ECOL-2001 for the species studied is shown in Table 3. Most of the species are recorded in some risk category in the Mexican official publication named Norma Oficial Mexicana NOM-059-ECOL-2001 (SEMARNAT 2002). In addition, 14 of them are recorded in the IUCN Red data List of Threatened Species (2009), whereas eight species are registered in the Appendices I and II of CITES (the Convention on International Trade in Endangered Species of Wild Fauna and Flora, http://www.cites.org/) (Table 3). *Nopalxochia phyllanthoides* is not included in the IUCN Red Lists, whereas *Bernardia mollis* and *Bouvardia xylostoides* are not included in CITES and IUCN Red Lists, but an extreme case is that of Ternstroemiaceae, because four species are not included in any list, these are *Cleyera theaeoides, C. velutina, Ternstroemia dentisepala* and *T. huasteca* (Table 3).

Table 3. Risk categories of the 31 species involved in this work

Species	NOM-059	CITES	IUCN	Main threats of the species	Conservation status
Acer negundo	Pr	--	VU	1	A, B
Aporocactus flagelliformis	Pr	Appendix II	--	1, 3	A, C
Bernardia mollis	A	--	--	1	A
Bouvardia xylosteoides	Pr	--	--	1, 7	A
Carpinus caroliniana	A	--	--	1, 2	A
Ceratozamia kuesteriana	Pr	Appendix I	CR	1, 3, 8	A
Ceratozamia Mexicana	A	Appendix I	VU	1, 3	E
Ceratozamia sabatoi	A	Appendix I	--	1, 3	A
Cleyera cernua	--	--	DD	1, 8	A
Cleyera theaeoides	--	--	--	1	A
Cleyera velutina	--	--	--	1, 8	A
Cupressus lusitanica	Pr	--	LR	1	A, B, F, G
Cyathea mexicana	P	Appendix II	--	1, 3	A
Deppea hernandezii	--	--	--	1, 8	A
Diospyros riojae	P	--	EN	1, 4	A
Juglans pyriformis	A	--	--	1, 2	A
Litsea glauscescens	Pr	--	--	1, 5	A
Magnolia dealbata	Pr	--	PN	1, 4	D
Magnolia schiedeana	A	--	EN	1, 6, 7	A
Matudaea trinervia	A	--	VU	1	A
Nopalxochia phyllanthoides	A	Appendix II	--	1	A
Ostrya virginiana	Pr	--	--	1, 2	A
Podocarpus reichei	Pr	--	DD	1,7	A
Symplocos coccinea	Pr	--	VU	1, 7	D
Taxus globosa	Pr	--	LR	1, 4	A
Ternstroemia dentisepala	--	--	--	1, 8	A
Ternstroemia huasteca	--	--	--	1	A
Tilia mexicana	P	--	--	1,7	A
Zamia vazquezii	A	Appendix II	EN	1,3	E
Zamia loddigesii	A	Appendix II	NT	1	A
Zinowiewia concinna	P	--	--	1	A

THREAT FACTORS AND STRATEGIES TO PREVENT THEIR EXTINCTION

In Mexico, habitat loss is a serious problem, mainly due to agricultural practices and deforestation; if these activities continue in temperate forests, a potential loss of many species, especially those scarcely represented, will occur in the near future (Luna et al. 2006a).

When we evaluated the threat factors for all the species studied, we noted that habitat destruction due to agricultural practices, forestry exploitation, and animal husbandry are the main problems, which are related with the predilection of humans for the establishment of settlements and towns and related activities in temperate forests (Luna et al. 2006a); other important threat factors are clandestine timber exploitation, illicit extraction, and traffic of plants (Table 3). If we consider that these factors occur within the NAPs, the conservation purposes of these areas will be more effective when the conservation programs minimize these threat factors.

Conservation strategies include: (1) demographic and ecological studies of the species analyzed herein; (2) collection of seeds and vegetative propagation for ex situ conservation in botanical gardens; (3) targeting of specific areas where these species inhabit, in order to include them in future conservation plans. Some of these strategies have been applied to several of the studied species (Table 3), but unfortunately in the case of cycads, for example, several problems with propagation by tissue culture techniques have persisted. In contrast, a few species herein studied are largely represented in botanical and public gardens and greenhouses, such as *Acer negundo* and *Cupressus lusitanica*. Also, some species are candidates for propagation by seeds such as *Ceratozamia mexicana*, *Ceratozamia sabatoi*, *Cupressus lusitanica*, and *Podocarpus reichei*, mainly in localities where they grow naturally.

FINAL CONSIDERATIONS

Regarding species geographic range based on the MER (SEMARNAT 2002), we propose that the Mexican government evaluate this criterion, because with it the distributional range of each species is roughly calculated, in absence of sufficient information and technology to objectively evaluate species geographic range. All Mexican temperate plant species studied herein belong to the category of very restricted geographic distribution (less than 5% of Mexican territory) established by the MER. We thus consider that the majority of the species restricted to specific localities are in imminent danger of extinction in short and medium term.

Repeatability of this methodology should be tested using independent data from different organisms, since areas of endemism are fundamentally the same for a wide range of taxa (Humphries and Parenti 1999; Crisp et al. 2001). In our study, some grid-cells might benefit from a detailed analysis based on other groups of the Mexican biota, such as birds, reptiles or mammals, because these taxa are well studied in this country; these grid-cells are Zacualtipán (Hidalgo-Veracruz), Jacala (Hidalgo-Querétaro), Misantla (Veracruz), Molango (Hidalgo), Pahuatlán (Hidalgo-Veracruz-Puebla), and Xalapa (Veracruz). Also, these six grid-cells need more detailed studies with the objective to be included in future conservation plans, because in our results these areas represent priority areas for conservation, considering that they contain most of the species and more restricted species studied herein and are threatened areas due to human impact.

Due to the fact that many challenges in biodiversity preservation are basically questions of spatial distribution (Peterson et al. 2000; Luna et al. 2009), our methodology offers an exciting potential for new insights and understanding of geographic distribution of Mexican biota, due that is more informative and operative to use small geographic units instead of using the Mexican states (Luna et al. 2006a), in order to detect areas with high values of richness and endemicity, particularly in the case of threatened taxa.

In the last years, a renewed enthusiasm on locating centers of richness or endemism is evident in attempts to optimize conservation strategies (Ceballos and Brown 1995; Linder 2001; Luna et al. 2004; Contreras-Medina and Luna 2007; Santa Anna del Conde et al. 2009), because the Mexican protected areas system was created based on a variety of factors, such as scenic beauty, historical significance, recreational use and watershed protection (Peterson et al. 2000), and biodiversity conservation was not considered in the management and planning of most of the protected areas, and therefore, is not optimized. Thus, key biodiversity features (Jackson et al. 2009) are almost missing from the Mexican protected areas system, and species with significant conservation concerns are underrepresented, and thus, the protected areas system is in consequence inefficient. Notwithstanding, as our analysis indicates, the present protected areas system does include some areas of importance, housing some of the species studied herein.

The grid system used in this study is an alternative method to perform a distributional analysis, and it can be helpful in understanding the distributional patterns of other groups of Mexican organisms. This procedure has the advantage to detect and locate with precision congruent patterns of distribution, and it objectively calculates the size of the area occupied by the species under study. Also, this system offers a standard format and minimal area units, which enable us to recognize better areas for conservation in situ. With this method, other biogeographical and biological criteria can be objectively included, and we can, thus, propose conservation areas in a more rigorous manner in order to preserve the biological diversity of Mexico.

Acknowledgments

Frank Columbus invited us to contribute with this manuscript. Juan J. Morrone, Jaime Jiménez and Tania Escalante made useful suggestions on an early draft of this manuscript. We are also indebted to the staff of the herbaria cited in the text for their courtesy and cooperation during our review of their specimens. Field assistance was provided by Armando Ponce, Alberto González, Hamlet Santa Anna del Conde, Jorge Escutia, Rogelio Aguilar, and Francisco Yberri. RCM dedicates this chapter to his daughter Regina Contreras Castañeda Support from projects PAPIIT IN206202, SEMARNAT-2004-01-311, and CONABIO W025 is gratefully acknowledged.

REFERENCES

Alcántara, O. & Luna, I. (1997). Florística y análisis biogeográfico del bosque mesófilo de montaña de Tenango de Doria, Hidalgo, México. *Anales Inst. Biol. UNAM (Ser. Bot.) 68*, 57-106.

Alcántara, O. & Luna. I. (2001). Análisis florístico de dos áreas con bosque mesófilo de montaña en el estado de Hidalgo, México: Eloxochitlán y Tlahuelompa. *Acta Bot. Mex*, *54*, 51-87.

Angulo, M. J. & Soto, M. (1990). Ebenaceae. *In* A. Gómez-Pompa, (Ed.). *Bioclimatología de Flora de Veracruz*. Instituto de Ecología, A. C. and University of California, Jalapa. *Fascicle*, *2*, 43.

Arreguín, S. M. L., Cabrera, L. G., Fernández, N. R., Orozco, L. C., Rodríguez, C. B. & Yepes, B. M. (1997). *Introducción a la flora del estado de Querétaro*. Consejo de Ciencia y Tecnología del estado de Querétaro, Instituto Politécnico Nacional y *Universidad Autónoma de Chapingo*. México, D. F. *361*.

Arriaga, L., Espinoza, J. M., Aguilar, C., Martínez, E., Gómez, L. & Loa, E. (coord.). (2000). *Regiones terrestres prioritarias de México*. Comisión Nacional para el Uso y Conocimiento de la Biodiversidad (CONABIO), México, D. F.

Borhidi, A. (2006). *Rubiáceas de México*. Akadémiai Kiadó, Budapest.

Bravo, H. (1978). *Las cactáceas de México*. Vol. 1. UNAM. México, D. F.

Bravo, H. & Sheinvar, L. (1995). *El interesante mundo de las cactáceas*. CONACYT-Fondo de Cultura Económica. México, D. F.

Breedlove, D. E. (1986). *Listados florísticos de México. IV. Flora de Chiapas*. Instituto de Biología, UNAM. México, D. F.

Cabrera, R. (1985). Aceraceae. *In* A. Gómez-Pompa, (Ed.). *Flora de Veracruz*. INIREB, Xalapa, Veracruz. México. *Fascicle*, *46*, 7.

Calderón de Rzedowski, G. (2001). Familia Aceraceae. *In* J. Rzedowski, & G. Calderón, (Eds.). *Flora del Bajío y de regiones adyacentes*. Instituto de Ecología. Pátzcuaro, Michoacán, México. *Fascicle*, *94*, 7.

Carranza, E. (2000). Familia Ebenaceae. *In* J. Rzedowski, & G. Calderón, (Eds.). *Flora del Bajío y de regiones adyacentes*. Instituto de Ecología. Pátzcuaro, Michoacán, México. Fascicle, *83*, 10.

Cartujano, S., Zamudio, S., Alcántara, O. & Luna, I. (2002). El bosque mesófilo de montaña en el municipio de Landa de Matamoros, Querétaro, México. *Bol. Soc. Bot. México*, *70*, 13-43.

Ceballos, G. & Brown, J. H. (1995). Global patterns of mammalian diversity, endemism, and endangerment. *Conserv. Biol.*, *9*, 559-568.

Cervantes, A. (2002). Revisión taxonómica de las especies mexicanas del género *Bernardia* Houst. ex Mill.: Euphorbiaceae: Bernardieae. Master in Sciences Thesis. *Facultad de Ciencias*, UNAM. México, D. F. *189*.

Chávez, V. M. & Rubluo, A. (1995). El cultivo de tejidos vegetales en la conservación. *In* E., Linares, P., Dávila, F., Chiang, R. Bye, & T. Elías, (Eds.). *Conservación de plantas en peligro de extinción: Diferentes enfoques. Instituto de Biología*, UNAM. México, D. F., 123-131.

Churchill, S. P., Balslev, H., Forero, E. & Luteyn, J. L. (Eds.). (1995). *Biodiversity and conservation of Neotropical montane forests*. The New York Botanical Garden, New York.

CITES (the Convention on International Trade in Endangered Species of Wild Fauna and Flora) in http://www.cites.org/.

CONABIO (Comisión Nacional para el Conocimiento y Uso de la Biodiversidad). (1998). "División política estatal". Escala 1:250,000 - 1000,000. México, D. F.

CONABIO (Comisión Nacional para el Conocimiento y Uso de la Biodiversidad). (1999). "Índice de cartas 1:50,000". Escala 1:50,000. Digitized from the Inventario de Información Geográfica, Instituto Nacional de Estadística, Geografía e Informática. (INEGI, 1992). México, D. F.

CONABIO (Comisión Nacional para el Conocimiento y Uso de la Biodiversidad). (2000). "Regiones Terrestres Prioritarias". Escala 1:1000,000. México, D. F.

CONANP (Comisión Nacional de Áreas Naturales Protegidas). (2007). http://www.conanp.gob.mx

Contreras-Medina, R. & Luna, I. (2001). Presencia de *Taxus globosa* Schltdl. (Taxaceae) en el estado de Chiapas, México. *Polibotánica*, *12*, 51-56.

Contreras-Medina, R., Luna, I. & Alcántara, O. (2001). Las gimnospermas de los bosques mesófilos de montaña de la Huasteca Hidalguense, México. *Bol. Soc. Bot. México*, *68*, 69-80.

Contreras-Medina, R., Luna, I. & Alcántara, O. (2003). Zamiaceae en Hidalgo, México. *Anales Inst. Biol. UNAM (Ser. Bot.)* 74, 289-301.

Contreras-Medina, R., Luna, I. & Alcántara, O. (2006). La familia Podocarpaceae en el estado de Hidalgo, México. *Rev. Mex. Biodivers*, *77(1)*, 115-118.

Contreras-Medina, R. & Luna, I. (2007). Species richness, endemism and conservation of Mexican gymnosperms. *Biodivers. Conserv*, *16*, 1803-1821.

Crisp, M. D., Laffan, S., Linder, H. P. & Monro, A. (2001). Endemism in the Australian flora. *J. Biogeogr*, *28*, 183-198.

Cuevas, R. & Núñez, N. M. (1994). *Cyathea mexicana* Schltdl. et Cham., en el estado de Jalisco, México. *Boletín IBUG*, *2*, 105-108.

Diego, N., Peralta, S. & Ludlow, B. (2001). El Jilguero. Bosque mesófilo de montaña. In N. Diego, & R. M. Fonseca, (Eds.). *Estudios florísticos en Guerrero, México*. Facultad de Ciencias, UNAM. México, D. F. 42.

Dieringer, G. & Espinosa, J. E. (1994). Reproductive ecology of *Magnolia schiedeana* (Magnoliaceae), a threatened cloud forest tree species in Veracruz. *Bull. Torrey Bot. Club*, *121*, 154-159.

ESRI (Environmental Systems Research Institute). (1999). *ArcView GIS* ver. 3.2. *Environmental Systems Research Institute Inc*. Redlands, USA.

Fonseca, R. M., Velázquez, E. & Domínguez, E. (2001). Carrizal de Bravos. Bosque mesófilo de montaña. In N. Diego, & R. M. Fonseca, (Eds.). *Estudios florísticos en Guerrero, México*. Facultad de Ciencias, UNAM. México. 41.

García, F. & Castillo, P. (2000). Aspectos ecológicos de *Taxus globosa* Schlecht. en las Mesas de San Isidro, municipio de río Verde, San Luis Potosí. *Biotam*, *11*, 11-18.

García-Franco, J. G., Castillo-Campos, G., Mehltreter, K., Martínez, M. L. & Vázquez, G. (2008). Composición florística de un bosque mesófilo del centro de Veracruz, México. *Bol. Soc. Bot. México*, *83*, 37-52.

Godínez-Ibarra, O. & López-Mata, L. (2002). Estructura, composición, riqueza y diversidad de árboles en tres muestras de selva mediana subperennifolia. *Anales Inst. Biol. UNAM (Ser. Bot.) 73*, 283-314.

Gómez, M. (1993). Juglandaceae. In A. Gómez-Pompa, (Ed.). *Bioclimatología de Flora de Veracruz*. Instituto de Ecología, A. C. and University of California, Xalapa. *Fascicle, 10*, 63.

Gutiérrez, L. & Vovides, A. P. (1997). An *in situ* study of *Magnolia dealbata* Zucc. in Veracruz state: An endangered endemic tree of Mexico. *Biodivers. Conser, 6*, 89-97.

Guzmán, U., Arias, S. & Dávila, P. (2003). *Catálogo de cactáceas mexicanas*. UNAM-CONABIO, México, D. F.

Hernández-Cerda, M. E. (1980). Magnoliaceae. In Gómez-Pompa, A. (ed.). *Flora de Veracruz*. INIREB, Xalapa, Veracruz. México. *Fascicle, 14*, 14.

Hilton-Taylor, C. (2000). 2000 IUCN Red list of threatened species. *IUCN*, Glanz, Switzerland.

Holmgren, P. K., Holmgren, N. H. & Barnett, L. C. (1990). *Index Herbariorum, part I, The Herbaria of the World*. International Association of Plant Taxonomy and New York Botanical Garden, Bronx.

Humphries, C. J. & Parenti, L. R. (1999). *Cladistic biogeography*. Oxford University Press, New York.

INEGI (Instituto Nacional de Estadística, Geografía e Informática). (1992). Inventario de Información Geográfica. México, D. F.

IUCN (Red List of Threatened Species). (2009). http:// www.iucnredlist.org.

Jackson, S. F., Walter, K. & Gaston, K. J. (2009). Relationship between distributions of threatened plants and protected areas in Britain. *Biol. Conserv, 142*, 1515-1522.

Juárez, A. K. (2008). Biodiversidad de la flora del bosque mesófilo de montaña del municipio de Huayacocotla, Veracruz. Bachelor Thesis. Facultad de Ciencias, *UNAM. México*, D. F. 72.

M. Kappelle, & A. D. Brown, (Eds.). (2001). *Bosques nublados del Neotrópico*. Instituto Nacional de Biodiversidad, Santo Domingo de Heredia, Costa Rica.

Linder, H. P. (2001). Plant diversity and endemism in sub-Saharan tropical Africa. *J. Biogeogr, 28*, 169-182.

Lozada, L., León, M. E., Rojas, J. & De Santiago R. (2003). Bosque mesófilo de montaña en el Molote. *In* Diego, N. & R.M. Fonseca (eds.). *Estudios florísticos en Guerrero, México*. Facultad de Ciencias, UNAM. México. *35*.

Luna, I., Almeida, L. & Llorente, J. (1989). Florística y aspectos fitogeográficos del bosque mesófilo de montaña de las cañadas de Ocuilan, estados de Morelos y México. *Anales Inst. Biol. UNAM (Ser. Bot.) 59*, 63-87.

Luna, I., Ocegueda, S. & Alcántara, O. (1994). Florística y notas biogeográficas del bosque mesófilo de montaña del municipio de Tlanchinol, Hidalgo, México. *Anales Inst. Biol. UNAM (Ser. Bot.) 65*, 31-62.

Luna, I., Alcántara, O., Morrone, J. J. & Espinosa, D. (2000). Track analysis and conservation priorities in the cloud forests of Hidalgo, Mexico. *Divers. distrib., 6*, 137-143.

Luna, I., Alcántara, O. & Contreras-Medina, R. (2004). Patterns of diversity, endemism and conservation: an example with Mexican species of Ternstroemiaceae Mirb. ex DC. (Tricolpates: Ericales). *Biodivers. Conserv, 13*, 2723-2739.

Luna, I., Alcántara, O., Contreras-Medina, R. & Ponce, A. (2006a). Biogeography, current knowledge and conservation of threatened vascular plants characteristic of Mexican temperate forests. *Biodivers. Conserv*, *15*, 3773-3799.

Luna, I., Alcántara, O., Ruiz, C. & Contreras-Medina, R. (2006b). Composition and structure of humid montane oak forests at different sites in central and eastern Mexico. *In* M. Kappelle, (Ed.). *Ecology and conservation of Neotropical montane oak forests. Ecological Series*, Vol. *185*, Springer Verlag, Heidelberg, 101-112.

Luna, I., Morrone, J. J. & Escalante, T. (2009). Conservation biogeography: a viewpoint from evolutionary biogeography. *In* F. Columbus, (Ed.). *Nature Conservation: Global, Environmental and Economic Issues*. Nova-Science Publishers, New York (in press).

Malda, G. (1990). Plantas vasculares raras, amenazadas y en peligro de extinción en Tamaulipas. *Biotam*, *2*, 55-61.

Martínez, L. & Chacalco, A. (1994). *Los árboles de la Ciudad de México*. Universidad Autónoma Metropolitana. México, D. F.

Mayorga, R., Luna, I. & Alcántara, O. (1998). Florística del bosque mesófilo de montaña de Molocotlán, Molango-Xochicoatlán, Hidalgo, México. *Bol. Soc. Bot. México*, *63*, 101-119.

McVaugh, R. (1992). Gymnosperms. *In* W. R. Anderson, (Ed.). *Flora Novo-Galiciana*. Vol. 17. The University of Michigan Herbarium, *Ann Arbor*, 4-119.

Meyrán, J. (1962). *Nopalxochia phyllanthoides* silvestre. *Cact. Suc. Mex*, *7*, 72-73.

Moretti, A., Sabato, S. & Vázquez-Torres, M. (1982). The rediscovery of *Ceratozamia kuesteriana* (Zamiaceae) in Mexico. *Brittonia*, *34*, 185-188.

Narave, H. V. (1983). Juglandaceae. *In* A. Gómez-Pompa, (Ed.). *Flora de Veracruz*. INIREB, Xalapa, Veracruz, México. *Fascicle*, *31*, 30.

Nee, M. (1981). Betulaceae. *In* A. Gómez-Pompa, (Ed.). *Flora de Veracruz*. INIREB, Xalapa, Veracruz, México. *Fascicle*, *20*, 20.

Niembro, A. (1990). *Árboles y arbustos útiles de México*. Ed. Limusa. México, D. F.

Osborne, R. (1995). The 1991-1992 world cycad census and a proposed revision of the threatened species status for cycads. *In* P. Vorster, (Ed.). *Proceedings of the Third International Conference on Cycad Biology*. Cycad Society of South Africa, Stellensboch, 65-83.

Pavón, N. & Rico-Gray, V. (2000). An endangered and potentially economic tree of Mexico: *Tilia mexicana* (Tiliaceae). *Econ. Bot*, *54*, 113-114.

Pacheco, L. (1981). Ebenaceae. *In* Gómez-Pompa, A. (ed.). *Flora de Veracruz*. INIREB, Xalapa, Veracruz. México. *Fascicle*, *16*, 21.

Pattison, G. (1986). *Magnolia dealbata*. *J. Magnolia Soc.*, *21*, 17-18.

Pérez-Escandón, B. E. & Villavicencio, M. A. (1995). *Listado de las plantas medicinales del estado de Hidalgo*. Universidad Autónoma del estado de Hidalgo. Pachuca, México.

Peterson, A. T., Egbert, S. L., Sánchez-Cordero, V. & Price, K. P. (2000). Geographic analysis of conservation priority: endemic birds and mammals in Veracruz, Mexico. *Biol. Conserv.*, *93*, 85-94.

Puig, H. & R. Bracho, (Eds.). (1987). *El bosque mesófilo de montaña de Tamaulipas*. Instituto de Ecología. México, D. F. *186*.

Ruiz-Jiménez, C. A., Meave, J. & Contreras-Jiménez, J. L. (2000). El bosque mesófilo de la región de Puerto Soledad, Oaxaca, México: análisis estructural. *Bol. Soc. Bot. México*, *65*, 23-37.

Rzedowski, J. (1981). *Vegetación de México*. Ed. Limusa. México, D. F.

Rzedowski, J. (1996). Análisis preliminar de la flora vascular de los bosques mesófilos de montaña de México. *Acta Bot. Mex*, *35*, 25-44.

Sánchez-González, A. & López-Mata, L. (2003). Clasificación y ordenación de la vegetación del norte de la Sierra Nevada, a lo largo de un gradiente altitudinal. *Anales Inst. Biol. UNAM (Ser. Bot.) 71*, 47-71.

Sánchez-Mejorada, H. (1982). *Las cactáceas de México*. Secretaría de Desarrollo Agropecuario del Gobierno del estado de México. Toluca, México.

Santa Anna del Conde, H., Contreras-Medina, R. & Luna, I. (2009). Biogeographic analysis of endemic cacti of the Sierra Madre Oriental, Mexico. *Biol. J. Linnean Soc.*, *97*, 373-389.

SEMARNAT (Secretaría de Medio Ambiente y Recursos Naturales). 2002. Norma Oficial Mexicana NOM-059-ECOL-2001, Protección ambiental-Especies nativas de México y de flora y fauna silvestres-Categorías de riesgo y especificaciones para su inclusión, exclusión o cambio-lista de especies en riesgo. *Diario Oficial de la Federación*. México, 6 de marzo, 1-80.

Sosa, V., Vovides, A. P. & Castillo-Campos, G. (1998). Monitoring endemic plant extinction in Veracruz, Mexico. *Biodivers. Conserv*, *7*, 1521-1527.

Soto, M. & Gómez-Pompa, A. (1990). Betulaceae. *In* A. Gómez-Pompa, (Ed.). *Bioclimatología de Flora de Veracruz*. Instituto de Ecología, A. C. and University of California, Xalapa. *Fascicle*, *1*, 45.

Standley, P. &. Steyermark, J. (1946-1976). Cactaceae, Euphorbiaceae and Lauraceae. *Flora of Guatemala*. Chicago Natural History Museum. *Fieldiana Botany*, Vol. *24*, Part VI.

Téllez-Valdés, O., Dávila-Aranda, P. & Lira-Saade, R. (2006). The effects of climate change on the long-term conservation of *Fagus grandifolia* var. *mexicana*, an important species of the cloud forest in eastern Mexico. *Biodivers. Conserv*, *15*, 1095-1107.

Van der Werff, H. & Lorea, F. (1997). Familia Lauraceae. *In* J. Rzedowski, & G. Calderón, (Eds.). *Flora del Bajío y de regiones adyacentes*. Instituto de Ecología. Pátzcuaro, Michoacán, México. *Fascicle*, *56*, 58.

Vázquez, J. (1994). *Magnolia* (Magnoliaceae) in Mexico and Central America: a synopsis. *Brittonia*, *46*, 1-23.

Vázquez, J., Cuevas, R., Cochrane, T. S., Iltis, H. H., Santana, F. J. & Guzmán, L. (1995). *Flora de Manantlán*. Botanical Research Institute of Texas, Forth Worth.

Vovides, A. P. (1981). Lista preliminar de plantas mexicanas raras o en peligro de extinción. *Biotica*, *6*, 219-228.

Vovides, A. P. (1985). Systematic studies on Mexican Zamiaceae II. Additional notes on *Ceratozamia kuesteriana* from Tamaulipas, Mexico. *Brittonia*, *37*, 226-231.

Vovides, A. P. (1988). Relación de plantas mexicanas raras o en peligro de extinción. *In* O. Flores, & P. Gerez, (Eds.). *Conservación en México: Síntesis sobre vertebrados terrestres, vegetación y uso del suelo*. INIREB. México, D. F., 289-302.

Vovides, A. P. (1995). Experiencias y avances en el conocimiento de las plantas mexicanas en peligro de extinción. *In* E., Linares, P., Dávila, F., Chiang, R. Bye, & T. Elías, (Eds.). *Conservación de plantas en peligro de extinción: Diferentes enfoques*. Instituto de Biología, UNAM. México, D. F., 139-144.

Vovides, A. P. (1999). Familia Zamiaceae. *In* J. Rzedowski, & G. Calderón, (Eds.). *Flora del Bajío y de regiones adyacentes*. *Fascicle* 71. Instituto de Ecología, A. C. Pátzcuaro, Michoacán, México. 17.

Vovides, A. P., Rees, J. D. & Vázquez-Torres, M. (1983). Zamiaceae. *In* Gómez-Pompa, A. (ed.). *Flora de Veracruz*. INIREB, Xalapa, Veracruz, México. *Fascicle, 26*, 31.

Yberri, F. G. (2009). Distribución geográfica de *Nopalxochia phyllanthoides* (DC) Britton et Rose (Cactaceae): modelos predictivos y conservación. Bachelor Thesis. Facultad de Ciencias, UNAM. México, D. F. 62.

Zamudio, S. (1992). Familia Taxaceae. *In* J. Rzedowski, & G. Calderón, (Eds.). *Flora del Bajío y de regiones adyacentes*. Instituto de Ecología. Pátzcuaro, Michoacán, México. *Fascicle 9*, 7.

Zamudio, S. (2002). Familia Podocarpaceae. *In* J. Rzedowski, & G. Calderón, (Eds.). *Flora del Bajío y de regiones adyacentes*. Instituto de Ecología. Pátzcuaro, Michoacán, México. *Fascicle, 105*, 7.

Zamudio, S. & Carranza, E. (1994). Familia Cupressaceae. *In* J. Rzedowski, & G. Calderón, (Eds.). *Flora del Bajío y de regiones adyacentes*. Instituto de Ecología. Pátzcuaro, Michoacán, México. *Fascicle, 29*, 21.

Zanoni, T. A. (1982). Cupressaceae. *In* A. Gómez-Pompa, (Ed.). *Flora de Veracruz*. INIREB, Xalapa, Veracruz, México. *Fascicle, 23*, 15.

Zulueta, R. & Soto, M. (1993). Magnoliaceae. *In* A. Gómez-Pompa, (Ed.). *Bioclimatología de Flora de Veracruz*. Instituto de Ecología, A. C. and University of California, Xalapa. *Fascicle, 6*, 38.

In: Species Diversity and Extinction
Editor: Geraldine H. Tepper, pp. 341-359

ISBN: 978-1-61668-343-6
© 2010 Nova Science Publishers, Inc

Chapter 11

GENETIC VARIABILITY IN *CAIMAN LATIROSTRIS* (BROAD- SNOUTED CAIMAN) (REPTILIA, ALLIGATORIDAE). CONTRIBUTIONS TO THE SUSTAINABLE USE OF POPULATIONS RECOVERED FROM THE RISK OF EXTINCTION

P.S. Amavet[*,a,b,d], *J.C. Vilardi*[c,d], *R. Markariani*[a], *E. Rueda*[a,d], *A. Larriera*[a,b] *and B.O. Saidman*[c,d]

[a]Departamento de Ciencias Naturales, Facultad de Humanidades y Ciencias, Universidad Nacional del Litoral, Ciudad Universitaria, 3000, Santa Fe, Argentina;
[b]Proyecto Yacaré, Laboratorio de Zoología Aplicada: Anexo Vertebrados (FHUC-UNL/ MASPyMA), Santa Fe, Argentina;
[c]Departamento de Ecología, Genética y Evolución, Facultad de Ciencias Exactas y Naturales, Universidad de Buenos Aires, Buenos Aires, Argentina;
[d]CONICET, Argentina

ABSTRACT

Genetic population analysis using molecular markers is probably the most important issue in conservation genetics and today it is a very useful tool for the study of species subjected to sustainable use. *Caiman latirostris* (broad-snouted caiman) is one of the two crocodilian species cited for Argentina. Their wild populations were drastically reduced in the 1950s and 1960s due to commercial hunting and intense alteration of their habitat, and *C. latirostris* was included in the Appendix I of CITES. Since 1990, management plans that use ranching system (harvest of wild eggs, captive rearing and reintroduction to nature) began in Argentina. Through these management activities, Argentine caiman populations were numerically increased and transferred to the Appendix II of CITES that allows the regulated trade of their products. Genetic population studies are being

---

[*] Corresponding autor: Tel.: +54 342 4575105 int. 128- Fax: +54 342 4575105 int.110. E-mail address: pamavet@fhuc.unl.edu.ar (PS. Amavet)

developed together with these sustainable use plans because genetic monitoring is considered essential in management program execution. This chapter includes genetic population studies about broad-snouted caiman in Santa Fe province, Argentina. Analysis related to variability, differentiation and genetic structure were carried out through isozyme electrophoresis, RAPD markers, and quantitative traits. Furthermore, paternity studies were conducted using microsatellite markers. The obtained results indicate that the broad-snouted caiman populations analyzed have low to intermediate genetic variability values, a significant population differentiation, and a high phenotypic variability for some of the morphometric traits studied. In addition, we found indications that *C. latirostris* mating system could include multiple paternity behavior, since we found more than one paternal progenitor in at least one of the families analyzed. Although the utility and broad applicability of molecular studies are widely accepted, this approach should be complemented by population analyses conducted by means of traditional methods such as morphometry, cytogenetics, ecology, and ethology to get a deeper biological knowledge of the species. To increase the efficiency in the use of natural resources the development of suitable legal guidelines as well as their effective implementation becomes very important to protect wildlife.

1. INTRODUCTION

Genetics Applied to Conservation

One of the main goals of conservation genetics is to preserve species as dynamic entities capable of evolving to adapt to environmental changes, it is therefore essential to understand the natural forces that determine evolutionary changes in populations and such information is essential for the sustainable use of natural resources.

Because evolution in its most basic level is a change in the genetic composition of a population, it only can occur if enough genetic diversity is present. Genetic variation may be expressed as differences among individuals at different levels, such as morphological traits, chromosomal structure and number, protein and DNA sequence polymorphisms.

Although the definition of gene is not simple, in its basic concept it is represented by a sequence of nucleotides in a particular segment (locus) of a DNA molecule. Genetic diversity can be represented by slightly different sequences in a gene. Many of these variants in DNA sequence result in differences in amino acid sequences of the protein encoded by this locus. Such a variation in a protein can result in structural changes in the molecule with no effect on its catalytic activity (isozymes), or cause morphological, biochemical or functional dissimilarities, which can give rise differences between individuals, which in time, may affect components of fitness such as the reproductive ability, survival rates or behavioral patterns [1].

The Species Analyzed in this Study

Caiman latirostris, commonly referred as the broad-snouted caiman (Fig. 1) is one of 23 species within Crocodylia Order, belonging to Alligatoridae family [2]. It is one of the two crocodilians cited in Argentina, whose geographical distribution covers the North of the country.

Figure 1. Adult 2.40 m length specimen of *C. latirostris*.

C. latirostris is a medium-sized crocodilian. Hatchlings weigh about 40 grams and measures 22 cm in total length approximately, whereas adult males can reach 2.60 m length [3]. Their wild populations are subject to management in Santa Fe Province, through "Proyecto Yacaré", a program of monitoring and restocking started in 1990 with the aim of species and environment conservation. This project was also proposed to ensure indirectly the conservation of the habitat that this species shares with a lot of birds, other reptiles and mammals associated with the wetlands. "Proyecto Yacaré" is based on the ranching system that consists of keeping the breeding stock in nature, harvesting the wild eggs for artificial incubation, captive rearing and reintroduction of a percentage of the animals to the wild [4, 5], for sustainable use and habitat conservation of this species. The initial objective was restoring broad-snouted caiman populations that were numerically reduced due to a drastic loss of their habitat. This was the consequence of several causes: advancing of agricultural frontiers, construction of canals and dams, and indiscriminate hunting [3, 6].

Thanks to these activities, the numerical recovery of broad-snouted caiman populations promoted the status change of *C. latirostris* in Argentina from Appendix I to Appendix II of CITES in the X Conference of the Parties - The Convention on International Trade in Endangered Species of Wild Fauna and Flora- of 1997 in Zimbabwe.

This status allows their commercial use by the ranching system within a legal framework that strictly regulates trade where a proportion of the animals born from eggs harvested from the wild can be derived to commercial utilization [7].

Currently, there are *C. latirostris* management programs using ranching system in other Argentinean provinces (Chaco, Corrientes and Formosa) of Northeast.

Genetic Markers used for Population Studies of Crocodilians

In wild population characterization it is essential the knowledge and measurement of the existent genetic variability because this is related to the dynamics of ecological and behavioral variables [8]. Genetic variation information allows to predict the behavior of local populations over time since a considerable level of variation increases the ability of the species to adapt to environmental changes [9], what is crucial in implementing management and conservation strategies. In variability analyses genetic markers are frequently used [10], which can be viewed as any phenotypic characteristic with single Mendelian inheritance, with negligible environmental effects on its expression, used to infer the genotype of an organism [11].

Crocodilians have shown relatively low values of variability [12 -14], depending on the taxa studied and the methodology used. Low levels of variation detected were attributed by various authors to one or a combination of factors. Among them, it has been suggested that the low variation found may be due to intrinsic characteristics of these species, such as large size, ectothermy, potentially long life span, and few changes in its external morphology from its earliest records. Further, genetic drift events, the relative stability of the environments they inhabit, and particular genetic systems were mentioned as possible causes of low genetic variation [12].

Initially, allozyme analyses were developed in *Alligator mississippiensis* wild populations [12, 15, 16, 17]. The authors suggested that low heterozygosity in the analyzed populations is due to important bottlenecks (genetic drive) underwent for several decades, added to particular characteristics of the species such as big size and long life span [12, 17]. These analyses were later applied on other crocodilian species [13, 18], and indicated that they show few allozyme polymorphic loci.

In relation to DNA studies, probes were initially used for fingerprinting in genealogy reconstruction and phylogenetic analyses in *A. mississippiensis* and several related crocodilians [19 - 21]. Additional studies on mitochondrial genes were conducted: Densmore and White [22] analyzed crocodilian phylogenetic relationships including genus *Caiman*. Later, Janke and Arnason [23] sequenced fully *A. mississippiensis* mitochondrial genome, Ray and Densmore [24] described the complete structure of mitochondrial control region (D-loop), and Glenn *et al.* [14] found low variation between populations of *A. mississippiensis*. By contrast, Ray *et al.* [25] and Farias *et al.* [26] estimated high variability on *Osteolaemus tetraspis tetraspis*, and good differentiation in populations of *Melanosuchus niger* and *Caiman crocodylus*. More recently, Yan *et al.* [27] identified different types of meat for consumption including *Alligator sinensis,* by mean multiplex PCR analyzing the citochrome b gene.

Concerning to application of RAPD technique (Random Amplified Polymorphic DNA) previous works are scarce in this reptile order, restricted to genus *Alligator* [28, 29] and *Crocodylus acutus* [30].

Some research groups started microsatellite analysis on wild crocodilian populations. These markers consist of motifs of one to five base pairs repeated in tandem widely distributed in genomes [31] which, thanks to their high polymorphism, are an important tool in the analysis of population variability, as well as kinship between individuals [32]. Applying these markers variability estimates were higher than values obtained using methods previously mentioned [28, 33, 34, 35, 36, 37]. Several authors have also used microsatellite amplification for genealogical and mating system studies on crocodilians [38 - 40].

In Argentina scarce antecedents of genetic works in *C. latirostris* are recorded. One of them is a comparison of karyotypic structure between this species and *Caiman yacare*, the other Argentine caiman species [41]. The results obtained showed a high similarity between them with a diploid number of 42 chromosomes, including 12 telocentric and 9 biarmed pairs, 2 of which are microchromosomes. Both species lack sexual chromosome differentiation and show very similar C and NOR banding patterns. These data agree with results obtained by simple banding and biometric analyses on *C. latirostris* in Brazil [42], but disagree with karyotypic structure studies conducted on all remaining crocodilian species, where no microchromosomes were observed [43].

Regarding to molecular techniques, population studies were made in Brazil using microsatellites developed for *C. latirostris* [44, 45]. Verdade *et al.* [46] applied these markers and found microgeographic variation among populations in the region of São Paulo state; Villela [47] found high levels of genetic diversity and heterozygosity in Brazilian populations. In contrast, in our country, data about specific differentiation, variability, genetic structure or mating systems on *C. latirostris* were not available.

Related to morphometric studies on this species, Verdade [48 - 50] analyzed body size and head measures in *C. latirostris*, as well as relative measures (ratio of absolute measures) at different ages. Besides, this author related morphometric measures between females and their offspring. In Argentina, Larriera *et. al* [51], and Piña *et al.* [52] carried out morphometric analyses using cranial measures on newborns of *C. latirostris*. The authors found sexual dimorphism evidence using multivariate morphometric analyses. Monteiro *et al.* [53] analyzed ontogenetic changes in the shape of skull of *C. latirostris* and other two caiman species. Other authors conducted similar analyses on other crocodilian species: Montague [54] presented a formula for predicting snout-vent length from 17 other body measurements and vice versa from data collected on wild *Crocodylus novaeguineae*. Also, the author suggested equations for predicting live body attributes from dried skulls. Miranda *et al.* [55], working with *Caiman crocodylus yacare* found that weight and 10 length measures in juveniles varied with egg incubation temperature. Later, Isberg *et al.* [56] presented a wide range of quantitative genetic analyses for economically-important traits in the production of saltwater crocodiles (*Crocodylus porosus*) in Australia, to propose a genetic improvement program. The authors employed several traits: reproductive traits, age at slaughter, survival and number of scales rows in univariate and multivariate analyses, and the results indicated that sufficient genetic variability was in their breeding stocks for a selection program to be successful.

The principal aim in our work was to know the structure and genetic variability existing within and among wild populations of *C. latirostris* in Santa Fe province in Argentina. We considered that obtaining such information is necessary and a priority because of the lack of genetic information about of this species in our country. Consequently, we started genetic studies of wild populations as a contribution to establish the best management, conservation and sustainable use strategy for this native species. Simultaneously, this work had the goal to increase the knowledge on the biology of this species and to contribute to the characterization, conservation and management of the environment that it inhabits.

2. POPULATION GENETICS STUDIES IN *CAIMAN LATIROSTRIS*

Genetic Variability Analysis

Allozyme Studies

The electrophoretic technique allows the separation of enzymatic proteins (allozymes) according to their net electric charge, molecular weight, and tridimensional structure. Because of its low cost, ease of use and the number of loci that can be examined, the protein polymorphism revealed by gel electrophoresis and specific staining techniques, was one of the most widely used tools to investigate levels of genetic variability in natural populations [57].The variation observed among band patterns from different individuals may be converted into genotype and allele frequencies, producing information that allows the estimation of the genetic variability of particular loci [1].

We studied four allozyme systems on 34 specimens of *C. latirostris* from four populations of Santa Fe province: "Estancia El Estero" (EEE) -Departamento San Javier (30° 29' S 59° 59' W), "Costa del Salado" (CSA) –Departamento San Cristóbal (29° 58'S 60° 50'W), "Estero del Paraje 114" (EDP)- Departamento San Javier (30° 43'S 60° 17'W), and "Arroyo El Espín" (AES) - Departamento Vera (29° 58'S 60° 04' W) (Fig. 2)

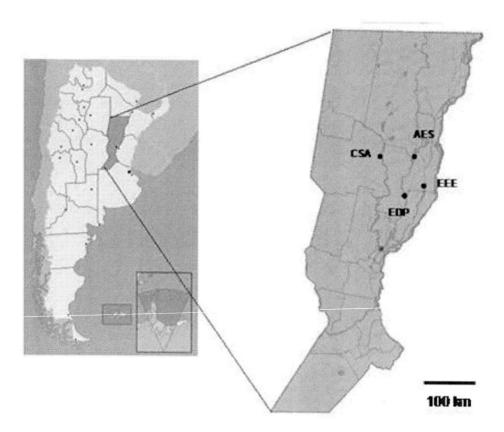

Figure 2. Map of Santa Fe Province showing the location of the four sampled populations.

All sampled individuals belonged to the group of animals intended for slaughter as part of activities of Proyecto Yacaré. Each animal was measured, weighed and sexed. From each individual portions of liver, kidney and heart tissues were dissected and preserved at -20°C.

For allozyme analysis 12.5% starch gels were prepared [13]. For sample preparation, of each individual a mix of tissue portions was homogenized in water following the Flint et al. method [13]. We revealed four different allozyme systems following Murphy et al. [11]: SOD (Superoxide dismutase), EST (Esterases), IDH (Isocitrate dehydrogenase) and MDH (Malate dehydrogenase).

Zymograms showed only one monomorphic band for each analyzed locus. Our data agree with isozyme studies by other authors who observed high levels of homozygosity in other species of crocodilians [12, 13, 15, 17].

RAPD Analysis

This technique uses primers of about 10 to 20 pb length to amplify genomic DNA regions. PCR products are solved on gels which typically detect multiple bands representing different fragments (loci) amplified. If there is variation in the primer annealing region, some bands will reveal patterns of presence / absence. This method analyzes many loci without requiring of sequencing the genome nor designing specific primers for the species under study. Its disadvantage is the dominant mode of inheritance and the considerable care in the protocols required to obtain repeatable results [58].

For RAPD analysis we used ten individuals coming from different nests of each studied population, making a total of 40 animals. DNA was extracted from blood samples obtained by spinal vein puncture [59, 60]. For DNA isolation we followed the method described in our previous work [61].

We amplified all samples using eight RAPD primers from Promega® (B050-10 and B051-10): A01, A02, A03, A05, A06, B04, B05 and B07. DNA extraction method and amplification protocols are described in our previous work [61]. PCR products were analyzed by electrophoresis on 4% polyacrylamide gels under denaturing conditions stained with silver nitrate solution [62].

The patterns observed were similar in all populations (Fig. 3), with the exception of AES which have two exclusive RAPD bands, A1-12 and A1-17, not present in the other populations.

Genetic variability levels in the studied population were estimated using Average number of alleles per locus (A), Percentage of polymorphic loci (P), and Expected Heterozygosity (H_e).

The Average number of alleles per locus is the sum of all alleles detected in all loci, divided by the total number of loci and provides information to the polymorphism [1]. The genetic polymorphism following Cavalli-Sforza and Bodmer [63] is the occurrence of two or more alleles in a locus in one population, each with significant frequency. In practice, it is decided arbitrarily an upper limit for the frequency of most common allele, typically 0.95 or 0.99; in this work we used 0.95 criterion. H_e represents the expected proportion of heterozygous individuals within a population for a diploid species. This is the most widely used measure of variability because it has a biologically meaning [64].

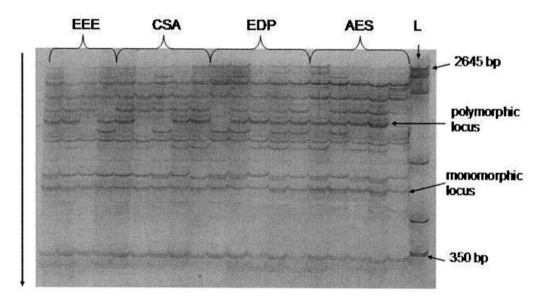

Figure 3. RAPD bands amplified with primer B05, visualized on 4% polyacrylamide gel stained with silver nitrate solution. EEE, CSA, EDP and AES refer to the populations analyzed. The vertical arrow shows electrophoresis direction, L: ladder pGem.

The estimated values of average number of alleles per locus (A), percentage of polymorphic loci (P) and expected heterozygosity (H_e) in each population are described in Table 1.

Table 1 Genetic variability measures at 32 loci in four populations of *Caiman latirostris*. N: sample size; A: average number of alleles per locus; P: percentage of polymorphic loci (5% criterion); H: expected heterozygosity; (Standard errors in parentheses)

POB	N	A	P	H_e
EEE	10	1.4 (0.1)	40.6	0.139(0.034)
CSA	10	1.3 (0.1)	31.3	0.130(0.038)
EDP	10	1.4 (0.1)	40.6	0.143(0.035)
AES	10	1.7 (0.1)	65.6	0.242 (0.037)

These variability values obtained were low to moderate contrasting with the results of isozymes where all loci were monomorphic

Taking into account the particular values of each population, Arroyo El Espín population (AES) have the highest variability values. The difference among this population and the remaining studied populations is also reflected in the phenogram (Fig. 4) based on Nei's genetic distances [65]. Nei's distance (D) is the most widely used measure of genetic distance. It is defined as $D = -\ln(I)$, where I (genetic identity) represents the probability of obtaining a homozygote after a crossing between individuals from different populations. A genetic distance $D = 0$ means absence of differences between two populations (i.e, $I = 1$).

By means of statistical pairwise comparisons (Wilcoxon tests) we could demonstrate that He estimates of AES were significantly different from those obtained in any of the remaining populations. The presence of bands exclusive of AES and the higher variability of this population may be related to the fact that EEE, CSA and EDP, share relatively stable environments while the population AES occupies a transitional environment that receives water flows (and expectedly gene flow) from two basins (Paraná and Salado rivers), and may be subject to higher levels of disturbances that can result in higher genetic variability values.

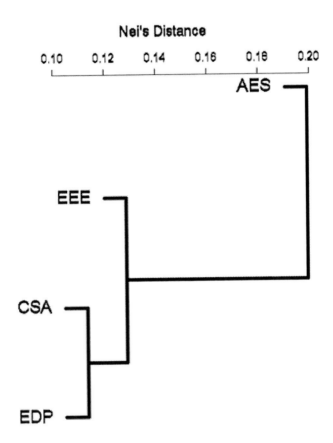

Figure 4. Phenogram based on Nei's (1978) genetic distances between populations of *C. latirostris*.

Population structure was analyzed by means of the F_{ST} statistics [66], which expresses the average reduction in heterozygosity of group of populations due to isolation among populations and random genetic drive within each population. This index varies from 0 to 1. The maximum F_{ST} value means a complete isolation (lack of migration) among populations. This index is estimated from the variation in allelic frequencies among populations.

In regard to the population structure, we found highly significant population differentiation among populations (F_{ST} =0.27). According to Wright [66] this high differentiation suggests that the gene flow among populations is approximately only one migrant every three generations. It may be argued that this low migration rate is an underestimation because in our case the genetic and geographic distance matrices are not correlated, what means that the balance among migration, mutation and genetic drift among these populations was not reached. However, the high genetic differentiation among population is consistent with the hypothesis of the occurrence of bottlenecks and genetic drift.

Phenotypic Variation Studies

The purpose of morphometric studies is to quantify the size and shape of living beings. Both size and shape are important biological properties of organisms, resulting from its genetic basis in complex association and interaction with internal and external environment, with important functional and ecological implications [67].

The differences in shape could indicate the existence of different functional roles performed by the same parts, different responses to the same selection pressures or differences in selective pressures, as well as differences in processes of growth and morphogenesis.

The shape analysis is an attempt to understand the various causes of variation and morphological transformation [68] and indirectly contributes to the study of genetic diversity.

From each of the four populations, four nests were random chosen, and from each one, 10 randomly selected newborns were measured. A total of 160 animals were measured within 48 hours after birth. The sex in newborns of *C. latirostris* was not taken into account because it can not be detected without sacrificing the animals.

Eleven morphometric measures suggested by Verdade [50] (Table 2, Fig. 5) were obtained from each individual.

Table 2. Description of allometric measurements obtained (Adapted from Verdade, 2001)

Acronym	Description
TTL	Total length: anterior tip of snout to posterior tip of tail.
SVL	Snout-vent length.
DCL	Dorsal cranial length: Anterior tip of snout to posterior surface of occipital condyle.
SL	Snout length: Anterior tip of snout to anterior orbital border, measured diagonally.
LCR	Length of the postorbital cranial roof: Distance from the posterior orbital border to the posterolateral margin of the squamosal.
CW	Cranial width: Distance between the lateral surfaces of the mandibular condyles of the quadrates.
SW	Basal snout width: Width across anterior orbital borders.
WN	Maximal width of external nares.
IOW	Minimal interorbital width.
ML	Mandible length: Anterior tip of dentary to the posterior tip of the retroarticular process.
BM	Body mass

In agreement with data from other authors [48, 69] who found low morphological variation; in this study the univariate tests (ANOVA) showed no significant differences among populations but highly significant differences in the quantitative traits among nests within populations. However, the overall multivariate test (MANOVA), showed highly significant differences between populations. This means that if we consider all the studied traits as a global phenotype, the results suggest that individual differences accumulate to show significant results.

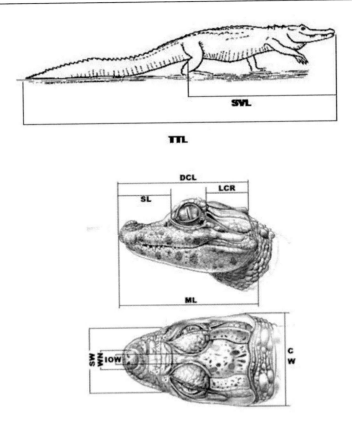

Figure 5. Allometric measurements obtained in individuals of *C. latirostris*. (Adapted from Verdade, 2001).

In a Principal Component Analysis (PCA) [70] we demonstrated that most of analyzed traits have a similar contribution to the first axis (CP1), which reflects differences in body size. In the CP2, three traits (CW, WN and IOW) have an important contribution with different signs. Similarly, the CP3 is also determined by three traits (WN, SVL and BM) that contribute with different signs. These results suggest that CP2 and CP3 are associated with differences in body shape.

The axis that showed the highest differences among populations was the CP2, and consistently the traits with the highest difference among populations were WN and IOW, which highly contribute to this axis.

It is noticeable that these measures of the head (WN and IOW) have major influence on the morphological variability, which could be a consequence of the important role of these traits on reproductive behavior and perhaps also in feeding.

High values of variation within populations of most traits were consistent with high heritability estimates. Heritability is one of the central concepts of quantitative genetics. In its simplest form, it is the proportion of total phenotypic variance in a population due to genetic differences among individuals. More specifically, the heritability is the proportion of phenotypic variance attributable to genetic variation that parents can transmit to their offspring. Therefore, the heritability of a character in a population determines their evolutionary potential [1].

In our case the high heritability of the analyzed traits is evidenced by the highly significant differences detected among nests (*i.e.* families or fraternal groups). The traits TL (total length), SVL (snout-vent length), and BM (body mass) are usually taken into account to measure the growth of animals for commercial use in breeding programs. Due to its high heritability values estimated in our work, a high response to directional selection on these traits is expected. In agreement, the high variability within populations detected for these traits indicates its great potential for improvement in breeding plans.

Although in our country all the commerce of *C. latirostris* products are based on the ranching system, which does not include the use of parents in captivity, the exponential growth of the sustainable utilization of species of crocodiles justifies a comprehensive analysis of these species in their biological (morphological, physiological, reproductive, ethological, etc.), as well as ecological aspects. In addition, information about the heritability of morphological characters is potentially useful for the development of close cycle systems (farming), as implemented in other countries like Brazil. This kind of systems will be soon developed experimentally in Argentina.

Paternity Analysis

The knowledge of reproductive strategies is an important contribution to the conservation of a species. Multiple paternity has been described in theory as an agent that increases the effective population size [71], potentially increasing the genetic diversity of populations that have experienced recent genetic drift events, such as population bottlenecks [40].

We used microsatellites as a tool to analyze this biological issue. Microsatellites have the advantage over other markers of being highly polymorphic and codominant. Because of that they have become the markers of choice in population studies. However, there are shortcomings that limit their use in species where the genomic information is scarce. In particular, their amplification requires the identification of the flanking sequences to obtain specific primers.

We sampled four adult *C. latirostris* females who were in the care of nests in populations of Santa Fe. Each female was assumed as the mother of the eggs in the corresponding nest. We obtained a blood sample from each female by puncturing the internal jugular vein [59, 60]. The eggs found in each nest were collected for artificial incubation at 31.5 °C, with 95% relative humidity in Proyecto Yacaré facilities in Santa Fe city. Five days after the birth of the offspring, 10 to 15 newborns from each nest were randomly chosen, and a blood sample of each specimen was obtained. For DNA isolation we used the method described in Amavet *et al.* (2007) [61]. We amplified the samples from the four families using four pairs of primers designed by Zucoloto *et al.* [44] for *C. latirostr*is: *Cla*μ 2, *Cla*μ 5, *Cla*μ 9, and *Cla*μ 10. PCR conditions are described in our previous work [72]. PCR products were analyzed by electrophoresis on 10% polyacrylamide gels under denaturing conditions stained with silver nitrate solution [62].

Through analysis of the bands obtained in polyacrylamide gels we determined the genotype of each locus in each individual analyzed. A paternity analysis was conducted assuming Mendelian segregation and independence of the loci studied. The possible paternal genotypes were proposed using three analysis systems: Single locus Minimum method [73], Cervus 3.0 program [74], and Gerud 2.0 program [75].

In the paternity analysis [72] we observed the existence of more than one father in at least one of the *C. latirostris* families, whose genotypes were assigned to the offspring with different probabilities.

Although the sample size was small, and the number of loci analyzed was low, we observed a possible breeding behavior of multipaternity in *C. latirostris*, in agreement with what Davis *et al.* [38] described for *A. mississippiensis*. The results obtained indicate that this reproductive behavior may be a feature shared by species of the family Alligatoridae.

3. Conclusion

The hunting pressure underwent by this species, combined with the effect of the changes and drastic reductions in their habitat (pipelines from rivers, deforestation, increased agricultural areas) might probably have created population bottlenecks by reducing their effective population size and causing the loss of alleles by genetic drift action. In determining the numerical increase and recovery of populations in recent years due to the implementation of management plans and conservation, we could consider that current populations have grown from a few surviving individuals, which could reveal the consequences of the effect founder.

With the data obtained in this work it could be argued that the values of variability depend largely on the analysis technique employed. The comparisons between species or populations with data from other authors are difficult to perform because of the great diversity of methods and parameters used.

Anyway, we can conclude that the variability values determined for the populations studied here for *C. latirostris* would be acceptable to keep population plasticity, what is reflected in the good levels of population viability and reproductive capacity in the environments they occupy.

The population-genetic analysis using molecular techniques is perhaps the most important issue of conservation genetics, and currently is very helpful for the study of commercially exploited populations or species subject to conservation plans.

Although its widely recognized usefulness and applicability of molecular studies, they should be complemented by population analyses carried out by traditional methods such as morphology, cytogenetics, ecology and behavior. In fact, these disciplines should be integrated with molecular studies to get a complete and integrative biological knowledge of native species of our region.

To increase the efficiency in the sustainable use of our natural resources, the development of suitable legal guidelines, as well as their effective implementation, becomes very important to protect wildlife. Therefore, the deeper understanding of the biology of these species can allow more appropriate human decisions that ensure a continued existence of our regional ecosystems.

ACKNOWLEDGMENTS

Our studies are supported by Universidad Nacional del Litoral (UNL) Santa Fe, Argentina. We would like to thank other members of Proyecto Yacaré- Laboratorio de Zoología Aplicada: Anexo Vertebrados (FHUC-UNL/ MASPyMA), Santa Fe, Argentina; especially Pablo Siroski and Gisela Poletta for their assistance in providing samples. PSA, BOS, JCV are members of Carrera del Investigador Científico, Consejo Nacional de Investigaciones Científicas y Técnicas (CONICET), Argentina.

REFERENCES

[1] Frankham, R; Ballou, JD; Briscoe, DA. *Introduction to Conservation Genetics*. Cambridge, UK: Cambridge University Press; 2002.
[2] Ross, P. *Status survey and conservation Action Plan:* Revised action Plan for Crocodiles. Gland, Switzerland: IUCN-The World Conservation Union; 1998.
[3] Larriera, A; Imhof, A. Proyecto Yacaré: Cosecha de huevos para cría en granjas, del género *Caiman* en la Argentina. In: Bolkovic, M; Ramadori, E, editors. *Manejo de fauna silvestre en la Argentina. Programas de uso sustentable. Dirección de Fauna Silvestre*. Buenos Aires, Argentina: Secretaría de Ambiente y Desarrollo Sustentable; 2006; 51-64.
[4] Larriera, A. *Caiman latirostris* ranching program in Santa Fe, Argentina, with the aim of management. In: *Crocodiles. Proceedings of the 12th Working Meeting of the CSG/IUCN*. Gland, Switzerland: The World Conservation Union; 1994; 188-198.
[5] Larriera, AL; Imhof, A; Siroski, P. Estado actual de los programas de conservación y manejo del género *Caiman* en Argentina. In: Castroviejo J, Ayarzagüena J, Velasco A, editors. *Contribución al conocimiento del Género Caiman de Suramérica*. Sevilla, España: Publicaciones Asociación Amigos de Doña Ana; 2008; 143-179.
[6] Ubeda, C; Grigera, D. Análisis de la evaluación más reciente del estado de conservación de los anfibios y reptiles de Argentina. *Gayana*, 2003, 67, 97-113.
[7] Larriera, A. The *Caiman latirostris* ranching program in Santa Fe, Argentina. The first commercial rearing. In: *Crocodiles. Proceedings of the 14th Working Meeting of the Crocodile Specialist Group of the Species Survival Commission of IUCN*. Gland, Switzerland and Cambridge, UK: The World Conservation Union; 1998; 379-385.
[8] Burke, JS. Influence of abiotic factors and feeding on habitat selection of summer and southern flounder during colonization of nursery grounds. PhD dissertation. Carolina State University, Raleigh, USA; 1991.
[9] Mayr, E. *Animal species and evolution*. Cambridge, USA: Belknap Press of Harvard University Press; 1963.
[10] Avise, JC. *Molecular Markers, Natural History and Evolution*. New York, USA: Chapman & Hall; 1994.
[11] Murphy, RW; Sites, JW; Buth, DG; Haufler, CH. Proteins: Isozyme electrophoresis. In: Hillis DM, Moritz C, Mable BK, editors. *Molecular Systematics*. (2nd Edition) Sunderland, Massachusetts, USA: Sinauer Associates Press; 1996; 51-120.

[12] Gartside, DF; Dessauer, HC; Joanen, T. Genic homozygosity in an ancient reptile (*Alligator mississippiensis*). *Biochemical Genetics*, 1976, 15, 655-663.

[13] Flint, NS; van der Bank, FH; Grobler, JP. A lack of genetic variation in commercially bred Nile Crocodiles (*Crocodylus niloticus*) in the North-West Province of South Africa. *Water SA,* 2000, 26, 105-110.

[14] Glenn, TC; Staton, JL; Vu, AT; Davis, LM; Alvarado Bremer, JR; Rhodes, WE; Brisbin Jr, L; Sawyer, RH. Low Mitochondrial DNA Variation among American Alligators and a Novel Non-Coding Region in Crocodilians. *Journal of Experimental Zoology*, 2002, 294, 312–324.

[15] Menzies, RA; Kushlan, J; Dessauer, HC. Low degree of genetic variability in the American alligator (*Alligator mississippiensis*). *Isozyme Bulletin*, 1979, 12, 61.

[16] Adams, SE; Smith, M.H.; Baccus, R. Biochemical variation in the American alligator (*Alligator mississippiensis*). *Herpetologica*, 1980, 36, 289-296.

[17] Lawson, R; Kofron, CP; Dessauer, HC. Allozyme variation in a natural population of the Nile Crocodile. *American Zoologist*, 1989, 29, 863-871.

[18] Jurgens, A; Kruger, J; van der Bank, FH. Allozyme variation in the Nile crocodile (*Crocodylus niloticus*) from Southern Africa. In: *Crocodiles. Proceedings of the 12th Working Meeting of the Crocodile Specialist Group, IUCN*. Gland, Switzerland: The World Conservation Union; 1994; 72-76.

[19] Aggarwall, RK; Lang, JW; Singh, L. Isolation of high –molecular-weight DNA from small samples of blood having nucleated erythrocytes, collected, transported, and stored at room temperature. *GATA*, 1992, 9, 54-57.

[20] Aggarwall, RK; Majumdar, KC; Lang, JW; Singh, L. Generic affinities among crocodilians as revealed by DNA fingerprinting with a Bkm-derived probe. *Proceedings of the National Academy of Sciences of USA*, 1994, 91, 10601-10605.

[21] Lang, JW; Aggarwall, RK; Majumdar, KC; Singh, L. Individualization and estimation of relatedness in crocodilians by DNA fingerprinting with a Bkm-derived probe. *Molecular and General Genetics*, 1992, 238, 49-58.

[22] Densmore, LD; White, PS. The systematic and evolution of the Crocodylia as suggested by restriction endonuclease analysis of mitochondrial and nuclear ribosomal DNA. *Copeia*, 1991, 3, 602-615.

[23] Janke, A; Arnason, U. The complete mitochondrial Genome of *Alligator mississippiensis* and the separation between recent Archosauria (Birds and Crocodiles). *Molecular Biology and Evolution*, 1997, 14, 1266-1272.

[24] Ray, DA; Densmore, LD. The crocodilian mitochondrial control region: General structure, conserved sequences and evolutionary implications. *Journal of Experimental Zoology*, 2002, 294, 334–345.

[25] Ray, DA; White, PS; Duong, HV; Cullen, T; Densmore, LD. High levels of genetic variability in west African dwarf crocodiles *Osteolaemus tetraspis tetraspis*. In: Grigg G, Seebacher F, Franklin CE, editors. *Crocodilian Biology and Evolution.* Chipping Norton, UK: Surrey Beatty and Sons; 2000; 58-63.

[26] Farias, IP; Da Silveira, R; de Thoisy, B; Monjeló, LA; Thonrbjarnarson, J; Hrbek, T. Genetic diversity and population structure of Amazonian crocodilians. *Animal Conservation*, 2004, 7, 265-272.

[27] Yan, P; Wu, XB; Shi, Y; Gu, CM; Wang, RP; Wang, CL. Identification of chinese alligators (*Alligator sinensis*) meat by diagnostic PCR of the mitochondrial cytochrome b gene. *Biological Conservation*, 2005, 121, 45-51.

[28] Glenn, TC; Dessauer, HC; Braun, MJ. Characterization of microsatellite DNA loci in American Alligators. *Copeia*, 1998, 3, 591-60.

[29] Wu, BX; Wang, YQ; Zhou, KY; Zhu, WQ; Nie, JS; Wang, CL; Xie, WS. Genetic variation in captive population of Chinese alligator, *Alligator sinensis*, revealed by random amplified polymorphic DNA (RAPD). *Biological Conservation*, 2002, 106, 435-441.

[30] Porras Murillo, LP; Bolaños, JR; Barr, BR. Variación genética y flujo de genes entre poblaciones de *Crocodylus acutus* (Crocodylia: Crocodylidae) en tres ríos del Pacífico Central, Costa Rica. *Revista de Biología Tropical*, 2008, 56, 1471-1480.

[31] Tautz, D; Schlötterer, C. Simple sequences. *Current Opinion in Genetics and Development*, 1994, 4, 832-837.

[32] Bruford, MW; Wayne, RK. Microsatellites and their application to population genetic studies. *Current Opinion in Genetics and Development*, 1993, 3, 939-943.

[33] Davis, LM; Glenn, TC; Elsey, RM; Brisbin Jr, IL; Rhodes, WE; Dessauer, HC; Sawyer, RH. Genetic structure of six populations of American alligators: a microsatellite analysis. In: Grigg G, Seebacher F, Franklin CE, editors. *Crocodilian Biology and Evolution*. Chipping Norton, UK: Surrey Beatty and sons; 2000; 38-50.

[34] Davis, LM; Glenn, TC; Strickland, DC; Guillette, LJ; Elsey, RM, Rhodes, WE; Dessauer, HC; Sawyer, RH. Microsatellite DNA analysis support an east-west phylogeographic split of American alligator populations. *Journal of Experimental Zoology*, 2002, 294, 352-72.

[35] Fitzsimmons, NN; Tanksley, S; Forstner, MRJ; Louis, EE; Daglish, R; Gratten, J; Davis, S. 2000. Microsatellite markers for *Crocodylus:* new genetic tools for population genetics, mating system studies and forensics. In: Grigg G, Seebacher F, Franklin CE, editors. *Crocodilian Biology and Evolution*. Chipping Norton, UK: Surrey Beatty and sons; 2000; 51-57.

[36] Fitzsimmons, N; Buchan, JC; Lam, PV; Polet, G; Hung, TT; Thang, NQ; Gratten, J. Identification of purebred *Crocodylus siamensis* for reintroduction in Vietnam. *Journal of Experimental Zoology*, 2002, 294, 373-38.

[37] Dever, JA; Strauss, RE; Rainwater, TR; mcMurry, ST; Densmore, LR. Genetic Diversity, Population Subdivision, and Gene Flow in Morelet's Crocodile *(Crocodylus moreletii)* from Belize, Central America. *Copeia*, 2002, 1078–1091.

[38] Davis, LM; Glenn, TC; Elsey, RM; Dessauer, HC; Sawyer, RH. Multiple paternity and mating patterns in the American alligator, *Alligator mississippiensis*. *Molecular Ecology*, 2001, 10, 1011-1024.

[39] Isberg, SR; Chen, Y; Barker, SG; Moran, C. Analysis of microsatellites and parentage testing in saltwater crocodiles. *Journal of Heredity*, 2004, 95, 445-449.

[40] McVay, JD; Rodriguez, D; Rainwater, TR; Dever, JA; Platt, SG; McMurry, ST; Forstner, MRJ; Densmore, LD. Evidence of multiple paternity in Morelet's Crocodile (*Crocodylus moreletii*) in Belize, CA, inferred from microsatellite markers. *Journal of Experimental Zoology*, 2008, 309 A, 643-648.

[41] Amavet, P; Markariani, R; Fenocchio, A. Comparative Cytogenetic Analysis of the American Alligators *Caiman latirostris* and *Caiman yacare* (Reptilia, Alligatoridae) from Argentina. *Caryologia*, 2003, 56, 489-493.

[42] Lui, JF; Tapia Valencia, EF; Boer, JA. Karyotypic analysis and chromosome biometry of cell cultures of the yellow throated Alligator (*Caiman latirostris* DAUDIN). *Revista Brasileira de Genetica*, 1993, 17, 165 - 169.

[43] Cohen MM; Gans C. The chromosomes of the order Crocodilia. *Cytogenetics*, 1970, 9, 81 -105.

[44] Zucoloto, RB; Verdade, LM; Coutinho, LL. Microsatellite DNA library for *Caiman latirostris*. *Journal of Experimental Zoology*, 2002, 294, 346–351.

[45] Zucoloto, RB. Desenvolvimento de seqüencias de DNA microssatélite para estudo de populações remanescentes de Jacaré-de-Papo Amarelo (*Caiman latirostris*), da região central do estado de São Paulo. Tese de Doutorado. Centro de Energia Nuclear na Agricultura, Universidade de São Paulo, Piracicaba, Brasil; 2003.

[46] Verdade, LM; Zucoloto, RB; Coutinho, LL. Microgeographic variation in *Caiman latirostris*. *Journal of Experimental Zoology*, 2002, 294, 387-396.

[47] Villela, PMS. Caracterização genética de populações de jacaré-de-papo-amarelo (*Caiman latirostris*), utilizando marcadores microssatélites. Dissertação de Mestrado. Escola Superior de Agricultura Luiz de Queiroz, Universidade de São Paulo, Piracicaba, Brasil; 2004.

[48] Verdade, LM. Morphometric analysis of the Broad-snout caiman (*Caiman latirostris*): An assessment of individual's Clutch, Body size, Sex, Age, and Area of origin. Ph. D Dissertation, Univ. of Florida, Gainesville, Florida, USA; 1997.

[49] Verdade, LM. Regression equations between body and head measurements in the broad-snouted caiman (*Caiman latirostris*). *Revista Brasileira de Biologia*, 2000, 60, 469-482.

[50] Verdade, LM. Allometry of reproduction in broad-snouted caiman (*Caiman latirostris*). *Brazilian Journal of Biology*, 2001, 61, 431-435.

[51] Larriera, A; Piña C. I.; Siroski P.; Verdade, L. Allometry of reproduction in wild Broad-Snouted Caimans (*Caiman latirostris*). *Journal of Herpetology*, 2004, 38, 301-304

[52] Piña, C; Larriera, A; Siroski, P; Verdade, LM. Cranial sexual discrimination in hatchling broad-snouted caiman (*Caiman latirostris*). *Iheringia Série Zoologia*, 2007, 97, 17-20.

[53] Monteiro, LR; Cavalcanti, MJ; Sommer, HJS. Comparative ontogenetic shape changes in the skull of *Caiman* species (Crocodylia, Alligatoridae). *Journal of Morphology*, 1997, 231, 53-62.

[54] Montague, JJ. Morphometric analysis of Crocodylus novaeguineae from the Fly River Drainague, Papua New Guinea. *Australian Wildlife Research*, 1984, 11, 395-414.

[55] Miranda, MP; Vanini de Moraes, G; Nunes Martins, E; Pinto Maia, LC ; Rus Barbosa, O. Thermic variation in incubation and development of Pantanal Caiman (*Caiman crocodilus yacare*) (Daudin, 1802) kept in metabolic box. *Brazilian archives of Biology and Technology*, 2002, 45, 333-342.

[56] Isberg, S; Thomson, P; Nicholas, F; Barker, S; Moran, C. Farmed saltwater crocodile. A genetic improvement program. Report for the Rural Industries Research and Development Corporation, Sydney, Australia; 2004.

[57] Lewontin, RC. Biology as ideology: The doctrine of DNA. New York, USA: Harper Collins; 1991.
[58] Hadrys, H; Balick, M; Schierwater, B. Applications of random amplified polymorphic DNA (RAPD) in molecular ecology. *Molecular Ecology*, 1992, 1, 55–63.
[59] Tourn, S; Imhof, A; Costa, AL; von Finck, MC; Larriera, A. Colecta de sangre y procesamiento de muestras en *Caiman latirostris* (Informe de avance). In: Larriera A, Imhof A, von Finck MC, Costa AL, Tourn SC, editors. *Memorias del IV Workshop sobre Conservación y Manejo del Yacaré Overo (Caiman latirostris)*. Santa Fe. Argentina: La Región, Fundación Banco Bica; 1993; 25-30.
[60] Zippel, KC; Lillywhite, HB; Mladinich, CRJ. Anatomy of the crocodilian spinal vein. *Journal of Morphology*, 2003, 258, 327–335.
[61] Amavet, PS; Rosso, EL; Markariani, RM; Larriera, A. Analysis of the population structure of Broad-Snouted Caiman (*Caiman latirostris*) in Santa Fe, Argentina, using the RAPD technique. *Journal of Herpetology*, 2007, 41, 285-295.
[62] Bassam, BJ; Caetano-Anolles, G; Gresshoff, PM. Fast and sensitive silver staining of DNA in polyacrylamide gels. *Analytical Biochemistry*, 1991, 196, 80-83.
[63] Cavalli–Sforza, LL; Bodmer, WF. The Genetics of Human Populations. San Francisco, USA: Freeman; 1971.
[64] Hedrick, PW. Genetics of populations. Boston, USA: Science books international; 1983.
[65] Nei, M. Estimation of average heterozygosity and genetic distance from a small number of individuals. *Genetics,* 1978, 83, 583-590.
[66] Wright, S. The genetical structure of populations. *Annals of Eugenics*, 1951, 15, 323-354.
[67] Marroig, G. When size makes a difference: allometry, life-history and morphological evolution of capuchins (*Cebus*) and squirrels (*Saimiri*) monkeys (Cebinae, Platyrrhini). *BMC Evolutionary Biology,* 2007, 7-20.
[68] Zelditch, ML; Swiderski, DL; Sheets, HD; Fink, WL. Geometric morphometrics for Biologist. A primer. San Diego, USA: Elsevier Academic Press; 2004.
[69] Wu, XB; Xue, H; Wu, LS; Zhu, JL; Wang, RP. Regression analysis between body and head measurements of Chinese alligator (*Alligator sinensis*) in the captive population. *Animal Biodiversity and Conservation*, 2006, 29, 65- 71.
[70] Amavet, P; Vilardi, JC; Rosso, E; Saidman, B. Genetic and morphometric variability in *Caiman latirostris* (broad-snouted caiman), Reptilia, Alligatoridae. *Journal of Experimental Zoology*, 2009, 311A, 258-269.
[71] Sugg, DW; Chesser, RK. Effective population size with multiple paternity. *Genetics,* 1994, 137, 1147-1155.
[72] Amavet, P; Rosso, E; Markariani, R; Piña, CI.. Microsatellite DNA markers applied to detection of multiple paternity in *Caiman latirostris* in Santa Fe, Argentina. *Journal of Experimental Zoology*, 2008, 309A, 637-642.
[73] Myers, EM; Zamudio, KR. Multiple paternity in an aggregate breeding amphibian: the effect of reproductive skew on estimates of male reproductive success. *Molecular Ecology*, 2004, 13, 1951-1963.

[74] Marshall, TC; Slate, J; Kruuk, L; Pemberton, JM. Statistical confidence for likelihood-based paternity inference in natural populations. *Molecular Ecology*, 1998, 7, 639-655.

[75] Jones, AG. GERUD 2.0: A computer program for the reconstruction of parental genotypes from half-sib progeny arrays with known or unknown parents. *Molecular Ecology Notes*, 2005, 5, 708-711.

In: Species Diversity and Extinction
Editor: Geraldine H. Tepper, pp. 361-381
ISBN: 978-1-61668-343-6
© 2010 Nova Science Publishers, Inc

Chapter 12

THE AQUATIC PLANT SPECIES DIVERSITY IN LARGE RIVER SYSTEMS

Dragana Vukov and Ružica Igić*
Department of Biology and Ecology, Faculty of Sciences,
University of Novi Sad, Serbia

ABSTRACT

This chapter relates to the survey of aquatic plants – hydrophytes in sections of Danube and Tisa rivers located in Central European part of Serbia, as well as in two small lowland rivers Zasavica and Jegrička. The methodology applied in the survey is in accordance to the Water Framework Directive of the European Council. The main aim of the survey was to list the plant species and to estimate their abundance according to the five-level descriptive scale, in each survey unit. The survey unit in studied rivers was the river kilometer, where the starting and ending point of river km is marked on the river bank with the navigation sign. In small rivers, which are not used for navigation, the survey units were of different length, and their beginning and ending points were the artificial objects on their banks (bridges, farms, etc.) The list of recorded plant species and their estimated abundance in survey unit was the base for calculation of diversity parameters: species richness, Shannon diversity index, and Evenness, in each survey unit. The next step in the study was the spatial analyzes of diversity parameters. Since the survey units are the continual sections along the river, it was possible to observe the longitudinal (upstream-downstream) as well as the latitudinal (backwaters-main channel) trends in diversity. The studied plants belong to the ecological group that contains limited number of species in the given eco-region; the main aim of the study was to test the indicative capacity of their diversity parameters. Analyzes of the aquatic plant species diversity showed that the species richness, diversity index, and evenness, have the great potential to indicate the hydrological conditions of the river.

---

* Corresponding author: Email: dragana.vukov@dbe.uns.ac.rs

INTRODUCTION

The EC Water Framework Directive (WFD – Directive 2000/60/EC) is the most comprehensive legislation ever enacted in Europe for addressing the integrity of fresh waters. For the purpose of this Directive the term "river" is defined as a body of inland water flowing for the most part on the surface of the land (Directive 2000/60/EC, article 2, paragraph 4). Among the environmental objectives of the WFD, the most important is the determination of the ecological status; which is an expression of the quality of the structure and functioning of aquatic ecosystems associated with surface waters (Directive 2000/60/EC, article 2, paragraph 21). One of the proposed biological quality elements for the classification of river's ecological status is the composition and abundance of aquatic flora (Directive 2000/60/EC, Annex V). In accordance to the WFD, the European Committee for Standardization (CEN) specified a method for surveying aquatic plants in running waters.

Aquatic plants, or hydrophytes (*sensu lato*), include different adaptive types which inhabit different water bodies. Their basic structural and physiological characteristics are in the accordance with the ecological conditions and resources of aquatic environment in which they develop and live. According to their eco-morphology and their position in the water body, hydrophytes are divided in to following groups: submerged plants, floating plants, and emergent plants, were the first two eco-morphological types are known under the term hydrophytes (*sensu stricto*), while the emergent plants are also called helophytes. Although the hydrophytes taxonomically include algae, mosses, ferns, and a variety of flowering plants, this study deals with aquatic vascular plants that belong to the divisions of higher plants, and to the group of hydrophytes (*sensu stricto*). Hydrophytes are important component of aquatic ecosystems and can be used to facilitate the monitoring of ecological status. In addition to their important ecological role, the use of hydrophytes as indicators of ecological quality in running waters is based on the fact that certain species and species groups are indicators for specific running water types and are adversely affected by anthropogenic impact. In certain situations the lack of hydrophytes is also characteristic for certain types of running waters habitats. For example, in deeper rivers hydrophytes may be absent due to habitat limitations imposed by water depth, current flow velocity, turbidity, etc. In large rivers, hydrophyte vegetation usually inhabits the littoral zone of riverbed, forming the narrow or wide belt along the bank.

Study Sites

In this chapter the results of the aquatic plants survey in rivers Danube and Tisa, were analyzed, and compared with the similar data obtained for rivers Zasavica and Jegrička.

Danube is the second largest, but the most important international river in Europe. It runs through ten European states, its river basin covers the surface of 817 000 km^2, and includes territory of 18 countries. It originates in Germany, and runs in to the Black Sea. Its flow is 2778 km long, and its average annual discharge is 6550 m^3/s. According to the typological requirements of the WFD, the Danube River has been divided into ten homogeneous section types (Robert et al, 2003). Studied part of the river is 588 km long (from river km 1433 to 846), and includes 358 km of the left and 451 km of the right river bank, and is placed in the

Republic of Serbia. It largely belongs to the Pannonian Plain Danube section type (river km 1433 – 1071), than to the Iron Gate Danube (river km 1071 – 931), while the shortest section of the studied part (river km 931 – 846) is characterized as Western Pontic Danube (Robert et al, 2003).

The Pannonian Plane Danube meanders through floodplain landscape. In this section Danube is lowland river, 400 to 1200 m (average 750m) wide, and up to 19 m deep (average 6m). The most frequent sediment is fine, mostly inorganic substrate with grain diameter less than 0.063 cm (pelal), and sand (psammal), while gravel (microlital, acal), and rocs (mega-, macro-, mesolithal) occur rarely. The average current velocity is 0.4m/s. The average slope is 0.04‰. The runoff character is Alpine, with the maximum water level and discharge in May and June, and minimum during the autumn and winter months (Robert et al., 2003). The water transparency varies between 40 and 120 cm (Vukov et al., 2006). The main anthropogenic pressure on the river's geomorphology in this portion is a numerous artificial structures used for flow regulation and bank protection, as well as a number of settlements of different size along the banks. The difference between left and right river bank is in its altitude: the left bank is lower, never exceeding 90 m above see level, while the right bank is sometimes even 289 m high (the loess plateau); in the diversity of sediment types present: along left riverside the only recorded substrates are pelal and psammal, while along the right riverside microlithal and megalithal also occur (Vukov et al., 2006). In this chapter data from 358 river km of left, and 225 river km of the right bank were analyzed.

The Iron Gate Danube runs through Đerdap/Iron Gates gorge, which is composed of four canyons (necks) and three valleys. The main channel has an average width of about 750 m. Slope values range from 0.04‰ and 0.25‰ (Robert et al., 2003). The dominant substrate are represented by cobbles, boulders and bedrocks (meso -, macro-, megalithal) and frequent coarse, medium and partial fine gravel interspersed with sand and mud in the slow-flowing parts. The main anthropogenic influence on the complex of hydrological, hydro-morphological, and by that on ecological conditions in this part of Danube was the construction of the dam and power plant Đerdap/Iron Gates I, which is on the river km 943, its construction was finished in 1972. This dam creates a reservoir that is usually some 200 km long, because the influence of the dam could be detected at the mouth of the Danube tributary Tamis (on the river km 1155) under conditions of the mean water level. Prior to the construction of the dams this part of the Danube had characteristics of mountain river, with the depth up to 82m, and current velocity of 50 m/s. Today, the average depth of the reservoir is 100 m, and current velocity is between 1.8 and 4 m/s. The main effect of the risen water level is wider littoral zone along the sections of river where the geo-morphology of the terrain allows it, which is potentially inhabited by hydrophytes, and the significantly slower current enabled higher sedimentation rate, which caused higher water transparency. In this section water transparency ranges from 70 to 150 cm (Vukov et al. 2006). In this chapter data from the right bank of the whole Iron Gates Danube were analyzed.

Western Pontic Danube is passing through a floodplain landscape, with right bank higher than left, mainly built of alluvial and diluvial sediments. The dominant substrate is gravel, with the small amounts of sand and mud (Vukov et al., 2006). Average width of main channel is 830 m, and depth is 8.5 m. The average slope is 0.04‰ (Robert et al., 2003). The main anthropogenic influence has the second dam - Đerdap/Iron Gates II, which is on the river km 863.4, and it was operational approximately ten years after the Đerdap/Iron Gates I. The upstream part of the Western Pontic Danube is the reservoir between two dams of hydro-energy complex Đerdap/Iron Gates, while the rest of this section is downstream of the second dam. This chapter deals with the data obtained during the survey of 85 river km along the right bank.

Tisa originates in Ukraine, and runs through Romania, Slovakia, Hungary and Serbia where it flows in to the Danube. Its basin covers the area of 157000 km2. The average slope of Tisa was 0.04‰, and its length was 1419 km. The consequence of its small slope and the adjacent lowland terrain was development of numerous meanders. After the catastrophic flood in 1830, the intensive and drastic hydro-technical measures were undertaken. Today most of the former meanders are cut of the main channel by flood protection dikes, and therefore the length of the river is now 966 km, also the average slope is higher, due to the construction of numerous dams, today it is 0.06‰. (Knežev ed., 2003). In this chapter data from the 150 river km of the Lower Tisa were analyzed, with the dam on the river km 63.

The Jegrička and Zasavica are two small lowland rivers. The length of Zasavica is 33 km, it is protected natural asset since 1997, and it has great value as a aquatic habitat of high naturalness. Jegrička is 27 km long, it was natural water body, but today it is canalized and it is part of large canal network (Vukov, 2003; Vukov et al., 2000; 2006; Borišev et al., 2004).

The Survey of Hydrophytes

The data analyzed in this chapter were collected during the survey of aquatic plants in time period between 1998 and 2007 (Janauer, 2003; Janauer et al., 2003; Vukov et al., 2000a; 2000b; 2003; 2004a; 2004b; 2005; 2006a; 2006b; 2008a; 2008b; Polić et al., 2004; Igić et al., 2000; Pajević et al., 2008). Investigations undertaken in above mentioned water bodies are methodologically compatible, and therefore comparable. The methodology applied in the survey is in accordance to the Water Framework Directive of the European Council and EU Standards (CEN). The main aim of the survey was to list the plant species and to estimate their abundance according to the five-level descriptive scale (Kohler, 1978; Kohler et al., 1971; Kohler & Janauer, 1995; Janauer, 2003; CEN Standard EN 1484, 2003; WFD/2000/60/EC), in each survey unit. The survey unit in studied rivers was the river kilometer, where the starting and ending point of river km is marked on the river bank with the navigation sign. In small rivers, and in side arms of Danube, which are not used for navigation, the survey units were of different length, and their beginning and ending points were the artificial objects on their banks (bridges, farms, etc.) The list of recorded plant species and their estimated abundances formed the base for calculation of diversity parameters: species richness, Shannon diversity index, and Evenness for each survey unit. The next step in the study was the spatial analyzes of diversity parameters. Since the survey units are the continual sections along the river, it was possible to observe the longitudinal (upstream-downstream) as well as the latitudinal (backwaters-main channel) trends in diversity.

Species Composition and Species Richness

The first step in the survey of aquatic plants is the inventory of plant species present in survey unit. That enables the creation of species composition list for each studied water body. In case of large rivers like Danube and Tisa left and right riverside were studied separately, and in case of Danube, which is characteristic by its numerous side arms and oxbows in its Pannonian section, those backwaters were also analyzed separately along the left and right riverside.

Biological diversity can be quantified in many different ways. The simplest one is species richness – the number of different species present. In all surveyed water bodies a total number of 40 vascular hydrophyte species was recorded. Among studied rivers a different number of aquatic species was found. The highest number of hydrophytes or the highest value of species richness (31) was found in side arms along left riverside of Danube (Table 1). It is followed by Zasavica River, with 29 species, than by main channel of Danube, where along right riverside 28, and along left bank 27 species were recorded. The number of species in side arms along right riverside is 27, in Jegrička – 22, while in Tisa, along both riversides, only ten vascular hydrophytes were recorded.

The species composition in each water body varies. Some hydrophytes are present in all surveyed water bodies, and they could be treated as typical for lowland running water bodies in the eco-region of Hungarian lowlands (WFD/2000/60/EC, Annex XI). Those species are: *Ceratophyllum demersum* L., *Hydrocharis morsus-ranae* L., *Lemna minor* L., *Najas marina* L., *Potamogeton pectinatus* L., *Salvinia natans* (L.) All. and *Spirodela polyrrhiza* (L.) Schleiden (Table 1). On the other side, there are species characteristic for the water body, where they exclusively ocurr. For example, *Aldrovanda vesiculosa* L., *Hottonia palustris* L., *Potamogeton acutifolius* Link, *Potamogeton trichoides* Cham. & Schlecht., *Ranunculus circinatus* Sibth., *Utricularia australis* R. Br. and *Wolffia arrhiza* (L.) Horkel. ex Wimm. were recorded only in Zasavica. Most of those species, except *Ranunculus circinatus* and *Wolffia arrhiza*, are natural rarities, and as such they contributet largely to valorization of Zasavica River as Special Nature Reserve. They witness about unique complex of ecological conditions and naturalness of this water body (Igić et al. 2000). This unique occurrence of hydrophytes could not be treated as typical, since it is known that aquatic environment tends to uniform habitat conditions, and that is, for example, the reason why the majority of aquatic species have wide, cosmopolitan, range of distribution. The species composition of Tisa and Jegrička are more typical for running waters in eco-region of Hungarian lowlands.

The composition of aquatic plant species in the Danube and in its side arms is unexpectedly rich for this large river, since the largest number of the surveyed units (river kms) belongs to the Pannonian Plane Danube. The environmental conditions of this part is not quite suitable for the successful development of aquatic vegetation: the sediment (mostly pelal and psammal) of littoral zone is very labile, because of its alluvial origin, also this labile sediment and current velocity enable low sedimentation rate, and that result in low water transparency. Also, in the other part of surveyed Danube (The Iron Gates section and the Western Pontic Danube), there are numerous reasons, such as high current velocity and rocky type of sediment, for example, why the occurrence of such diversity of hydrophytes is not to be expected.

Table 1. The list of recorded species

		DmcL	DmcR	DsaL	DsaR	TL	TR	Z	J
1.	*Aldrovanda vesiculosa* L.							+	
2.	*Azolla filiculoides* Lam.	+	+	+	+				
3.	*Ceratophyllum demersum* L.	+	+	+	+	+	+	+	+
4.	*Ceratophyllum submersum* L.			+	+			+	+
5.	*Elodea canadensis* Michx	+	+	+	+	+	+		+
6.	*Elodea nuttallii* (Planchon) St John	+	+	+	+				
7.	*Hottonia palustris* L.							+	
8.	*Hydrocharis morsus-ranae* L.	+	+	+	+	+	+	+	+
9.	*Lemna gibba* L.	+	+	+	+			+	
10.	*Lemna minor* L.	+	+	+	+	+	+	+	+
11.	*Lemna trisulca* L.	+	+	+	+			+	+
12.	*Myriophyllum spicatum* L.	+	+	+	+			+	+
13.	*Myriophyllum verticillatum* L.		+		+			+	+
14.	*Najas marina* L.	+	+	+	+	+	+	+	+
15.	*Najas minor* All.	+	+	+				+	
16.	*Nuphar lutea* (L.) Sibth. & Sm.	+	+	+	+			+	
17.	*Nymphaea alba* L.	+		+				+	+
18.	*Nymphoides peltata* (S.G.Gmelin) O.Kuntze	+	+	+	+			+	+
19.	*Potamogeton acutifolius* Link							+	
20.	*Potamogeton crispus* L.	+	+	+	+			+	+
21.	*Potamogeton gramineus* L.	+	+	+	+				
22.	*Potamogeton lucens* L.	+	+	+	+			+	+
23.	*Potamogeton natans* L.	+	+	+	+				+
24.	*Potamogeton nodosus* L.	+	+	+	+	+	+		+
25.	*Potamogeton pectinatus* L.	+	+	+	+	+	+	+	+
26.	*Potamogeton perfoliatus* L.	+	+	+	+				+
27.	*Potamogeton pusillus* L.	+	+	+	+			+	
28.	*Potamogeton trichoides* Cham. & Schlecht.							+	
29.	*Potamogeton zizii* Koch ex Roth	+	+	+	+				
30.	*Ranunculus circinatus* Sibth.							+	
31.	*Ranunculus trichophyllus* Chaix in Vill.			+					
32.	*Salvinia natans* (L.) All.	+	+	+	+	+	+	+	+
33.	*Spirodela polyrrhiza* (L.) Schleiden	+	+	+	+	+	+	+	+
34.	*Stratiotes aloides* L.			+				+	
35.	*Trapa natans* L.	+	+	+	+	+	+		+
36.	*Utricularia australis* R. Br.							+	
37.	*Utricularia vulgaris* L.			+				+	+
38.	*Vallisneria spiralis* L.	+	+	+	+				+
39.	*Wolffia arrhiza* (L.) Horkel. ex Wimm.							+	
40.	*Zannichellia palustris* L.		+					+	
	Total number of species	27	28	31	27	10	10	29	22

Legend: DmcL – Danube main channel left bank; DmcR – Danube main channel right bank; DsaL – Danube side arms left bank; DsaR – Danube side arms right bank; TL – Tisa left bank; TR – Tisa right bank; Z – Zasavica; J – Jegrička.

Table 2. Floristic similarity of surveyed water bodies

	DmcL	DmcR	DsaL	DsaR	TL	TR	Z	J
DmcL		0.95	0.93	0.92	0.54	0.54	0.61	0.78
DmcR			0.88	0.94	0.53	0.53	0.63	0.76
DsaL				0.90	0.49	0.49	0.67	0.79
DsaR					0.54	0.54	0.61	0.82
TL						1	0.36	0.62
TR							0.36	0.62
Z								0.63
J								

Legend: DmcL – Danube main channel left bank; DmcR – Danube main channel right bank; DsaL – Danube side arms left bank; DsaR – Danube side arms right bank; TL – Tisa left bank; TR – Tisa right bank; Z – Zasavica; J – Jegrička.

Floristic Similarity

The surveyed water bodies differ more or less in species composition and species richness. One of the most frequently used ways of quantifying the similarity of species composition (or floristic composition) is by calculating the Sørensen index of similarity. The Sørensen index, also known as Sørensen's similarity coefficient, is a statistic used for comparing the similarity of two samples (Sørensen, 1948). It can be calculated using following formula:

$$QS = \frac{2C}{A+B},$$

where A and B are the species numbers in samples A and B, respectively, and C is the number of species shared by the two samples.

The highest similarity is recorded between species composition along left and right bank of Tisa River. To be more specific, these two water bodies are identical floristically since their similarity index has maximal value – 1 (Table 2). It is of course, expected because their species lists are the same. The lowest similarity is between Tisa (both riversides), and Zasavica. In general, the highest similarity is between analyzed water bodies of Danube, where similarity index ranges between 0.88 and 0.95, than between floristic composition of Jegrička and water bodies of Danube (0.76 – 0.82), and between Zasavica and side arms surveyed along left riverside of Danube (0.67). The similarities between flora of Zasavica and floristic composition of other Danube backwaters is on the same level as similarity between Jegrička on one side, and Tisa and Zasavica on the other (0.61 – 0.63). The low values of similarity index clearly separate Tisa, from other water bodies (0.49 – 0.54).

The surveyed water bodies are rivers in the same eco-region, but they differ in their habitat conditions. This differences influence the different species composition. Since the Sørensen index could be used as a distance, or a dissimilarity measure, $(1-QS)$, it could also be a measure of β diversity, which describes alternations in species number in comparison of habitats in particular areas (Schulze, et al., 2005). If applied to the data on floristic similarity (Table 2), index calculated as a measure of distance show the opposite picture (Table 3). The higher values indicate the higher diversity, and indirectly, indicate the similarities and differences, as well as complexity of habitat conditions in surveyed rivers.

The Spatial Distribution of Species Richness

The spatial distribution of species richness is the graphical display of the number of recorded species (S) per survey unit in downstream direction. For the Danube River, on the same diagram both, the values gained for each river kilometer and for side arms along adequate riverside are shown. The surveyed section of the left riverside of Danube is 358 river km long, and it ranges between river km 1433 and 1073. The number of species recorded in those river kilometers rises in downstream direction (Figure 1). The highest leap could be noticed around river km 1160. The number of recorded species in side arms is usually higher than in adequate river kilometer of main channel. The right riverside covered with this investigation is 451 river km long, and it stretches between river km 1296 and 846. The number of recorded species rises in downstream direction and at river km 1078 it has the highest recorded value (18), while further downstream in gradually drops (Figure 2). In side arms the species richness is usually higher than in the part of the main channel they run in to.

Although the species composition and the totally recorded species number are the same when comparing left and right riverside of the Tisa River, the spatial distribution of species richness in surveyed river km differs. In the upstream part of the studied left riverside the species number is uniform downstream to the river km 90, and after that it begins to vary in wider range (Figure 3). Generally, the species richness does not show the trend of dropping or rising. Along the right riverside, the number of recorded species rises in the downstream direction, and in river km 72 it has its maximal value. Further downstream the trend of dropping could be observed (Figure 4).

Table 3. Floristic dissimilarity of surveyed water bodies

	DmcL	DmcR	DsaL	DsaR	TL	TR	Z	J
DmcL		0.05	0.07	0.08	0.46	0.46	0.39	0.22
DmcR			0.12	0.06	0.47	0.47	0.37	0.24
DsaL				0.1	0.51	0.51	0.33	0.21
DsaR					0.46	0.46	0.39	0.18
TL						0	0.64	0.38
TR							0.64	0.38
Z								0.37
J								

The Aquatic Plant Species Diversity in Large River Systems 369

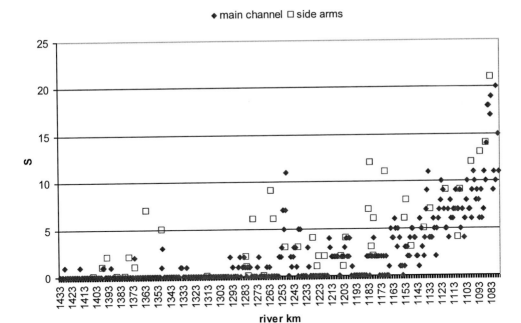

Figure 1. The spatial distribution of species richness along the left riverside of Danube.

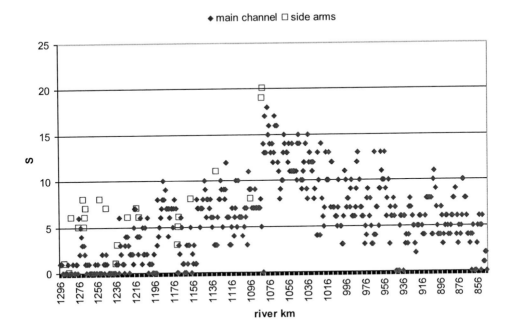

Figure 2. The spatial distribution of species richness along the right riverside of Danube.

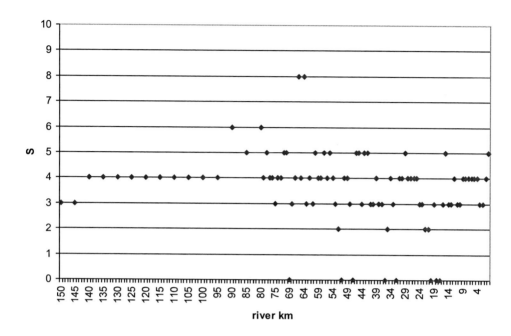

Figure 3. The spatial distribution of species richness along the left riverside of Tisa.

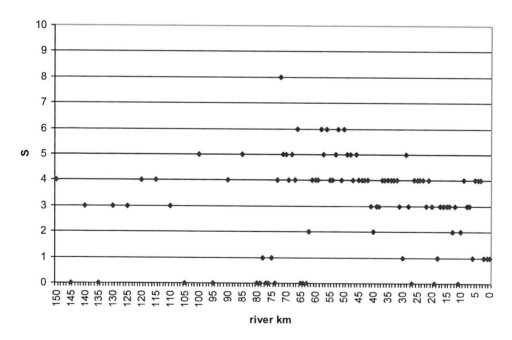

Figure 4. The spatial distribution of species richness along the right riverside of Tisa.

The number of recorded species in survey units of the Zasavica River rises in the downstream direction, although this trend breaks in survey units Poljansko and Batve, and in the last section Modran (Figure 5).

The Jegrička River is characteristic because the number of species recorded in its survey units does not indicate any regularity. If observed in continual downstream direction values are very variable (Figure 6).

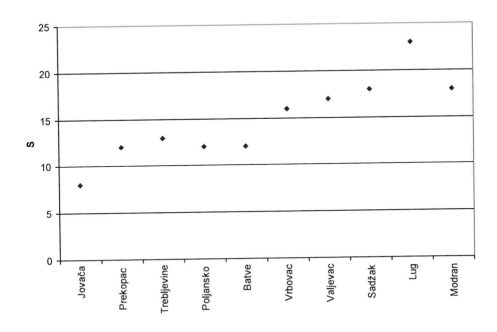

Figure 5. The spatial distribution of species richness along the Zasavica River.

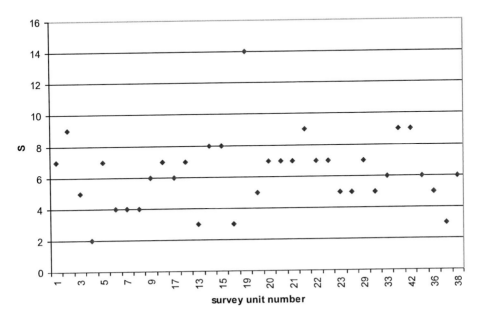

Figure 6. The spatial distribution of species richness along the Jegrička River.

The Diversity Index and Evenness

The diversity index is a mathematical measure of species diversity in a community. Diversity indices provide more information about community composition than simply species richness, because they also take the relative abundances of different species in to account. By taking the relative abundances in to account, a diversity index depends not only on species richness but also on the evenness, or equitability with which individuals are distributed among different species.

The one of the best known and most frequently used index is Shannon diversity index (Schultze et al., 2005):

$$H' = -\sum_{i=1}^{S} p_i \times \ln p_i,$$

where the Hs is diversity index, S the number of species, pi the relative frequency per area of the i-th species measured from 0-1.

In the study described in this chapter the parameter pi needed to be adjusted to the type of the data on species abundance gathered in the field. The abundance of the each species recorded in survey unit was estimated according to the five level descriptive scale (1-rare, 2-occasional, 3-frequent, 4-abundant, 5-very abundant). The proportion of i-th species in k-th survey unit was calculated as:

$$p_i = \frac{a_i}{a_{\max k}},$$

where a_i is the estimated abundance of the species i, and a_{maxk} is the maximal potential abundance of all species recorded in the survey unit k, and that could be calculated as:

$$a_{\max k} = 5 \times S_k,$$

S_k is the number of species recorded in the survey unit k, it is multiplied with 5, because 5 is the highest estimated abundance according to the already mentioned five level descriptive scale.

The H_s value rises with increasing number of species and increasing uniformity of distribution of relative abundance of individual species. A maximum H_s (H_{smax}) value indicates that all species in habitat are evenly distributed (Schulze et al., 2002). This value was introduced in order to describe the uniformity of species distribution in a habitat, or the evenness (E). Evenness in the survey unit k is calculated as:

$$E_k = \frac{H'_k}{\ln S_k}$$

where H'_k is the diversity index gained for the survey unit k, S_k is number of species in survey unit k, and lnS_k is the way to calculate H_{smax}.

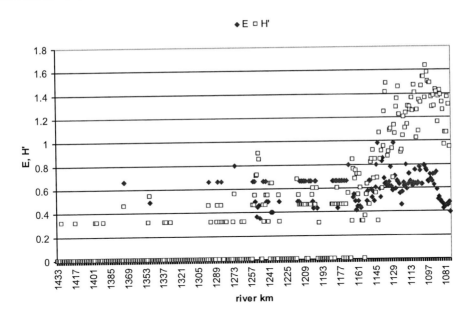

Figure 7. Evenness and diversity index, left bank, Danube, main channel.

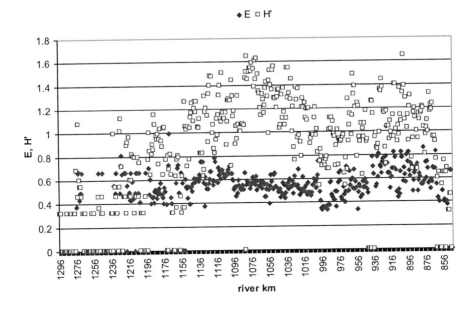

Figure 8. Evenness and diversity index, right bank, Danube, main channel.

The Spatial Distribution of Diversity Indices and Evenness

The values of evenness and diversity indices for each survey unit are presented on the same figure. Along the left riverside of Danube River, the values of diversity index grow in the downstream direction. The maximum value was recorded in river km 1098, and after that the values are dropping. According to the spatial distribution of diversity index, two sections could be identified along the left side of the Danube main channel: the upstream one (1433-1155), with values that range between 0 and 0.9; and the downstream section (1155-1076), with values that vary between 0.3 and 1.64, but is never 0 (Figure 7). The evenness does not show such regularity, and its value is in majority of cases lower than the value of diversity index. The highest evenness was recorded in river km 1143 and 1128, and is near 1 (0.98). Spatial distribution of diversity index values has two peeks: one on the river km 1080 (1.65), and the other on river km 902 (1.65). In downstream direction values are rising up to the first peek, than they are drooping on 0.56 (river km 969), after that they are rising again up to the second peek, and after that they are falling again (Figure 8). Again, due to the spatial distribution of the diversity index, the segmentation of Danube flow could be observed, this time in to three or maybe four sections. The upstream section (1296-1155), with diversity index very variable, and the 0 as frequent value; the second section (1155-965) with high values that quite rarely reaches down to 0; the third section has similar pattern as the previous one, only it is shorter it starts approximately at river km 960 and ends at 860; the fourth section that could be included is the shortest one, between river km 860 and 846. The values of evenness follow similar pattern that could be observed not by rising and fowling, but with grouped and ungrouped or more and less variable values in these sections. Evenness is very variable from river km 1296 to 1155, values vary between 0 and 1; further downstream they are grouped mainly between 0.4 and 0.6, and after that they vary again but in narrower range 0.46 and 0.86; in the fourth section values are dropping, and vary between 0 and 0.68. The values of evenness and diversity index gained for side arms were put at the adequate river km. In the side arms that are situated along the left riverside, the similar trend like in the main channel - the growth of diversity index - could also be observed. Unlike the diversity index, evenness is quite low and uniform, with values usually between 0 and 0.2 (Figure 9). In side arms along the right riverside the similar trend could be observed, the only significant difference is that evenness has higher values and varies between 0 and 0.6 (Figure 10).

The spatial distribution of evenness and diversity index in Tisa River also follows some pattern. Both, left and right riverside in the upstream part have relatively uniform values, while in the downstream section values are very variable and scattered (Figure 11; Figure 12). The highest diversity index was recorded in river km 66 along left, and in river km 72 along right riverside.

In Zasavica River, spatial distribution of evenness and diversity index follows the same pattern (Figure 13). First two survey units are in fact two streams with confluence before the third survey unit (Trebljevine). From Trebljevine to Vrbovac the diversity grows, and downstream of Vrbovac slightly decrease, after Lug it rises again, and in the last survey unit both diversity index and the evenness have the highest values.

Jegrička River is characteristic, because the spatial distribution of recorded evenness and diversity index does not show any pattern observed in previous cases. Both parameters have, in general, stagnant trend but with wade range of variation (H'- between 0.64 and 1.42; E- between 0.46 and 0.93; Figure 14).

The Aquatic Plant Species Diversity in Large River Systems

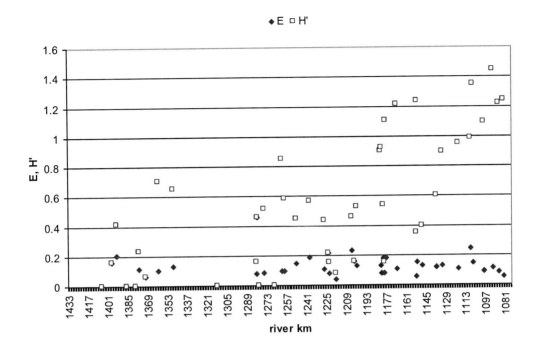

Figure 9. Evenness and diversity index, left bank, Danube, side arms.

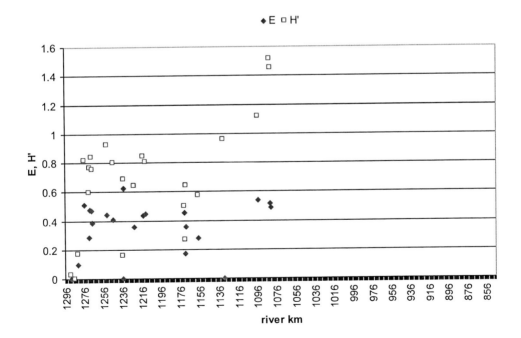

Figure 10. Evenness and diversity index, right bank, Danube, side arms.

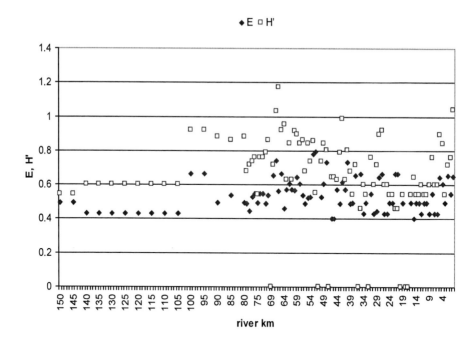

Figure 11. Evenness and diversity index, left bank, Tisa.

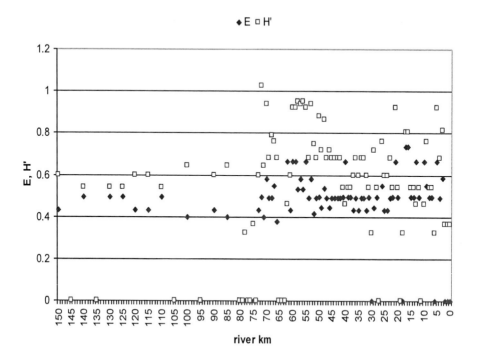

Figure 12. Evenness and diversity index, right bank, Tisa.

The Aquatic Plant Species Diversity in Large River Systems

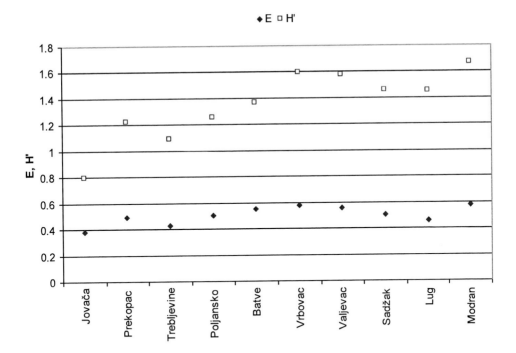

Figure 13. Evenness and diversity index, Zasavica.

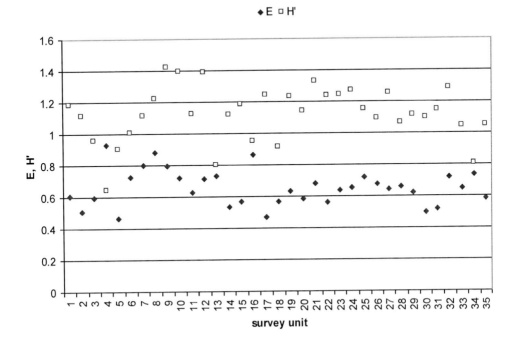

Figure 14. Evenness and diversity index, Jegrička.

Conclusion

In the running waters in eco region of Hungarian lowlands (Central Europe), grow approximately 40 to 50 plant species that belong to the group of hydrophytes (sensu stricto), which includes submerged and floating eco-morphological types. They have important ecological role in aquatic ecosystems, and can be used as parameters in the monitoring of ecological status, as well as the indicators of certain habitat types. The lack of hydrophytes in rivers also has indicative value. The data analyzed in this chapter originates from two large lowland rivers – Danube and Tisa, and two small lowland rivers – Zasavica and Jegrička. The largest human influence on hydrology and hydromorphology of Danube has the contruction of dams, and in the case of Tisa River the intensive and drastic hydro-technical measures, that cut of the river from its natural flood plane, and have shortened its flow length. Zasavica is protected natural asset with well preserved naturalness, while Jegrička used to be small natural lowland river with inflow to Tisa, and now its waterbed is canalized and it is part of the large network of canals, without connection to its source or mouth. The large rivers were divided in to different water bodies. Water bodies analyzed in the case of Danube River were: the left riverside of main channel, the right riverside of main channel, the side arms that are situated along the left riverside, the side arms that are situated along right riverside. Tisa was divided in to two water bodies: the left riverside, the right riverside. Small rivers were treated as single water body each. Since these waterbodies are different in their habitats conditions, primarely hidrology and hydromorphology, the floristic composition differ between them, too.

The total species number recorded in all of them toagether is 40. The highest number of hydrophytes was recorded in side arms along left riverside of Danube. The water body with the highest number of unique species (the species that have not been recorded in other surveyed water bodies) is Zasavica, and majority of those species are rare and endangered and therefore protected. This was expected, since Zasavica is protected natural asset. The composition of aquatic plant species in the Danube and in its side arms is unexpectedly rich. The floristic similarity between surveyed water bodies indicate that the most similar are the water bodies of Danube, while the Tisa is the least similar to other water bodies. The similarity index, and dissimilarity index that could be also used, show that although the surveyed water bodies belong to the same type – all of them are rivers – there is significant diversity between them as they are complexes of different habitat types. The measures of diversity used in this chapter – species richness, diversity index and evenness were calculated for every survey unit in each water body in order of their spatial distribution display. The spatial distribution of these diversity parameters in Danube is quite indicative. As it was mentioned before, the highest impact on the hydrological conditions of surveyed part of Danube River have two large dams constructed in 1970-es and 1980-es. Their impact on habitat conditions of the Danube flow is mostly noticeable in the sections upstream of them, where the accumulations are now situated. In those two accumulations, or reservoirs, the water level raised and the current velocity is slower. The main effect of the risen water level is wider littoral zone along the sections of river where the geo-morphology of the terrain allows it, which is potentially inhabited by hydrophytes, and the significantly slower current enabled higher sedimentation rate, which caused higher water transparency. These factors enabled the growth of aquatic vegetation in Danube. This is the case explained by intermediate disturbance hypothesis, because the environmental conditions in reservoirs of Danube became more diverse and favorable because of disturbance produced by construction of dams and they enabled the high diversity of hydrophytes in water body that otherwise would not support their growth in such number and abundance. This raises some questions: Is this diversity sustainable? Will the natural succession lead these new aquatic habitats to extinction? Is it justified to protect these areas as natural assets, even if that is not sustainable? The spatial distribution of diversity parameters also indicate that generally accepted division of the Danube River on sections, based on hydro- and geo-morphological characteristics, is not supported by the distribution and abundance of aquatic vegetation. The results of this spatial analyzes show the segmentation, and according to them the surveyed part of Danube has the: upstream section that still has the characteristics of Pannonian Plane Danube, and it stretches downstream to the (approximately) river km 1155; the first Đerdap/Iron Gates reservoir, between river km 1155 and the first dam (river km 943); the second Đerdap/Iron Gates reservoir, between two dams (river km 943 – 863); the Western Pontic Danube, which is downstream of second dam. This division of Danube river is based only on diversity of hydrophytes as one of the elements for the estimation of ecological status and must be supplemented with data on other quality elements.

The spatial distribution of diversity parameters along Tisa River also shows the segmentation of flow. The upstream part, which is not disturbed by dam, has low diversity, which is expected if the habitat conditions are took in to account; the part of Tisa upstream of dam, which is 10 to 15 river km long, and has higher diversity; and the downstream section, from dam to the mouth in Danube, which has a spatial distribution pattern like reservoir. In this case Danube, the large water mass that Tisa flows in to, has a role of a natural obstacle that is slowing the current velocity of Tisa, and that form the environmental conditions of the reservoir in the downstream section.

The spatial distribution of diversity parameters in Zasavica River show the regular growing trend in downstream direction, and could be characterized as referent for small lowland river. The situation in Jegrička is quite opposite. The distribution of diversity does not show any expected regularity observed in other rivers studied in this chapter. Only the detail analyses of patterns in diversity reveal some fragments along this water body, but they are, probably the result of the presence of different habitat types along it. According to diversity distribution pattern, it could be concluded that this water body is not a river at all, because it lacks the longitudinal gradient in diversity observed in Zasavica, which could be the referent case for small water bodies.

At the end, it could be concluded that even a small ecological group of species, like hydrophytes, growing in large rivers, like Danube and Tisa, with environmental conditions that usually do not support the successful development of aquatic vegetation could be the source of important information on habitat conditions that occur in those rivers. The ongoing implementation of Water Framework Directive in majority of European countries will among all, provide a large amount of data on hydrophyte species composition and abundance in whole variety of water bodies across continent. As it is shown in this chapter, that type of data could be successfully used to measure diversity.

Acknowledgments

Basic data for this study were collected during the realization of the MIDCC project funded by the Austrian Federal Ministry of Education, Science and Culture (bm:bwk, www.midcc.at), while further analyses was done in frame of the project No. 143037, financed by the Ministry of Science and Environmental Protection, Republic of Serbia.

References

Borišev, M., Igić, R. & Vukov, D. (2004). Aquatic macrophytes of Jegrička River on two connected sections (Zmajevo-Sirig; Sirig-Temerin). *IAD Limnological Reports Vol, 35*, 491-496.

Igić, R., Stojšić, V., Vukov, D. & Panjković, B. (2000). Retke i zaštićene biljke Zasavice. In D. Stevanović, (Ed). Zasavica. Sremska Mitrovica, PMF, Institut za biologiju u Novom Sadu, Goransko-ekološki pokret, *Sremska Mitrovica*, 42-48.

Janauer, G. A. (2003). Methods. Arch. Hydrobiol. (Suppl.) 147, *Large Rivers, 14,* 9-16.

Janauer, G., Vukov, A. & Igić, D. R. (2003). Aquatic macrophytes of the Danube River near Novi Sad (Yugoslavia, river-km 1255-1260), Archiv für Hydrobiologie, Suppl. – *Large Rivers, Vol, 147,* No. 1-2, 195- 203.

Kneževv, M. (2003). Nautičko ekološko istraživanje Tise. *Tiski cvet, Novi Sad*.

Kohler, A. (1978). Methoden der Kartierung von Flora und Vegetation von Süßwasserbiotopen. *Landsch. Stadt, 10*, 73-85.

Kohler, A. & Janauer, G. A. (1995). Zur Methodik der Untersuchungen von aquatischen Makrophyten in Fließgewässern, In: Handbuch Angewandte Limnologie (C. Steinberg, H. Bernhardt, & H. Klapper, Eds.), 1-22. Ecomed Verlag, Lansberg-Lech.

Kohler, A., Vollrath, H. & Beisl, E. (1971). Zur Verbreitung, Vergesellschaftung und Ökologie der Gefäß-Makrophyten im Fließwassersystem Moosach (Münchner Ebene). Arch. *Hydrobiol, 69*, 333-365.

Pajević, S., Igić, R., Stanković, Ž., Vukov, D., Krstić, B. & Rončević, S. (2008). Chemical composition of aquatic macrophytes from the Danube River and their role in biomonitoring and bioremediation, Fundam.Appl.Limnol./Arch.Hydrbiol.Suppl.162, No.1-2 - *Large Rivers. Vol. 18*, No. 1-2, 351-360.

Polić, D., Igić, R., Anačkov, G., Janauer, G. & Vukov, D. (2004). Aquatic vegetation in Dubovačka ada oxbow, 35th IAD Confernce, *Limnological Reports, Vol. 35,* IAD *Limnological Reports, Vol. 35*, 2004, 457- 462.

Robert, S., Birk, S. & Sommerhauser, M. (2003). Definition of Reference Conditions for the Section Types of the Danube River. In Sommerhauser, M., Moog, O. Developing the *Typology of Surface Waters and Defining the Relevant Reference Conditions*. UNDP/GEF Danube Regional Project.

Schulze, E. D., Beck, E. & Müler-Hohenstein, K. (2005). Plant Ecology. *Berlin Heildelberg*, Springer.

Sørensen, T. (1948). A method of establishing groups of equal amplitude in plant sociology based on similarity of species and its application to analyses of the vegetation on Danish commons. *Biologiske Skrifter, 5(4)*, 1-34.

Vukov, D. (2003). Aquatic Macrophytes in Zasavica River. MSc. thesis, *PMF Novi Sad*.

Vukov, D. (2008). Floristic and Ecological analyses of aquatic vascular macrophytes in Danube River (Serbia). *PhD Dissertation*, PMF, Novi Sad.

Vukov, D., Anačkov, G., Boža, P. & Janauer, G. (2004). Genus Potamogeton L. 1753 in the Danube River Corridor (rkm 1296 – 1076), IAD *Limnological Reports, Vol. 35*, 413- 419.

Vukov, D., Anačkov, G., Igić, R. & Janauer, G. (2004). The Aquatic macrophytes of »Mali Đerdap« (Danube, rkm 1039 – 999). *IAD Limnological Reports, Vol. 35*, 421-426.

Vukov, D., Boža, P., Igić, R. & Anačkov, G. (2008). The distribution and the abundance of hydrophytes along the Danube River in Serbia, *Central European Journal of Biology, Vol. 3(2)*, 177-187.

Vukov, D. & Igić, R. (2005). Aquatic Plants in Tisa River. In Knežev, M. ed. Ecological Research of Tisa River. *Tiski Cvet, Novi Sad.*

Vukov, D., Igić, R., Anačkov, G. & Boža, P. (2000). Akvatične makrofite Prekopca i Jovače. In Stevanović, D. ed. Zasavica. Sremska Mitrovica, PMF, Institut za biologiju u Novom Sadu, Goransko-ekološki pokret, *Sremska Mitrovica*, 49-56.

Vukov, D., Igić, R., Borišev, M. & Janauer, G. A. (2006). Quantitative Ecological analyses of aquatic vegetation in the Jegricka river, Archiv für Hydrobiologie, *Suppl. – Large Rivers, Vol. 158*, No. 16/3, 419-426.

Vukov, D., Igić, R., Boža, P., Anačkov, G. & Butorac, B. (2000). Survey into the aquatic macrophytes of the Zasavica Natural Reservation (Yugoslavia). In: L. Kórmóczy, & L. Gallé, (Ed). *Ecology of river vallyes*, Szeged: University of Szeged, Department of Ecology, 183-187.

Vukov, D., Igić, R., Boža, P., Anačkov, G. & Janauer, G. A. (2006). Habitat and Plant Species Diversity along the River Danube in Serbia, 36th International Conference IAD, *Vienna*, Austria: AC-IAD, 4-8 September, 127-131.

Water Framework Directive 2000/60/EC

Water Quality – Guidance standard for surveying of aquatic macrophytes in running water. EN 14184. CEN, 2003.

In: Species Diversity and Extinction
Editor: Geraldine H. Tepper, pp. 383-404

Chapter 13

CHANGES IN PLANT SPECIES DIVERSITY AROUND THE COPPER PLANT IN SLOVAKIA AFTER POLLUTION DECLINE

Viera Banásová*, Anna Lackovičová‡ and Anna Guttová†
Institute of Botany, Slovak Academy of Sciences, Bratislava, Slovakia

ABSTRACT

Air pollution is a factor that modifies the naturalness of ecosystems, including species diversity. The greatest source of emissions in the region of Krompachy (East Slovakia) was the copper smelter, which has been functioning more than 100 years. Air pollution induced large degradation process, e.g. vegetation decline or die back of a number of plant species. The long-term monitoring of vascular plant, bryophyte and lichen diversity around the smelter started in 1986. The plant diversity decreased rapidly due to the extreme pollution, mostly during the peak in 1987–1992. Over the last 15 years the amount of the pollution decreased. Currently the levels of SO_2 decreased under the limit, high concentrations of Cu, Zn, As and low pH is persisting in the soil. Achieved air quality standards due to the reduction of pollutants in the last years reflect in the first positive changes in the environment. Regeneration trend implies successful secondary succession towards increased species diversity and cover.

INTRODUCTION

Development and/or evolution is in the nature of every living and functioning ecosystem. These processes result into permanent changes of biodiversity of an ecosystem, which have natural character. An important component of all ecosystems in the Earth is man with all his activities. In many cases it is complicated to resolve, which changes of biodiversity were

---
* viera.banasova@savba.sk
‡ anna.lackovicova@savba.sk; † anna.guttova@savba.sk

induced only by man and which were determined by natural agent. The study of vegetation succession in man induced influences can help to understand complicated interactions within the ecosystem. Species diversity is not a stabile feature of the plant communities. It is responding to both global and local environmental condition. The qualitative and quantitative species composition and structure sensitively reacts to all changes of ecosystem under human impact. Knowledge of succession in plant communities in man-influenced habitat helps to understand overall biodiversity changes.

Biodiversity is often used as a "measure of the health" of ecosystems. Air pollution is an important factor that modifies the natural characteristics of the ecosystems in an industrial area. Habitat destruction due to the human activities and their implications (e.g. pollution) is the major cause of extinction of sensitive species. It can induce large degradation process in plant communities including the species diversity. The dramatic effects on vegetation often being detected several kilometers from the point source (e.g. Seaward, Richardson 1989).

The air pollution causes direct and indirect toxic effects on the plant communities. One of the most dangerous phytotoxic gaseous pollutants is SO_2, penetrating mostly into the plant tissue. It causes acute injury likes foliar necrosis as well as chronic injury, manifested in reduced growth and /or yield accompanied by foliar chlorosis and premature senescence (Wilson, Bell 1986). Except gaseous pollutants also heavy metal particles were distributed in the area of smelter.

This contribution presents results from the study carried out in the surroundings of a copper smelter situated in Slovakia (Central Europe). The long term monitoring of vegetation structure, species composition and diversity of vascular plants, bryophytes and lichens around the copper plant was carried out from 1987 to 2009. The study of epiphytic lichens initiated earlier (1983-1984). Main object of the research was the air pollution impact on the plant diversity and structure. This chapter can provide unique information as to the ambient air quality changes within particular grassland vegetation in the period of highest pollution level (1986-1992) to the period with decreasing of pollution when the air quality achieved standards (from 2003 on). The aims of the study are to explain the changes in plant species richness and diversity and regeneration capability of the grassland plant communities in relation to the intensity of the pollution.

MATERIAL AND METHODS

Study Area

The investigated area of the town Krompachy (22.9 km², alt. 379 m a.s.l.) is situated in the east Slovakia, in the basin of the river Hornád (Hornádske Podolie) surrounded by Branisko Mts (with Sľubica Mt.) from the north, Šarišská vrchovina Mts (with Roháčka Mt.) from the East and Hnilecké vrchy Mts from the south (Fig. 1). The climate is moderately warm with mean annual temperatures of -5 °C in January and 18 °C in July. Average annual rainfall is 550 mm. Prevailing winds blow in SW – NE and W – E directions (Mazúr et al. 1980). The study area is populated by ca 8700 people. Major stationary pollution source is smelter producing technical-purity copper. The plant in Krompachy was established in 1804. However, the history of copper productions dates back to the 14th century, when the term "cuprum Crumbasis" (copper of Krompachy) was known in European markets. During the past 100 years the smelter dispersed high amount of airborne pollutants. The principal pollutants were SO_2 and heavy metals (Cu, Zn, As).

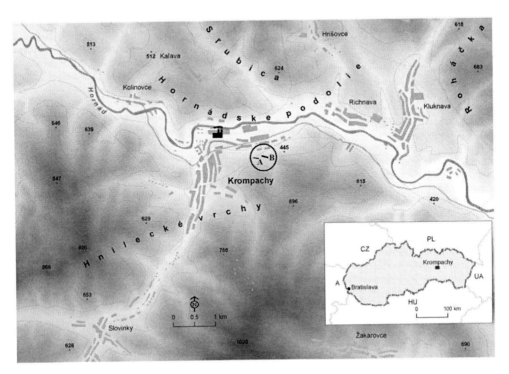

Figure 1. Map of investigated area. Localization of plant communities in site A (*Nardo-Agrostidion tenuis*) and site B (*Arrhenatherion*).

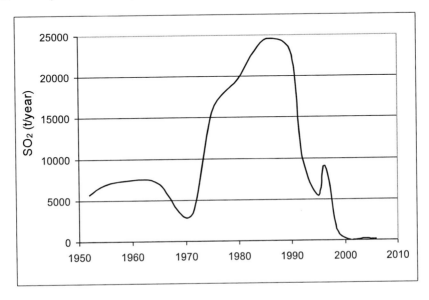

Figure 2. Annual SO_2 emissions (Hronec 1996; www.env.gov.sk).

Regarding the dispersal of air pollutants, the situation of the town was unfavorable. The atmospheric deposition influenced by climatological factors, topographical conditions, often at the microenvironmental level (Seaward, Richardson 1989). Prevailing winds and geomorphology of the terrain cause inversions, mainly in spring and winter (Babušík et al. 1984). In last 15 years the amount of the pollution decreased. Currently the levels of SO_2 decreased under the limit (Fig. 2), high concentrations of Cu, Zn, As and low pH remains in the soil.

Field Work

The field investigations were carried out in SW slope opposite of the smelter. The data sampling in the permanent plots started in 1987 in the period of the highest smelter pollution. Uneven pollution causes different degradation levels of the vegetation. This is why two grassland communities were chosen for monitoring on two sites (Fig. 1). At the beginning of the research the site A, situated on the chimney level, there hosted weakly damaged natural grown grassland of the union *Nardo-Agrostidion tenuis* Sillinger 1933. The site B, located lower, was hosted by the grassland of the union *Arrhenatherion* Koch 1926. Within either site three representative permanent plots were established for the long-term observations. The species composition and cover were sampled using phytocoenological relevés on the plots by Zürich-Montpellier approach (Braun-Blanquet 1964).

The species diversity index (H) was calculated from transformed percentage cover data using Shannon's formula. The cover data were transformed into mean percentage values. The similarity of plant community was calculated using Sørensen index (S) with the reference to the first year (c.f. Begon et al. 1990).

The analysis of *Avenella flexuosa* population was carried out in 1989 in the site A. The size of the flowering shoots of 25 plants was measured. The control population originated from non-contaminated soil in the Borská nížina lowland (SW Slovakia). The data were arranged into the size classes. The nomenclature of the vascular plants and mosses follows Marhold et Hindák (1998); the nomenclature of lichens is according to Bielczyk et al. (2004).

Elemental Analyses and Soil pH

For the investigation of the pollution distribution in the soil the samples were taken from rhizosphere (depth 0-10 cm) in both sites. Besides this, the soil samples were taken up to the depth of 60 cm, in the distance of each 5 cm, from the soil pit dug out in 1989. The total contents of metals were measured by flame AAS (atomic absorption spectroscopy). The exchangeable metals were estimated using extraction in 0.05 M EDTA (Žemberyová et al., 2007). The pH of the soil was measured potentiometrically.

RESULTS

The long term pollution with copper induced large degradation of vegetation including forest decline. The results of our study are commented in three time periods. They deal with the impact of emissions on soil, vegetation (two types of plant communities and terrestrial and epiphytic lichens), similarity and diversity of plants.

Soil Contamination

Continual long-term environmental burden of the area lead to the deposition of copper, zinc and arsenic in the soil. High metal concentrations in comparison with non-contaminated soils were detected in both sites (both plant communities). Their concentrations slightly increased since 1987 to 1992. It is known, that the forces may be very different by aerial deposition of metal particulates and other pollutants (e.g. Baker 1987). The atmospheric situation and the landscape configuration caused spatial variation in distribution of the pollution and soil contamination also in the study area. Soil in the site with *Nardo-Agrostidion* community was strongly attacked. Soil in the site with *Arrhenatherion* community was slightly contaminated. The metal concentrations in the soil increased with increasing the intensity of the pollution (Tab. 1). Although the pollution level after 1992 decreased we recorded high metal contents in the soil even in 2003. The total metal concentrations are still higher at present in comparison with the amounts characteristic for uncontaminated soil. Also the exchangeable fractions were high, which show the analyses taken in 2008 (Tab. 2).

Table 1. Contents of metals in the soil (mg.kg^{-1}) in the site A and the site B

A) *Nardo-Agrostidion*														
	1987			1989			1991			2003			2008		
	Cu	Zn	As	Cu	Zn	As	Cu	Zn	As	Cu	Zn	As	Cu	Zn	As
Min	245	600	80	1259	650	105	444	n.d.	130	n.d.	n.d.	n.d.	602	219	n.d
Max	350	760	240	3090	1079	205	2378	200	680	n.d.	n.d.	n.d.	607	591	n.d
Mean	267	650	160	2175	865	178	1145	198	347	1310	374	652	605	347	n.d
StDe	54	95	80	1295	303	103	1071	n.d.	293	n.d.	n.d.	n.d.	3	211	n.d

B) *Arrhenaterion*														
	1987			1989			1991			2003			2008		
	Cu	Zn	As	Cu	Zn	As	Cu	Zn	As	Cu	Zn	As	Cu	Zn	As
Min	182	215	80	103	186	56	444	n.d.	54	n.d.	n.d.	n.d.	568	266	196
Max	435	755	200	340	278	92	2378	179	114	n.d.	n.d.	n.d.	586	288	234
Mean	267	398	160	213	239	73	466	n.d.	91	1294	306	237	577	277	215
StDe	145	309	69	120	186	18	102	n.d.	32	n.d.	n.d.	n.d.	9	11	6

*n.d. – not detected

The analyses of soil test pit in 1989 showed that large amount of metals accumulated in the upper horizons (0-20 cm). The highest values for Cu 2159 mg.kg^{-1}, Zn 1079 mg.kg^{-1} and As 252 mg.kg^{-1} found on the soil surface (0-5 cm). The metal concentrations slightly decreased with increasing in the soil depth. The layers deeper than 20 cm showed the metal levels characteristic for natural non-polluted soils. (Fig. 3). Even high concentration of Cu 3260 mg.kg^{-1}, and Zn 1530 mg.kg^{-1} we found in litter.

Table 2. Metal concentration in the soil of the site B in 2008

Element	Total contents (mg.kg^{-1})	0.05 M EDTA (mg.kg^{-1})
Cu	577 ± 9	314 ± 5.0
Zn	277 ± 11	81.9 ± 3.0
Pb	104 ± 6	62.6 ± 4.2

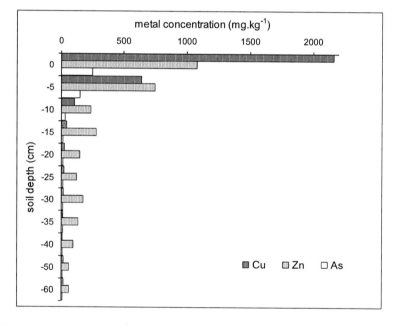

Figure 3. Distribution of metals (mg.kg^{-1}) along the soil profile in the site A.

SO$_2$ pollution leads to the decrease in soil pH. The soil pH is an important factor determining the species richness. Our data taken in the area opposite of the smelter showed that species richness decreased with increase of the soil acidity (Fig. 4). Due this we measured soil pH on permanent plots during the monitoring. The relation is rapid decrease in soil pH due to SO$_2$ pollution. In the plot A in *Nardo-Agrostidion* community pH quickly dropped in the short time: from 5.1 in 1987 to 3.9 in 1993. The decreasing of soil pH was not so dramatic in the plot B (*Arrhenatherion* community).

Vegetation Changes

The environmental pollution affected the species composition of the original vegetation during long time. The resistance to the aerial pollutants and heavy polluted metals as the solid particles are important trait of plants for its distribution. Plant species could switch from non-metalliferous substrates and recolonize the new modified by pollution metalliferous sites (Ernst 1990).

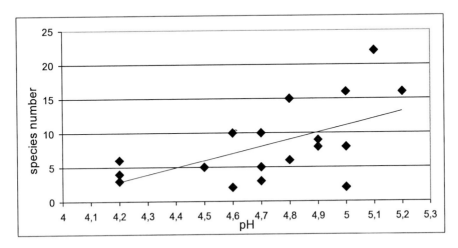

Figure 4. Relation of soil pH to the species richness in the surrounding of the smelter.

We observed several changes in plant communities in long time, which can divided to three time period. Several injuries were visible in the plant communities at the beginning of our research in 1987.

1. PERIOD OF THE AIR POLLUTION CULMINATION (BETWEEN 1980-1992)

1.1. *Nardo-Agrostidion* Community in the Site A

Mosaic grassland with prevailing of *Nardus stricta, Festuca rubra*, and *Agrostis capillaris*. The species such as *Avenella flexuosa, Briza media, Luzula luzuloides, Dianthus carthusianorum* and *Vaccinium myrtillus* occurred with high abundance in 1987. *Galium verum* and *Gentiana asclepiadea* occurred with low frequency. The rapid decrease in cover of dominants such as *Nardus stricta* happened in the next years. Deep rooting plant *Carlina acaulis* occupied the free space. Sensitive plants e.g. *Arrhenatherum elatius, Dactylis glomerata*, and *Galium verum* disappeared. They were replaced by tolerant species *Avenella flexuosa, Festuca rubra*, and *Holcus lanatus* later also by *Luzula luzuloides* which achieved also higher cover. Although was *Avenella flexuosa* the tolerant species, their population exposed to the high pollution level in situ showed reduction of shoot growth. The morphometric analyses show that the short plants prevailed in comparison with control population from uncontaminated site (Fig. 5).

The bare lands started to develop. There was a litter of non-destructed biomass from the last years which indicated the inhibition of soil microorganisms responsible for humus creation. Species richness of vascular plants dropped from 18 in 1987 to 11 in 1992.

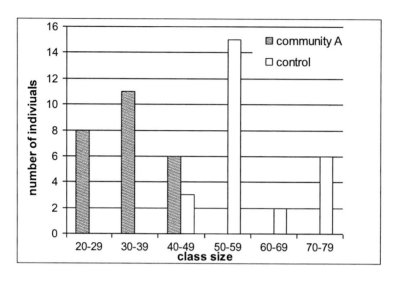

Figure 5. Growth parameters of the grass *Avenella flexuosa* (class size in cm).

1.2. *Arrhenatherion* Community in the Site B

The investigations in 1987 showed that the grassland vegetation was composed of 19 vascular plants. The tall grass such as *Arrhenatherum elatius, Dactylis glomerata, Festuca pratensis, Festuca rubra*, were dominat. Great portion of dicotyledonous (74 %) was detected. The species e.g. *Trifolium pratense, Campanula patula, Dianthus carthusianorum, Galium album, Knautia arvensis, Pimpinella saxifraga, Silene alba, Daucus carota, Ranunculus acris* were present. The plant community of this cutting meadow reached 90 % of total cover in 1987.

Qualitative and quantitative changes were observed already in 1988. Total plant cover reached 80 %. The presence and cover of dicotyledonous decreased and gradually increased the abundance of tolerant grasses. Sensitive species such as *Dactylis glomerata, Festuca pratensis, Poa pratensis* disappeared and were replaced by most tolerant grasses *Agrostis capillaris, Holcus lanatus, Luzula luzuloides*. Also some forbs disappeared e.g. *Centaurea jacea, Cruciata glabra, Galium album, G. aparine, Knautia arvensis, Ranunculus acris*. From dicotyledons could survive *Acetosa vulgaris, Dianthus carthusianorum, Daucus carota, Lotus corniculatus, Pimpinela saxifraga*. Metal tolerant plant *Silene vulgaris* and *S. alba* increased their cover. The occurrence of tolerant but not typical meadow species such as *Avenella flexuosa, Luzula luzuloides* indicated the previous changes induced by pollution.

Species richness of vascular plants dropped from 21 in 1987 to 12 in 1992.

1.3. Terrestrial Cryptogams in the Sites A and B

No epigeic lichens were observed on both permanent plots in 1987. Only one moss species *Ceratodon purpureus*, was recorded with low cover of 15 % in the spaces open to competition in the site with *Nardo-Agrostidion* community and in 1990 also in the site with *Arrhenatherion* community, with the cover of 5 %.

1.4. Epiphytic Lichens in the Region of Krompachy

The study of epiphytic lichens in 1980s showed the existence of „lichen desert" for foliose and fruticose species in the area of Krompachy. None of these species was occurring in the both monitoring plots, neither in the radius of ca 2 km around the smelter. Only the three toxitolerant species were recorded – *Lecanora conizaeoides*, *Scoliciosporum chlorococcum* and *Lepraria incana*; *L. conizaeoides* featured the highest cover and abundance on the tree trunks. A few thalli smaller than 1 cm^2 of the foliose species *Hypogymnia physodes* indicating the limits of the „lichen desert" zone, were recorded only in the lee of the slopes, ca 1.7 km SW of the pollution source. First well developed thalli of *Hypogymnia physodes* were found in the distance 4.2 km SE from the pollution source (skiing resort Plejsy), in the SW direction in the distance of 5.5 km (the town Slovinky) and ca 6 km NE from the smelter, further behind the village Hrišovce situated in the sheltered valley (Fig. 6a). Another foliose species *Parmelia sulcata*, was recorded in the last mentioned locality, too. However, the thalli of this lichen were not in good condition, showing typical pinkish to blackish, dying out patches. The area where no epiphytic macrolichens were observed was covering ca 12 km^2 around the smelter, including the areas of permanent plot (Lackovičová 1985, 1995).

2. PERIOD OF THE POLLUTION DECREASING (1993-2003)

The period the pollution peak was followed by pollution decline by 50% (1992-1995) due to both the recession and the purchase of a new technology. This trend is persisting up-to-now. In this time, first positive changes in both sites were observed: degradation of the soil was halted and the existing barren land became covered by protective cover of pioneer cryptogams. This is the process of Signal Reconvalescent Changes in the environment (Banásová et al., 2006).

2.1. *Nardo-Agrostidion* Community in the Site A

Although the amount of pollution decreased after 1992 the vegetation reflected this situation with time shift. Strong injury of vegetation was found also in 1993. Cu concentrations in the soil increased, pH was low (3.8). We noticed a huge decline and die-back of cover and diversity of vascular plants in 1992, after the period of extremely high intensity of pollution. Also the cover of tolerant species *Agrostis capillaris* and *Avenella flexuosa* dropped down (Fig. 7), whereas *Luzula luzuloides* temporarily occupied the free space. The forbs *Galium mollugo* and *G. verum* found sparsely. The total plant cover was only 10 %. A huge area of the bare lands occured rapidly.

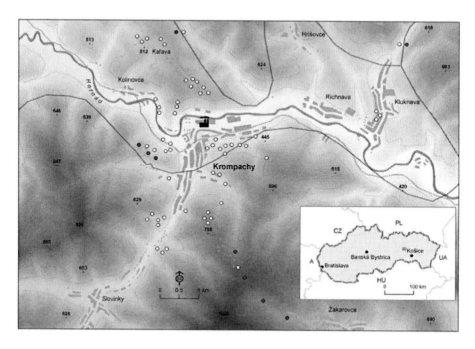

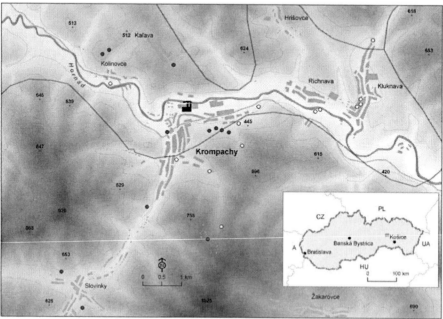

Figure 6. Occurrence of the lichen *Hypogymnia physodes* in the area of Krompachy in the period of pollution culmination 1983-1984 (a) and of low emission levels 2007-2009 (b). Legend: ● - presence, o – absence.

In 2003 no changes in vascular plant cover and diversity was observed. Only three grasses *Agrostis capillaris, Avenella flexuosa, Luzula luzuloides*, and young trees *Betula pendula* and *Populus tremula* were present, all with low cover (up to 30-50%). Species richness dropped to 4 vascular plants in 2003.

2.2. *Arrhenatherion* Community in the Site B

The total vegetation cover was only 50 %. The bare land increased their space. Humus horizon was washed out, a thick layer of undecomposed litter found on the soil surface. Only 13 plant species were able to resist the contamination stress.

Grasses prevailed and the portion of dicotyledons dropped to 46 %. Most tolerant grasses *Agrostis capillaris, Avenella flexuosa, Holcus lanatus,* and *Luzula luzuloides* remained in plant community (Fig. 7). Species *Agrostis capillaris* developed edge stand on the bare land. We found also in preserved grasses some injury. The former dominant grass *Arrhenatherum elatius* with low abundance in this time, had yellowish, degenerate spikes. The tussocks of *Luzula luzuloides* were low, chlorotic, the black and red spots occurred on the leaves. Species richness started to increase to17 species of vascular plants in 2003.

2.3. Terrestrial Cryptogams in the Sites A and B

After the onset of the period of low SO_2 emission levels significant changes in cover and diversity of terrestrial cryptogams were recorded in both types of plant communities. Already in 1998 a terrestrial lichen *Cladonia rei* was found along two axes of maximum impact of emissions – SSE from the smelters and ENE along the river Hornád, northwards of the village Richnava (cf. Hajdúk & Lisická 1999), falling within the „lichen desert" zone of foliose and fruticose epiphytic lichens delineated by Lackovičová (1995). This cup lichen is a subneutrophilous – moderately acidophilous species with wide ecological amplitude, often recorded in disturbed habitats. It was covering extensive carpets of bare lands together with the pioneer moss *Ceratodon purpureus*. This species was already recorded in the plots in the 1980s, in 2003 however it reached lower cover. *C. purpureus* is a common and widespread species with quite broad ecological tolerances. It grows in wide range of habitats, disturbed either naturally or anthropogenically (Jules & Shaw 1994). Its relatively rapid colonization after long-term disturbance helped prevent soil erosion and its growing abundance promotes an accumulation of organic matter, which favors the development of other elements of biodiversity, e.g. invertebrates. This is an initial stage of bringing the severe, barren land habitat back into life.

Tolerant, pioneer terrestrial lichen species *Cladonia rei, C. fimbriata,* and the bryophyte *C. purpureus* occupied a free niche on bare soil of the sampling plots after retreated vascular plants (30-50 % in site A, 10 % in site B). Out of these cryptogams, the highest values of cover featured corticolous lichens *Diploschistes muscorum* and *Placynthiella icmalea*, the carpets of which fragmented into patches (0.1-0.4 mm thick) were covering not only the barren land and hardly decomposed plant debris, but also other living cryptogams (e.g. thalli of *Cladonia* sp. div.).

A) Nardo-Agrostidion

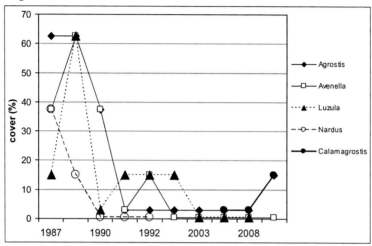

B) Arrhenaterion

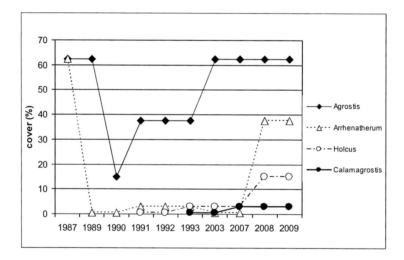

Figure 7. Cover changes of important grass species in the site A (a) and the site B (b).

2.4. Epiphytic Lichens in the Region of Krompachy

We observed only toxitolerant corticolous lichens *Lecanora conizaeoides*, *Lepraria* sp. and *Scoliciosporum chlorococcum* growing directly on the trees or shrubs on the permanent plot A in 2003. The phorophytes were pioneer woody species *Salix caprea*, *Corylus avellana* and *Populus tremula*. However, for the first time we recorded a juvenile thallus of a foliose lichen *Hypogymnia physodes* in the forested area neighboring the permanent plot. The tiny thallus of moderately toxitolerant lichen was growing on the bark of *Pinus sylvestris*. On the

3. PRESENT TIME (2007-2009)

The levels of emissions around the smelter have been under the limits for the past years. While the atmospheric SO_2 levels decreased under the limit (Fig. 2), high concentrations of heavy metals and low pH are remaining in the soil. In 2007 the cover and the diversity of vascular plants started to increase. The most interesting is comeback of *Agrostis capillaris* on bare lands. Young, native pioneer woody species *Salix caprea*, *Corylus avellana* and *Populus tremula* occurred abundantly. Hence the terrestrial lichens and mosses play an importnt role in covering of the bare soils.

3.1. *Nardo-Agrostidion* Community in the Site A

The air pollution induced the drastic changes in vegetation composition and species diversity. The development of the plant community progresses slowly. The total cover persists low. The cover of vascular plants became higher. Recently only 5 vascular plants were found. Vegetation is composed only from three grass species: *Agrostis capillaris*, *Avenella flexuosa*, *Calamagrostis epigejos*. Also young trees *Betula pendula* and *Populus tremula* occurred. Open soil was colonized by non-vascular species. According to the increasing of *Calamagrostis epigejos* cover (Fig. 7) and its intensive clonal growth we suggest that it will occupy large space in the future.

3.2. *Arrhenatherion* Community in the Site B

The total cover increases to 80 %. The species *Agrostis capillaris* is a dominant in the whole area. Other grasses e.g. *Arrhenatherum elatius*, *Holcus lanatus* and *Calamagrostis epigejos* are abundant (Fig. 7) also the portion of dicotyledonous increased. The presence of heavy metal tolerant species (e. g. *Acetosa pratensis*, *Agrostis capillaris*, *Arabidopsis halleri*, *Campanula patula*, *Equisetum arvense*, *Holcus lanatus*, *Silene vulgaris*, *Plantago lanceolata*, *Viola tricolor* agg.) recorded. Recently 27 vascular plants were found.

3.3. Terrestrial Cryptogams in the Site A and B

On-going colonization of disturbed and polluted soil by terrestrial cryptogams was observed in this period. In 2007 another copper-tolerant bryophyte *Pohlia drumondii* was found, together with persisting *Ceratodon purpureus*. Terrestrial, heavy metal tolerant lichens *Cladonia rei*, *C. fimbriata*, *C. subulata*, *Diploschistes muscorum* and *Placynthiella icmalea* spread from the edges of the barren lands towards their centers and significantly increased the cover up to 50-80% in site A (*Nardo-Agrostidion*), while to 5 % in site B (*Arrhenatherion*).

3.4. Epiphytic Lichens in the Region of Krompachy

Number of epiphytic species in 2007 was almost two times than in the previous data gathering campaign in 2003. On *Salix caprea, Corylus avellana* and *Populus tremula* abundantly occurred 8 species: corticolous *Lecanora conizaeoides, Lepraria* sp., *Soliciosporum chlorococcum*, foliose *Hypogymnia physodes, Parmeliopsis ambigua, Melanelia fuliginosa, Xanthoria parietina* and for the first time from the period also fruticose lichen *Evernia prunastri*.

In 2009 the total number of epiphytes increased up to 17, so as their frequency and abundance increased also on thicker branches of shrubs and tree trunks. Ecologically wide ranging species were represented by *Amandinea punctata, Melanelia fuliginosa, Scoliciosporum chlorococcum*. Species of subneutral to basic substrates were represented by *Caloplaca* sp., *Lecanora hagenii, Physcia adscendens* and *Xanthoria parietina*. The largest species group is that of acid to subneutral substrates: *Cladonia* cf. *coniocraea, Evernia prunastri, Hypogymnia physodes, Hypogymnia tubulosa, L. symmicta* agg., *Lepraria* sp., *Parmelia sulcata, Pseudevernia furfuracea* and *Usnea* sp. (a small, juvenile up to 1 cm long thallus). The acidophilous species *Lecanora conizaeoides* is retreating. Currently, there is no „lichen-desert" zone in Krompachy town. The smaller desert zone occurs now in the E of the smelter in the Hornád valley (Fig. 6b). The trees around the smelter, in the built-up areas of the town or the town outskirts support again crustose, foliose and fruticose epiphytic lichens. The recolonisation event is in line with the same trend, regarding the species composition, as is evident in other urban environments in Europe (e.g. Dymytrova 2009, Gombert et al. 2005, Isocrono et al. 2007, Larsen et al. 2007, Munzi et al. 2007)

SIMILARITY AND SPECIES DIVERSITY

The qualitative and quantitative changes in plant communities can be expressed by similarity (S) and species diversity (H) indices. The level of vegetation changes expressed by these indices was connected with the intensity of the air pollution stress. We calculated Sorensen's similarity index (S) to compare qualitative vegetation changes.

The S started decrease in 1990 and was the lowest in 2008 in *Nardo-Agrostidion* community (site A). This shows that the former plant community completely changed. The species e.g. *Dianthus carthusianorum, Galium verum, G. album, Nardus stricta, Vaccinium myrtillus* abundant in *Nardo-Agrostidion* disappeared quickly. The grass *Agrostis capillaris* preserved as the singleton species of the former community while *Calamagrostis epigejos* and epigeic lichens are immigrant species recently (Fig. 8a).

The S index of *Arrhenatherion* community (site B) showed also the important changes after pollution increasing. Species richness and cover decreased and the open place arised. The similarity dropped down. Some species such as *Galium aparine, Gypsophilla muralis, Urtica dioica* occupied the free space only for one vegetation season. The invading garden plant *Lysimachia punctata* occurred only in two season 1990-1993. Some scattered and short time occurrence of *Lysimachia punctata* may be caused by anthropogenic distribution. Regress to the former plant community started in 2007 however the similarity index reached only 76.9 % now (Fig. 8b). Some species such as *Festuca pratensis, Rumex acetosella*

a) *Nardo – Agrostidion*

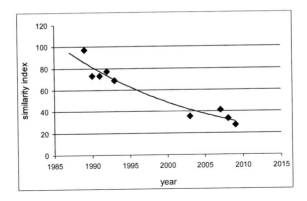

b) *Arrhenaterion*

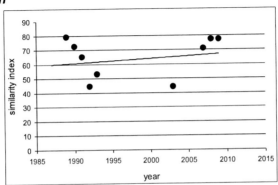

c) non-vascular plants

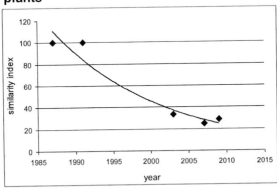

Figure 8. Sorensen's index of similarity – all plants in the site A (a), all plants in the site B (b), non-vascular plants in the site A (c).

occurred in the former plant community are still missing. In comparison with former community "new" species *Calamagrostis epigejos*, *Equisetum arvense* and terrestrial lichens are important recently.

Cover of the former dominant *Arrhenatherum elatius* dropped down in the period of high level of pollution and retrieved higher part in the community recently.

If we look at the S index calculated exclusively for terrestrial lichens in the site A (*Nardo-Agrostidion*) (Fig. 8c). It is evident that since the baseline investigation in 1987 up till now completely new species composition has been formed in the monitoring site. In 1987 a free niche of barren land was covered exclusively by one bryophyte species – *Ceratodon purpureus*, whereas after 2000 the cryptogamic species composition is becoming diverse, with 6 species, lichens dominating. All the species overgrowing the toxic soil feature significantly broad ecological amplitude (cf. Palice & Soldán 2004, Pohlová 2004). Generally, the diversity is still quite poor comparing to the results e.g. from Cd, Pb and Zn contaminated soils in France (Cuny et al. 2004).

Another important aspect of plant communities is information on species diversity (H). When only species number is taken into consideration, it misses the information if some species are rare or common (Begon et al., 1990). The community with the same species richness can differ in the abundance of the species. The character of the community expressed by diversity index takes into account both the abundance patterns and species richness. We found the relation of species diversity (H) to increasing of the pollution in both studied communities. The amount of pollution increased rapidly in period from 1987 to 1990. The response of diversity had the time shifting. The diversity index dropped rapidly in 1990 in the case of *Nardo-Agrostidion* community (site A). We calculated low species diversity index also in 1993 even in 2003. The diversity starts to increase in last two years. In the case of *Arrhenatherion* community (site B) the diversity index was also low in years 1992 and 1993. Since 2003 we observed the trend of diversity increasing (Fig. 9).

DISCUSSION

Spatial and vertical distribution of the pollutants depends on several factors. Various pollutants were deposited along the direction of prevalent winds in amounts varying with their contribution and the distance from emitting source. The impact of combined emissions (mostly SO_2 and heavy metals) was visible in various injury, trend of secondary succession (similarity), and development of species diversity of vegetation during the monitoring.

Degradation of soils is an important problem in industrial area. The loss of soil organic matter was proceeding rapidly. Therefore some signs of natural revegetation now is a good assumption for amelioration e.g. developing new organic layer. Giocchiio (2000) suggests that the soil pH and metal concentrations were main factors explaining the low level of plant species abundance.

SO_2 induced direct injury such as the foliar chlorosis and necrosis caused also the decrease of soil pH. The pH between 4 and 8 does not injure of plants. If the pH values decreases below 4 the physiological functions are restricted e.g. absorption of important elements e.g. calcium and magnesium and nitrogen fixation decreased, the solubility and availability of the heavy metals increased. The result of an extreme pH level may show

a) *Nardo - Agrostidion*

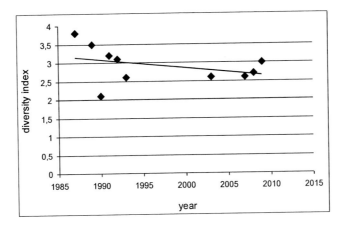

b) *Arrhenaterion*

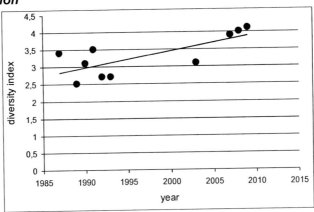

c) non-vascular plants

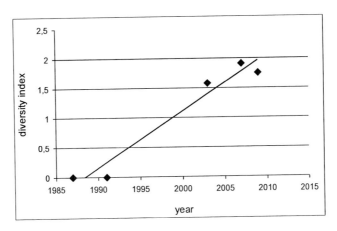

Figure 9. Shannon`s index of diversity – all plants in the site A (a), all plants in the site B (b), non-vascular plants in the site A (c).

nutrient deficiency symptoms in plants. Due to the low soil pH also the density of the vegetation decreases (Grime 1979) which was evident mainly in *Nardo-Agrostidion* vegetation type in study site A. Although the typical acido-tolerant species composed this community, probably the combination of acidity, heavy metals and SO_2 pollution have led to the extinction of several species and the decreasing of diversity and cover.

Decrease in the atmospheric SO_2 concentrations immediately induces decrease of tree bark acidity that hosts epiphytic lichens. In 1984 extremely low values of bark pH were measured on surviving *Pinus sylvestris* trees ranging from 2.8 to 3.2 (Lackovičová 1985). In 2003, after decrease in emission load, bark pH raised to 4.1. This event significantly contributed to recolonisation of epiphytes.

We found that the heavy metals retained in the rhizosphere (0-20 cm). The plants rooted in this layer were affected by such toxicity of the soils. After the extinction of sensitive plants and decreasing of the vegetation cover proceeded the loss of soil organic matter rapidly in most polluted sites. Highest metal concentrations occurred also in litter, which can be also another source of contamination. Although emission level decreased in this time the soils remained rich in Cu, Zn and As. The new technology in the smelter signified that SO_2 pollution do not affect the decreasing of soil pH more but the soil contamination by heavy metals persist and reflect in plant species composition. Heavy metal rich soils were the suitable habitat for the high diversity of plants which have genes coding metal resistance in their genome (Ernst 2004). Therefore metal-resistant plant species can survive at the investigated site with high abundance. They are the species mainly of family *Poaceae* (*Agrostis capillaris*, *Avenella flexuosa*, *Festuca rubra*, *Holcus lanatus*), *Brassicaceae* (*Arabidopsis halleri*), *Caryophyllaceae* (*Silene vulgaris*, *S. alba*) which grow around the smelter recently.

Species richness and resident species can indicate the sensitivity of ecosystems. We detected that the long-term pollution has resulted in extinction of several species. It is known that the basic physiological functions such as photosynthesis and respiration were blocked due to the toxicity of the environment. The remaining of plants in the sites depended on their resistance to the gaseous pollutants and their genetic properties to be able to adapt to the pollution stress (Ernst 1974).

Dueck et al. (1987) suggest that some grasses are able to evolve the tolerance to the air pollution conditions. Although grasses showed the high capability to tolerate pollution we found that some of them e.g. *Arrhenatherum elatius*, *Dactylis glomerata* react by decreasing of their frequency, abundance, and cover in the period of the highest emission concentration in the study area. A lower vitality of *Arrhenatherum elatius* and *Calamagrostis epigejos* was demonstrated by degeneration of their reproductive organs. We observed that site with the highest pollution level however was too toxic also for surviving of tolerant grass such as *Avenella flexuosa*. In one hand, *Avenella flexuosa* increased the abundance when most of other species passed away. In other hand the pollution attack brought low vitality and the growth reduction of *Avenella flexuosa*. The analyze showed that in the natural population exposed to the high pollution level in situ the short plants prevailed in comparison with control population from uncontaminated site.

We found that *Agrostis capillaris* achieved high tolerance to the environmental pollutants while *Nardus stricta* appears to be less tolerant. We observed highest capability of *Agrostis capillaris* to live there in the period of the highest emission concentration. Meanwhile the abundance of *Agrostis capillaris* increased the massive extinction of *Nardus stricta* after rising of the SO_2 pollution was observed. Good vitality of *Agrostis capillaris* demonstrated also the the occupation of the former bare lands when the intensity of pollution dropped. In most affected sites *Agrostis capillaris* created a monodominant stand.

Area with highest pollution level however was too toxic also for surviving of tolerant plants. An example was tolerant plant *Avenella flexuosa*.

The damage of vegetation manifested by decreasing in species richness and diversity and resulted to poor species composition of the plant communities. The visible poorness of species diversity and cover occurred in the period of the highest pollution. The total species richness in *Nardo-Agrostidion* (site A) dropped from 19 to 12 during five years (1987-1992). The total cover decreased from 90 % to 45 % in the same period. The low number of species (only 10 including epigeic lichens) preserved in this time, while total cover increased to 70 %.

The total species richness in *Arrhenatherion* community (site B) drooped also rapidly from 21 to 12 during five years. The total cover decreased from 90 % in 1987 to 50 % in 1992. Recently was the total species richness 30 species and the total cover reached 80 %, but in species composition prevailed plants with wide ecological amplitude and capability to evolve heavy metal tolerant ecotypes.

A diversity index is a mathematical measure of species diversity in a community. The ability to quantify diversity can be the way to understand changes in the community structure. The insidious build-up of the pollutants has the potential to affect the species diversity and stability of future plant communities (Seaward, Richardson 1989). We found a lower species diversity index (H) in the *Nardo-Agrostidion* community even in 2003 in comparison with *Arrhenatherion* community. A closer look at the H index calculated exclusively for terrestrial lichens in the site A (*Nardo-Agrostidion*), shows increase in species diversity in the period of decrease in emission levels. This trend however slows down recently, as the barren lands are being covered by vascular plant vegetation to much greater extent, thus closing the space for cryptogams to develop.

Various degree of injury of grassland vegetation suggested also the species diversity and similarity index (S) in 2003, when amount of pollution was under the limit. The similarity index showed that community *Nardo-Agrostidion* was highly affected by emissions than *Arrhenatherion* community. According this index we found that the community *Nardo-Agrostidion* completely changed. It was characteristic by occurrence of two grasses and epigeic lichens and mosses. In the case of *Arrhenatherion* community the regress to the former plant community started in 2007. We can carefully predict that the similarity index indicated the tendency to come back to the former status of *Arrhenatherion* community.

CONCLUSION

Our results showed the viability of natural vegetation in high damaged habitats due to the environmental pollution. The cost of the toxicity was dramatic changes in species composition and diversity. The development of the species-poor plant communities with low diversity was the results on the highest atmospheric and soil pollution. Sensitive plants, mostly dicotyledonous, disappeared quickly and were replaced by tolerant plants. The most tolerant grass species, e.g. *Agrostis capillaris*, *Holcus lanatus*, *Luzula luzuloides* survived also in the period of the highest pollution intensity most intensive pollution.

Currently, in line with the event of the pollution decreased under limits, we observed the increasing of the recovery of former bare lands by spontaneous vegetation. The rate of this recovery is becoming faster every year. The typical dominant species of meadow community *Arrhenatherum elatius* increased their abundance. The dicotyledonous started to comeback with low frequency. In recent time take an important role also aggressive grass *Calamagrostis epigejos*. Due its clonal growth it can occupy mainly the open, moderate polluted soils. The most striking is fast comeback of epiphytic lichens to the plots with more than 30-year lichen desert. Along with the toxitolerant species *Lecanora conizaeoides*, *Parmeliopsis ambigua* and *Scoliciosporum chlorococcum*, numerous thalli of moderately sensitive species *Hypogymnia physodes*, *H. vittata*, *Parmelia fuliginosa* and *Platismatia glauca*, even the moderately sensitive species *Evernia prunastri* and *Xanthoria parietina* were observed. This process is currently on-going in many of European towns after the decrease in air pollution levels. Our observations of progressing recolonisation are surprising, because it happened in relatively short period – shorter than 15 years. This shows that epiphytic lichen vegetation is capable of rapid regeneration also in heavily polluted areas.

The regeneration trend shows that secondary succession both vascular plants and epiphytic lichens will successfully continue in the area. We can suppose, that diversity and cover of organisms in the area will increase.

ACKNOWLEDGMENT

This work was supported by Slovak Research and Development Agency (No. APVV-51-040805; No. APVV-0432-06) and Grant Agency VEGA (No. 2/0149/08, No. 2/0071/10). We acknowledge Dušan Senko for elaboration of the maps.

REFERENCES

Babušík, I., Schlosser, L., Ronchetti, L., Szabó, G. Výskum znečisteného ovzdušia vybraných oblastí SSR. [*Research of air pollution in selected region of SSR.*] SHMÚ, Bratislava, 1984.

Baker, JM. Metal tolerance. *New Phytol.,* 1987, 106, (Suppl.), 93-111.

Banásová, V ; Guttová, A; Lackovičová, A. Signálne zmeny diverzity cievnatých rastlín a lišajníkov v okolí kovohuty v Krompachoch (východne Slovensko). [Signal changes of vascular plant and lichen diversity in the vicinity of the copper smelter in Krompachy (East Slovakia)]. In: Kontrišová, O; Marušková, A; Váľka, J. (eds.): *Monitorovanie a hodnotenie stavu životného prostredia* VI. Zvolen; Vyd. Fakulta ekológie a environmentalistiky TU vo Zvolene a Ústav ekológie lesa SAV vo Zvolene, 2006; 65-71.

Begon, M; Harper, J.I; Townsend, C.R. *Ecology: individuals, populations and communities.* Blackwell Sci. Publ.; 1990.

Bielczyk, U; Lackovičová, A; Farkas, E; Lőkös, L; Liška, J; Breuss, O; Kondratyuk, S.Ya. *Checklist of lichens of the Western Carpathians.* W. Szafer Institute of Botany, Polish Academy of Sciences. Kraków; 2004.

Braun-Blanquet, J. *Pflanzensoziologie.* Wien-New York; Springer; 1964.

Cuny, D; Denayer, F.O; de Foucault, B; Schumacker, R; Colein, P; van Hauwyn, C. Patterns of metal soil contamination and changes in terrestrial cryptogamic communities. *Environmental Pollution*, 2004, 129, 289-297.

Dueck, TA; Dil, EW; Pasman, FJM. Adaptation of grasses in the Netherlands to air pollution. *New Phytol.* 1987. 108, 167-174.

Dymytrova, L. Epiphytic lichens and bryophytes as indicators of air pollution in Kyiv city (Ukraine). *Folia Cryptog. Estonica*, 2009, 46, 33-44.

Ernst, W. Schwermetallvegetation der Erde. Stuttgart; Gustav Fischer Verlag; 1974.

Ernst, WHO. Mine vegetation in Europe. In: Shaw AJ. (ed): *Heavy metal tolerance in Plant Evolutionary aspects.* CRC Press, 1990, 22-33.

Ernst, WHO; Knolle, F; Kratz, S; Schnug, E. Aspect of ecotoxicology of heavy metals in the Harz region. *Landbauforschung Völkenrode*, 2004, 2, 33-71.

Ginocchio, R. Effects of a copper smelter on a grassland community in the Puchuncaci Valey, Chile. *Chemopshere,* 2000, 41, 15-23.

Gombert, S; Asta, J; Seaward, MRD. The use of autecological and environmental parameters for establishing the status of lichen vegetation in a baseline study for a long-term monitoring survey. *Environmental Pollution* 2005,135, 501-514.

Grimme, JP. *Plant strategies and vegetation processes.* Viley; Chichester; 1979.

Hronec, O. *Exhaláty – pôda – vegetácia.* – [Emissions – soil – vegetation], TOP; Prešov; 1996.

Hajdúk, J; Lisická, E. *Cladonia rei* (lichenizované askomycéty) na stanovištiach kontaminovaných imisiami z Kovohút Krompachy (SV Slovensko). [Cladonia rei (lichenized Ascomycotina) on haevy metal contaminated habitats near copper smelters at Krompachy (NE Slovakia).]. *Bull. Slov. Bot. Spoločn., Bratislava, 1*999, 21, 49- 51.

Isocrono, D; Matteucci, E; Ferrarese, A; Pensi, E; Piervittori, R. Lichen colonisation in the city of Turin (N Italy) based on current and historical data. *Environmental Pollution* 2007, 145, 258-265.

Jules, ES; Shaw, AJ. Adaptation to metal-contaminated soils in populations of the moss, *Ceratodon purpureus*: vegetative growth and reproductive expression. *American Journal of Botany,* 1994, 81,791-797.

Lackovičová, A. Diverzita epifytických lišajníkov v oblasti Krompách [Diversity of epiphytic lichens in Krompachy region] - In: Topercer, J. (ed.), Diverzita rastlinstva Slovenska, Zborn. zo VI. zjazdu SBS pri SAV, Blatnica 6.-10. 6. 1994, *Nitra*; 1995; 158-163.

Lackovičová, A., Flora epifitnych lišajnikov v oblasti goroda Krompachy. *Ekol. Kooperacija, Bratislava*, 1985, 4, 93-94.

Larsen, RS; Bell, JNB; James, PW; Chimonides, PJ; Rumsey, FJ; Tremper, A; Purvis, OW. Lichen and bryophyte distribution on oak in London in relation to air pollution and bark acidity. *Environmental pollution* 2007,146, 332-340.

Marhold, K; Hindák, F. (eds.) Zoznam nižsich a vyšších rastlin Slovenska. *[List of non-vascular and vascular plants of Slovakia]*. Bratislava; VEDA; 1998.

Mazúr , E. et al. *Atlas Slovenskej socialistickej republiky.* [Atlas of Slovak socialistic republik]. Bratislava; Slovenská akadémia vied, Slovenský úrad geodézie a kartografie. Slovenská kartografia; 1980:

Munzi, S; Ravera, S; Caneva, G. Epiphytic lichens as indicators of environmental quality in Rome. *Environmental Pollution*, 2007, 146, 350-358.

Palice, Z; Soldán, Z. Lichen and bryophyte species diversity on toxic substrates in the abandoned sedimentation basins of Chvaletice and Bukovina. In: Kovář, P. (ed.), *Natural recovery of human-made deposits in landscape (Biotic interactions and ore/ash-slag artificial ecosystems)*. Academia, Praha; 2004; 200-221.

Pohlová, R. Changes on microsites of the moss *Ceratodon purpureus* and lichens *Peltigera didactyla* and *Cladonia* sp. div. in the abandoned sedimentation basin in Chvaletice. In: Kovář, P. (ed.), *Natural recovery of human-made deposits in landscape (Biotic interactions and ore/ash-slag artificial ecosystems)*. Academia, Praha; 2004; 222-234.

Seaward, MRD; Richardson, DHS. Atmospheric sources of metal pollution and effects on vegetation. In: Shaw AJ. (ed): *Heavy metal tolerance in Plant Evolutionary aspects*. CKC Press; 1990; 75-92.

Wilson, GB; Bell, JN. Studies on the tolerance to sulphur dioixide of grass populations in polluted areas. *New Phytol.,* 1986, 102, 563-574.

Žemberyová, M; Barteková, J; Závadská, M; Šišoláková, M. Determination of bioavailable fractions of Zn, Cu, Ni, Pb and Cd in soils and sludges by atomic absorption spectrometry. *Talanta*, 2007, 71, 1661- 1668.

http://www.env.gov.sk

In: Species Diversity and Extinction
Editor: Geraldine H. Tepper, pp. 405-413

ISBN: 978-1-61668-343-6
© 2010 Nova Science Publishers, Inc.

Chapter 14

CRUSTACEAN ZOOPLANKTON BIODIVERSITY IN CHILEAN LAKES: TWO VIEW POINTS FOR STUDY OF THEIR REGULATOR FACTORS

Patricio De los Ríos[*1], *Luciano Parra and Marcela Vega*

Universidad Católica de Temuco, Facultad de Recursos Naturales, Escuela de Ciencias Ambientales, Casilla 15-D, Temuco, Chile

ABSTRACT

The crustacean zooplankton in Chilean lakes is characterized mainly by their low species number and high predominance of calanoids copepods in comparison to daphnids cladocerans. This is a different pattern in comparison to North American lakes that have abundant cladocerans populations and high number of crustacean species.

The aim of the present study is do an analysis of published information for Chilean lakes and ponds. The information was analyzed using two view points: a) the first step consisted in a Principal Component Analysis considering geographical features, conductivity, trophic status, crustacean species number and calanoid copepod relative abundances. b) the second step consisted in an application of null model co-ocurrence for determine the existence of potential structures or random distribution in species for different groups of water bodies.

The results revealed that the main regulator factors would be the oligotrophy and conductivity, because the high species number are observed in mesotrophic status and low to moderate conductivity, that corresponded to Patagonian oligo-mesotrophic lakes and ponds. In another side, low species number was observed in oligotrophic and/or high conductivity lakes and ponds. The results of null models revealed the presence of regulator factors in one simulation, whereas the other two simulations denoted random in species associations, although this result does not agree with PCA analysis, the cause would be the presence of species repeated in many sites. These results about regulators factors of species diversity are similar with Argentinean Patagonian lakes.

Keywords: copepods, cladocerans, zooplankton, species number, oligotrophy, conductivity.

---

* Corresponding author: E-mail: prios@uct.cl; E-mail(2): patorios@msn.com

INTRODUCTION

The zooplankton assemblages in Chilean lakes is characterized by the presence of low species number and marked predominance of calanoids copepods, specifically of *Boeckella* and *Tumeodiaptomus* genus that is due mainly to the oligotrophy of these environments (Soto & Zúñiga, 1991; De los Ríos & Soto, 2007). A different situation would success in northern hemisphere lakes where there is high species richness (Dodson et al., 2009).

Also, the current literature indicates a potential effect of natural ultraviolet radiation that has an increase in its penetration to southern latitudes, due ozone depletion (Villafañe et al., 2001; Marinone et al., 2006). In this scenario, the ultraviolet radiation can penetrate into water column and due this condition only the tolerant species can occurs (Marinone et al., 2006; De los Ríos et al., 2008a). In another hand, in northern Chile, there are many saline lakes associated with saline deposits of volcanic origin, and in this scenario, the conductivity is other important regulator factor of species richness and dominance (De los Rìos & Crespo, 2004).

The aim of the present study is analyze the available information of zooplankton crustacean species diversity in Chilean lakes along a wide geographical gradient (23-51° S), with the aim of determine the existence of regulator factors, with emphasis in the role of conductivity and chlorophyll concentration. For this purpose, it will use a traditional statistical exploratory analysis and null model because this last kind of analysis determine it is possible found random in species associations.

MATERIALS AND METHODS

It was revised the literature about zooplankton assemblages in Chilean lakes between 23-51° S (De los Rios, unpublished data; Schmid-Araya & Zúñiga, 1992; Campos et al., 1983, 1988, 1988, 1990, 1992a,b; 1994a,b; Villalobos, 1999; Soto & De los Ríos, 2006). To the available data was considered the conductivity, latitude, surface, maximum depth, chlorophyll "a" concentration and species number, and to these data was applied a Principal Component Analysis (PCA) using the software Xlstat 5.0.

The comparison of the dataset gathered is useful to test the hypothesis that species reported are non randomly associated. For this, we use the "C score" index (Tiho & Johens, 2006; Tondoh, 2007), which determines the co-occurrence based on presence (1)-absence (2) matrices for a zooplankton species in the sample. Following Gotelli (2000) and Tiho & Johens (2007) the presence/absence matrix was analysed as follows: (a) fixed-fixed: In this algorithm,, the row and the column sums of the original matrix are preserved. Thus each random community contains the same number of species as the original community (fixed column), and each species occurs with the same frequency as in the original community (fixed row). In this instance, it is not prone to type I errors (falsely rejecting the null hypothesis) and has a good power for detecting non-randomness (Gotelli, 2000; Tiho & Johens, 2006). (b) fixed-equiprobable: In this simulation, only the row sums are fixed, whereas the columns are treated as equiprobable. This null model treats all the samples (columns) as equally suitable for all species (Tiho & Johens, 2006; Gotelli, 2000).(c) fixed-proportional In this algorithm, the species occurrence totals are maintained as in the original community, and the probability that a species occurs in a sample (= column) is proportional to the column total for that sample (Gotelli, 2000; Tiho & Johens, 2006; Tondoh 2007). Data were analysed with Ecosim program version 7.0 (Gotelli & Entsminger 2009).

RESULTS AND DISCUSSION

The results of PCA indicates that the most important variables contributes to the 64.13 %, for the first axis the latitude, surface and maximum depth 36.97 %, whereas for the second axis, the conductivity, chlorophyll *a* and species richness contributes with a 27.16 % (Figure 1a). The results indicate the existence of two sites (Miscanti and Miniques) that have high conductivity, low chlorophyll concentration and low species number, whereas in a second group there are two sites Peñuelas and Rungue that are with small surface, moderate chlorophyll concentration and high species number (Figure 1b; Figure 1b; Tables 1 and 2). Finally a third group is joined by Patagonian lakes, that have large surface and depth, low conductivity, oligotrophic and oligo-mesotrophic status and low species number (Figure 1b; Tables 1 and 2). The results of null model revealed the existence of regulator factors only for fixed-fixed simulation, whereas the fixed-proportional and fixed-equiprobable simulations denoted absence or regulator factors, this is that species associations are random (Table 2). This difference in the results of both simulations is probable due many species present in many of the studied sites that simultaneously have low species number (Table 3).

The exposed results revealed that the main regulador factors of species assemblages in Chilean lakes are conductivity and chlorophyll concentration (Soto & De los Ríos, 2006; De los Ríos & Soto, 2009). The role of conductivity or salinity as main regulator factor in zooplankton assemblages in northern Chile was described for De los Ríos & Crespo (2004), who indicated an inverse association between salinity and species richness, because when the salinity increases until 90 g/l the halophilic copepod *B. poopoensis* can be an exclusive component in zooplankton assemblages. This zooplankton assemblage with marked dominance of halophilic copepod *B. poopoensis* was described for saline lakes of Bolivian and Peruvian Altiplano (Hurlbert et al., 1986; Williams et al., 1995).

Table 3. Results of null model analysis for studied sites. "P" values lower than 0.05, denoted presence of non random factors as regulator of species association

Model	Observed index	Mean index	Standard Effect Size	P
Fixed-Fixed	7.809	6.928	5.406	0.000
Fixed-Proportional	7.809	7.019	1.257	0.107
Fixed-Equiprobable	7.809	8.526	0.194	0.940

In central Chilean water bodies and Patagonian lakes, the main regulator factor would be the trophic status, that is denoted in the exposed results, where the central Chilean water bodies had high species number associated to high chlorophyll concentration, whereas the low species number was reported in Patagonian lakes with oligotrophic and/or oligomesotrophic status (Soto & Zúñiga, 1991; De los Ríos & Soto, 2007; 2009).

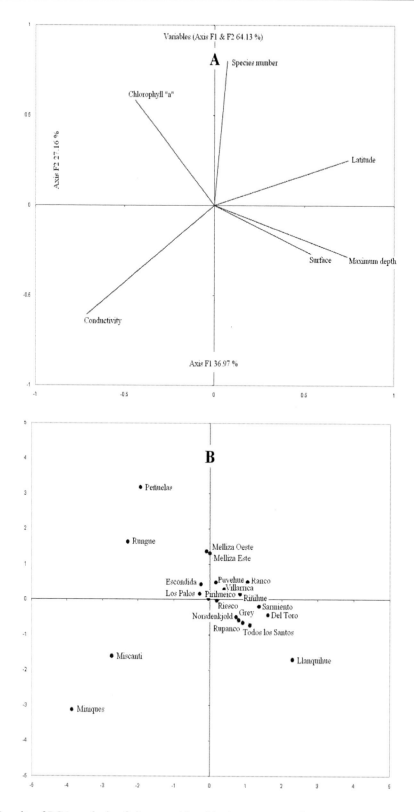

Figure 1. Results of PCA analysis of sites considered in the present study.

Table 1. Geographical location, surface, maximum depth, chlorophyll "a", conductivity, species number for studied sites

	Latitude	Surface (km2)	Maximum depth (m)	Chlorophyll "a" (mg/l)	Conductivity (mS/cm)	Species number	Reference
1- Miscanti	23°44' 67°46'	13.40	9.00	1.00	7.73	3	De los Ríos (unpublished data)
2- Miniques	23°45' 67°44'	1.60	5.00	1.00	15.42	1	De los Ríos (unpublished data)
3- Rungue	33°00' 70°53'	0.48	15.00	65.90	0.33	4	Schmid-Araya & Zuñiga (1992)
4- Peñuelas	33°10' 71°29'	19.00	15.00	57.40	0.12	11	Schmid-Araya & Zuñiga (1992)
5- Villarrica	39°18' 72°07'	175.80	185.00	0.40	0.06	7	Campos et al., (1993)
6- Pirihueico	39°50' 71°48'	30.50	145.00	0.60	0.05	5	Wölfl (1996)
7- Riñihue	39°50' 72°20'	77.50	323.00	1.20	0.05	7	Wölfl (1996)
8- Ranco	40°13' 70°22'	442.60	199.00	0.80	0.07	9	Campos et al., (1992a)
9- Puyehue	40°40' 72°30'	165.40	123.00	2.10	0.08	7	Campos et al. (1989)
10- Rupanco	40°50' 72°30'	235.00	273.00	1.20	0.06	4	Campos et al., (1992b)
11- Llanquihue	41°08' 72°50'	870.50	317.00	0.50	0.11	3	Campos et al., (1988)
12- Todos los Santos	41°08' 72°50'	178.50	335.00	0.40	0.04	4	Campos et al., (1990)
13- Los Palos	45°19' 72°42'	59.00	5.00	0.80	0.02	4	Villalobos, (1989)
14- Riesco	45°39' 72°20'	14.70	130.00	0.90	0.02	4	Villalobos, (1989)
15- Escondida	45°49' 72°40'	43.00	7.00	0.48	0.01	5	Villalobos, (1989)
16- Nordensjold	51°01' 72°56'	25.00	200.00	0.25	0.13	2	Soto & De los Ríos, (2006)
17- Melliza Oeste	51°03' 72°57'	0.13	25.00	3.24	0.66	8	Soto & De los Ríos, (2006)
18- Melliza Este	51°03' 72°56'	0.12	16.00	5.32	0.80	8	Soto & De los Ríos, (2006)
19- Sarmiento	51°03' 72°37'	86.00	312.00	0.34	0.85	5	Campos et al., (1994a)
20- Grey	51°07' 72°56'	15.00	200.00	3.18	0.20	2	Soto & De los Ríos, (2006)
21- Del Toro	51°12' 72°38'	196.00	300.00	0.35	0.07	4	Campos et al., (1994b)

Table 2. Crustacean zooplanctonic species reported for studied sites

	1	2	3	4	5	6	7	8	9	10	11	12	13	14	15	16	17	18	19	20	21
Boeckella gracilipes (Daday, 1902)					X	X	X	X		X	X	X	X	X			X	X	X	X	X
B. michaelseni (Mrázek, 1901)															X	X	X	X	X	X	X
Boeckella poopoensis (Marsh, 1906)	X	X																			
Tumeodiaptomus diabolicus (Dussart, 1979)				X	X		X	X	X												
Parabroteas sarsi (Mrázek, 1901)													X	X							
Mesocyclops longisetus (Thiebaud, 1914)					X	X	X	X	X												
Mesocyclops araucanus (Pilati & Menu-Marque, 2003)																	X	X	X		X
Tropocyclops prasimus (Fisher, 1960)			X	X	X	X	X	X	X	X	X	X	X	X	X	X			X	X	X
Metacyclops mendocinus (Wierzerkski, 1892)			X																		
Acantocyclops vernalis (Fisher, 1853)																	X	X			
Daphnia ambigua (Scourfield, 1967)				X	X		X	X	X												
Daphnia pulex (De Geer, 1877)		X						X				X					X	X			
Ceriodaphnia dubia (Richard, 1894)				X		X			X								X	X			
Diaphanosoma chilense (Daday, 1902)				X	X		X	X	X												
Eubosmina hagmanni (Stingelin, 1903)			X		X	X	X	X	X	X	X	X	X	X	X		X	X	X		
Scapholeberis expinifera (Nicolet, 1848)								X									X				
Chydorus sphaericus (O.F. Müller 1785)		X								X											
Alona affinis (Leydig, 1860)				X																	
Alona guttata (Sars, 1862)				X																	
Alona pulchella (King, 1863)				X																	
Moina micrura (Kurz, 1874)			X	X																	

These results are similar with descriptions for New Zealand lakes, where it is possible found more species associated in mesotrophic status (Jeppensen et al., 1997, 2000). The low species number associated for low chlorophyll concentration in Chilean Patagonian lakes is similar with descriptions for Argentinean Patagonian lakes that are unpolluted (Modenutti et al., 1998).A partially different situation would success in northern Chilean Patagonian lakes that have high species number due the oligo-mesotrophic status (De los Ríos & Soto, 2007), this is due the increase of nutrients imputs from the surrounding basin due the change of native forest to towns, agricultural and industrial zones (Soto, 2002; Woelfl et al., 2003). Other important factor for Patagonian lakes, would be the role of natural ultraviolet radiation exposure, because in the Patagonian lakes, the radiation can penetrate into water column (Morris et al., 1995). Then in this scenario of penetration of natural ultraviolet radiation only the tolerant species can occurs, and under oligotrophic status, the species occurrence would be markedly low (Marinone et al., 2006). Then in Patagonian deep lakes only with mesotrophic status can be high species number (De los Ríos & Soto, 2007), because the species can protect against ultraviolet radiation damage migrating to deep zones without exposure to natural ultraviolet radiation (Villafañe et al., 2001).

The results of null models, agree partially with PCA, in spite of the previous antecedents exposed in the present study, the results of null model that indicate the absence of regulator factors in species assemblages are similar with descriptions for other Patagonian lakes (De los Ríos et al., 2008a,b; De los Ríos & Soto, 2009). This information apparently contradictory with multivariate statistical analysis, is due the presence of numerous species repeated in many sites, or environmental homogeneity (Tondoh, 2007). Nevertheless, it is necessary more studies that include more variables, and more sites in regions that were not included in the present study, for understand the species richness regulators mechanisms.

ACKNOWLEDGMENTS

The present study was founded by project DGI-CDA 2007-01 of the Research Direction of the Catholic University of Temuco.

REFERENCES

Campos, H., Arenas, J., Steffen, W., Román, C. & Agüero, G. (1982). Limnological study of lake Ranco (Chile): morphometry, physics and plankton. *Archiv für Hydrobiologie*, *94*, 137-171.

Campos, H., Soto, D., Steffen, W., Agüero, G., Parra, O. & Zúñiga, L. (1994a). Limnological studies in lake del Toro, Chilean Patagonia. *Archiv für Hydrobiologie*, *99*, 199-215.

Campos, H., Soto, D., Steffen, W., Agüero, G., Parra, O. & Zúñiga, L. (1994b). Limnological studies in lake Sarmiento, a subsaline lake from Chilean Patagonia. *Archiv für Hydrobiologie*, *99*, 217-234.

Campos, H., Steffen, W., Agüero, G., Parra, O. & Zúñiga, L. (1983). Limnological studies in lake Villarrica. Morphometry, physics, chemistry and primary productivity. *Archiv für Hydrobiologie, (Supplement)*, *71*, 37-67.

Campos, H., Steffen, W., Agüero, G., Parra O. & Zúñiga, L. (1987a). Limnology of lake Riñihue. *Limnológica*, *18*, 339-357.

Campos, H., Steffen, W., Agüero, G., Parra, O. & Zúñiga, L. (1988). Limnological study of lake Llanquihue (Chile): morphometry, physics, chemistry and primary productivity. *Archiv für Hydrobiologie, (Supplement)*, *81*, 37-67.

Campos, H., Steffen, W., Agüero, G., Parra, O. & Zúñiga, L. (1989). Estudios limnológicos en el lago Puyehue (Chile): morfometría, factores físicos y químicos, plancton y productividad primaria. *Medio Ambiente*, *10*, 36-53.

Campos, H., Steffen, W., Agüero, G., Parra, O. & Zúñiga, L. (1990). Limnological study of lake Todos los Santos (Chile): morphometry, physics, chemistry and primary productivity. *Archiv für Hydrobiologie, (Supplement)*, *117*, 453-484.

Campos, H., Steffen, W., Agüero, G., Parra, O. & Zúñiga, L. (1992a). Limnological study of lake Ranco (Chile). *Limnológica*, *22*, 337-353.

Campos, H., Steffen, W., Agüero, G., Parra, O. & Zúñiga, L. (1992b). Limnological studies of lake Rupanco (Chile): Morphometry, physics, chemistry and primary productivity. *Archiv für Hydrobiologie, (Supplement)*, *90*, 85-113.

De los Ríos, P. & Soto, D. (2009). Estudios Limnológicos Lagos y Lagunas del Parque Nacional Torres del Paine (51° S, Chile). *Anales Instituto de Patagonia*, *37*, 63-71.

De Los Ríos, P., Acevedo, P., Rivera, R. & Roa, G. (2008a). Comunidades de crustáceos litorales de humedales del norte de la Patagonia Chilena (38°S): rol potencial de la exposición a la radiación ultravioleta. En: Efectos de los Cambios Globales sobre la Diversidad. A. Volpedo, & L. Fernández, (Eds). 209-218.

De los Ríos, P., Rivera, N. & Galindo, M. (2008b). The use of null models to explain crustacean zooplankton associations in shallow water bodies of the Magellan region, Chile. *Crustaceana*, *81*, 1219-1228.

De los Ríos, P. & Soto, D (2007). Crustacean (copepoda and cladocera) zooplancton richness in Chilean Patagonian lakes. *Crustaceana*, *80*, 285-296.

De Los Ríos, P. & Crespo, J. (2004). Salinity effects on the abundance of *Boeckella poopoensis* (Copepoda, Calanoida) in saline ponds in the Atacama desert, northern Chile. *Crustaceana*, *77*, 417-423.

Dodson, S., Newman, A., Wolf, S., Alexander M. & Woodford, M. (2009). The ralationship between zooplankton community structure and lake characteristics in temperate lakes (Nothern Wisconsin, USA). *Journal of Plankton Research*, *31*, 93-100

Gotelli, N. J. (2000). Null models of species co-occurrence patterns. *Ecology.*, *81*, 2606-2621.

Gotelli, N. J. & Entsminger, G. L. (2001). EcoSim: *Null models software for ecology*. Version 7. Acquired Intelligence Inc. & Kesey-Bear. Jericho, VT 05465.

Jeppensen, E., Lauridsen, T. L., Mitchell, S. F. & Burns, C. W. (1997). Do zooplanktivorous fish structure the zooplankton communities in New Zealand lakes? *New Zealand Journal of Marine and Freshwater Research*, *31*, 163-173.

Jeppensen, E., Lauridsen, T. L., Mitchell, S. F., Chirstofferssen, K. & Burns, C. W. (2000). Trophic structure in the pelagial of 25 shallow New Zealand lakes: changes along nutrient and fish gradients. *Journal of Plankton Research*, *22*, 951-968.

Marinone, M., S Menu-Marque, C., Añón-Suarez, D., Diéguez, M. C., Pérez, A., De los Ríos, P., Soto, D. & Zagarese, H. E. (2006). UV radiation as a potencial driving force for zooplankton community structure in Patagonian lakes. *Photochemistry and Photobiology*, *82*, 962-971.

Modenutti, B. E., Balseiro, E. G., Queimaliños, C. P., Añón Suárez, D. A., Dieguez, M. C. & Albariño, R. J. (1998). Structure and dynamics of food webs in Andean lakes. *Lakes and Reservoirs Research and Management, 3*, 179-189.

Morris, D. P., Zagarese, H. E., Williamson, C. E., Balseiro, E. G., Hargreaves, B. R., Modenutti, B. E., Moeller, R. E. & Queimaliños, C. P. (1995). The attenuation of solar UV radiation in lakes and the role of dissolved organic carbon. *Limnology & Oceanography, 40*, 1381-1391.

Schmid-Araya, J. M. & Zuñiga, L. R. (1992). Zooplankton community structure in two Chilean reservoirs. *Archiv für Hydrobiologie, 123*, 305-335.

Soto, D. (2002). Oligotrophic patterns in southern Chile lakes: the relevance of nutrients and mixing depth. *Revista Chilena de Historia Natural, 75*, 377-393.

Soto, D. & De los Ríos, P. (2006). Trophic status and conductivity as regulators of daphnids dominance and zooplankton assemblages in lakes and ponds of Torres del Paine National Park. *Biologia, Bratislava, 61*, 541-546.

Tiho, S. & Johens, J. (2007). Co-occurrence of earthworms in urban surroundings: a null models of community structure. *European Journal of Soil Biology, 43*, 84-90.

Tondoh, J. E. (2006). Seasonal changes in earthworm diversity and community structure in central Côte d'Ivoire. *European Journal of Soil Biology, 42*, 334-340.

Villafañe, V. E., Helbling, E W. & Zagarese, H. E. (2001). Solar ultraviolet radiation and its impact on aquatic ecosystems of Patagonia. *Ambio, 30*, 112-117.

Williams, W. D., Carrick, T. R., Bayly, I. A. E., Green, J. & Herbst, D. B. (1995). Invertebrates in salt lakes of the Bolivian Altiplano. *International Journal of Salt Lake Research, 4*, 65-77.

Wölfl, S., Villalobos, L. & Parra, O. (2003). Trophic parameters and method validation in a Lake Riñihue (North Patagonia, Chile) from 1978 to 1997. *Revista Chilena de Historia Natural, 75*, 459-474.

In: Species Diversity and Extinction
Editor: Geraldine H. Tepper, pp. 415-430
ISBN: 978-1-61668-343-6
© 2010 Nova Science Publishers, Inc.

Chapter 15

RESTORATION OF PROPAGATION AND GENETIC BREEDING OF A CRITICALLY ENDANGERED TREE SPECIES, *ABIES BESHANZUENSIS*

Shunliu Shao[1] and Zhenfu Jin[*2]

[1]Institute of Ecology, Zhejiang Forestry Academy, Hangzhou, Zhejiang, China
[2]School of Engineering, Zhejiang Forestry University, Lin'an, Zhejiang 311300, China

ABSTRACT

Abies beshanzuensis, Pinaceae, is a geographically distinct tree species in China, which had grown widely at middle to south coastal mountainous area in China during the Riss Ice Age (130–180kBP). However, the population of the species reduced drastically being due to climate change, natural disaster and human activities after the last ice age (Würm: 15–70kBP). There were only 7 wild individuals in 1963, while in 1998 only 3 individuals were discovered at about 1700 m elevation of Mt. Baishanzu in Qingyuan, Zhejiang province, China. *A. beshanzuensis* was approved as one of 12 critically endangered plant species in the world in 1987 by the Species Survival Commission (SSC) of the International Union for Conservation of Nature and Natural Resources (IUCN). Though the efforts to conserve the species by people and many scientists have been performed, it has been recognized that the species has lost natural reproductive ability due to the loss of genetic diversity. It is urgent subject in particular to develop techniques for restoration of propagation and genetic breeding based on the genetic diversity of remaining trees. We developed techniques to produce seeds and seedlings with restoration of high propagation by finding the most suitable tree species for grafting based on taxonomic and genetic information of neighborhood species, and promotion of efflorescence using the most suitable gibberellin.

---

* Corresponding Author: Zhenfu Jin, Ph.D, School of Engineering, Zhejiang Forestry University, 88 North Huancheng Road, Lin'an, Zhejiang 311300, China, Tel & Fax: +86-571-63741607, e-mail: jinzhenfu@yahoo.com.cn

INTRODUCTION

Abies beshanzuensis M. H. Wu, Pinaceae, is a geographically distinct tree species in China, which had grown thickly at middle to south coastal mountainous area of Zhejiang, Jiangsu, Anhui and Fujian provinces in China on the Riss (the Illinoian or Saale: 125-200 kBP) glacial period [1]. However, the population of the species reduced drastically due to climate change, natural disaster and human activities after the last (the Würm or Wisconsin: 15-70 kBP) glacial period. There were only 7 wild individuals in 1963 [1], while in 1998 only 3 individuals were discovered at about 1700 m elevation of Mt. Baishanzu in Qingyuan, Zhejiang province, China [2,3] (Fig. 1). *A. beshanzuensis* was listed as one of 12 critically endangered plant species in the world in 1987 by Species Survival Commission (SSC) of International Union for Conservation of Nature and Natural Resources (IUCN) [4,5]. In 1992, *A. beshanzuensis* stamp was issued, and in 1999, *A. beshanzuensis* was also listed as the primary protection tree species by the State Council of China (Fig. 2) [2,6].

Figure 1 *A. beshanzuensis* growing at 1700 m elevation of Mt. Baishanzu.

Figure 2 *A. beshanzuensis* in a stamp of Chinese post.

In the past years, researches involving hybridization between natural *A. beshanzuensis* species (1976), grafting *A. beshanzuensis* onto Japanese *Abies firma* as stock (1978-1990), cutting experiments with ABT (rooting hormones developed by Chinese Academy of Forestry) treatment during 1978-1980, as well as hybridization among grafts, have been conducted [3,6]. However, all these efforts were not successful. The reason of *A. beshanzuensis* becoming endangered and difficult to propagate would be due to the long-term interval of flowering, the flowering in the rain season, and/or weak vitality of the pollen-cones [7,8]. It has been recognized that *A. beshanzuensis* has lost natural regeneration ability due to the loss of genetic diversity.

The absence of natural regeneration of *A. beshanzuensis* would be due to insufficient supply and/or low quality of the seeds produced by mother trees. However, failure to produce seeds may result from limited pollination or resources, and there is little consensus regarding their relative importance in natural systems. Techniques to promote production of cone improvement of seed quality and seedlings with high reproduction ability by selecting the most suitable parent species for grafting based on taxonomic, genetic and chemical information of neighborhood species should be urgently developed.

This chapter, we review the researches on the community structure of *A. beshanzuensis*; rejuvenation by grafting *A. beshanzuensis* onto *A. firma*; lignin characteristics of *A. beshanzuensis*; promotions of adventitious rooting of *A. beshanzuensis* cuttings by rooting hormones such as RTN (rooting hormones developed by the University of Tokyo), ABT, α-naphthalene acetic acid (NAA), and indole-3-butyric acid (IBA); flowering promotion by application of gibberellin $A_{4/7}$ ($GA_{4/7}$) and progeny vitality of *A. beshanzuensis* as shown in Fig. 3.

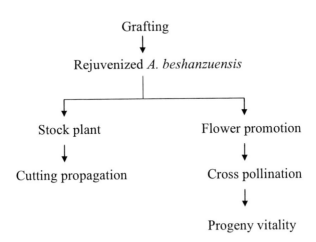

Figure 3. Scheme of the study.

1. FOREST COMMUNITY STRUCTURE OF *A. BESHANZUENSIS*

A. beshanzuensis has gained considerable importance as one of 12 critically endangered plant species in the world [3,4]. *A. beshanzuensis* was found in 1963, while was published as a new species of Abies in 1976 [1]. *A. beshanzuensis* distributed in subtropical zone of Mt. Baishanzu in Qingyuan (119.3-119.6 E, 27.4-27.5 N), southern Zhejiang province, China, at altitudes about 1700 m (Fig. 4). The site belongs to the subtropical monsoon climate, where is abundant precipitation, and four distinct seasons. However, the site shows special alpine climate, namely the annual average temperature is 12.8°C, average annual precipitation is 2,342 mm, average annual relative humidity is 84%, extreme low temperature is -13.2°C, extreme high temperature 30.1°C, frost-free season is 187 day, because of high elevation [1,9]. Compare with the Qingyuan County locates at 350 m elevation, the site conditions of *A. beshanzuensis* forest plant communities are distinguished by lower temperature, higher rainfall, and longer frost day.

Figure 4. Location of *A. beshanzuensis* habitat, Mt. Baishanzu, Qingyuan, Zhejiang province, China (119.3-119.6 E, 27.4-27.5 N).

A. beshanzuesis commonly occurs in forestry together with evergreen *Cyclobalanopsis multinervis* and deciduous *Fagus lucicla*, which is a natural broadleaf mixed forest. About 8-9 species form the high canopy, and the high canopy divided into 3 sub-layers, the first sub-layer composed of *A. beshanzuensis* with 13 m height, the second dominant community of *C. multinervis* and *F. lucicla* with 8-12 m height, and the third one by young trees of the second layer with 4-8 m height [9,10]. The diversity of the community has 17-19 species and the diversity index of high canopy is higher than shrub and ground layer.

2. REJUVENATION BY GRAFTING

Cuttings collected from natural *A. beshanzuensis* branches treating with rooting hormones (ABT) were not successful in 1978 [6]. The failure of root regeneration in cuttings might be due to the physiological age of plant. The physiological age of natural *A. beshanzuensis* branches was too old to regenerate adventitious roots in cuttings. Grafting is a method of asexual plant propagation where the tissues of stock are encouraged to fuse with scion. A shoot of a desired selected plant cultivar is grafted onto the stock plant. An easily rooted plant is used to provide regeneration of roots, which is called the stock, and the scion contains the desired genes to be duplicated in future production by the stock/scion plant. For taking place of successful grafting, the vascular cambium tissues of the rootstock and the scion plants must be placed in contact with each other to ensure the vascular connection between the two tissues. The physiologically young and vigorous *A. firma* was selected as a rootstock. Grafting of *A. beshanzuensis* onto *A. firma* was conducted to rejuvenation of *A. beshanzuensis* during 1978-1990, because the *A. beshanzuensis* was difficult to propagate by other methods, such as cuttings [7,8]. The grafts grew well and the average height of grafts was investigated in 2005 as shown in Fig. 5 [7,8,11].

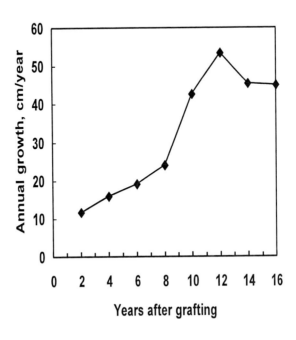

Figure 5. Average annual height growth of grafted trees of *A. beshanzuensis*.

3. LIGNIN CHARACTERISTICS OF *A. BESHANZUENSIS*

Lignin is one of major component of plant cell walls, distributes mainly throughout the secondary xylem. As integral cell wall components, functions of lignin are mechanical support of plants, long-distance water conduction, impenetrable barriers, and defense from biotic attacks [12]. Lignin contents and monomeric aromatic compositions as well as their intermolecular linkage patterns vary predictably among different species. Lignin is derived from the phenylpropanoid (sikimic acid) pathway [13]. Lignin macromolecule is formed by the dehydrogenative polymerization of three monolignols, coniferyl, sinapyl, and *p*-coumaryl alcohols, which differ only in the degree of methoxylation of the aromatic ring. Dehydrogenative polymerization gives varieties of substitution patterns and intermolecular linkages, mainly non-condensed ether linkage and condensed type C-C linkage [14].

Root development of cuttings taken from natural *A. beshanzuensis* was not successful, because of physiologically old of stock plant, while cuttings from grafts grew well [7]. These facts suggest that *A. beshanzuensis* rejuvenated by grafting and their branches recovered vigor and repropagation ability. It is well known that nutrients and plant hormones would be exchanged between stock and scion, while genetic basis of scion would not be affected by the stock. In other words, *A. forma* would not affect on the genetic basis of *A. beshanzuensis*. *A. forma* was selected as stock not only physiologically young and vigor, but also genetically the closest species of *A. beshanzuensis* and the similarity of lignin characteristics, which is one of the most important constitution of vascular system of gymnosperm plants, tracheids. The similarity in the genetical and chemical status of the stock would result good affinity between scion and stock.

Table 1. Klason lignin contents and yield of alkaline nitrobenzene oxidation and ozonation products

	A. beshanzuensis	*A. firma*
Lignin content %		
KR	39.2	33.3
ASL	0.4	0.4
Total	39.2	33.7
Alkaline nitrobenzene oxidation products (mmol·(200g lignin)$^{-1}$)		
Total yield	264.9	280.6
H/G	0.067	0.091
Ozonation products (mmol·(200g lignin)$^{-1}$)		
E	77.5	85.1
T	67.8	75.1
Total yield	145.2	160.2
E/T	1.14	1.13

KR: Klason residue; ASL: acid soluble lignin;
H/G: H/V: molar ratio of *p*-hydroxybenzaldehyde to vanillin
E: *erytro* type of β-O-4 intermonomer linkage;
T: *threo* type of β-O-4 intermonomer linkage;
E/T: molar ratio of *erythro* type to *threo* type.

Lignin content of *A. beshanzuensis* was higher comparing with *A. firma* (Table 1) and other coniferous species (25-35% of cell walls). The high lignin content may be due to the adaptation of *A. beshanzuensis* to environmental stresses in surviving from the Riss glacial period. The composition of non-condensed linkage, total yield of alkaline nitrobenzene oxidation products, based on lignin of *A. beshanzuensis* was almost the same with that of *A. firma* (Table 1), and good agreement with gymnosperm wood lignins [15,16]. The lignin of both species was characterized with significantly high molar ratio of *p*-hydroxyphenyl nuclei to guaiacyl nuclei (H/G ratio) of alkaline nitrobenzene oxidation products. These findings agree with the results reported for other Abies species, *A. concolor* [17], *A. fraseri* [18], *A. balsamea* [19], *A. sibirica* [20] and *A. sachalinensis* [21].

Alkaline nitrobenzene oxidation of the Björkman lignins of both *A. beshanzuensis* and *A. firma*, which were isolated by neutral solvent extraction after finely ground extract free wood meal with Björkman's procedure [22], gave quite similar results with those of the extract-free samples (Table 2). The H/G ratio of Björkman lignin of moso bamboo, *Phyllostachys pubescens*, which is known to have *p*-hydroxyphenyl nuclei involved in both esterified and arylglycerol-β-aryl intermonomer linkages [23], gave considerably high value (Table 2). The presence of arylglycerol-β-aryl ether intermonomer linkages is essential to distinguish lignin from other polyphenols [24]. The acidolysis products of the Björkman lignins were identified by gas chromatography-mass spectroscopy (GC-MS) as trimethylsilyl derivatives. Although H/G ratio of *P. pubescens* was considerably high (Table 2), the relative intensities of *p*-hydroxyphenyl propanones of *P. pubescens* Björkman lignin were smaller than those of *A. beshanzuensis* and *A. firma* [25]. The molar ratio of *p*-hydroxyphenyl-type acidolysis products to the corresponding alkaline nitrobenzene oxidation products (H_{acid}/H_{nb}) was calculated and the results showed that the values of *A. beshanzuensis* and *A. firma* were higher than that of *P. pubescens* (Table 2). These results suggest that the most of *p*-hydroxybenzaldehyde and *p*-hydroxybenzoic acid in alkaline nitrobenzene oxidation products of *P. pubescens* would originate from *p*-coumaric acid esterified to lignin [26]. In contrast, lignin of *A. beshanzuensis* was characterized by the presence of *p*-hydroxyphenylglycerol-β-aryl ether intermonomer linkages with the same rate of *A. firma* lignin.

Table 2. Alkaline nitrobenzene oxidation products from Björkman lignin (mmol·(200g lignin)$^{-1}$)

	A. beshanzuensis	*A. firma*	*P. pubescens*
Total yield	263.5	260.1	325.1
H/V	0.066	0.086	0.206
S/V	-	-	1.56
H_{acid}/H_{nb}	0.18	0.19	0.12

H/V: molar ratio of *p*-hydroxybenzaldehyde to vanillin;
S/V: molar ratio of syringaldehyde to vanillin:
H_{acid}/H_{nb}: molar ratio of *p*-hydroxyphenyl-type acidolysis products to *p*-hydroxybenzaldehyde in alkaline nitrobenzene oxidation products.

The arylglycerol-β-aryl ether intermonomer linkage is an important characteristic of lignin relating with the definition [27,28], and can be either *erythro*- or *threo*-forms, and the ratio of *erythro*- to *threo*-forms of arylglycerol-β-aryl ether intermonomer linkage (*E/T* ratio). The molar ratio of erythronic acid to threonic acids in ozonation products of the extract-free sample of *A. beshanzuensis*, which reflects the *E/T* ratio of arylglycerol-β-aryl ether intermonomer linkage, was exactly the same with that of *A. firma* (Table 1). The values were in the range of some gymnosperm wood lignins reported by Akiyama et al. [28]. The ^{1}H NMR spectra of acetylated Björkman lignins of both *A. beshanzuensis* and *A. firma* exhibited similar strength of signals at δ 6.01 and δ 6.06 ppm [25]. These results were in agreement with *E/T* ratios of *A. beshanzuensis* and *A. firma* found by ozonation (Table 1). The *p*-hydroxybenzoic acid, *p*-coumaric acid, and ferulic acid have been reported to enhance rooting of stem cuttings [29]. The high level of *p*-hydroxyphenylglycerol-β-aryl ether intermonomer linkages in *A. beshanzuensis* and *A. firma* may promote adventitious rooting of cuttings.

Wang et al. [30] investigated genetic phylogenies and divergence times in Pinaceae using three genomes, the chloroplast *matK* gene, the mitochondrial *nad5* gene, and low-copy nuclear gene encoding 4CL (4-coumarate CoA ligase), which is one of the genes concerned deeply in lignin biosynthesis [14]. They concluded that *A. firma* is genetically the closest species of *A. beshanzuensis* [30]. It was confirmed that structural feature of lignin of *A. beshanzuensis* was quite similar with that of *A. firma* [25]. All these results suggest that *A. firma* would be the best species as the rootstock for grafting *A. beshanzuensis*, and the similarity between *A. beshanzuensis* and *A. forma* could have good affinity between scion and stock. In addition, physiologically young and vigorous *A. forma* supply enough water, nutrients, minerals resulting rejuvenation of *A. beshanzuensis*.

4. ROOT PROMOTION OF *A. BESHANZUENSIS* CUTTINGS

Propagation through stem cuttings in woody plants is used to capture specific genetic combinations (phenotypes) and to provide superior cultivars for planting. The formation of adventitious roots of cuttings is affected by plant growth regulators, stock plant age, season of cutting collection and shoot position [29,31]. Especially, the formation of adventitious root in shoot cuttings was strongly dependent on aging of the donor tree [32,33]. Cutting propagation of *A. beshanzuensis* conducted in 1978 could not succeed. The reason of failure would be due to the physiological age of natural *A. beshanzuensis* branches, which was too old to regenerate adventitious root in cuttings [6]. In Feb. 2004, cutting experiment was conducted with cuttings taken from 2.5 m height, healthy, disease-free, the year's growth branches of grafts. Cuttings were treated with 0.02% RTN, ABT, NAA, or IBA solution, and adventitious roots were propagated in the Forest Nursery, Lishui (119°54' W, 28°27' N), Zhejiang province, China [7]. RNT and NAA exhibited the most outstanding promotion effects on the formation of adventitious roots (Fig. 6). The cuttings treated with RNT, NAA, ABT, IBA and control (the cuttings without any hormone treatment) had average rooting percentages of 79.2%, 70.8%, 62.5%, 58.3% and 54%, respectively (n=3) (Table 3). The highest rooting rate (79.2%) was observed in the cuttings treated with 0.02% RNT solution, which showed significantly differences with control and the treatment with IBA solution [7]. There was no significant difference among other treatment (Table 3). The cuttings treated with 0.02% RNT solution had 10 adventitious roots per cutting, and the average length of adventitious roots was 28 cm, which was significantly higher than those of control and other treatments (Table 4). Cuttings taken from natural grown *A. beshanzuensis* branches could not form adventitious roots, while cuttings collected from grafts grew well, suggesting that the branch of *A. beshanzuensis* was rejuvenated by grafting and recovered vigor and repropagation ability [7].

The effects of cutting on adventitious rooting were examined by the $L_8(2)^7$ orthogonal experiment (Table 5). There were 12 cuttings in every treatments, and every variables took 2 levels, namely length of cutting (5 and 10 cm length), with or without top bud, leaf size (whole and half leaf), the way of base incision (basal and mallet ends of cutting), and the interaction of cutting length×top bud and cutting length×leaf size were investigated [7]. Among those 6 variables, cutting length and top bud were two important factors affecting on rooting rates of *A. beshanzuensis* cuttings (Table 6). Cuttings with top bud and 10 cm length cuttings showed better rooting rate than those of without top bud and 5 cm cuttings. The average rooting rates of 5 cm and 10 cm length cuttings were 5.5 and 8.5, and the longest adventitious roots were 14 cm and 25 cm, respectively (Table 7). Rooting rate of cuttings with top bud was higher than cuttings without top bud. There were no significant differences between whole and half leaf, or basal and mallet ends of cuttings.

Figure 6. Adventitious roots of *A. beshanzuensis* cuttings.

Table 3. Effect of hormone treatment on rooting rate

Hormones	Rooting rate (%)	x_i-x_5	x_i-x_4	x_i-x_3	x_i-x_2
RTN	79.2 (x_1)	25.2*	20.9*	16.7	8.4
NAA	70.8 (x_2)	16.8	12.5	8.3	
ABT	62.5 (x_3)	8.5	4.2		
IBA	58.3 (x_4)	4.3			
Control	54.0 (x_5)				

note: *significant difference.
RTN: rooting hormones developed by the University of Tokyo, ABT: rooting hormones developed by Chinese Academy of Forestry, NAA: α-naphthalene acetic acid, IBA: indole-3-butyric acid, Control; without any hormones (n=3).

Table 4. Effect of hormone treatment on adventitious root development

Hormones	Adventitious root		The longest adventitious root (cm)	
	Average	Range	Average	Range
RTN	10.0	7.5-12.0	28.0	23.5-32.0
NAA	8.0	6.5-9.5	21.5	17.0-23.5
ABT	7.5	5.0-8.5	21.0	16.5-22.0
IBA	6.5	5.5-7.5	20.0	15.5-23.0
CK	6.5	5.5-7.0	21.0	16.0-22.0

Abbreviations: refer footnotes of Table 3. (n=3)

Table 5. Four cutting type of the $L_8(2)^7$ orthogonal experiment

Level	Length (A)	Top buds (B)	A×B	Leaf size (C)	A×C	Residual (e)	Incision (D)
1	5 cm	With	1	Whole	1	1	Mallet
2	10 cm	Without	2	Half	2	2	Basal

A×B and A×C: Interactions between factors A and B, and A and C, respectively.

Table 6. Analysis of variance of the $L_8(2)^7$ orthogonal experiment of rooting rate

Variables	Degree of freedom (df)	Sum of deviation square	Mean square	Mean square ratio	F-test	
A	1	701.3	701.3	81.4*	$F_{0.05}$=18.5	
B	1	705.0	705.0	81.9*	$F_{0.01}$=98.5	
A×B	1	217.4	217.4	25.2*		
A×C	1	78.8	78.8	9.1*		
D	1	217.4	217.4	25.2*		
Residual (C、e)	2	17.2	8.6			

Note: *significant difference with 5% level.

Table 7. The results of adventitious roots of different cuttings

Cutting type		Number of roots		The longest root (cm)	
		Average number	Range	Average number	Range
Cutting length (cm)	10	8.5	6.5-11.0	25.0	21.5-30.5
	5	5.5	4.5-6.0	14.0	10.5-17.0
Top bud	With	10.0	7.5-13.5	26.5	21.5-31.5
	Without	4.5	4.0-6.0	12.0	10.0-15.5
Leaf size	Whole leaf	6.5	4.5-7.0	24.5	19.5-28.0
	Half leaf	7.5	6.5-8.5	23.0	20.0-26.5
Incision	Mallet	7.5	5.0-8.5	24.5	22.5-28.5
	Basal	7.0	5.5-8.5	23.5	18.5-25.5

The formation of adventitious root of *A. beshanzuensis* was investigated by Olympus BH2 Microscope (Olympus Co. Ltd., Tokyo). Callus growth was very slow at the beginning, only a tine of callus was observed until the early June. Callus formation could be observed near the cortex and vascular cambium of incision of the cuttings. During middle June to early July, the callus grew rapidly, and quickly covered the xylem. The hard cuttings of *A. beshanzuensis* had no latent root primordial. The root primordial was induced from a group of parenchyma cells originating from vascular cambium and phloem, and inner callus. Those parenchyma cells have thick cytoplasm, big nuclear and closely-arranged, which was different obviously with other cells. Root primordial differentiation into primordial roots, developed finally to adventitious roots. It was found that 80% of adventitious roots were originated from callus [7].

5. PROMOTION OF CONE PRODUCTION *A. BESHANZUENSIS* BY $GA_{4/7}$ APPLICATION

The successful promotion of flowering in a forest tree species by gibberellins (GA's) was first reported by Kato et al. [34] for a member of the Taxodiaceae family. However for the Pinaceae family conifers seemed to be unresponsive to GA's until 1973 when a mixture of less-polar GA_4 and GA_7 and occasionally also GA_9 were found to promote flowering in Douglas-fir seedlings and mature propagate [35]. The first positive results on Douglas-fir were quickly extended to a variety of other Pinaceae species. Now it is well known that gibberellin $A_{4/7}$ ($GA_{4/7}$) is effective in promoting pollen- and seed-cone productions, especially for the Pinaceae family [36,37]. Grafts of *A. beshanzuensis* were sprayed with $GA_{4/7}$ [8]. The two-way analysis of variance (n=3) showed highly significant ($\alpha=0.01$) differences among variations for the $GA_{4/7}$ concentration and spraying timing, and significant ($\alpha=0.05$) differences among variations for the interaction of $GA_{4/7}$ concentration × spraying timing (Table 8).

Table 8. Analysis of variance of $GA_{4/7}$ promoted pollen-cones

Variations	Degree of freedom (df)	Sum of deviation square	Mean square	Mean square ratio	F-test
Concentration (A)	2	63,494	31,747	11.18**	$F_{0.01(2,18)}=6.01$
Spraying time (B)	1	37,203	37,203	13.10**	$F_{0.01(1,18)}=8.29$
A×B	2	30,808	15,404	5.42**	$F_{0.05(2,18)}=3.55$
Residual (e)	18	51,120	2,840		
Sum (T)	23	182,625			

Note: **: significant difference with 1% level, *: significant difference with 5% level.

The highest pollen-cone, 93.1 per branch (150 cm^3), was recorded in the treatment with $GA_{4/7}$ concentration of 500 mg/L, followed by 250 mg/L, and the lowest was in the control (Table 9). It was suggested that high $GA_{4/7}$ concentration within the 0-500 mg/L affect strongly on the promotion of the pollen-cone production. The results of Q test showed highly significant difference ($\alpha=0.01$) between 500 mg/L concentration and control, and significant difference ($\alpha=0.05$) between 250 mg/L concentration and control (Tables 9, 10). However, the difference between 500 mg/L and 250 mg/L concentration was not significant, suggesting that the effect of 250-500 mg/L concentration on the promotion of pollen cone production of *A. beshanzuensis* was clear (Tables 9, 10) [8].

Table 9. Multiple comparison of different concentration of $GA_{4/7}$ promoted pollen-cones

Concentration (mg/L)	Pollen-cones per branch (150 cm^3)	x_i-x_3	x_i-x_2
500	93.14 (x_1)	89.61**	51.95
250	51.19 (x_2)	47.66	
0	3.53 (x_3)		

$GA_{4/7}$ sprays were applied at different timings during the growing season to promote pollen-and seed-cone production of *A. beshanzuensis*. The average numbers of pollen-cone were 15.4 and 54.8 per branch (150cm^3) sprayed with $GA_{4/7}$ during May-June ("Early") and June-September ("Late"), respectively. Treatment in June-September significantly increased pollen-cone production. Interaction between 500 mg/L × Late spraying significantly increased pollen-cone production, the average number of pollen-cone reached 126.5/branch (150cm^3) (Table 10, Fig. 7). The result of *A. beshanzuensis* was different from that of eastern white pine [36]. For the eastern white pine, early $GA_{4/7}$ application by either spraying or injection during the period of rapid shoot elongation (May-June) promoted pollen-cone production, while the late applications (August-September) did not increase the production [36]. The late $GA_{4/7}$ spraying was more effective, and early treatment in *A. beshanzuensis* may be counting on environment at high altitude habitat of *A. beshanzuensis*. Results indicate that the best treatments for pollen- and seed-cone productions should be carried out in June-September with the concentration of 250-500 mg/L of $GA_{4/7}$ spraying. The $GA_{4/7}$ application could promote pollen-cone production. Although the pollen-cone production did not closely relate to increase of seed-cone, the application of $GA_{4/7}$ is important for genetic resource preservation. The highest seed-cone production was recorded at the $GA_{4/7}$ concentration of 250 mg/L, followed by 500 mg/L, and the control was the lowest (Tables 11, 12). The Q test results showed significant difference ($\alpha=0.05$) between 250 mg/L concentration and control, while there was no significant difference among other concentrations. Similar to pollen-cone promotion by $GA_{4/7}$, the 250-500 mg/L concentration was also appropriate for seed-cone production (Tables 11, 12). Different from pollen-cone, the seed-cone influenced directly to the seed production. The effect of $GA_{4/7}$ application on promotion of seed-cone was not significant compare with pollen-cone production [35,36]. This result was in good agreement with previous papers [35,36]. Spraying of $GA_{4/7}$ increased the number of pollen-cone, and also seed-cones with some extent [37].

Figure 7 Pollen-cone production promoted by $GA_{4/7}$ spray.

Table 10. $GA_{4/7}$ promoted pollen-cone of concentration × spraying timing

Concentration × spraying timing	Pollen-cones per branch (150cm^3)	x_i-x_6	x_i-x_5	x_i-x_4	x_i-x_3	x_i-x_2
500 mg/L × Late	126.48(x_1)	123.33*	122.26*	106.68	101.01	69.58
250 mg/L × Late	56.90(x_2)	53.75	52.68	37.1	31.43	
500 mg/L × Early	25.47(x_3)	22.32	21.25	5.67		
250 mg/L × Early	19.80(x_4)	16.65	15.58			
0 mg/L × Late	4.22(x_5)	1.07				
0 mg/L × Early	3.15(x_6)					

Table 11. Analysis of variance of $GA_{4/7}$ promoted seed-cones

Variations	Degree of freedom (df)	Sum of deviation square	Mean square	Mean square ratio	F-test
Concentration (A)	2	36.4	18.2	5.17*	$F_{0.05}(1,18)$=4.41
Spraying time (B)	1	14.3	14.3	4.06	$F_{0.05}(2,18)$=3.55
A×B	2	22.58	11.3	3.21	
Residual (e)	18	63.36	3.52		
Sum (T)	24	161.03			

Table 12. Multiple comparison of different concentration of GA$_{4/7}$ promoted seed-cones

Concentration (mg/L)	Seed-cones per branch (150cm^3)	x_i-x_3	x_i-x_2
250	3.70 (x_1)	2.49*	1.65
500	2.05 (x_2)	1.84	
0	1.21 (x_3)		

Figure 8. *A. beshanzuensis* recovering vigor and repropagation ability by grafting.

6. PROGENY HEREDITARY VITALITY OF *A. BESHANZUENSIS* BY CROSS-POLLINATION

Sexual reproduction is the basic method to conserve *A. beshanzuensis*. The level of genetic variation in population originated from cross-pollinated seedlings was higher than that in population of old trees and population of graft-originated, suggesting that cross-pollination increased genetic diversity of *A. beshanzuensis* [38-40]. The absence of seed vitality may be due to the lack of genetic diversity and low quality of the seeds produced by mother trees. Low quality of seeds may result from limited pollen- and seed-cones and self pollination.

It is important to note that the grafts of *A. beshanzuensis* were from a few natural *A. beshanzuensis* trees, the genetic diversity may not be improved. Therefore even the cross-pollination among pollen-cone and seed-cone produced by GA$_{4/7}$ application was performed, significant increase in the vitality of seeds would not be expected. The weight of 1,000 seeds increased from 39.3 g of natural pollination to 44.7 g of cross-pollination of *A. beshanzuensis* grafts [8], namely the quality of seeds from cross-pollination was obviously improved. In addition, the vitality of seeds was increased by cross-pollination, especially the vitality of seeds was increase 24-28% by cross-pollination in the early May (Fig. 8).

CONCLUSIONS

1) The dominant community of *A. beshanzuensis* phytogroup is a natural broadleaf mixed forest, in which about 8-9 species form the high canopy divided into 3 sub-layers, and the first sub-layer composed of *A. beshanzuensis* with 13 m height.
2) *A. beshanzuensis* rejuvenated by grafting onto *A. firma* as a stock and the branch recovered vigor and repropagation ability.
3) Structural feature of lignin of *A. beshanzuensis* was investigated comparing with that of *A. firma* to be used as a stock for grafting. The lignin of both species was characterized with the presence of significant amount of *p*-hydroxyphenyl nuclei. Structural feature of lignin of *A. beshanzuensis* was quite similar with that of *A. firma*, suggesting that *A. firma* would be the best species as the mother tree for grafting.
4) The cuttings taken from grafts of *A. beshanzuensis* onto *A. firma* treated with 0.02% RNT had the highest average rooting rate of 79.2%, and had higher adventitious rooting formation. The root primordium and adventitious root developed during middle of June to early July.
5) The $GA_{4/7}$ application could promote seed- and pollen-cone production, and the concentration of 250-500 mg/L was appropriate for the production of seed- and pollen-cone. And late (June-September) $GA_{4/7}$ spraying increased significantly pollen-cone production. The effect of cross-pollination of *A. beshanzuensis* grafts on restoration of filial generation hereditary vitality of *A. beshanzuensis* was made clear.

ACKNOWLEDGMENTS

This study was financially supported by the Toyota Foundation (Grant D06-R-351), Japan, and by the Natural Science Fund of Zhejiang Province (Grant Y304441), China.

REFERENCES

[1] Wu, MX. *Acta Phytotaxonamica Sinica* 1976, 14, 15-22.
[2] Kong, S. *Plants* 1999, 1, 8-10.
[3] Ye, Z; Shen, L; Zhou, R. *Science World Magazine* 2007, 6, 57.
[4] Wu, MX. *Plants*, 1999, 1, 6-8
[5] Lin, X. *Human and Biosphere*, 1999, 3, 1-4.
[6] Cai, Z. *Journal of Botanic*, 1992, 5, 6.
[7] Shao, S; Qian, H; Jin, Z; Wang, B; Liu, B; Bai, M; Chen, X. *Bulletin of North-east Forestry University*, 2006, 34(5), 47-48.

[8] Shao, S; Chen, X; Tang, L; Chen, D; Fang, J; Liu, B. *Journal of Zhejiang Forestry & Technology*, 2007, 27(5), 21-24.
[9] Yu, J; Yao, F; Chen, X; Zhou, R; Cheng, Q; Ding, B. *Journal of Tropical and Subtropical Botany*, 2003, 11(2), 93-98.
[10] Wu, H; Chen, D; Yu, J. *Journal of Zhejiang Forestry College*, 1997, 14(1), 22-28.
[11] Hu, B; Shao, S; Qian, H; Zhou, Q. *Journal of Zhejiang Forestry Science & Technology*, 2004, 24(3), 12-17.
[12] Wardrop, AB. In Sarkanen, KV.; Ludwig, CH. Eds.; *"Lignins: Occurrence, Formation, Structure and Reactions"*. Wiley-Interscience: New York, 1971, p 19-42.
[13] Higuchi, T. *Proceeding of Japan Academy Series B - Physical and Biological Sciences*, 2003, 79(8), 227-236.
[14] Iiyama, K; Lam, TBT; Meikle, PJ; Ng, K; Rhodes, D; Stone, BA. In Jung, HG; Buxton, DR; Hatfield, RD; Ralph, J. Eds.; *"Forage Cell Wall Structure and Digestibility"*. American Society of Agronomy, Madison, USA, 1993, p 621-683.
[15] Creighton, RHJ; Gibbs, RD; Hibbert, H. *Journal of American Chemical Society*, 1944, 66, 32-37.
[16] Chen, CL. In Lin, SY.; Dence, CW. Eds.; *"Methods in Lignin Chemistry (Springer Series in Wood Science)"*. Springer-Verlag, Heidelberg, 1992, p 301-321.
[17] Bicho, JG; Zavarin, E; Brink, DL. *Tappi*, 1966, 49, 218-226.
[18] Balakshin, MY; Capanema, EA; Goldfarb, B; Frampton, J; Kadla, JF. *Holzforschung*, 2005, 59, 488-496.
[19] Donaldson, LA. *Phytochemistry*, 2001, 57, 859-873.
[20] Kuznetsov, BN; Efremov, AA; Levdanskii, VA; Kuznetsova, SA; Polezhayeva, NI; Shilkina, TA; Krotova, IV. *Bioresource Technology*, 1996, 58, 181-188.
[21] Ozawa, S; Sasaya, T. *Mokuzai Gakkaishi*, 1991, 37, 847-51.
[22] Björkman, A. *Svensk Papperstidning*, 1956, 59, 477-485.
[23] Higuchi, T; Tanahashi, M; Sato, A. *Mokuzai Gakkaishi*, 1972, 18, 183-189.
[24] Adler, E; Pepper, JM; Erikson, E. *Industrial & Engineering Chemistry*, 1957, 49, 1391-1392.
[25] Shao, S; Jin, Z; Weng, Y. *Journal of Wood Science*, 2008, 54, 81-86.
[26] Lam, TBT; Iiyama, K; Stone, BA. *Phytochemistry*, 1990, 29, 429-433.
[27] Brunow, G; Karlsson, O; Lundquist, K; Sipilä, J. *Wood Science & Technology*, 1993, 27, 281-286.
[28] Akiyama, T; Goto, H; Nawawi, DS; Syafii, W; Matsumoto, Y; Meshitsuka, G. *Holzforschung*, 2005, 59, 276-281.
[29] Bhardwaj, DR; Mishra, VK. *New Forests*, 2005, 29, 105-116.
[30] Wang, XQ; Tank, DC; Sang, T. *Molecular Biological Evolution*, 2000, 17, 773-781.
[31] Rowe, DB; Blazich, FA; Goldfarb, B; Wise, FC. *New Forests*, 2002, 24, 53-65.
[32] Kaul, K. *New Forests*, 2008, 36, 217-224.
[33] Husen, A; Pal, M. *New Forests*, 2007, 31, 57-73.
[34] Kato, Y; Fukuhara, N; Kobayashi, R. In *"Transaction of the 2nd Meeting of the Society for the Study of Gibberellins"*. Tokyo. Kyowa Hakko Kogyo, Tokyo, 1958, p 67-68.
[35] Pharis, RP. *Annual Review of Plant Physiology*, 1980, 36, 517-568.
[36] Ho, RH; Eng, K. *Forest Ecology and Management*, 1995, 75, 11-16.
[37] Ho, RH; Schnekenburger, F. *Tree Physiology*, 1992, 11, 197-203.
[38] Pharis, RP; Webber, J; Ross, SD. *Forest Ecology and Management*, 1987, 19, 65-84.

[39] Wesoly, W; *Forest Ecology and Management*, 1987, 19, 121-127.
[40] Ai, J; Qiu, Y; Yu, J; Chen, X; Ding, B. *Journal of Zhejiang University (Agricultural & Life Sciences)*, 2005, 31(3), 277-283.

INDEX

A

absorption spectroscopy, 386
acclimatization, 52, 294
accommodation, 256
accounting, 246, 252, 253
accuracy, 197
acetic acid, 417, 424
acetylcholine, 12
achievement, 37
acid, 10, 11, 32, 33, 41, 53, 54, 117, 131, 132, 142, 269, 342, 396, 417, 420, 421, 422, 424
acidity, 8, 388, 400, 403
acrosome, 118, 135
acrylate, 237
activation, 12, 35, 36, 37, 38, 44, 48
adaptability, 99, 288
adaptation, 42, 206, 219, 221, 224, 227, 298, 305, 312, 421
additives, 47
adductor, 90
adipose, 157
adjustment, 115
adults, 306, 307
Africa, 113, 131, 160, 162, 299, 336, 355
age, xiv, 7, 8, 31, 32, 49, 59, 66, 73, 80, 83, 86, 87, 90, 91, 92, 105, 106, 192, 201, 220, 290, 345, 415, 419, 422
agent, 11, 12, 172, 204, 207, 352, 383
aggregates, 185, 188
aggregation, 189, 261
aggression, 210
aging, 6, 116, 422
agriculture, vii, 1, 20, 51, 145, 150, 154, 166, 216, 257, 298
agrosystem, 44
air pollutants, 386

air quality, xiv, 383, 384
Alaska, 70, 71, 72, 81, 98, 101
albumin, 11, 120, 121
alcohol, 188
alcohols, 420
algae, 248, 250, 253, 256, 257, 258, 259, 260, 262, 362
algorithm, 222, 406
alkaloids, 12, 42, 43, 46, 48
allele, 346, 347
allometry, 358
alternative, vii, 2, 26, 42, 57, 119, 121, 133, 140, 175, 177, 312, 333
ambient air, 384
amino acids, 10, 34, 121
ammonia, 220, 235, 264
amphibia, 358
amphibians, 147, 153
amplitude, 237, 380, 393, 398, 401
Amur River, 288, 290
analgesic, 11, 12
anatomy, 14, 46, 55, 58, 115
androgens, 135, 137
aneuploid, 14
animal husbandry, xii, 317, 332
animals, ix, 6, 17, 20, 30, 112, 115, 116, 117, 118, 121, 122, 124, 125, 136, 140, 146, 152, 153, 155, 172, 195, 200, 202, 204, 216, 245, 260, 262, 267, 298, 314, 343, 347, 350, 352
annealing, 237, 347
annihilation, 25
ANOVA, 187, 192, 193, 197, 225, 302, 307, 350
anther, 6
antioxidant, 10, 13, 58
apoptosis, 12, 136
appendix, 17, 25
aquatic habitats, 379

aquatic systems, 146, 167, 168
Argentina, vii, xii, 1, 14, 341, 343, 345, 352, 354, 357, 358
argument, xii, 298, 312
arsenic, 387
arson, 122
artery, 128
arthropods, x, 184, 192, 273
ascorbic acid, 10, 13
ash, 404
Asia, 62, 98, 104, 156, 160, 162, 179, 180, 181, 282, 314
aspiration, 115, 125
assessment, 45, 139, 216, 222, 230, 234, 261, 273, 311, 313, 357
assets, 24, 379
assimilation, 41
assumptions, x, 215, 217, 225, 228, 229, 230
astringent, 11
asymmetry, 133
atmospheric deposition, 386
atoms, 13
atrium, 280, 281
attachment, 285, 290
attacks, 13, 168, 420
attractiveness, 31
Australia, 131, 241, 315, 345, 357
Austria, 21, 278, 294, 381
authority, 22, 24, 25, 174
availability, 8, 19, 21, 31, 33, 145, 151, 169, 257, 263, 279, 398
averaging, 75
awareness, 174, 208, 220
axilla, 6

B

BAC, 233
backwaters, xiii, 361, 364, 365, 367
bacteria, 216, 217, 219, 221, 225, 231, 233, 235, 236, 237, 238
bacterium, 42, 219
banking, 131, 132, 134, 176
banks, viii, xiii, 8, 26, 111, 113, 129, 134, 141, 176, 177, 361, 363, 364
barriers, 34, 205, 217, 221, 237, 244, 305, 420
base pair, 219, 221, 344
basic research, 117
beetles, 212
behavior, x, xiii, 33, 51, 52, 55, 59, 133, 135, 182, 184, 186, 188, 191, 199, 200, 342, 344, 351, 353
Beijing, 179
Belgium, 21

beneficial effect, 20, 118
Bengal, Bay of, ix, 143, 144, 147, 149
bias, 221, 227, 228, 237
bioavailability, 290
biogeography, 92, 93, 234, 336, 337
bioindicators, 102
biological processes, 240
biomass, 8, 41, 216, 257, 258, 264, 390
biomechanics, 46
biomonitoring, 380
biopsy, 133, 134
bioremediation, 13, 235, 238, 380
biosphere, 221, 233, 234, 237
biosynthesis, 422
biotechnology, 34, 41, 43, 114
biotic, 42, 62, 99, 242, 420
birds, 6, 20, 270, 332, 337, 343
birth, 12, 350, 352
births, 116, 139
Black Sea, 362
bladder, 115, 157, 158
blocks, 128
blood, 10, 12, 46, 52, 347, 352, 355
body composition, 314
body shape, 351
body size, 345, 351
Bolivia, 14
bonds, 118
Bosnia, 268, 286, 292
Brazil, x, 14, 183, 184, 203, 207, 212, 213, 290, 345, 352
breeding, viii, x, xiv, 52, 55, 111, 112, 144, 145, 147, 150, 151, 152, 153, 154, 157, 167, 171, 174, 175, 176, 178, 194, 196, 343, 345, 352, 353, 358, 415
Britain, 286, 314, 336
Brno, 294, 295
bryophyte, xiii, 383, 393, 395, 398, 403, 404
buffalo, 126, 129, 136
buffer, 117
Bulgaria, 268, 286, 288, 291
Byelorussia, 268, 286

C

Ca^{2+}, 305
calcium, 11, 12, 13, 35, 44, 80, 122, 302, 304, 305, 312, 315, 398
calcium carbonate, 80, 305
cambium, 419, 425
Cambodia, 156
Cambrian, 66
Cameroon, x, 239
Canada, vii, 1, 14, 45, 269

canals, ix, 143, 145, 146, 148, 172, 173, 174, 343, 378
cancer, ix, 59, 111, 123, 216, 235
cancerous cells, 13
candida, 37, 42, 49, 57
candidates, 332
capillary, 220
capsule, 158
carapace, 64, 66, 70, 84, 86, 88, 99
carbohydrates, 10, 11
carbon, 34, 40, 41, 103, 119, 222, 238, 413
cardiotonic, 11
Caribbean, 29, 237
carotene, 118
carotenoids, 10
case study, 104, 210, 300, 316
castration, 115
catabolism, 237
catalytic activity, 342
catastrophes, 113
catchments, 150
catfish, 170, 171, 275
cattle, 11, 20, 51, 122, 136, 318
causal relationship, 63, 83, 97, 98, 99
C-C, 420
cell, ix, 8, 12, 13, 30, 34, 35, 36, 40, 48, 112, 117, 123, 129, 131, 132, 134, 322, 324, 357, 420, 421
cell culture, 13, 357
cell line, ix, 112, 129, 131
cell membranes, 132
Central Asia, 289
Central Europe, xiii, 280, 286, 288, 361, 378, 381, 384
cestodes, 267, 269, 286, 287, 291, 294
changing environment, 100
channels, 150, 153
cheese, 10
chemoprevention, 59
Chile, 14, 403, 405, 406, 407, 411, 412, 413
China, ix, xiv, 62, 90, 93, 143, 170, 179, 268, 286, 415, 416, 418, 422, 429
chlorophyll, 40, 406, 407, 409, 411
chloroplast, 14, 46, 422
cholesterol, 10, 12, 118, 137, 138
chromatography, 421
chromosome, 123, 135, 231, 345, 357
class size, 390
classes, 232, 386
classification, 3, 209, 235, 362
cleaning, 159
climate change, xiv, 14, 16, 62, 96, 99, 102, 216, 318, 338, 415, 416
clone, 216, 220, 221, 222, 223, 225, 226, 232

cloning, 129, 132, 222
closure, viii, 61, 80, 81, 172, 173, 177
cluster analysis, 227, 245
clusters, 256
CO_2, 4, 41, 54, 96
coccus, 11
codes, 250
coding, 235, 400
codominant, 352
collenchyma, 4
Colombia, 11, 57
colonisation, 403
colonization, x, 184, 192, 206, 256, 267, 268, 279, 280, 282, 288, 354, 393, 395
combined effect, 16
commerce, 352
commercial hunting, xiii, 341
communication, 166, 173, 237, 263
compensation, 80, 309
competence, 121
competition, 46, 171, 184, 227, 252, 258, 263, 287, 391
complement, 112, 145, 222
complexity, 262, 368
complications, ix, 143
components, 10, 43, 117, 118, 119, 227, 229, 238, 242, 342, 420
compounds, 11, 12, 47, 52, 118
computing, 211
concentration, xi, 25, 36, 41, 96, 116, 117, 118, 130, 140, 186, 218, 239, 245, 257, 258, 262, 267, 302, 304, 305, 312, 322, 326, 388, 400, 406, 407, 411, 425, 426, 427, 428, 429
concrete, 207
conduction, 420
conductivity, xiv, 302, 304, 305, 310, 405, 406, 407, 409, 413
confidence, 225, 229, 247, 307, 359
confidence interval, 225, 247
configuration, 387
conflict, 174, 210
conflict of interest, 174
Congress, 211
connectivity, 207, 208, 256, 299, 311
consensus, 191, 217, 417
constipation, 13
construction, ix, 28, 91, 104, 143, 151, 153, 154, 166, 221, 227, 293, 298, 303, 312, 313, 343, 363, 364, 379
consumers, 261, 262
consumption, 10, 59, 344
contaminant, 35
contaminated soils, 235, 387, 398, 403

contamination, 35, 115, 176, 387, 393, 400, 403
control, 12, 17, 24, 32, 33, 123, 153, 154, 166, 178, 225, 241, 262, 286, 295, 299, 344, 355, 386, 389, 400, 422, 425, 426
conversion, 224
cooling, 68, 119, 120, 125, 126, 127, 128, 129, 130, 132, 134
copper, xiii, 120, 128, 234, 383, 384, 387, 395, 402, 403
corolla, 7
correlation, 31, 78, 106, 109, 116, 192, 246, 256, 257, 302, 304, 305, 312
correlation coefficient, 246, 302, 305
cortex, 4, 127, 425
Costa Rica, 11, 14, 336, 356
costs, viii, 6, 11, 40, 111, 120, 121, 129, 176, 221
Côte d'Ivoire, 244, 413
cotton, 6, 50
cotyledon, 3
covering, 24, 150, 153, 174, 222, 391, 393, 394, 395
credit, 177
creep, 3, 5, 27
Croatia, 282, 291, 297, 301, 302, 303, 304, 313, 316
crocodile, 104, 355, 357
crops, 73, 167, 168, 170
crying, 177
cryobanking, 126
cryopreservation, viii, ix, 111, 117, 118, 119, 120, 121, 122, 123, 124, 125, 127, 128, 130, 132, 133, 134, 135, 138, 139, 140, 141, 176
crystal growth, 119
crystalline, 120
crystallization, 120, 126
crystals, 119, 126
Cuba, 14
cues, 167
cultivation, 167, 172, 219
culture, vii, 2, 11, 12, 34, 35, 36, 37, 40, 41, 42, 47, 49, 51, 54, 56, 113, 128, 144, 145, 154, 206, 216, 332
culture media, 34, 36, 40
cumulative percentage, 250
currency, 145
cuticle, 4, 35, 40
cyanide, 266, 269, 278, 288, 290, 294
Cyanophyta, 250
cycles, viii, 8, 41, 61, 93, 94, 97, 100, 120, 138, 198, 216
cystic fibrosis, 238
cytochrome, 356
cytogenetics, xiii, 342, 353
cytology, 136
cytoplasm, 425

Czech Republic, 265, 268, 270, 280, 282, 283, 284, 286, 287, 289, 291, 293

D

danger, 145, 268, 288, 332
Danube River, xi, 265, 282, 288, 291, 295, 362, 368, 373, 378, 380, 381
data analysis, 231, 233
data gathering, 193, 195, 396
data set, 222, 225, 226, 229, 230
database, 70, 219, 269, 320
dating, 73
death, viii, 41, 83, 111, 113, 125, 139
decay, 11
decisions, 224, 353
decomposition, 222, 238
defense, 41, 267, 420
defense mechanisms, 41, 267
deficiency, 400
definition, 216, 217, 342, 421
deforestation, 20, 205, 318, 332, 353
degenerate, 393
degradation, xiii, 150, 166, 167, 181, 189, 330, 383, 384, 386, 387, 391
degradation process, xiii, 383, 384
dehydration, 8, 121, 138, 141
delivery, 242
demography, 55
denitrifying, 233, 236
density, 19, 20, 26, 31, 119, 186, 201, 222, 232, 247, 248, 249, 250, 252, 256, 306, 312, 400
Department of the Interior, 58
deposition, 78, 108, 192, 387
deposits, 73, 108, 109, 404, 406
derivatives, 12, 227, 421
destruction, vii, xii, 1, 18, 25, 35, 150, 167, 184, 298, 300, 310, 313, 317, 332, 384
detection, 50, 169, 222, 225, 234, 261, 358
developing countries, 312
deviation, 424, 425, 427
devolution, 22
dew, 6, 191
diet, 118, 145, 171, 202, 203, 210
dietary fat, 131, 132
dietary supplementation, 140
differentiation, xiii, 34, 35, 36, 40, 84, 205, 212, 342, 344, 345, 349, 425
diffusion, 27
diluent, 116
dimorphism, 66, 212
diploid, 14, 185, 345, 347
disaster, xiv, 267, 273, 415, 416

discharges, 256, 308
discrimination, 227, 245, 357
disinfection, 35
disorder, 232
dispersion, 7, 15, 30, 186, 207, 210, 282
disposition, 5
disseminate, 287
dissolved oxygen, viii, xi, 61, 82, 83, 93, 94, 96, 98, 100, 239, 245, 252, 253, 256, 257, 258, 302
distillation, 225
distress, 35
diuretic, 11
divergence, 217, 226, 228, 234, 422
diversification, 2, 16
division, x, 71, 184, 194, 210, 379
division of labor, 210
DNA, 14, 118, 122, 123, 131, 134, 136, 137, 217, 218, 219, 221, 222, 223, 232, 233, 234, 235, 236, 237, 238, 342, 344, 347, 352, 355, 356, 357, 358
DNA polymerase, 221, 237
docosahexaenoic acid, 118
dogs, 129, 134
dominance, xi, 30, 185, 211, 226, 228, 234, 240, 245, 248, 249, 253, 256, 258, 261, 275, 304, 406, 407, 413
donors, 134
dosage, 171
draft, 333
drainage, 41, 144, 149, 222, 287, 300, 301, 302
dressings, 13
Drosophila, 136
drought, 4, 5, 15, 34
dry ice, 119
drying, viii, 111, 121, 135, 136, 139, 140, 141, 151, 155, 173
duration, 48, 127, 150

E

E.coli, 222
early warning, 293
earnings, 145
earth, 62, 64, 96, 99, 100, 101, 147
earthworms, 413
East Asia, 88, 92, 93, 96, 108, 267
East China Sea, 62, 90
eating, 270
economic activity, 206
ecosystem, xi, 2, 19, 150, 151, 168, 169, 172, 179, 216, 217, 219, 234, 240, 241, 257, 261, 266, 270, 288, 293, 295, 299, 301, 311, 314, 322, 383
Education, 101, 103, 108
effluents, 167, 178, 257

egg, 66, 118, 119, 132, 139, 162, 185, 194, 315, 345
Egypt, 43
ejaculation, 115, 136, 140
elaboration, 402
electric charge, 346
electricity, 299, 310
electrolyte, viii, 111, 121
electrophoresis, xiii, 216, 220, 234, 235, 238, 342, 346, 347, 348, 352, 354
elephants, 115, 118, 131, 133, 139, 141
elk, 123, 139
elongation, 426
e-mail, 297, 415
embryo, ix, 3, 8, 39, 40, 112, 113, 126, 130, 138
embryogenesis, 35, 37, 38, 39, 40, 49, 51, 54, 58
emission, 393, 395, 400, 401
employment, 145, 155
encoding, 234, 237, 422
endocrine, ix, 111, 137
endonuclease, 355
energy, 112, 117, 121, 138, 139, 258, 298, 364
England, 268, 287
enteritis, 285
enthusiasm, 333
environmental change, 63, 70, 71, 77, 93, 102, 105, 189, 264, 342, 344
environmental conditions, 7, 8, 91, 98, 220, 242, 256, 266, 365, 379
environmental factors, x, 63, 70, 98, 135, 239, 240, 241, 242, 256, 257, 262, 302
environmental issues, 24
enzymes, 13, 119
epidemic, 113, 138
epidermis, 6, 31
epididymis, 115, 116, 125
epithelial cells, 137
epithelium, 282
equilibrium, 22
erosion, 26, 261
erythrocytes, 355
essential fatty acids, 132
EST, 347
estimating, 225, 234, 236, 241
ethanol, 35, 59, 269, 302
ethology, xiii, 342
ethylene, 117, 127
ethylene glycol, 117, 127
EU, 364
eucalyptus, 205, 206
Eurasia, 160
Europe, 11, 98, 162, 176, 212, 267, 268, 269, 278, 282, 286, 287, 292, 293, 294, 295, 300, 303, 314, 362, 396, 403

evaporation, 173
evapotranspiration, 4, 7
evolution, vii, viii, 2, 14, 44, 99, 104, 106, 108, 111, 209, 212, 219, 233, 238, 257, 342, 354, 355, 358, 383
examinations, 282
excision, 123
exclusion, 311
execution, xiii, 342
experimental condition, 8
expertise, 322
exploitation, vii, xii, 1, 147, 151, 166, 178, 317, 332
explosives, 172
exposure, 125, 169, 240, 322, 411
external environment, 350
extraction, xii, 20, 22, 54, 116, 167, 173, 205, 219, 317, 332, 347, 386, 421
extrapolation, 122

F

facies, 103
failure, 117, 140, 295, 417, 419, 422
family, vii, 1, 2, 3, 17, 41, 43, 44, 45, 55, 114, 145, 158, 159, 160, 161, 162, 163, 164, 195, 196, 217, 222, 232, 342, 353, 400, 425
famine, 234
farmers, 154, 168, 169, 170, 177
farms, xiii, 145, 146, 208, 289, 361, 364
fatty acids, 118, 140
feces, 115
federal law, xii, 318, 322
females, x, 66, 113, 115, 123, 124, 176, 183, 185, 195, 196, 307, 345, 352
fermentation, 52
fertility, ix, 55, 111, 120, 122, 127, 132, 133, 141, 150
fertilization, 42, 113, 116, 117, 118, 134, 136, 181, 233
fiber content, 10, 13
fibroblasts, 132, 136
fibrosis, 216, 237, 238
fidelity, 201, 210
financial support, 289
financing, 204
fingerprints, 219, 229, 231
h and Wildlife Service, 47
fisheries, 145, 147, 149, 150, 154, 166, 167, 169, 172, 173, 174, 177, 178, 179, 180, 181, 282, 290
fishing, 145, 150, 151, 153, 155, 166, 172, 173, 174, 177, 178
fission, x, 184, 188, 189, 194, 195, 204
fitness, 342

fixation, 4, 269
flame, 386
flavonoids, 13, 43
flexibility, 118, 176, 207, 229
floating, 88, 362, 378
flood, 146, 150, 152, 153, 154, 172, 180, 181, 299, 364, 378
flooding, ix, 143, 146, 147, 151, 167, 173, 269
flora, 23, 47, 57, 58, 59, 112, 153, 155, 207, 233, 294, 330, 334, 335, 336, 338, 362, 367
flora and fauna, 153, 155, 294, 330
fluctuations, viii, 61, 62, 63, 79, 80, 81, 91, 94, 96, 97, 98, 99, 100, 105, 107, 240
fluid, 118, 119, 120
focusing, 186
follicle, 136, 284
follicles, 124, 127, 128, 132, 133, 280, 281, 283, 284, 285
food, vii, 1, 6, 10, 11, 13, 48, 54, 57, 58, 83, 93, 112, 144, 150, 152, 167, 168, 169, 171, 173, 179, 184, 185, 188, 192, 196, 201, 202, 203, 204, 242, 258, 262, 264, 306, 413
forbs, 390, 392
forecasting, 243
forest fragments, 186, 205, 206, 208
forests, xii, 152, 205, 206, 207, 208, 212, 317, 318, 320, 322, 332, 335, 336, 337
fossil, viii, 14, 61, 63, 64, 65, 66, 67, 70, 73, 77, 78, 82, 84, 86, 87, 88, 91, 92, 93, 98, 99, 100, 102, 103, 104, 105, 107, 109
fractures, 12
France, 21, 210, 398
free radicals, 13
freedom, 112, 424, 425, 427
freedom of choice, 112
freezing, ix, 15, 111, 117, 118, 119, 120, 121, 123, 125, 126, 127, 128, 130, 132, 133, 134, 135, 138, 139, 140, 141
frost, 418
fruits, 3, 4, 7, 10, 11, 12, 13, 48, 54, 57, 59, 171, 203, 204
fuel, 216
functional analysis, 221, 238
funding, 174, 316
fungi, 42, 233

G

gamete, 130, 134
garbage, 204
gastric mucosa, 59
gastritis, 13, 59
GDP, 145

gel, 118, 216, 219, 235, 238, 346, 348
gender, 123, 187
gene, 26, 113, 116, 175, 176, 185, 217, 219, 221, 222, 223, 225, 229, 230, 232, 236, 237, 264, 342, 344, 349, 356, 422
generation, 37, 38, 42, 145, 181, 230, 235, 298, 299, 429
genetic alteration, 36
genetic diversity, xiv, 113, 116, 130, 177, 208, 217, 218, 233, 342, 345, 350, 352, 415, 417, 428
genetic drift, 116, 205, 206, 344, 349, 352, 353
genetic information, xiv, 205, 222, 345, 415
genetics, xii, 46, 235, 341, 342, 351, 353, 356
genome, viii, 111, 113, 132, 217, 218, 222, 238, 280, 344, 347, 400
genomics, 237
genotype, 344, 346, 352
Georgia, 212
Germany, 21, 111, 131, 215, 268, 286, 362
germination, 4, 8, 31, 32, 33, 34, 39, 44, 49, 50, 53, 57, 58, 203
gibberellin, xiv, 415, 417, 425
gill, 269
gland, 6, 131
glass transition, 122
glasses, 131, 140
glucose, 10, 12, 46, 121
glucoside, 13
glutathione, 118
glycerin, 269
glycerol, 117, 120
glycol, 117
glycolysis, 121
goals, 40, 342
gold, 219, 266
gonads, 134
government, 11, 21, 24, 26, 152, 167, 168, 170, 172, 173, 174, 177, 332
government policy, 167, 174
gracilis, 158, 159
grains, 13
graph, 71, 80, 98
grass, 221, 235, 268, 275, 278, 286, 390, 393, 394, 395, 396, 400, 401, 402, 404
grasses, 390, 393, 395, 400, 401, 403
grazing, vii, 1, 147, 152
Great Britain, 268, 286
Greece, 300
Green Revolution, 180
grids, xii, 317, 321
gross domestic product, 145
groundwater, 167

groups, xiv, 14, 20, 23, 42, 43, 67, 71, 96, 99, 113, 117, 156, 157, 170, 195, 206, 217, 222, 225, 227, 237, 243, 245, 249, 253, 273, 278, 299, 311, 332, 333, 344, 352, 362, 380, 405
growth rate, vii, 1, 8, 20, 43, 172, 257
Guatemala, 338
guidelines, xiii, 174, 342, 353
guiding principles, 179
Guinea, x, 159, 239, 244, 259, 357
Guinea, Gulf of, x, 239, 244, 259
gut, 220, 222, 232, 233
gymnosperm, 420, 421, 422

H

haploid, 185
hardening process, 40
hardness, 306
harvesting, 6, 150, 343
health, 10, 112, 169, 172, 232, 282, 288, 293, 384
heat, 62, 120, 128, 129, 173
heat transfer, 120, 128
heating, 128
heavy metals, xi, 265, 266, 267, 292, 385, 395, 398, 400, 403
height, 26, 187, 191, 192, 227, 260, 419, 422, 429
height growth, 419
hemisphere, 108, 406
herbicide, 181
heritability, 351, 352
hermaphrodite, 6
heterogeneity, 16, 220, 225, 237, 241, 261, 311
highlands, 152
homogeneity, 260, 411
homozygote, 348
Hong Kong, 170, 241
hormone, 31, 41, 127, 422, 423, 424
host, xi, 196, 265, 267, 269, 270, 271, 272, 274, 275, 276, 277, 278, 279, 280, 282, 285, 286, 288, 290, 293, 294, 295
households, 145
housing, vii, 1, 152, 333
human activity, ix, 112, 143, 322
human development, vii, 1
humidity, 33, 40, 41, 42, 144, 200, 352, 418
humus, 390
Hungary, 266, 268, 269, 270, 273, 280, 282, 283, 286, 287, 288, 289, 291, 293, 364
hunting, 212, 343, 353
hybridization, 14, 46, 58, 234, 417
hydroelectric power, 154
hydrophyte, 362, 365, 379
hypothesis, 46, 82, 202, 234, 242, 349, 379, 406

hypothesis test, 46
hypoxia, 82

I

identification, 66, 135, 187, 188, 231, 237, 270, 294, 311, 312, 313, 318, 352
identity, 217, 223, 348
ideology, 358
illumination, 124, 125
illusion, 260
image analysis, 270
images, 68, 74, 77, 84, 85, 89, 90
imbibition, 33
immersion, 31
immigration, 227
immune response, 169
impact assessment, 174
implementation, xiii, 172, 179, 342, 353, 379
in vitro, vii, ix, 2, 12, 13, 26, 34, 35, 36, 39, 40, 41, 42, 47, 48, 52, 53, 56, 57, 59, 111, 113, 116, 121, 124, 125, 126, 128, 132, 135, 138, 139, 140, 293
in vivo, 52, 131
inbreeding, 116, 133
incidence, 15, 188
inclusion, 11, 255
income, 11, 152, 155, 177, 181
incompatibility, 34, 46
incubation time, 116
independence, 352
India, ix, 143, 144, 148, 156, 158, 170, 179, 181, 182
indication, 249, 266, 272
indicators, 108, 273, 290, 293, 362, 378, 403
indices, x, 66, 215, 216, 218, 220, 224, 225, 226, 228, 229, 231, 245, 252, 282, 304, 318, 372, 373, 396
indigenous, ix, 9, 10, 143, 150, 152, 153, 154, 155, 167, 169, 171, 172, 173, 176, 178, 267, 269, 288
Indonesia, 156
induction, 141
industry, ix, 99, 123, 143, 166, 298
infancy, x, 215
infection, 269, 270, 274, 276, 277, 280, 282, 283, 293, 294
infertility, 117
inflammation, 12
inflammatory bowel disease, 236
inflation, 234, 246
infrared spectroscopy, 118
inheritance, 344, 347
inhibition, 13, 390
inositol, 34
insecticide, 150, 154, 181

insects, 7, 171, 202
insight, 221
instability, 267
institutions, 22, 113
insulin, 12
insurance, 113, 261
integration, 177
integrity, ix, 36, 112, 122, 125, 128, 130, 131, 134, 135, 167, 322, 362
interactions, 59, 169, 185, 211, 212, 240, 241, 242, 243, 252, 255, 260, 261, 262, 263, 313, 314, 384, 404
interface, 119, 122, 128, 293
interference, 188, 270
internode, 5
interrelationships, 227
interval, 64, 81, 140, 417
intestine, 12, 282, 291
invertebrates, 153, 155, 202, 246, 248, 253, 255, 258, 259, 273, 299, 311, 393
ions, 302, 305, 312
iron, 269, 291
ischemia, 52, 127
Islam, 43
isolation, 16, 193, 206, 347, 349, 352
isotope, 80, 81, 92, 109
isozyme, xiii, 342, 347
isozymes, 342, 348
Israel, 111, 113, 119, 140, 289
Italy, 21, 300, 403

J

Japan, Sea of, 102, 106, 107, 108, 109
Java, 101, 161
jaw, 162
joints, 30
Jordan, 15, 52
justice, 22
justification, 141
juveniles, 302, 345

K

K^+, 12
Kazakhstan, 268, 286, 289
kidney, 46, 347
killing, 35, 169, 174
kinetics, 218, 223
Korea, 288
Krebs cycle, 121

L

lactose, 117
lakes, ix, xiv, 64, 143, 145, 146, 153, 154, 305, 405, 406, 407, 411, 412, 413
land, 2, 11, 15, 62, 63, 112, 144, 146, 147, 208, 216, 233, 238, 244, 258, 298, 299, 311, 362, 391, 393, 394, 398
land use, 216, 238
Land Use Policy, 173
landscapes, xii, 261, 263, 298, 311, 316, 317, 323
land-use, 298
larva, 187, 194, 260
Latin America, 131, 141, 299
Latvia, 268, 286, 288, 294
laws, 22, 177
LDL, 12
learning, 169
legislation, 25, 177, 207, 362
Lepidoptera, 203
levees, 147, 152
liberation, 12
lichen, xiii, 383, 391, 393, 395, 396, 402, 403
life cycle, vii, 1, 8, 31, 43, 270, 273, 288, 290, 292, 293
life span, 46, 344
lifespan, 113, 127
lifetime, 315
light conditions, 201
lignin, 10, 417, 420, 421, 422, 429
likelihood, 26, 359
limitation, ix, 43, 112, 246
linkage, 420, 421
links, 216, 237, 259, 262
lipids, 10, 11, 52, 117, 118, 125, 203
liposomes, 118, 140, 142
liquid chromatography, 50
Lithuania, 268, 286
liver, 12, 137, 347
livestock, 2, 11, 113, 117, 123
living environment, 70
local community, 241
locus, 342, 347, 348, 352
logistics, 204
Louisiana, 131
low temperatures, 126, 138
luciferase, 221
lumen, 285
lying, 149

M

macroalgae, x, 239, 244, 245, 246, 248, 255, 257, 258, 259, 260, 261, 263
magnesium, 11, 13, 306, 398
Malaysia, 156, 241
males, 66, 115, 185, 193, 195, 213, 307, 343
mammal, 136, 189
management, ix, xiii, 26, 45, 123, 138, 141, 143, 166, 167, 172, 173, 174, 177, 178, 179, 180, 181, 236, 243, 264, 310, 311, 313, 314, 315, 319, 333, 341, 343, 344, 345, 353, 354
MANOVA, 350
manure, 30, 237
mapping, 186
marine environment, 71, 84, 86, 87, 88, 96, 103, 263
market, 20, 21, 31, 119
marketing, 145, 155
markets, 25, 158, 384
marsh, 233
matrix, 51, 54, 206, 224, 227, 230, 406
maturation, 31, 113, 123, 125, 128, 194
measurement, vii, x, 53, 215, 217, 225, 234, 236, 238, 263, 293, 344
measures, xii, 172, 173, 178, 206, 207, 208, 226, 227, 230, 234, 298, 300, 302, 310, 311, 313, 343, 345, 348, 350, 351, 364, 378
meat, 121, 344, 356
media, 34, 35, 40, 116, 260, 267, 389
median, 88, 90
Mediterranean, 209, 231
medusa, 28
membrane permeability, 118
membranes, 118, 125, 132
memory, 8, 48
men, 132, 133
menopause, 12
mercury, 220
meristem, 4, 36, 48
meta-analysis, 217
metabolism, 3, 53, 54, 121, 169, 238
metabolites, 42, 57, 115, 121
metals, 267, 290, 386, 387, 388, 389, 400
Mexico, vi, vii, xii, 1, 2, 3, 4, 9, 10, 11, 14, 15, 22, 25, 27, 29, 30, 44, 45, 47, 49, 50, 51, 52, 56, 57, 59, 212, 317, 319, 320, 321, 322, 323, 326, 330, 332, 333, 336, 337, 338
Mg^{2+}, 306
mice, 121, 122, 123, 128, 129, 132, 137, 140, 141
microbial communities, 216, 217, 219, 220, 221, 222, 225, 226, 227, 229, 230, 231, 233, 234, 235, 236, 237

microbial community, x, 215, 216, 218, 221, 227, 230, 231, 233, 235, 236, 237
microclimate, 33
microenvironments, 44
microhabitats, 105, 301, 302, 306, 307, 309, 310, 312
micrometer, 302
micronutrients, 10, 34
microorganism, 30, 34, 35
microsatellites, 345, 352, 356
microscope, 128, 270, 302
middle class, 145
Middle East, 300
migration, 62, 93, 98, 99, 104, 145, 166, 167, 172, 189, 196, 212, 314, 349
milk, 13
mining, 20, 292
Ministry of Education, 380
Miocene, 83, 88, 99, 102, 103, 105, 108
mixing, 413
MMA, 204
mobility, 128, 186
models, xiv, 52, 126, 216, 218, 223, 226, 231, 240, 241, 242, 246, 253, 260, 264, 318, 405, 411, 412, 413
moisture, 8, 242, 322
molecular biology, 14, 230
molecular mobility, 131
molecular weight, 346
molecules, 13, 121
momentum, 170
money, 311
morning, 6, 198
morphogenesis, 350
morphology, xi, 2, 6, 14, 44, 45, 46, 56, 58, 66, 88, 102, 119, 120, 152, 169, 212, 240, 265, 269, 282, 289, 292, 294, 310, 344, 353, 362, 363, 378
morphometric, xiii, 342, 345, 350, 358, 389
mortality, 20, 40, 169, 171, 209, 282
mortality risk, 209
morula, 121
mosaic, 238
motion, 112, 131, 242
mountains, 318, 320
movement, 108, 118, 206, 242
multidimensional, 227
multiple factors, ix, 111
multiple regression, 246, 252
multiplication, vii, 2, 25, 34, 35, 36, 40, 41, 129
mutation, 349
Myanmar, 144, 148, 156, 158

N

naphthalene, 417, 424
narcotic, 12
nares, 350
nation, 175
National Park Service, 52, 56, 58
native species, x, xi, 16, 183, 297, 298, 299, 310, 311, 345, 353
natural food, 171
natural resources, xiii, 2, 22, 330, 342, 353
nature conservation, 315
NCA, 319, 320
Nd, 9
necrosis, 384, 398
negative relation, 252
nematode, 141, 266, 270, 273, 280
Nepal, 156, 158, 170
Netherlands, 231, 403
network, 144, 146, 147, 152, 266, 364, 378
New South Wales, 264
New Zealand, 291, 411, 412
NGOs, 24, 172, 173, 174, 177, 178
niacin, 10
Nicaragua, 11
Nile, 138, 355
nitrifying bacteria, 236
nitrobenzene, 420, 421
nitrogen, viii, ix, 13, 103, 111, 112, 119, 120, 121, 126, 127, 128, 129, 136, 138, 139, 142, 220, 234, 238, 258, 259, 263, 264, 398
nitrogen fixation, 234, 398
NMR, 422
North Africa, 300, 314
North America, xiv, 14, 15, 54, 55, 56, 98, 116, 131, 141, 162, 176, 268, 270, 282, 291, 295, 299, 314, 315, 405
North Sea, 290
Norway, 280, 314, 315
nuclei, 135, 137, 207, 421, 429
nucleotides, 218, 342
Nuevo León, 15, 16, 17, 18, 23, 28, 52, 58, 327
null hypothesis, 245, 406
nursing, 20, 42
nutraceutical, 55
nutrients, xi, 19, 34, 41, 83, 146, 204, 239, 240, 242, 245, 253, 257, 258, 411, 413, 420, 422

O

obesity, 13
observations, xiv, 188, 189, 192, 194, 196, 197, 198, 200, 223, 231, 257, 285, 287, 386, 402

obstruction, 166
occlusion, 285
oceans, 80, 98
oil, 66, 99, 102, 137, 138, 269
Oklahoma, 221, 222
oocyte, 113, 124, 130, 138
organ, ix, 31, 34, 36, 55, 111, 127, 128, 130
organic matter, 41, 306, 393, 398, 400
organism, 171, 217, 242, 344
orientation, 42
oscillation, 200
osmotic pressure, 258
ovariectomy, 130, 132
ovaries, 125, 126, 127, 128, 130, 136, 185
overgrazing, 20
ovulation, 113, 119, 138
ovum, 113, 126
oxidation, 41, 118, 420, 421
oxidation products, 420, 421
oxidative stress, 131
oxygen, xi, 80, 81, 82, 92, 96, 109, 240, 252, 255, 257, 258
ozonation, 420, 422
ozone, 406

P

Pacific, viii, 61, 62, 63, 68, 69, 71, 81, 86, 87, 88, 89, 90, 91, 92, 94, 95, 96, 98, 100, 102, 103, 104, 106, 107, 108, 109, 179, 180, 259, 260
packaging, 55
Pakistan, 156, 170
paleontology, 105, 109
Panama, 262
Paraguay, 14
parameter, 372
parasite, xi, 207, 265, 266, 269, 270, 271, 273, 275, 278, 279, 282, 286, 287, 288, 290, 292, 293, 294, 295
parenchyma, 4, 6, 30, 425
parentage, 356
parents, 351, 352, 359
parthenogenesis, 291
particles, 64, 302, 384, 389
pasture, 44, 146, 233
pastures, 205, 206, 235, 318
pathogens, 41, 279, 288
PCA, xiv, 227, 229, 351, 405, 406, 407, 408, 411
PCR, 216, 219, 221, 222, 232, 237, 238, 344, 347, 352, 356
penis, 115
perfusion, 128
permeability, 126

permit, 257, 269
personal communication, 278, 287
Peru, 14
pesticide, 168, 169, 170, 171, 180
pests, 43, 168
pH, xiv, 116, 117, 135, 139, 235, 253, 255, 257, 302, 306, 314, 383, 386, 388, 389, 391, 395, 398, 400
pharmaceuticals, 121
phase transitions, 132, 136, 142
phenotype, 42, 350
phenylalanine, 9
Philippines, 170
phloem, 6, 425
phosphates, 257
phosphatidylcholine, 118, 139
phospholipids, 117
phosphorous, 263
photographs, 321
photosynthesis, 14, 30, 41, 44, 257, 400
physical environment, 243, 261
physical properties, 141
physics, 260, 411, 412
physiology, 14, 19, 113, 133, 169, 171, 260, 264
phytoplankton, 169, 262
pigs, 122, 125, 129, 134, 138
pilot study, 138
planning, 121, 136, 207, 208, 333
plasma, 116, 117, 118, 119, 129, 137, 138
plasticity, 99, 353
Platyhelminthes, 269, 295
Pliocene, 70, 77, 78, 79, 83, 86, 87, 90, 94, 98, 102, 103, 104, 105, 106, 108, 109
ploidy, 14, 55
Poland, 268, 286, 288, 294
policy makers, 178
pollen, 131, 417, 425, 426, 427, 428, 429
pollination, 7, 34, 59, 417, 428, 429
pollinators, 7
pollutants, xiv, 181, 252, 267, 290, 295, 383, 384, 385, 387, 389, 398, 400, 401
polyacrylamide, 347, 348, 352, 358
polymerase, 235
polymerase chain reaction, 235
polymerization, 420
polymers, 10, 117
polymorphism, 220, 234, 237, 344, 346, 347
polymorphisms, 342
polyphenols, 44, 421
polyunsaturated fat, 117, 132, 140, 142
pools, 64, 244, 262
poor, x, 116, 118, 124, 144, 145, 155, 166, 167, 220, 221, 222, 273, 275, 285, 311, 398, 401
population density, 20, 273

population size, 169, 187, 352, 353, 358
positive correlation, 312
positive relationship, 14, 306
potassium, 11, 121
poverty alleviation, 145, 181
power, 226, 298, 363, 406
precipitation, 150, 198, 305, 418
prediction, 225
predictors, 253
preference, 157, 187, 191, 253
pregnancy, 115, 142
pressure, 21, 26, 127, 153, 166, 172, 173, 243, 260, 266, 298, 353, 363
prevention, 179
prices, 22
primate, 122, 137, 139
probability, xii, 30, 126, 205, 206, 225, 297, 348, 407
probe, 115, 139, 355
producers, 13
productivity, 147, 167, 241, 242, 263, 264, 282, 411, 412
profit, 22, 25, 43
program, xiii, 24, 29, 183, 236, 245, 270, 342, 343, 345, 353, 354, 357, 359, 407
prokaryotes, 217, 234, 235
proliferation, 15, 36, 39, 49, 51, 55, 288
propagation, vii, xii, xiv, 2, 22, 23, 24, 26, 29, 30, 31, 32, 34, 35, 36, 37, 38, 40, 41, 43, 47, 48, 49, 50, 51, 52, 53, 54, 55, 56, 58, 59, 119, 126, 318, 332, 415, 419, 422
propylene, 117
protected areas, 24, 43, 207, 311, 315, 318, 328, 329, 330, 333, 336
proteins, 10, 11, 223, 346
protocol, 41, 42, 127, 128, 176
public awareness, 177
public support, 177
pulp, 11, 49, 52
pulse, 180, 181
pumps, 147
pupa, 187
purification, 138
PVP, 117

Q

quadriceps, 184, 189, 191, 192, 193, 194, 195, 196, 198, 204, 209, 211, 213
quality control, 221

R

radiation, 54, 102, 406, 411, 413
radius, 391
rain, x, 5, 167, 183, 196, 199, 200, 207, 212, 417
rainfall, 6, 52, 144, 146, 244, 384, 418
Ramadan, 11, 56
RDP, 219
reactive oxygen, 118, 131
reality, 141, 221
recession, 391
recognition, 216, 217, 318
recombination, 217
reconstruction, 84, 238, 344, 359
recovery, 25, 130, 138, 189, 206, 207, 208, 343, 353, 402, 404
recruiting, 189
rectum, 115
reflection, 220
regenerate, 419, 422
regeneration, 35, 36, 40, 49, 50, 55, 205, 206, 207, 208, 384, 402, 417, 419
regression, 246, 252, 257, 258, 304
regulation, 22, 25, 211, 241, 262, 315, 363
regulations, 22, 172, 173, 178
regulators, xiv, 34, 405, 411, 413, 422
rejection, 128
relationship, xii, 20, 59, 70, 84, 97, 98, 216, 233, 252, 257, 261, 297, 302
relationships, 58, 69, 84, 102, 106, 209, 216, 246, 252, 257, 264, 301, 304, 305, 344
relevance, 413
relict species, 70
remediation, 175, 222
replication, 197, 222
reproduction, x, 4, 7, 28, 45, 46, 59, 65, 112, 129, 138, 141, 169, 176, 183, 185, 188, 195, 210, 217, 357, 417, 428
reproductive organs, 137, 400
reptile, 344, 355
reserves, 56, 186, 208
residuals, 246
residues, 41, 171
resistance, 5, 6, 43, 220, 234, 241, 389, 400
resolution, 217, 220, 321
resource management, 177
resources, ix, 24, 138, 141, 143, 144, 150, 152, 155, 166, 172, 177, 178, 179, 184, 242, 244, 257, 312, 318, 362, 417
respiratory, 237, 238
restriction enzyme, 220
restriction fragment length polymorphis, 216, 231, 232, 233, 234, 235, 236

retail, 155
retention, 5, 154, 184
returns, 68, 188, 201, 202
Rhodophyta, 250
rhodopsin, 222, 231
rhythm, 200
ribosomal RNA, 231
rice, 64, 145, 167, 168, 169, 170, 181
rice field, 64
risk, xii, 17, 19, 43, 51, 117, 150, 176, 206, 256, 288, 311, 317, 318, 321, 330
river systems, 147, 298, 299, 312, 315
RNA, 218, 219, 236
rods, 6
Romania, 266, 267, 268, 282, 286, 287, 288, 293, 294, 364
room temperature, ix, 32, 41, 112, 121, 122, 129, 355
root cap, 4
Royal Society, 133, 263, 291, 315
runoff, 168, 171, 363
rural people, 147
Russia, 98, 268, 286

S

safety, 267
sales, 168
salinity, viii, xi, 8, 61, 62, 67, 68, 69, 70, 71, 72, 79, 80, 81, 82, 83, 86, 87, 88, 91, 93, 94, 98, 100, 107, 239, 240, 242, 245, 252, 253, 256, 257, 258, 407
saliva, 115
salmon, 280
salt, 233, 258, 286, 413
salts, 11
sample, 73, 75, 99, 115, 119, 120, 121, 123, 126, 138, 188, 189, 218, 224, 225, 226, 228, 229, 230, 232, 233, 236, 245, 302, 318, 347, 348, 352, 353, 406, 422
sampling, 219, 221, 226, 232, 243, 249, 257, 266, 273, 301, 302, 303, 306, 386, 393
sanctuaries, x, 144, 173, 174, 175, 178, 179
satellite, 270, 275, 278
saturated fatty acids, 118
saturation, 30, 118, 248
savannah, 244
scaling, 227, 230, 234
scarcity, 151
scolex, 279, 280, 282, 284, 285, 288
scores, 254
sea-level, 62, 80, 81, 86, 91, 94, 100, 105, 107, 109
search, 29, 51, 54, 186, 201

seasonality, 114
Second World, 205
sectoral policies, 173
security, 112, 152, 179
sediment, 63, 64, 67, 101, 107, 152, 229, 234, 292, 363, 365
sedimentation, 258, 363, 365, 378, 404
sediments, 63, 66, 83, 89, 99, 102, 106, 220, 230, 231, 232, 236, 238, 257, 273, 292, 363
seed, 7, 8, 11, 15, 20, 26, 31, 32, 33, 34, 35, 46, 49, 53, 55, 57, 145, 172, 203, 207, 417, 425, 426, 427, 428, 429
seedlings, xiv, 5, 19, 26, 33, 34, 35, 41, 48, 208, 212, 415, 417, 425, 428
segregation, 352
selecting, 174, 417
selectivity, 16, 178
semen, viii, 111, 113, 115, 116, 117, 119, 121, 123, 130, 131, 132, 133, 134, 135, 136, 137, 138, 139, 140, 141, 176, 181
senescence, 384
sensitivity, 12, 125, 130, 139, 142, 220, 293, 310, 400
separation, 118, 137, 219, 261, 346, 355
sequencing, x, 215, 216, 220, 221, 222, 223, 229, 230, 235, 236, 238, 347
Serbia, xiii, 282, 287, 288, 361, 363, 364, 380, 381
serum, 120, 121, 137
serum albumin, 120, 121
sex, 123, 131, 132, 133, 134, 137, 141, 290, 350
sex ratio, 123, 134
sexual dimorphism, 107, 345
sexual reproduction, 184, 212
shape, 3, 6, 7, 15, 40, 66, 88, 157, 158, 231, 284, 345, 350, 357
shaping, 224
shares, 343
sharing, 91, 178
sheep, ix, 111, 124, 125, 126, 127, 129, 130, 133, 136, 137
shelter, 150, 152, 204
shock, 40, 132
shoot, 3, 5, 6, 36, 37, 39, 40, 51, 389, 419, 422, 426
shores, x, 207, 239, 240, 241, 244, 245, 246, 248, 249, 250, 252, 255, 256, 258, 259, 260, 261
shrimp, 145
shrubs, 3, 394, 396
sieve plate, 90
sign, xiii, 27, 361, 364
signals, 422
silver, 275, 347, 348, 352, 358

similarity, xi, 227, 228, 229, 230, 234, 237, 245, 265, 270, 272, 275, 278, 345, 367, 368, 378, 380, 386, 387, 396, 397, 398, 401, 420, 422
simulation, xiv, 234, 405, 406, 407
Singapore, 170, 241, 261
skin, 52, 163
slag, 404
Slovakia, v, xi, xiii, 265, 266, 267, 268, 269, 270, 271, 273, 274, 275, 276, 277, 278, 280, 281, 282, 283, 284, 285, 286, 287, 288, 289, 291, 292, 294, 364, 383, 384, 386, 402, 403
social context, 173
social group, 123
social relations, 112
sodium, 10, 35
software, 229, 303, 321, 406, 412
soil erosion, ix, 143, 393
soil pollution, 401
solidification, 126
solubility, 398
somatic cell, ix, 36, 112, 129, 130, 131, 132, 133, 134, 135, 136, 138, 140
South Africa, 337, 355
South Asia, 144, 154
South China Sea, 93
Southeast Asia, 86, 91, 94, 98, 158
Soviet Union, 268
specialization, 19
speciation, 16, 62, 98, 99, 108, 227
species, vii, viii, ix, x, xi, xii, xiii, xiv, 1, 2, 3, 5,
specificity, 19, 279, 295
spectroscopy, 421
spectrum, 266, 278, 286
speed, 32, 33, 128
sperm, viii, 111, 113, 115, 116, 117, 118, 119, 120, 121, 123, 124, 125, 126, 128, 130, 131, 132, 133, 134, 135, 136, 137, 138, 139, 140, 141, 175, 176, 177, 182
sperm function, 130
spermatheca, 185
spermatogonial stem cells, 122, 123, 134
spine, 30, 158, 162
sprouting, 20
SPSS, 303
Sri Lanka, 256, 260
stability, 112, 233, 234, 256, 257, 344, 401
stages, 33, 35, 39, 40, 63, 66, 80, 81, 86, 118, 174, 205, 206, 207, 208, 219, 224, 270
stakeholders, 174, 178
stamens, 7
standards, xiv, 189, 383, 384
starch, 4, 347
statistics, 169, 216, 230, 245, 246, 349

stem cells, 122, 123
sterile, 194
steroids, 12
sterols, 11, 12, 47, 57
stigma, 6
stimulant, 12
stock, 4, 153, 167, 168, 172, 173, 177, 178, 343, 417, 419, 420, 422, 429
storage, viii, ix, 111, 112, 119, 121, 129, 130, 131, 135, 136
strain, 43
strategies, xii, 22, 23, 24, 25, 31, 173, 181, 217, 219, 267, 311, 315, 317, 332, 333, 344, 352, 403
strength, 30, 118, 262, 422
stress, 42, 113, 133, 196, 242, 248, 262, 263, 293, 393, 396, 400
structural changes, 342
students, 26, 178
subsistence, 145, 151, 153, 298
substitution, 420
substrates, 121, 363, 389, 396, 404
success rate, viii, ix, 111, 112, 129
sucrose, 34, 40, 117, 120, 134
sugar, 41, 140
sulphur, 404
summer, 7, 63, 68, 69, 70, 71, 72, 82, 96, 97, 98, 194, 305, 308, 312, 354
Sun, 122, 136, 140
superiority, 185
supervision, 174
supply, vii, 2, 19, 83, 93, 145, 152, 153, 177, 191, 201, 258, 298, 306, 417, 422
surface area, 154, 298
survival rate, 26, 41, 342
susceptibility, 41, 42, 50, 173, 237
sustainability, 172
sustainable development, 22, 179
swamps, 144, 152, 160
Sweden, 21, 280
Switzerland, 21, 53, 134, 336, 354, 355
symmetry, 3
symptoms, 400
synchronization, 113
syndrome, 42
synthesis, 13, 221, 262

T

T cell, 44
tapeworm, 268, 270, 275, 278, 280, 282, 285, 286, 288, 290, 291, 292, 293, 294, 295
targets, 315
taxonomy, 2, 66, 88, 101, 103

teaching, 27
technical assistance, 289
teeth, 158, 162, 171
tenure, 174
terpenes, 12
territory, xii, 22, 144, 201, 204, 274, 276, 277, 278, 282, 288, 298, 317, 321, 324, 330, 332, 362
test statistic, 243
testicle, 116
testosterone, 137
Thailand, 156, 170, 179, 180
therapy, 216
threat, ix, xii, 112, 144, 165, 167, 184, 206, 207, 267, 317, 332
threats, 172, 257, 260, 299, 314, 331
threshold, 224, 245
tides, x, 239, 252
timber, 205, 332
time periods, 198, 387
timing, 30, 99, 146, 425, 427
tissue, vii, ix, 2, 35, 36, 37, 38, 39, 46, 47, 49, 54, 112, 123, 127, 128, 132, 133, 135, 139, 140, 332, 347, 384
Togo, 244
tonsillitis, 12
toxic effect, 219, 222, 384
toxicity, 117, 222, 233, 273, 290, 292, 400, 401
Toyota, 429
trade, xiii, 17, 45, 56, 57, 170, 209, 217, 341, 343
trade-off, 209
tradition, 144, 145
traffic, xii, 317, 332
training, 115, 141
traits, xiii, 133, 169, 242, 261, 312, 342, 345, 350, 351, 352
transformation, 43, 229, 245, 299, 350
transgression, 93, 104
transition, 49, 118, 122, 125, 130, 133, 139, 301
transition temperature, 118, 122, 130
translation, x, 215
translocation, 290
transmission, 12, 50
transparency, 171, 363, 365, 378
transplantation, ix, 41, 112, 122, 123, 124, 127, 128, 129, 130, 131, 134, 136, 140
transport, 242
trees, xiv, 3, 191, 205, 393, 394, 395, 396, 400, 415, 417, 419, 428
tribes, 4, 14
triggers, 98
triploid, 14, 280, 292
tropical forests, 299
tumor growth, 13
tundra, 299
turnover, 208
tyramine, 12
Tyrosine, 9

U

U.S. Geological Survey, 101
Ukraine, 266, 267, 268, 273, 282, 284, 286, 287, 364, 403
ultrasonography, 115
ultrastructure, 56, 136
uniform, 184, 282, 365, 368, 374
United States, 45, 52, 56, 131, 135, 137, 140, 232, 262, 320
universities, 27
uranium, 223, 232
urban areas, 145
urine, 115
Uruguay, 45
USSR, 142, 289, 292, 294
uterus, 126, 283
UV radiation, 412, 413

V

vacuum, 121
vagina, 115
Valencia, 357
validation, 413
validity, 226
vapor, 119, 120, 134
variability, x, xiii, 46, 53, 68, 176, 233, 239, 240, 241, 243, 247, 259, 311, 342, 344, 345, 346, 347, 348, 351, 352, 353, 355, 358
variation, 25, 36, 40, 44, 46, 186, 208, 228, 241, 248, 257, 258, 259, 260, 261, 262, 292, 315, 342, 344, 345, 346, 347, 349, 350, 351, 355, 356, 357, 374, 387, 428
vascular system, 420
vector, 221, 246
vegetation, xiii, 2, 33, 55, 147, 150, 155, 186, 187, 189, 205, 207, 210, 237, 261, 322, 362, 365, 378, 379, 380, 381, 383, 384, 386, 387, 389, 390, 391, 393, 395, 396, 398, 400, 401, 402, 403, 404
vein, 347, 352, 358
velocity, 119, 120, 128, 137, 263, 302, 309, 362, 363, 365, 378, 379
vertebrates, 155, 204
vesicle, 12, 125, 142
Vietnam, 11, 156, 170, 356
village, 269, 391, 393
viruses, 233

viscosity, 126
vision, 154
vitamin C, 13, 118
vitamin E, 11, 13, 118
vitamins, 10, 34
vulnerability, vii, 2, 322
vulnerable people, 154

women, ix, 111
wood, 238, 421, 422
workers, x, 28, 70, 183, 184, 185, 186, 188, 189, 191, 192, 193, 194, 195, 196, 198, 199, 201, 202, 204, 206, 207, 212, 218, 221
World Wide Web, 291
worms, 171, 272, 280, 290

W

Wales, 287, 290
waste water, 216, 259
water quality, 108, 273, 310
water resources, 155, 180
watershed, 150, 301, 311, 333
waterways, ix, 143
weakness, 173
web, 27, 262
welfare, 112
West Africa, v, x, 239, 240, 241, 244, 259, 260, 261
wetlands, ix, 143, 144, 153, 172, 266, 343
wheat, 54, 235
wild animals, 126
wild type, vii, 2
wilderness, 6, 8, 42, 262
wildlife, xiii, 17, 22, 24, 25, 26, 56, 113, 116, 123, 134, 138, 141, 266, 293, 342, 353
wildlife conservation, 123, 134
windows, 263
winter, viii, 7, 62, 68, 69, 70, 71, 72, 81, 82, 96, 98, 100, 149, 194, 235, 305, 308, 310, 363, 386

X

xenografts, 142
xenotransplantation, 135
xylem, 4, 420, 425

Y

Y chromosome, 134
yeast, 121, 136
yield, 7, 119, 181, 384, 420, 421
yolk, 118, 119, 132
Yugoslavia, 292, 380, 381

Z

Zimbabwe, 343
zinc, 387
zoogeography, 102
zooplankton, xiv, 169, 272, 405, 406, 407, 412, 413